T0329328

Seismic Safety of High Arch Dams

Seismic Safety of High Arch Dams

Authors:
Chen Houqun
Wu Shengxin
Dang Faning

Translators:
Kang Shusen
Zhou Jikai

CHINA ELECTRIC POWER PRESS

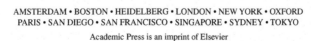
AMSTERDAM • BOSTON • HEIDELBERG • LONDON • NEW YORK • OXFORD
PARIS • SAN DIEGO • SAN FRANCISCO • SINGAPORE • SYDNEY • TOKYO
Academic Press is an imprint of Elsevier

British Library Cataloguing-in-Publication Data
A catalogue record for this book is available from the British Library

Library of Congress Cataloging-in-Publication Data
A catalog record for this book is available from the Library of Congress

ISBN: 978-0-12-803628-0

For information on all Academic Press publications
visit our website at http://store.elsevier.com/

Typeset by Thomson Digital

Printed and bound in the United States

Working together
to grow libraries in
developing countries

www.elsevier.com • www.bookaid.org

Publisher: Joe Hayton
Acquisitions Editor: Margaret Zhao
Editorial Project Manager: Hong Jin
Production Project Manager: Lisa Jones
Designer: Greg Harris

Cover picture provided by Liansheng Wang from China Three Gorges Corporation

Contents

Synopsis xv

Biographies xvii

Preface xix

CHAPTER 1 General Description 1

   1.1 Construction and Seismic Safety of High Arch Concrete Dams in China 1

      1.1.1 The General Conditions of High Concrete Arch Dam Construction
in Our Country 2

      1.1.2 Seismic General Conditions in Dam Site of Concrete High Dams in China 3

   1.2 The Basic Concept and its Seismic Case Enlightenment Evaluation
of Seismic Safety of High Concrete Arch Dam 4

      1.2.1 The Basic Concept of Seismic Safety Evaluation of High Concrete
Arch Dam 4

      1.2.2 Evaluation of Seismic Risk of High Concrete Arch Dams 6

PART I SEISMIC INPUTS AT SITE OF HIGH ARCH DAM

**CHAPTER 2 Outline of Bases of Seismic Fortification and Seismic
Hazard Analysis at Dam Site** 27

   2.1 Bases of Seismic Fortification Design and Prevention and General
Seismic Danger Analysis of Engineering Worksite 27

      2.1.1 Bases for Design and Prevention 27

      2.1.2 General Description of Earthquake Hazard Evaluation
in Engineering Site 34

   2.2 Aseismic Gradation Design and Prevention Level and Determination
of Corresponding Performance Objective of High Arch Dams 38

      2.2.1 Aseismic Design and Prevention Level and Corresponding
Performance Objective of Foreign Dams 38

      2.2.2 The Guiding Thinking for Working Out Aseismic Design
Prevention Level or Standard Framework 42

      2.2.3 Suggestions on Revision of Aseismic Design and Prevention
Level Framework in China 46

   2.3 Reservoir Earthquake 49

**CHAPTER 3 Determination of Correlation Design Seismic Motion
Parameters on Dam Site** 53

   3.1 Design of Peak Ground Acceleration 53

   3.2 Response Spectrum Attenuation Relations and its Design Response Spectrum 58

             3.2.1 Response Spectrum Attenuation Relation 58

             3.2.2 Design Response Spectrum 73

CHAPTER 4   **Design Acceleration Time Process**... **83**

        4.1   Amplitude and Frequency Nonsmooth Acceleration Time Process 83

             4.1.1 Seeking a Solution to the Evolutionary Power Spectrum..........................84

             4.1.2 Objective Evolutionary Power Spectrum Empiric Model Fitting................85

             4.1.3 Artificial Fitting Amplitude and Frequency Nonsmooth Seismic
                   Motion Acceleration Time Process Genesis 87

        4.2   Adopting Random Finite Fault Method to Directly Generate Acceleration
           Time Process.. 90

             4.2.1 Engineering Background of Adopting Random Finite Fault
                   Method to Directly Generate Acceleration Time Process........................90

             4.2.2 Random Finite Fault Method Behaviors, Basic Thinking Ways
                   and Effects 91

             4.2.3 Steps of Random Finite Fault to Generate Site Seismic Motion
                   Time Process...92

CHAPTER 5   **Dam Site Seismic Motion Input Mechanism** **97**

        5.1   Basic Concept of Site Design Seismic Motion Peak Ground Acceleration Input.....97

        5.2   Dam Site Seismic Motion Input Mode 99

        5.3   Free-Field Incident Seismic Motion Input Mechanism 100

             5.3.1 Artificial Homological Boundary Method 102

             5.3.2 Substructure Method Based on Dynamics...104

        5.4   Several Problems Need to be Clarified and Further Discussed in Seismic
           Motion Input Mode 106

        5.5   Suggestions 108

PART 2   **HIGH ARCH DAM BODY – RESERVOIR WATER –**
              FOUNDATION SYSTEM SEISMIC RESPONSE ANALYSIS
              AND SEISMIC SAFETY EVALUATION

CHAPTER 6   **High Arch Dam Body–Foundation System Three-Dimensional**
             Contact Nonlinear Dynamic Analysis Method **115**

        6.1   Performance and Engineering Background of High Arch Dam Engineering
           Seismic Response Analysis in Strong Seismic Areas 115

        6.2   Contact Problem Handling or Solving Method 117

             6.2.1 Normal Contact Condition...118

             6.2.2 Tangential Contact Conditions...118

        6.3   High Arch Dam Body–Foundation System Seismic Response Analysis Method 119

             6.3.1 High Arch Dam Body–Reservoir Water–Foundation System
                     Numerical Model...119

6.3.2 Dynamic Equation Discretion...122

6.4 Artificial Homology Boundary Implementation Method 124

6.4.1 Homology Formula Based on Finite Element Mesh Nodal Point 124

6.4.2 Computation Method of Boundary Nodal Point Displacement.................126

6.4.3 Concrete Computation Steps of Artificial Boundary Nodal Point
Displacement when Seismic Wave is Incident 127

6.4.4 Numerical Simulation of Arch Dam Contact Nonlinear Problem.............127

6.4.5 Static and Dynamic Combination Computation Method of Arch
Dam–Foundation System Contact Nonlinearity 134

**CHAPTER 7 Dam Abutment and Arch Support Rock Block Stability
and Seismic Safety Evaluation of High Arch Dam 137**

7.1 Aseismic Stability Analysis Behaviors of Arch Dam Abutment
and Earthquake Disaster Enlightenment.................................. 137

7.2 Basic Concept of Dam Abutment Rock Block Instability Safety Coefficient
of Arch Dam 139

7.3 Problems of Current Arch Dam Aseismic Stability Analysis Method.................141

7.4 Arch Dam Aseismic Stability with Deformation as its Core and Aseismic
Safety Evaluation.. 142

7.5 Practical Examples of Engineering Applications 144

**CHAPTER 8 Research on Parallel Computation of High Arch Dam
Structure Seismic Motion Response.....................................155**

8.1 Research and Significance of Large-Scale Structure Response.........................155

8.1.1 Developing Situation of High Performances Parallel Computation
in Hydraulic Structure Field in Our Country 155

8.1.2 The Significance of Research on High Dam Structure Seismic Motion
Parallel Computation 156

8.2 The Development and Existing Conditions of Finite Element Parallel
Computation 157

8.2.1 Brief Description of Parallel Finite element.............................157

8.2.2 Development Conditions of Domestic Finite Element
Parallel Computation 158

8.3 Dynamic Explicit Computation Format and Dynamic Contact Problem
Handling Method 160

8.3.1 Finite Element Explicit Computation Format in Structure Seismic
Response Analysis 160

8.3.2 Influence Factors on Explicit Integrated Computation Format
Numerical Stability 161

8.3.3 Lagrange's Method Of Multipliers of Motion Contact Nonlinearity 166

8.3.4 Point-by-Point Multiplier Method for Off-Diagonal Additional
Mass Matrix 168

8.4 FEPG System and Finite Element Method Based on FEPG 169

8.4.1 FEPG System Program Structure Performances 169

8.4.2 The Finite Unit Element Method on FEPG 170

8.5 Artificial Boundary Realization in FEPG 170

8.5.1 Input Formula of Seismic Wave in Artificial Viscoelastic Boundary 170

8.5.2 Realization of Artificial Viscoelastic Boundary in FEPG.......................180

8.5.3 Realization of Artificial Homology or Transmission Boundary in FEPG 188

8.6 Parallel Computation Program Development Based on PEFPG System

High Arch Dam Seismic Response...192

8.6.1 Message Passing Programming Interface 192

8.6.2 PFEPG System, its Structure and Work Mode 194

8.6.3 Explicit Format Parallel Program Development Based on FEPG

System Dynamic Equation 200

8.6.4 High Arch Dam Structure Dynamic Parallel Computation Stream 203

CHAPTER 9 Engineering Real Example Analysis of Parallel Computation...........207

9.1 Parallel Computation Analysis of Seismic Motion Responses

to Xiaowan Arch Dam 207

9.1.1 Xiaowan Arch Dam Computation Model 207

9.1.2 Parallel Computation Under the Condition of Transmitting Boundary 210

9.1.3 Parallel Computation Under Artificial Viscoelastic Boundary Conditions......215

9.2 Parallel Computation Analysis of Seismic Motion Dynamic Response

of Xiluodu Arch Dam 217

9.2.1 Basic Data for Computation 219

9.2.2 Parallel Computation of Xiluodu Arch Seismic Dynamic Response 220

9.3 Parallel Computation Analysis of Baihetan Left Bank Side Slope Sliding

Block Stability 223

9.3.1 Engineering General Conditions..224

9.3.2 Computation of Basic Data...224

9.3.3 Static Stability Computation 228

9.3.4 Dynamic Stability Computation 229

9.3.5 Analysis of Parallel Computation Efficiency............................232

PART 3 DYNAMIC TESTING, NUMERICAL SIMULATION AND MECHANISMS FOR DAM CONCRETE

CHAPTER 10 Research Progress on Dynamic Mechanical Behavior of High Arch Dam Concrete...239

10.1 Dynamic Mechanical Behavior of Normal Concrete 239

10.1.1 Dynamic Strength, Deformation, and its Influencing Factors

of Concrete 239

10.1.2 Experimental Techniques for Dynamic Mechanical Characteristics
of Concrete 245
10.1.3 Mesomechanical Numerical Analysis of Concrete 247
10.1.4 Discussion on the Mechanisms for Strain Rate Effect of Concrete 249
10.2 Dynamic Mechanical Characteristics of Dam Concrete....................251
10.2.1 The Main Features of Dam Concrete 251
10.2.2 Dynamic Tensile Strength of Dam Concrete..............................252
10.2.3 Dynamic Elastic Modulus of Dam Concrete 254
10.3 The Key Problem and Technical Way in the Dynamic Mechanical
Characteristic Study.. 255

CHAPTER 11   Dynamic Flexural Experimental Research on Dam Concrete............257
11.1 Testing Methods for Dynamic Flexural Characteristics of Dam Concrete..........257
11.1.1 Dynamic Flexural Instruments, Fixtures, and the Requirements.............258
11.1.2 Dynamic Loading Mode, Data Acquisition, and Processing 258
11.1.3 Casting of Dam Concrete Specimen 262
11.2 Testing on Dynamic Mechanical Characteristics of Dam Concrete 263
11.2.1 Experimental Materials 263
11.2.2 Loading Scheme..264
11.2.3 Experimental Results 265
11.3 Discussions on Experimental Results...288
11.3.1 The Ratio for Static Strength of Full Graded Concrete
to Wet-Sieved Concrete 288
11.3.2 The Ratio of the Static Flexural Tensile Strength to Cubic
Compressive and Splitting Tensile Strength.............................291
11.3.3 Strain Rate Effects on Flexural Strength and the Effects
of Dynamic Loading Mode 292
11.3.4 The Effect of Initial Loads on Dynamic Flexural Strength....................292
11.3.5 The Influence of Age on Dam Concrete Strength 293
11.3.6 The Analysis of the Characteristics of Flexural Deformation
and the Value of Dynamic Elastic Modulus 294

**CHAPTER 12   The Experimental Research on the Dynamic and Static
Mechanical Characteristics of Dam Concrete and
the Constitutive Materials 297**
12.1 The Experimental Techniques for the Dynamic and Static Direct Tensile
Stress–Strain Curves of Concrete 298
12.1.1 The Stable Fracture Conditions for Quasibrittle Materials 298
12.1.2 The Experimental Techniques for the Dynamic Uniaxial Tensile
Stress–Strain Curves of Concrete Materials.............................298
12.2 The Experimental Research on the Dynamic Tensile Characteristics
for Cement Mortar 304

12.3 The Experimental Research on Dynamic Direct Tensile Characteristics
of Aggregates 309

12.4 The Direct Tensile Tests on the Dynamic Mechanical Behavior
of the Interface between the Mortar and Aggregate 312

12.5 The Experimental Research on the Dynamic Tensile Characteristics
of Concrete for Dagangshan Arch Dam.. 312

12.6 Experimental Research on the Dynamic Tensile Stress–Strain Curves
of Concrete.. 316

CHAPTER 13 **Experimental Study of Dynamic and Static Damage Failure
of Concrete Dam Based on Acoustic Emission Technology**.............**321**

13.1 Research on Monitoring of Acoustic Emission Technology
of Concrete Materials 322

    13.1.1 Experimental Study of the Correlation Among the Feature
Parameters of Acoustic Emission..322

    13.1.2 Improvement of AE Source Location Technology...................................324

13.2 Experimental Study of Acoustic Emission Characteristics in the Damage
Process of Dam Concrete Under Dynamic and Static Flexural Failure 327

    13.2.1 Settings of Acoustic Emission Acquisition System 327

    13.2.2 Static Bending Test of Full-Graded and Wet-Sieving Arch Dam
Concrete Specimens...329

    13.2.3 Dynamic Flexural Test of Full-Graded and Wet-Sieving Arch
Dam Concrete Specimens...338

    13.2.4 Impact Dynamic Bending Test of Wet-Sieved Concrete
of Arch Dams with Different Loading Rates 344

13.3 Axial Tension Test of Concrete and its Constituents 351

    13.3.1 Acoustic Emission Collection System Settings.......................................351

    13.3.2 Concrete 353

    13.3.3 Aggregates..357

    13.3.4 Mortar..367

    13.3.5 Aggregate–Mortar Interface 371

    13.3.6 Comparative Analysis of Concrete and its Components.........................375

13.4 Experimental Study on AE Characteristics in Complete Uniaxial Tensile
Failure Process of Concrete Containing Softening Stage...................................377

    13.4.1 AE Acquisition System Setting 377

    13.4.2 Monotonic Uniaxial Load 377

    13.4.3 Cyclic Loading in Softening Stage 383

CHAPTER 14 **Testing Research on Large Dam Concrete Dynamic-Static
Damage and Failure Based on CT Technology** **395**

14.1 Application of CT Scanning Technology in Concrete Material Tests.................395

    14.1.1 General Description of CT Scanning Image Principle............................396

    14.1.2 A Brief Introduction to CT Scanning Equipment 398

14.2 Portable Type Dynamic Test Loading Equipment in Realignment with Medical-Use CT Machine.. 400

14.2.1 Technical Requirement and Structure of Test Loading Equipment 400

14.2.2 Specimen Installation and Load...403

14.3 Concrete CT Results and Initial Analysis 404

14.3.1 CT Scanning Test of Concrete Uniaxial Dynamic and Static Compression Destructive Process 405

14.3.2 CT Scanning Tests of Concrete Uniaxial Static, Dynamic Tensile Failure Process..413

14.3.3 Analysis of CT Test Results of Concrete Dynamic and Static Tensile and Compression Crack Process 423

14.4 Concrete subzone breaking theory...425

14.4.1 Existing Conditions and Problems of Concrete Research.......................425

14.4.2 Basic Assumption of Concrete Subzone Damage Theory 426

14.4.3 Degree of Perfect and Degree of Failure of Concrete.............................426

14.4.4 Concrete λ-Level Perfect Field and λ-Level Damage and Fracture Field 428

14.4.5 Concrete λ-Level Perfect Field and λ-Level Damage and Fracture Field Measure...429

14.4.6 Concrete $\lambda 1 - \lambda 2$ Cutoff or Intercept Joint...431

14.4.7 Concrete Perfect Space and Damage and Fracture 436

14.4.8 Concrete λ-Level Damage and Fracture Ratio and λ-Level Damage and Fracture Rate..437

14.4.9 Concrete $\lambda 1$-$\lambda 2$ Intercept Joint Ratio and Intercept Joint Rate...............438

14.4.10 Relation Among Concrete λ-Level Damage and Fracture and CT Number and Density..438

14.4.11 Concrete Damage and Fracture Generated Positions..............................439

14.4.12 Criterions for Concrete Damage and Fracture 440

14.4.13 Concrete Safety Zone, Damaging and Fracturing Zone, and Damaged and Fractured Zone 440

14.4.14 Concrete Damaging and Fracturing Zone Lower Limitation and Upper Limitation Boundary Surface 441

14.4.15 Concrete Weakening–Strengthening Norms 442

14.4.16 Subzone Description of Concrete Constitutive Relation.........................442

14.4.17 Constitutive Theory of Concrete Damage and Fracture Space 443

14.4.18 Summary 444

14.5 Classification of Support Vector Machine for Concrete CT Images.....................445

14.5.1 Optimum Classification Faces..445

14.5.2 Broad Sense Optional Classification Face 447

14.5.3 Classification Method of Support Vector Machine Image 448

14.5.4 Classification of Concrete Component Support Vector Machine...449

14.6 Fractal Dimension Computation and Analysis of Concrete CI Image.................452
    14.6.1 Fractal Theory and Dimensional Computation Method..........................452
    14.6.2 Fractal Dimensionality Computation and Analysis
           of Concrete Uniaxial Compression CT Images 454
    14.6.3 Fractal Dimensionality Computation and Analysis of Concrete
           Uniaxial Compression CT Image Under Dynamic Loading Action 457
    14.6.4 Comparison of Concrete Fractal Characteristics Under Static
           and Dynamic Actions 459
14.7 Concrete Damage Evolution Equation and Constitutive Relation Based
    on CT Images.. 460
    14.7.1 Geotechnical Damage Equation and Constitutive Model Based
           on CT Tests..460
    14.7.2 Concrete Statistic Damage Evolution Equation Based on CT Tests........462
    14.7.3 Concrete Average CT Number Damage Evolution Equation
           Based on CT Tests..464
    14.7.4 Concrete Fracture Dimensionality Damage Evolution
           Equation Based on CT Tests 468
14.8 Concrete 3-Dimensional Mesomechanics Analysis Based on CT Tests..............474
    14.8.1 Three Kinds of Concrete 3-Dimensional Computation Models 474
    14.8.2 Application of Damage Constitutive Relation Based on CT Images
           to Carry Out Concrete Damage Process Numerical Simulation 478
14.9 Three-Dimensional Cartoon Demonstration of Concrete Loading
    Damage Process 486

**CHAPTER 15 Research on Numerical Analysis of Full Gradation Large
               Dam Concrete Dynamic Behaviors...491**
15.1 Mesomechanism Numerical Method of Full-Grade Large Dam Concrete..........491
    15.1.1 Three-Dimensional Random Aggregate Model 492
    15.1.2 Finite Element Grid Profile Separation Program for Spherical
           Aggregate Model..502
    15.1.3 Mesofinite Element Profile Separation of Convex Polyhedron
           Aggregate Model..503
15.2 Damage and Failure Numerical Simulation Finite Element Equation
    of Concrete Test Specimens... 507
    15.2.1 Concrete Material Strain Rate Effect Enhancement Relations 507
    15.2.2 Concrete Constitutive Relation and Damage Evolution Model 508
    15.2.3 Virtual Work Equation in Static–Dynamic System 511
    15.2.4 Prestatic Loading Dynamic Equation...513
15.3 Concrete Numerical Simulation test FEPG Document 515
    15.3.1 GIO Command Stream Document...516
    15.3.2 GCN Command Document..516

15.3.3 NFE Algorithms Document...517

15.3.4 Scenario Document of Different Equations 521

15.4 Strain Rate Effects Upon Concrete Dam Dynamic Flexural Strength.................529

15.4.1 Numerical Computation Model...529

15.4.2 Numerical Computation Analysis 530

15.5 Physical Significance of Concrete Material Enhancement Parameters 536

15.6 Effects of Concrete Material Nonuniformity Upon Dynamic Flexural Strength........539

15.6.1 Random Aggregate Parameter Model (RAPM) 1 539

15.6.2 Mechanics Parameter Discreteness Affecting Dynamic Flexural
Tensile Strength...541

15.6.3 Mesoanalysis of Wet-Sieved Concrete Test Specimens 543

15.6.4 Effect of Concrete Gradation Upon Dynamic Flexural Tensile Strength 545

15.6.5 Effect of Selection of Mesoprofile Separation Zones Upon
Computation Results 547

15.6.6 Concrete Test Specimen Static–Dynamic Comprehensive
Flexural Tensile Mesodamage Mechanism 547

15.6.7 Effect of Aggregate Morphology Upon Concrete Flexural
Tensile Strength...551

15.7 Multiscale Algorithm to Predict Concrete Material Parameter Method 557

15.7.1 Multiscale Algorithm Basic Theory 557

15.7.2 Multiscale Algorithm Based on Double-Scale Method 560

15.7.3 Application of Multiscale Algorithm to Complete Gradation
Concrete Mesoparameter Prediction 561

15.7.4 Mesoflexural Tensile Numerical Simulation of Complete Gradation
Concrete Test Specimen 564

15.8 Research on Concrete Mesoanalysis and Parallel Computation
Based on PFEPG.. 567

15.8.1 Mesomodel Selection and Parallel Program Generation 567

15.8.2 Three-Dimensional Mesomodel Parallel Computation Analysis Under
Prestatic Loading Action 571

15.8.3 Postprocessing of Computation Results...574

15.8.4 Parallel Computation Efficiency 575

References..577

Index...589

Synopsis

This book is a nationally aided project by the National Nature Science Foundation, Sci-tech Innovation, and public welfare special funds of the Ministry of Water Resources. As viewed from the engineering application angle, this book makes a systematic and complete introduction of high arch dam behaviors, safety abundance, destruction mechanism, and other related knowledge concerning the seismic safety evaluation method of hydroelectric construction engineering in a macroseismic area. Its main contents are divided into the earthquake motion input of a high-arch dam site, high-arch dam body–reservoir water–seismic response analysis and seismic safety evaluation in the foundation system, large dam concrete dynamic resistance test, numerical simulation, and mechanism of three sections covering 15 chapters in total. They cover the setting-up defense norms in grades for high arch dam earthquake resistance and the determination of corresponding performance objects, the dam site seismic motion input mechanism, and presents research on the parallel computation of the seismic motion response of the high arch dam structure. Testing research on dynamic flexural tensile mechanics behaviors of large dam concrete and numerical analysis research on full gradation large dam concrete dynamic behaviors are also presented.

This book can provide a valuable reading reference for research workers who are engaged in hydraulic structure earthquake resistance and large dam concrete dynamic behaviors. This book can also serve as a teaching material for postgraduates who are engaged in the water resources and hydroelectric power professions.

Biographies

CHEN HOUQUN

Born on May 3, 1932. Currently positioned as an Academician of Chinese Academy of Engineering (CAE) and is also a member of Presidium of CAE (2000–2010). He is also a Research Professor of China Institute of Water Resources and Hydropower Research (IWHR). His major subject of specialization is hydraulic structures and earthquake engineering. He is associated with the following professional societies: Head of Expert Group of Quality Inspection for Three Gorges key project, State Council; Chairman of Expert Committee on South-to-North Water Diversion project, State Council; Member of Commission on Science and Technology of the China Ministry of Water Resources; Member of Standing Committee of Chinese National Committee on Large Dams; Honorary member of International Committee on Large Dams. Major areas of experience: he is one of the chief researchers and consultants on seismic safety of a series of critical hydraulic projects in China; he leads his team to be engaged in earthquake research on high-concrete dams for a long time. Their research activities include the following: to develop software package of seismic input of dam site using "stochastic finite fault method" and nonlinear dynamic finite element analysis of damage-rupture process of dam-foundation–reservoir system using high-performance parallel computation at supercomputer TH-1A; to take field test and modal test on shaking table as well as material test of dam concrete; to compile "Seismic Design Code of Hydraulic Structures" in China, etc.

WU SHENGXING

Born on September 1, 1963. Currently positioned as a Professor and Doctoral Mentor of College of Civil and Transportation Engineering, Hohai University, Nanjing, China. He is also a Standing Member of CPPCC Jiangsu Provincial Committee. His major subjects of specialization are as follows: early crack mechanism of concrete structures and complete set of control technology research; dynamic performance study of concrete; and simulation analysis of reinforced concrete structures. He is associated with the following professional societies: Secretary-General of Teaching Guidance Committee of Water Conservancy Subject, the Ministry of Education; member of Education Work Committee of China Civil Engineering Society; member of Committee of Structural Engineering, Chinese Society of Theoretical and Applied Mechanics; member of Concrete Structure Committee of China Association for Engineering Construction Standardization; and Executive Director of Jiangsu Civil Engineering and Architectural Society. Major areas of experience: having graduated in East China University of Water Resources and Electric Power in 1984 (now Hohai University), he received his MS in 1987 and PhD in 1994, both from Hohai University. He remained in school after graduation and was engaged in the teaching and research work of reinforced concrete structures. In 1998, he was judged to be a professor. He has been

a former vice president of the Institute of Civil Engineering, chief of Academic Affairs and deputy principal of Hohai University. He enjoyed the special allowance of the State Council in 2010.

DANG FANING

Born on October 4, 1962. Currently positioned as a member of Xi'an University of Technology Academic Committee. He is also a Leader of Geotechnical Engineering. His major subject of specialization is numerical analysis of hydraulic structure and numerical analysis of geotechnical engineering. He is associated with the following professional societies: Vice President of the Shaanxi Provincial Society of Rock Mechanics and Engineering, the Executive Director of the Shaanxi Provincial Mechanics Society, editorial board member of "Chinese Journal of Rock Mechanics and Engineering." Major areas of experience: with the CT tests as the base, he has used the fuzzy set to define such physical state new  indexes as rocks, concrete complete field, failure field, intercept theory and measures, etc., established the spatial evolution theory of geotech failure, and studied the damage evolution process of concrete materials. Also, he has suggested the rate effect principle of the static failure of inhomogeneous brittle material development along the structure of the weakest surface and (the dynamic failure of inhomogeneous brittle material development along) the fastest way of energy release. He has sponsored over 60 scientific research projects of various types and presented the optimal design suggestions for the seepage flow control and aseismic measures for the earth-rock filled dams of 36 large-scale hydroelectric power stations.

Preface

Nowadays, a series of high arch dam construction in western China is of great irreplaceable strategic significance for the rational utilization of water resources allocation, drought resistance, flood control, disaster mitigation, coping with the current climate changes, developing renewable clean hydropower energy, and promoting mineral energy resources for "emission reduction or mitigation." High arch dam engineering sites in the west regions of our country are located in the macroseismic area, and most of them are in the steeper river valleys related with regional geologic structure. For this reason, it is difficult for high arch dam engineering in hydroelectric construction in western China to avoid the complicated seismic safety problems. Once a high arch dam large reservoir sustains serious disaster change, it will lead to the secondary disaster consequences with seriously important effects upon society. These have clearly and apparently indicated the strategic importance of seismic safety evaluation of high arch dams and the prevention of seismic disaster changes. At present, a series of high dams of 300-m high are being constructed in the western regions of our country. There have been few precedent examples in these engineering projects so that there is a lack of engineering practical experiences and practical disaster examples for experiencing strong earthquakes. In order to ensure the engineering seismic safety, many key technically difficult problems confronted in engineering must be urgently solved, whereas at present, sci-tech support of high dam earthquake resistance cannot be completely adaptable to the requirement of engineering construction scale and speed. For this reason, it is necessary to make a breakthrough in the traditional concept and method, as well as to make intensive efforts to carry out and strengthen scientific interdisciplinary crossing and research on basic theoretical innovation of hi-tech applications in high dam seismic safety analysis. This is an important content closely related to the important task of disaster prevention and disaster mitigation in the overall development of the economic society in our country, and a challenge confronted in water resources and hydroelectric construction in our country as well as a development opportunity to accelerate the discipline frontier project.

For this reason, the National Natural Science Foundation Commission has listed "Research on the Problem of Frontier Basic Science Concerning High Arch Dam Seismic Safety Evaluation in Western China" as the Priority Research Project in its important planning project of "several key problems concerning energy resources utilization and environment protection in western China." The academician Chen Houqun is the sponsor of this research project. Under the unified coordination of the Chinese Academy of Water Resources and Hydropower, Hohai University, and Xi'an University of Technology, and closely in combination with the practical high arch dam engineering works in the macroseismic area in western China and entrusted by the research team organic integration of "production, learning or study, and research," the strong advantages of "unified planning, close cooperation, mutual advantage compensation, and resource co-sharing" should be brought into full play to fulfill this priority research project.

The contents in this book cover the major research achievements of the mentioned priority research project. These research achievements have provided the theoretical base, analysis method, and scientific foundation for the improvement of seismic safety design of high arch dam engineering works in the macroseismic areas, depending on the realization of practical natural behaviors, safety abundance, and destruction mechanism of high arch dam under the seismic actions, and probed into the working out of relational evaluation approaches of seismic safety of hydroelectric construction engineering works in the macroseismic areas in western China, and at the same time, boosted the development of hydraulic engineering seismic safety science.

Seismic safety evaluation of engineering structures should include earthquake motion input, structure seismic response, and structure resisting force of three elements. They are inseparable and intercoordinative components, being particularly important to the important high arch dam engineering works of this kind with the rapid development in computation technology. There occurs to have a growing meticulous research on structure seismic response in the seismic safety research and concrete large dams, while research on seismic motion input and structure resisting force is rather coarse appearing to have the situation of "being small in two heads and large in the middle." In fact, "two heads" can control the accuracy and level of seismic safety evaluation. These "two heads" have involved many interdisciplinary problems, which are difficult, with little research so that they need to be further strengthened. For this reason, in research on this priority project, the priority should be given to research on following frontier key technical problems with the urgent engineering problems to be solved in high arch dam seismic safety evaluation in our country as the background. It is necessary to determine and study the multiple seismic safety design and prevention norms and the corresponding quantitative performance objectives, as well as rationally selecting the seismic motion parameters and input mechanism related to dam site geology, earthquake, and topography or landform.

It is essential to study the mathematic model for the near-field and far-field ground geologic and seismic conditions of various kinds, and the integrating system of dam body–reservoir water–foundation of high arch dams and the effective solution method for nonlinear fluctuating problems, and also to make research on the seismic destruction mechanism and seismic safety quantitative judgment guideline.

It is important to study the full gradation dam concrete dynamic behavior test under the seismic loading actions, three-dimensional mesomechanics analysis, and CT scanning technology on its damaging mechanism.

Research is made on high arch dam nonlinear dynamic analysis and full gradation dam concrete three-dimensional mesomechanics dynamic analysis parallel computing.

This book consists of an introduction and high arch dam site seismic motion inputs, high arch dam dam body–reservoir water–foundation system earthquake response analysis and seismic safety evaluation, large dam concrete dynamic resisting force test, numerical simulation, and its mechanism of three sections covering 15 chapters in total.

Part I aims at a series of key technical problems of general characters and the foundation of fundamentals existing in the seismic motion inputs in earthquake resistance design of the current Chinese high arch dams. Overall, systematic, and in-depth research can be made on the appropriate determination of design and prevention standard framework, rational selection of seismic motion parameters, and correct understanding of the input mechanisms, and so on. Much more confusion of the many basic concepts existing in engineering application have been analyzed and clarified, and the approaches to solving these problems and theoretical bases are suggested and discussed. Also, a whole set of new thinking ways and new methods for the dam site seismic motion inputs are provided in close combination with our country's conditions and high arch dam characteristics, which are more rational and operable. The corresponding software development has been carried out so as to provide the basic data and scientific bases for the modification of hydraulic structure seismic safety codes.

Part II deals with the research and development work around seismic response to high arch dam systems, including a complete set of new analytical methods and software for high arch dam body–foundation system seismic response, being near engineering a new way of thinking and method of "arch dam seismic safety stability and seismic safety evaluation with deformation as the core," as well as the corresponding software system. These analytical methods and independent research and development software have been used in all the important water resource and hydropower projects, preferably with arch dams of above 300-m high in the hydropower development and construction in western China. These methods have become the important reference bases for the seismic safety design. Particularly after the Wenchuan earthquake, they have been extensively used in carrying out seismic safety reexamination of high arch dams in the important hydropower engineering works in accordance with the requirements by the National Development and Reform Commission, and also they have provided scientific bases for the revision of hydraulic engineering seismic safety codes that are underway. The parallel computation platform structured with high performance and the software researched and developed based on finite element language have obtained the obvious results and expended the depended research space of high arch dams, holding the vast application potential in the future.

Part III introduces research achievements obtained from carrying out the dynamic resisting force behaviors tests, numerical simulation method, and dynamic damage mechanism, whose contents include three aspects: (1) the large dam concrete dynamic resisting force aspect – mainly aiming at the urgent problems to be solved in the present high arch dam engineering seismic safety design, that is, large dam concrete full gradation material dynamic flexure tensile behaviors, and particularly the effect of prestatic loading upon the large concrete dynamic flexure tensile behaviors; the introduction of the material testing results in combination with the practical engineering working research on the concrete tensile of constitutive relation within the softening section included, and probing into the dynamic sound emission detection in concrete tests; (2) Application of CT technology to studying the internal damage spread of concrete testing samples or specimen and testing research on the damage process – dealing with the developing of dynamic–static tensile pressure loading adaptable to CT testing online. CT experiment technology concerning concrete damage and destruction, data acquisition, imagine processing and analysis method, and three-dimensional image reconstruction and carton display of space crack morphology and extension process; and (3) Digital concrete numerical simulation has introduced the following aspect – the multigradation aggregates to satisfy the requirements by the real engineering matching ratio and the space random input method for each media performance, the model establishment of three-dimensional mesomechanics of full-gradation large dam concrete materials, and the profile division of mesofinite element grid, the research on the strain rate effect relation between the counted media strength and elastic modulus as well as on the failure process of concrete materials in damaged variable evolution, the implementation of the series, parallel computation method and program research and development of the nonlinear dynamic numerical analysis, the comparison made of the results of tests and calculations of dynamic regulations of large dam concrete under the seismic conditions and the discussion of mechanism of fracture.

The general description and Part I of this book are written by the academician Chen Houqun of The Chinese Academy of Water Resources and Hydropower, including part of the contents of the doctoral student thesis of Zhang Cuikan under his guidance. Part II is written by academician Chen Houqun, Ma Huaifa, and Tu Jin of the Chinese Academy of Water Resources and Hydropower, including part of the contents of the doctoral thesis of Wang Litao under the guidance of his instructors or tutors Chen Houqun and Ma Huaifa. Part III is written by academician Chen Houqun, Ma Huaifa of the Chinese Academy of Water Resources and Hydropower, Wu Shengxin and Zhou Jikai of Hohai University, Dang Faning and Ding Weihua of Xi'an University of Technology with joint effort, including part of the contents of the doctoral thesis of Wang Yan under the guidance of Wu Shengxin. The whole book was planned and unified by the academician Chen Houqun.

This book consists of 15 chapters in three parts, translated from Chinese into English by senior translator and professor Kang Shusen of Xi'an University of Technology and Professor Zhou Jikai of Hehai University, of which Chapter 1–9, 14, and 15 are translated by Kang Shusen, and Chapters 10–13 are translated by Zhou Jikai. The whole translation version is unified by Kang Shusen.

As viewed from the engineering viewpoint, the purpose of the research contents in this book is to solve the fundamental key technical problems confronted in the safety evaluation of high arch dam earthquake resistance in the current strong seismic areas of our country. Largely owing to the fact that there are few precedented examples in constructing high arch dams of 300 m high in strong seismic areas, there has been a lack of engineering practices and seismic damage practical examples to follow as references. For this reason, it is bound to involve a series of scientific frontier problems to break through the traditional ideas and current methods. Accordingly, some problems are still in the stages of continuing to probe into and undergoing the engineering practice identification so that the viewpoints and the initial conclusions in this book need to be further developed and deepened. We expect to further probe into it with the broad masses of readers wholeheartedly, and we would like the readers to criticize and point out mistakes to be corrected.

In addition, the main research work in the book has obtained the financial support from the National Natural Science Foundation Project (90510017, 91010006, 51079164), Sci-Tech Innovation Project (SCX2002-02), the Public Welfare Special Project of the Ministry of Water Resources (200701004), and at the same time, also gained the support from Kongming Survey and Design Academy of Sinohydro Consultant Group, and Chengdu Survey and Design Academy of Sinohydro Consultant Group. We would like to extend our heartfelt thanks to them.

GENERAL DESCRIPTION

1.1 CONSTRUCTION AND SEISMIC SAFETY OF HIGH ARCH CONCRETE DAMS IN CHINA

Water and energy resources are the important material base for human society development, being directly concerned with the sustainable development of society and the national economy, the raising and improvement of the people's material and spiritual lives, as well as the important constrainable factors affecting economic and social development in China. In the course of realizing the grand objective of modernization and a moderately prosperous society in all aspects, China is facing the great challenges of population, resources, and environment.

Water resources serve as the basic natural resources and strategic economic resources are the important component of national comprehensive strength. Per capita water resources in China are extremely short, only accounting for 1/4 of the world per capita water resources. Based on the statistics by the World Bank, China among 153 countries in the world is ranked in the 88th place. It is predicted that by the year 2030, when the population in China is expected to increase to 1.6 billion, per capita water resources will be only 1760 m^3, being less than the internationally recognized water use standard for the countries with water sress. However, being subject to the influence of monsoon climate conditions, runoff amount distribution in time is so extremely uneven that runoff amount of flood water in the flood season within the year accounts for about two-thirds of the total. Accordingly, there are violent changes from year to year so that there frequently occur flood and drought disasters, thus seriously constraining social and economic development and affecting the ecological environment. The special distribution of water resources in China is also extremely uneven, successively decreasing from the southeast to the northwest and not matching with land resources. The construction of reservoir dams can regulate the utilization of floodwater in the flood season as possible, which is of great strategic significance for the rational disposal, utilization and drought resistance, flood prevention, and mitigation of disasters of water resources.

Energy resources are one of the important strategic resources in economic and social development. Energy resource security is widely concerned by every country in the world and plays a decisive role in completing the building of a moderately prosperous society in all aspects and the middle- to long-term strategic objectives of marching forward the completion of modernization. Electric power particularly served as advance of the national economy and basic industry has become even more important than any other secondary energy. The total sum of the national economy in China is large, but energy resources are relatively few so that per capita energy resources only equal to half of the world average level. Coal-fired electric power in particular consists of nearly 76% of the secondary energy structure, which is very hard to sustain because of its being subject to constraints by the environment and water resources capacity. Although the energy structure with coal-fired electric power as the main source is difficult to change in the middle-to-long-term, it is urgent to decrease its amount of contribution. China holds great potential for developing nuclear energy and renewable energy such as wind power energy, solar power energy, and bioenergy. Largely owing to the mixture causes of quality characteristics, economy, technology, resources, and so on, it is predicted that the percentage of wind power, solar power, biopower, and nuclear power energy in the power energy installation capacity cannot be large at least prior to 2030. Water energy resources in China occupy the first place in the world. The exploitable capacity by hydropower technology in mainland China will be 0.54 billion kW. Water energy resources account for about 40% of the exploited conventional energy resources based on the calculation of the 100-year utilization. In the middle- to long-term, water power energy, including pumping water for energy storage, occupies 22.3% of total energy, which has been the main energy in dealing with climate changes, reducing harmful emissions by coal-fired power plants, and improving the secondary energy structure. Great efforts made to develop renewable water power clean energy resources can be a sustainable engineering project to concentrate the national land realignment, river exploitation, flood prevention and drought resistance, coping with climate changes, optimization of energy structure, renewal of regional economy, poverty alleviation, and eco-improvement into one entity. High dams and big reservoirs characterized by good regulation performances, large installation capacity, and highly comprehensive economic returns can play the important and nonreplaceable role in hydropower engineering construction.

With a growing enhancement in the global sense and sustainable development demand, the realization of high-dam and big-reservoir effects and functions by the international community continues to be deepened. In the prerequisite of full attention paid to the settlement of migrants and the influence of the eco-environment, the construction of reservoirs and dams should be carried out in an active and orderly way, with high arch concrete dams built in particular, which is very suitable to the national conditions in our country and the urgent needs for socioeconomic development. Accordingly, the construction of reservoirs and big dams has become the indispensable and important component in the infrastructure construction in our country.

1.1.1 THE GENERAL CONDITIONS OF HIGH CONCRETE ARCH DAM CONSTRUCTION IN OUR COUNTRY

Since 1949 when the new China was established, our country has actively carried out the construction of big dams and reservoirs. There are 130 dams of higher than 100 m in China, out of which more than half are concrete dams, which is roughly the same percentage of concrete dams of higher than 100 m among 800 dams in the world. The percentage of concrete dams grows with a gradual increase in dam height. In China, in the dams with a height of 150 and 200 m, the concrete dams reach 58% and 78%, respectively. Particularly, the river heads of the big rivers and water energy resources in the country are concentrated in cliff river valleys of high mountains in the west part of the country, where the landforms and geological conditions are favorable for the construction of high concrete arch dams in relatively less inundated lands and with better regulation performances. Recently, a series of critical high dams of 200–300 m were constructed and will be constructed in the western part of China; most of which have adopted the schemes of concrete arch dams. The Xiaowan concrete double-curvature arch dam with a height of 294.5 m completed and impounded in the initial stage is the highest arch dam at present in the world. The Jinping Grade 1 concrete double-curvature arch dam with a height of 305 m is to be constructed. After this double-curvature arch dam is completed, it will replace the Xiaowan engineering project and will become the highest arch dam in the world. China is at present regarded as the largest country to build large dams internationally, and is building a series of rare precedent concrete arch dams, belonging to the frontier of international first-rate concrete high dam engineering projects.

1.1.2 SEISMIC GENERAL CONDITIONS IN DAM SITE OF CONCRETE HIGH DAMS IN CHINA

The Chinese mainland is in the clumping of several big plates in the earth's crust and is located in the crossing of two of the most active seismic belts in the world, bordering around the west branch of the Pacific earthquake belt in the east and the Eurasian earthquake belt passing through the west and southwest parts so that China is a country with more earthquakes. The Chinese mainland falls into the east part of the Eurasian plate so that the occurrence of earthquakes is characterized by the inner plate of the mainland inner part. The plate inner earthquake with high intensity can often aggregate large amounts of energy, so as to lead to the great intensity of earthquakes because of the thick earth crust and old rocks. Hence, the seismic source lies mostly in the depth of 10–30 km, but the seismic disasters caused by the inland shallow seismic source are always highly destructive. For this reason, China has the vast seismic regions but dispersed as well as frequent occurrence of earthquakes with violent intensities. Only within the twentieth century, there were earthquakes with the seismic intensity of no less than 8 scales, reaching over 10 times as many. Based on the data from the China Seismic Bureau, earthquakes of over 5-degree intensity took place in every province of China's mainland, of which earthquakes of over 6-degree intensity occurred in 29 provinces except for Zhejiang and Guizhou provinces; earthquakes of over 7-degree intensity happened in 20 provinces. There are two apparent earthquake belts in China's mainland: one earthquake belt passes through the middle part of national territory from the north to the south, and another earthquake belt passes through the northern regions in China from the east to the west. The seismic disasters have seriously threatened the safety of the lives and properties of the people, and at the same time, the seismic disasters are one of the important factors that constrain economic construction and social development. China is a country suffering from the most severe seismic disasters in the world. It has been recorded in human history that there were six strong earthquakes causing the death of over 200,000 persons, for which China accounted for four times. In the twentieth century, the Haiyuan earthquake in Ningxia took place in 1920, in which over 200,000 persons were killed, and the wounded in the earthquake were innumerable; the number of deaths in the 1976 Tangshan earthquake reached over 240,000 persons, and the losses were very extremely disastrous. Based on the statistics by the China Seismic Bureau, and since the twentieth century, the total number of deaths has reached nearly 1.2 million persons in the whole globe, of whom there have been nearly 600,000 persons in China, accounting for about 40%. From 1949 to 2000, the death of persons caused by various kinds of natural disasters in China reached about 550,000 persons, of whom the deaths due to the seismic disasters accounted for half, reaching 280,000 persons. Accordingly, the seismic situations and seismic disasters in China are one of the basic national situations of the country.

The western areas in China are the main seismic areas of the country. The seismic intensity, whether distributed in time and in space or in the west geosynclinal area, is larger than that in the platform area. In the modern times, 82% of strong earthquakes in the mainland take place in this region, while about 80% of water energy in China is concentrated on the high seismic area in western China. There are many dam sites in the high and steep gorges in the western areas where the concrete high arch dams are suitably constructed, but the regional formation by these landform and geological conditions can just be the conditions that carry the strong earthquakes. Therefore, the association and connection of dam sites and regional geological formation make it difficult to avoid high seismic areas so that keeping the seismic functions can often become the controlling engineering condition in design. For instance, a series of concrete high arch dams are being constructed in the strong earthquake areas in western China; the maximum dam height of the Jinping one-stage arch dam is 305 m, and the designed seismic acceleration is 0.2g (g = 9.8 m/s^2); the maximum dam height of the rock-filled dam at the mouth of Shangjing River is 314 m, and the designed seismic acceleration is 0.205g; the maximum dam height of the Dagang mountain arch dam is 210 m, and the designed seismic acceleration is 0.557g; the maximum dam height of the Xiluodu ferry arch dam is 278 m, and the designed seismic acceleration is 0.32g; the maximum dam height of the Ertan arch dam is 240 m, and the designed seismic acceleration is 0.2g; the maximum dam height of the Baihetan arch dam is 275 m, and the designed seismic acceleration is 0.325g; and the maximum dam height of the Xiaowan arch dam is 294.5 m, and the designed seismic acceleration is 0.31g.

As a result, severe seismic situations that are difficult to avoid are the severe challenges the country must face in building high concrete arch dams.

It is well-known that earthquakes are a very complicated phenomenon of natural disasters. Since earthquakes take place in the deep part of the crust, it is very difficult to directly observe and measure the genesis of seismic occurrence and its mechanism, whereas the seismic wave transmission approaches and medium conditions are very complicated. Also, there is a long interval of occurrence of strong earthquakes so that statistical data are only few but discrete, the understanding of the laws of the seismic occurrence is not enough, and there has been a big random time, location, and intensity in strong seismic occurrences. For this reason, seismic prediction has been a technically difficult problem that has not been well solved in the world hitherto. When many strong earthquakes take place, there are no obvious forewarning phenomena so that it is difficult to do early warning of seismic occurrences. As far as high arch dams and large reservoirs are concerned, once they suffer from destruction caused by strong earthquakes, the artificial peak flow that results in the dam failure will cause a serious secondary catastrophe in the downstream areas, which can be much heavier than the losses caused by the engineering itself. The high arch dam and large-scale reservoir in particular serve as the dragon-head reservoir in the cascade reservoirs, and the severe earthquake disaster change can result in the secondary catastrophe, whose consequence is so unthinkable that measures must be adopted to prevent them from happening.

Therefore, ensuring the seismic safety of high concrete dams and large-scale reservoirs with concrete dams in the main to prevent severe seismic disaster change is the serious challenge that must be encountered in water conservancy and hydropower construction in China, which has been of high concern by the country and society in such a way that great attention must be paid to it. What is particularly important is that the science and technology support for seismic safety of high concrete arch dams at present in the country cannot satisfy the requirements of the scale and speed of high arch dam construction so that intensive efforts must be made to deepen research on seismic safety of high concrete arch dams.

1.2 THE BASIC CONCEPT AND ITS SEISMIC CASE ENLIGHTENMENT EVALUATION OF SEISMIC SAFETY OF HIGH CONCRETE ARCH DAM

1.2.1 THE BASIC CONCEPT OF SEISMIC SAFETY EVALUATION OF HIGH CONCRETE ARCH DAM

The seismic problem of high concrete arch dams is a multidisciplinary crossing and a very complicated problem that involves hydraulic engineering structure, hydraulic mechanics, concrete materials, seismology, geology, and so on. In the research on the evaluation of seismic safety of high concrete arch dams, the following basic concepts must be implemented from the beginning to the end:

1. *Stressing the comprehensive evaluation*: The evaluation of seismic safety of any engineering structures should include seismic dynamic input, structure seismic response, and structure material resistance force of three basic elements. They are the indispensable and intermatching component parts, and they are particularly important for the vital engineering of high concrete arch dams. At present, research on seismic resistance of high concrete arch dams depends on the rapid development of computation technology. Research on seismic response of high concrete arch dams is gradually becoming finer and finer. Many problems difficult to analyze in the past are being gradually solved. Research on seismic dynamic input and structure material resistance force has still been rough, while seismic dynamic input and dam concrete dynamic resistance force have had an important effect upon the evaluation of seismic safety of engineering structures. For this reason, the situation of "small two ends, big intermediary" in the evaluation of seismic safety cannot be changed, and it is in fact that "the two ends" can only be used to control the accuracy and level of the safety evaluation of seismic resistance. Nevertheless, the "two ends" involved with the multi-interdisciplinary crossed problems with greater difficulty, and less research work has been done on this respect so it is urgent to be strengthened.

2. *Protruding engineering viewpoint*: Research on seismic safety of high concrete arch dams is to start from solving the urgent needs of current engineering construction to an objective to solve the problem of seismic resistance in actual engineering construction. Since the problem involves the crossing of multiple disciplines, it is necessary to surround the central objective to trace the developing frontier of some related sciences to which attention must be paid for their absorption and digestion and integrated application to the reinnovation of research on seismic safety of high concrete arch dams. However, the seismic safety of high concrete arch dams in engineering design belongs to the special working conditions, which must be required to be coordinated with the design method of basic working conditions, parameters, norms and accuracy, and being subject to their constraints. Seismic working conditions can only be based on the basic working conditions to carry out seismic resistance check and if necessary, the appropriate engineering seismic resistance measures must be adopted. As viewed from this sense, the so-called "optimum design of seismic resistance" is used to optimize the clam type and design parameters of high concrete arch dams, being irrational, for the earthquake is only the occasional function of miniprobability so that seismic working conditions cannot be used as the optimized objective. In addition, dynamic analysis in seismic resistance design is, in general, more complicated than static analysis, but in the physical concept, it is required to approach near reality, and the calculation accuracy must be appropriately matching with the basic working conditions. In taking the uncertainty of seismic function itself into account, it is essential to grasp the essence to simplify the secondary factors as possible so as to apply them to engineering.

3. *Paying attention to practical test*: Up to now, there has been little practical experience in constructing high concrete arch dams in the strong seismic areas abroad, so that there are only a very few high concrete arch dams suffering from the strong seismic functions in the world, and the seismic catastrophic cases are very few. In the recent 20 years, the construction of high concrete arch dams in China has developed rapidly. There have been few precedented cases and qualified engineering experiences that can be borrowed. Some of the traditional concepts and methods based on the past engineering practices are difficult to be used completely. In order to ensure seismic safety of engineering structures, there must be the innovative spirit to make the breakthrough, suggest new concepts, and adopt new materials and calculation methods. Largely

owing to seismic motion uncertainty and the complexity of seismic resistance of high concrete arch dams, although much progress has been made in present calculation tools and methods, a certain hypothesis and simplicity must be introduced into the analysis of dynamic response to seismic resistance of high concrete arch dams, whether the new concept is suggested or more complicated and rational mathematical models are established, or the parameters are determined. For this reason, great attention should be paid to the in-depth analysis of earthquake disaster real cases and observed records of strong earthquakes from which the enlightenments can be obtained and the tests can be accepted.

1.2.2 EVALUATION OF SEISMIC RISK OF HIGH CONCRETE ARCH DAMS

1.2.2.1 Typical seismic cases of high concrete arch dams

"Practice is only one standard for testing the truth." The concept of seismic safety of high concrete arch dams, theoretical analysis and test results, design, and engineering measures are subjected to tests in the seismic cases, particularly in the strong seismic cases, so as to probe into the enlightenments for their improvement and to let them fit the reality. The seismic response to high concrete arch dams is very complicated, whose main behaviors of seismic resistance dynamics can first be obtained from the investigations and summarization of earthquake disaster cases in which enlightenments can be obtained. Up to now, there have been very few cases of big dams (high concrete arch dams in particular) suffering from earthquake disasters in strong earthquakes. Accordingly, their seismic resistance dynamics can still be worth continuing to probe deeply into the cases. Over 40 seismic cases of high concrete arch dams experiencing earthquakes with different intensities in each country concerned in the world have been reported, of which there have been seven cases of high concrete arch dams with a height of over 100 m clearly suffering from the strong earthquake with 7-degree intensity. For instance, the Naber arch dam with a height of 112 m completed in Chile in 1968 experienced the 1985 earthquake with 7.7-degree intensity; earthquake epicenter distance was 80 km; the dam site intensity was 8-degree magnitude. The Lumiei arch dam with a height of 136 m completed in Italy in 1947 experienced the 1976 earthquake with 6.5-degree intensity; earthquake epicenter distance was 43 km. The Pieve di Cadore arch dam with a height of 112 m completed in Italy in 1949 suffered from an earthquake with 8-degree intensity. The Vid.Arges arch dam with a height of 167 m completed in Romania in 1965 experienced the 1977 earthquake with 7-degree intensity; the Pacoima arch dam with a height of 113 m in the United States experienced the 1971 San Fernando earthquake with 6.6-degree intensity and the 1994 Northridge earthquake with 6.8-degree intensity. The Sharpei rolled concrete arch dam with a height of 130 m completed in Sichuan Province in China in 2001, and the Deji double-curvature arch dam with a height of 181 m completed in 1974 in Taiwan experienced the 2008 Wenchuan earthquake with 8-degree intensity and the Titi strong earthquake with 7.6-degree intensity, respectively. There exist arch dam seismic cases including the high concrete arch dams with a height of over 100 m that have experienced through the strong earthquake with 8-degree intensity, most of which have had no seismic disasters or only had some seismic disasters such as seepage volume enlarged after the earthquake and local cracking to be repaired. Also, there are the various disaster cases of dam failures in an earthquake. For this reason, as far as the general condition of arch dams, seismic risks are concerned. High concrete arch dams that are designed in accordance with the requirements of seismic resistance specifications and constructed in terms of ensuring construction qualities, all are of rather strong seismic resistance capacity. In fact, the countries with more earthquakes, including Japan,

Iran, Turkey, Italy, the United States, Chile, Mexico, and so on, have constructed a large number of high concrete arch dams.

In the existing seismic cases of high concrete arch dams, it is worth pointing out that there are two seismic cases of high concrete arch dams with heights of over 100 m having experienced through a strong earthquake: (1) the Sharpei rolled concrete arch dam in China, having experienced through the 2008 Wenchuan strong earthquake tests, has remained perfect without any damage; and (2) the US Pacoima arch dam, having experienced through two strong earthquakes successively, has shown similar damages on the gravity pillar of the arch sent on the left bank.

The Shapai rolled concrete arch dam was completed in 2001. The dam height is 132 m. It is a gravity arch dam with a three-center arch. There are two horizontal cracks and two induced cracks set in the dam body. The radial longitudinal crack spacing is 45 m. A 12.5-m-thick concrete cushion is set in the dam bottom. The 90-day-old no. 900 concrete is adopted for the dam body. The dam foundation is on the stratum, consisting of granite and granodiorite as well as the later-intruded diorite. The Shapai rolled concrete arch dam was the highest rolled concrete dam at that time. Geological formation in the dam site zone is so complicated that seismic activities are very frequent. The terms of the analytical results of special seismic risk for engineering site made by the Seismic Department indicate that the seismic basic intensity in the dam site with 50-year transcendental probability of 10% is 7-degree magnitude, whose horizontal peak acceleration of land surface bedrock is $0.138g$. Based on the provisions in the current hydraulic seismic design specifications, peak acceleration and standard design response spectrum for the Shapai arch dam are designed in accordance with the horizontal $0.138g$ and vertical $0.92g$. The two kinds of dynamic analysis methods, the safety arch beam and finite element methods, of nonquality foundation are adopted to carry out the seismic resistance design of two kinds of working conditions in operation at normal water storage level and low water level. Meanwhile, based on the traditional rigid limit balance method, the analysis is made on the stability against sliding on the dam shoulders, and the dam strength is rechecked without any defects detected.

The Shapai arch dam during the Wenchuan earthquake was 36 km away from the epicenter, being located in the seismic intensity area with 9-degree magnitude, and the reservoir water storage level was close to the normal water storage level. Based on the photo taken by the helicopter soon after the earthquake, the initial survey by Chengdu Survey and Design Academy of Sinohydro Consultation Group indicated that the dam body structure was perfect and complete after the earthquake (see Fig. 1.1). The reservoir bank neighboring dam body was treated via the anchored resisting force body so that the side slopes are all still stable (see Fig. 1.2).

The US Pacoima arch dam has experienced through two strong earthquakes during the past 23 years. After the first strong earthquake, the damaged left dam shoulder part was consolidated in seismic resistance. In the second strong earthquake, similar damage appeared in the same part, with the strong seismic records of dam site and dam body actually measured, and there is a detailed survey report after the earthquake. Accordingly, the dam can be the rare seismic disaster case where an arch dam has suffered two strong earthquakes. For this reason, the Pacoima arch dam has aroused extensive attention. It is worth conducting an in-depth analysis and discussion of this rare seismic case of a high arch dam suffering from two strong earthquakes. Accordingly, in accordance with some related documents, a detailed introduction can be given to the engineering conditions and seismic disasters.

The Pacoima arch dam was built in 1928. Its main purpose is for flood prevention and water supply. It is a constant angle double-curvature arch dam with a height of 113 m, being the highest dam at that time in the United States but having the reservoir capacity of 7.5 million m^3. The ratio between

FIGURE 1.1 A Bird's-Eye-View of Shapai Arch Dam After Earthquake

FIGURE 1.2 Resisting Force Body Side Slopes on the Right and Left Banks of the Sharpei Arch Dam After Earthquake

the width and height in the river valley is 1.5; the length of the dam crest is 180 m; the thickness of the dam crest is 3.2 m; the thickness of the dam base bottom is 30.2 m; the ratio between the thickness and height of the dam body is 0.27; the elevation of the dam crest is 614.58 m above sea level. The keyways are set in the horizontal crack. Figures 1.3 and 1.4 are the plane figure and the arch crown section figures of the dam, respectively. The dam site bedrock is gneissic quartz diorite with the joint

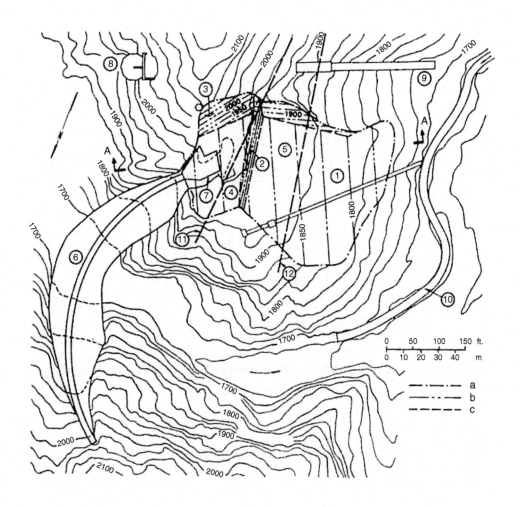

——— - ——— Broken surface

——— - - ——— Broken surface contour line

━ ━ ━ ━ ━ ━ Broken surface cross-cutting

FIGURE 1.3 Pacoima Arch Dam Plane

1, Broken face 1; 2, broken face 2; 3, broken face 3; 4, rock body "B"; 5, rock body "A"; 6, arch dam; 7 gravity pier on the left bank; 8 flood discharge tunnel inlet; 9, flood discharge tunnel outlet; 10, traffic way; 11, survey line "B"; 12, survey line "C".

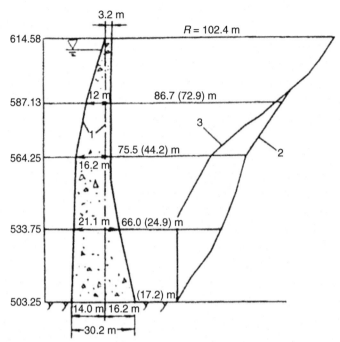

FIGURE 1.4 Pacoima Arch Dam Arch Crown Section (Numbers in the Bracket are the Values of Inner Radium)

1, Dam; 2,3, the upstream circular-arc central connective line.

and shearing belt development. A gravity pier with the height of 18 m is set up because of the poor geological conditions on the upper part of the left bank. The keyway is set in the joint seam between the pier and dam body, and can be used for grouting. The dam is designed using the arch crown beam method and without taking seismic function into account. In 1978, after the arch dam was in operation for 38 years, the trial-load method was used to carry out recheck, and the level acceleration of $0.15g$ evenly distributed in terms of static method is accounted into the seismic inertia functions. Based on the testing results of the dam body drilling hole for taking cores, the concrete compressive strength and tensile strength were to take 24.6 MPa and 12.23 MPa, respectively, in calculating the compressive and tensile safety coefficients with their results, reaching as high as 6.8 and 5.9 respectively. It has been proved through survey that rock body joints on the left dam shoulder well develop and one shearing belt is discovered again, and that the safety coefficient of stability against sliding on the checked left dam shoulder is only 1.0 or so. It is considered that consolidation treatment needs to be done.

On February 9, 1971, a strong earthquake with 6.6-degree intensity took place in San Fernando. The seismic source was about 13 km, and the dam site was only 6.4 km away from the seismic center. The seismic San Fernando fault is located 5 km downstream. Dam shoulder bedrock rock surface on the left bank 37 m away from the dam site is 15 m higher than the dam crest. The actually measured acceleration level and vertical component to the peak values are $1.27g$ and $0.27g$, respectively, and sustained time is about 10 s. Water depth in the reservoir was only 65% or so of full water storage in the reservoir when the earthquake took place and water level in the reservoir was under 45 m of the dam crest. The researchers estimated after the earthquake that the acceleration level component under the peak value of the dam bottom bedrock was $0.6g–0.8g$.

Earthquakes have caused the obvious broken phenomena of the land surface. The detailed survey results after the earthquake indicated that part of the mountain body at the dam site has been upgraded by 1.28 m, and levelly moved 2 m toward the south and the west, leading to a little contraction in the width of the river valley at the dam site so that the distance between the arch piers of both banks was shortened by 23.9 mm. The dam body was subjected to pressure, dam axis rotated 30 s clockwise, and dam crest on the right bank settled down by 17.3 mm in comparison with that on the left bank.

The main seismic disasters in the dam body in this earthquake are the vertical and radial horizontal cracks between the dam body and gravity pier on the left bank opening by 6.35–9.7 mm, extending 13.7 m in length, and ceasing at the intersection with one construction seam. The gravity pier on the left bank cracks along the horizontal construction seam, whose crack extends 1.5 m afterward, and goes down with an inclined angle of $35°{\sim}55°$ and intersects with the downstream bedrock of the arch pier at an elevation of 598 m above sea level. This again intersects with the bedrock below the gravity pier at an elevation of 596 m above sea level. The dam crest has a 15 mm upstream displacement so that no crack can be found in the dam body of this arch dam. Figure 1.5 are the upstream and downstream views of the gravity pier cracks on the left dam shoulder of this arch dam, respectively.

The grouting rock surface on the upper part of the left dam shoulder appears to have a large area crack and subsidence of about 8000 m^3, and the sloughing takes place in the downstream slope on the dam shoulder. The mountain body on the left bank dam shoulder meets with two air faces because the downstream river bends to the left. Therefore, in 2–3 m depth below the lower part of the gravity pier, there is a slip under the less inclined angle bottom sliding surface of the rock body along the inclined river bed. Figure 1.6 is an A–A profile figure in Fig. 1.3, indicating the situations of local ruptures of lower bedrocks of the gravity pier on the left bank. The sliding body is divided into two parts by the steeper-inclined angle rupture surface. The upper rock blocks A slip largely, reaching 200 mm vertically and 200–250 mm horizontally. The mountain edges on the left bank are damaged seriously so that rock blocks slide downward along the slopes, thus leading to blockage of the intake of the flood discharge tunnel in which there are four cracks occurring on the concrete lining face. The discharge flow steep flume is damaged. Apart from the upper part, the rock body on the left dam shoulder has had no damages so that the overall rock body is stable. The fine cracks and the opening of existing cracks, as well as the denudation of the rock face grouting layers, are detected on the arch piers and rock bed on the right bank.

A drilling hole for taking core tests and an inner hole photo indicate that the concrete within the dam body is good in quality in general and that any disturbances and relative motions have not been detected in the dam foundation interface and neighboring bedrocks.

The amount of seepage from a drainage hole on the left dam shoulder increased from 0.16 L/s to 1.89 L/s, and after 4 days decreased to 0.64 L/s, but this is still higher than the value prior to the earthquake. The records of six piezometers indicate that the underground water level suddenly increased when the earthquake took place, and decreased sharply a little later. After the earthquake, when the curtain grouting to consolidate core taking and pressure water tests were carried out, it was found that there were new cracks and reopening of the existing cracks on the rock body within the range of over 7.6 m and below 8.5 m of the contacting face between the dam body and bedrock.

The temporary consolidation was carried out in order to ensure the safety of tiding over the flood season after the earthquake. High-pressure water jet was used to clean away the disturbed rock blocks. The consolidated grouting was conducted to treat the disturbed rock body on the left dam shoulder up to the depth of 1.5 m below the disturbed zone, and the grouting enclosure must be carried out to treat the original grouting cracks. Seepage control grouting must be done within the range of 20 m upstream

FIGURE 1.5 The Cracks on Gravity Pier of the Left Dam Shoulder of Pacoima Arch Dam in the 1971 Earthquake

(a) Upstream face and (b) downstream face.

on the gravity pier, and the opened seams between arch dams and the gravity pier should be filled with soft filling materials.

In the permanent consolidation afterward, 35 pieces of post-tensioned prestressing anchorage cables, with each bearing 3100 KN tensile force, were used to anchor the sliding rock blocks on the upper part below the gravity pier into the lower part of the bedrock. The opened seams between the arch dam

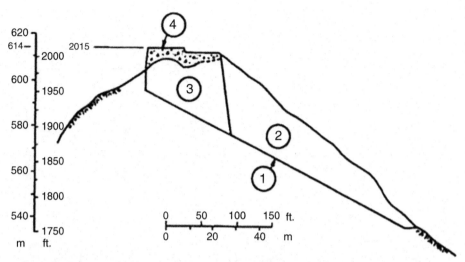

FIGURE 1.6 Local Broken Rock Blocks Below the Left Bank Gravity Pier (Fig. 1.5 A–A Profile Figure)

1, Broken crack face; 2, rock block A; 3, rock block B; 4, gravity pier.

and gravity pier must be sealed with epoxy resin from the upper to the lower. One row of the whole system curtain grouting should be made between the bedrock and dam shoulder up to the depth of one-third of the dam height. The consolidated grouting must be carried out to treat the damaged rock body on the left bank dam shoulder at an elevation of 518 m above sea level.

On January 17, 1994, an earthquake with 6.8-degree intensity took place 18 km of the southwest dam site. This was the Northridge earthquake with a depth of seismic focus of 17 km. When the earthquake took place, water level in the reservoir was 40 m lower than the dam crest, water depth was 71 m deep, and reservoir storage capacity was about two-thirds of full storage capacity. Particularly in this strong earthquake, the acceleration stations set in many locations of the dam body and the downstream valley land surface as well as the upper part of the bedrock on the left dam shoulder obtained the records of the full duration of time. Although the records on the dam crest, dam shoulder, and the upper rock bed have surpassed the limitation because of over shocks, they are still the hard-to-obtain and valuable data (CSMIP, 1994). Figure 1.7 is the layout figure of acceleration observation and measurement stations of the dam, of which nos. 1~17 stations are within the dam body. Table 1.1 shows the peak values of acceleration recorded by each station on January 17, 1994 when the earthquake took place.

The detailed investigation into the engineering seismic disasters indicates that though the cross-cracks between the left bank dam body and gravity pier were anchored using the prestressed anchorage cables after the earthquake in 1971, the cracks were pulled down again after this earthquake. The dynamometers set in the prestressed steel anchorage cables were damaged when the earthquake took place. This cross-crack top part opened with 50 mm, but the cross-crack opening at 3 m above the pier bottom was reduced to 6 mm; the gravity pier settled down slightly. There was a large crack inclined downward to extend to the pier bottom bedrock in the rock mass of the cross-crack end of the pier bottom. All these reveal that there were disturbed slips in gravity piers and their bottom local bedrock and in-between the piers. Also, there were microfine cracks declining upward to extend 15 m below the dam crest to the level construction seam in the dam body downstream face on the left bank arch pier

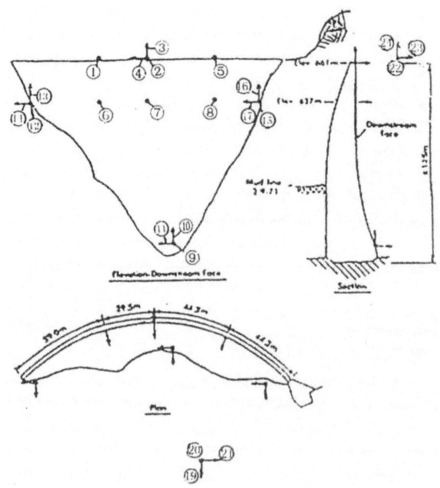

FIGURE 1.7 Layout of Pacoima Arch Dam Acceleration, Observation, and Measurement Stations

(a) Vertical face to downstream face, (b) profile of downstream face, and (c) plane of downstream face.

cross-seam bottom end. There were disturbances in the dam body in the neighboring cross-seams along the construction seams, whose upper part slipped 10–12 m downstream obviously. There are no apparent damages detected in the rest of the dam body. The joint gauge of the dam body cross-seam indicates that the gravity pier relative to the left bank arch abutment has moved 46 mm tangentially to the left and 12 mm to the radial downstream. The rest of the cross-seams had the signs of opening and closing in the earthquake. Particularly, it is difficult to measure the residual open degree closed after the earthquake in the arch crown position because of too-small openness. Also, there are several millimeters of settlement of the right side of the dam section in each cross-seam, being relative to the left dam section, but there is a shortage of comparable information with the data prior to the earthquake. It is likely that the keyway

Table 1.1 Actual Measurement Values of Each Acceleration Observation and Measurement Station of Pacoima Arch Dam in Earthquake in January 17, 1994

Stations	1	2	3	4	5	6	7	8	9
Directions	Parallel river	Parallel river	Vertical	Cross	Parallel river	Parallel river	Parallel river	Parallel river	Parallel river
Peak-value acceleration	>1.5g	>2.0g	>1.5g	–	>2.3g	0.98g	>1.5g	1.59g	0.49g
Stations	10	11	12	13	14	15	16	17	
Directions	Vertical	Cross	Parallel river	Vertical	Cross	Parallel river	Vertical	Cross	
Peak-value acceleration	0.43g	0.54g	1.03g	>1.7g	–	1.54g	>1.3g	>2.0g	

Note: g = 9.8 m/s².

function between cross-seams in each dam section makes the residual position disturbances become so small that the dam body can still maintain the entirety to a rather good extent.

Though the rock mass grouting layer on the upper part of the left dam shoulder stays mostly perfect, there has been a larger area cracking and a small local area grouting layer collapsed and fell with the horizontal loosened rock mass falling. The nine joint gauges were set in the existing cracks of the grouting layer prior to the earthquake to carry out monitoring every week. The results after the earthquake indicate that the existing cracks open obviously, and the measured values are correspondent to the rock mass motion morphology. The high and steep bank slopes 300 m away from the left bank downstream dam body have collapsed and slipped. The reinforced concrete protective facing in the discharge flume of the flood discharge tunnel was completely damaged, and the forward passages were blocked by fallen rocks. The 6-m-long lining within the flood discharge tunnel was seriously cracked and translocated, but the rest of the parts remained good.

The rock mass in the dam crest downstream on the right arch seat has an open fissure along the unloading joint of the parallel bank slope. No damages were found on the interface of the rock mass and bedrock. There is an unstable phenomenon on the rock side slope of the dam crest. There is a larger rock mass and landslide 60 m in the dam downstream and 40 m away from the dam crest to block the Pacoima gorge below.

Water ejection from the fissures of the river bottom can be seen from 459 m away from the downstream dam body, indicating that the underground water level is rising.

Earthquakes make the total amount of seepage water increase obviously. The largest amount was 70 L/m in the first month after the earthquake, being very obvious on the left bank, but gradually decreased after the earthquake. The underground water level recorded by the piezometer rose suddenly when the earthquake took place; the maximum value on the left bank was 3.9 m; the maximum value on the rear bank was 3.0 m, and they were gradually weakened. The variation in rock mass pore pressure indicates that the seepage curtain grouting in the rock mass suffered from damage during the earthquake, and the cracks in the rock mass of the dam foundation were opened, but the total amount of seepage was not large and gradually decreased, demonstrating that the damage of seepage prevention curtain grouting and drainage system was not serious.

In general, the arch dam in the two earthquakes displaced to the upstream; the length of the chord arch was shortened, and the joint seam between the left dam and the gravity pier was drawn apart and the rock mass moved. All these phenomena were roughly the same, but the degree of severity in the 1994 earthquake was much more serious than those in the 1971 earthquake.

The result of precision measurement of the clam crest of the arch dam after the earthquake shows that after the arch dam was displaced 15 mm to the upstream in the 1971 earthquake, this arch dam again was displaced 12 mm to the upstream in the succeeding earthquake, displaying that the seismic dynamic acceleration along the cross-river direction can result in the rock mass of the left dam shoulder to have a compressive effect upon the dam body, whose level component values at the location of the dam crest and dam bottom reached $2.0g$ and $0.5g$, respectively. The rock mass on the left dam crest slipped toward the downstream and river bed directions along the "B" surveyed line in Fig. 1.3, reaching 50–70 mm in level direction, and the vertical settlement of 50 mm. The very small phenomenon of opening degree of the cross-seam lower part between the left bank dam body and gravity pier indicates that rock block B was not out of stability completely and that the prestressed anchorage cables played an important role after the 1971 earthquake, but some anchorage cables suffered from damages because they were lengthened by overloading. The rock mass A slipped toward the downstream and river bed direction along the "C" surveyed line, reaching 400–480 mm in level direction and vertical settlement of 300 mm. Obviously, the rock mass A had experienced through slips, but there were no slope slides in its deep layer, so it recovered the balance after the earthquake. If there were no treatments of the cracks on the surface grouting layer, the rock mass A would be out of balance after suffering from a rainstorm intrusion and being likely to have strong after-quake.

Based on the records by the laser-reversed pendulum set in the arch crown within the dam in 1983, the relative dislocation 3 days after the earthquake was compared with the actually measured results 10 days before the earthquake. It has been found that the radial dislocation of the dam body was 46 mm to the upstream, and the tangential dislocation was 5 mm to the left bank.

The Sharpei Rcc arch dam in China and the Pacoima arch dam in the United States are the only global seismic cases that have been subjected to unpredictable extreme strong earthquakes with 9-degree intensity at present. The dam height of the former is subjected to extreme strong earthquakes, whose reservoir water level was over that of the latter, when the earthquake took place. Also, the rolled concrete arch dam is subjected to the temperature effect much larger than that of conventional concrete arch dams. After the earthquake, the former dam body remained basically complicated while the latter has had the severe seismic disasters. Accordingly, further in-depth analysis and comparison of strong seismic cases of these two high arch dams being subjected to severe strong earthquakes are of important significance for studying the seismic safety of high concrete arch dams. In addition, the seismic case of the Sharpei Rcc arch dam subjected to the extreme strong earthquake is of important reference significance for the seismic safety of this new type of rolled concrete arch dam.

The record of the dam body in a strong earthquake is of vital importance for understanding the actual seismic shock dynamics of arch dams. It is regrettable that apart from the Pacoima arch dam, there have been extremely few complicated and actually measured strong seismic cases of high arch dams in a strong earthquake. In this respect, it is worth noticing that the literature has reported the seismic cases of high arch dams of over 100 m in height in Switzerland. Although three arch dams were subjected to weak earthquakes with less than 5-degree intensity, the strong seismic observation and measurement stations were set

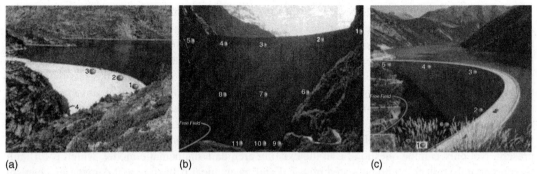

(a)   (b)   (c)

FIGURE 1.8 Locations of Three Strong Seismic Observation and Measurement Stations of Arch Dam Bodies in Switzerland

(a) Emosson arch dam, (b) Mauvoisin arch dam, and (c) Punt-Dal-Gall arch dam.

up in each location of the dam body and dam site downstream river valley bedrock (see Fig. 1.8) and the completed seismic record has been obtained. For instance, the Emosson arch dam with a height of 180 m was completed in 1974, being subjected to the 2001 earthquake with 3.6-degree intensity and epicenter distance was 10 km from the earthquake. The Mauvoisin arch dam completed in 1957 and heightened to 250 m in 1968 was subjected to a 1996 earthquake with 4.6-degree intensity. Epicenter distance was 12 km from the earthquake. The Punt-Dal-Gall arch dam with a height of 130 m completed in 1968 was subjected to a 1999 earthquake with 4.6-degree intensity, and epicenter distance was 12 km from the earthquake.

The *in situ* vibration tests for the three arch dams were conducted before the earthquake. The dam body-damping ratio measured via the land surface environment vibration (land surface impulsion) and forced vibration is 2–3%. The actually measured acceleration time scale of the dam downstream bedrock of each arch dam can be used as the seismic dynamic input. Also, the arch dam mathematic model of the present frequently used nonquality foundation is adopted to carry out the analysis of seismic dam body response. The results indicate that the calculated values of acceleration responses near the dam crests of the three dams are much larger than the actually measured values of the corresponding locations. If the calculated values are made to close the actually measured values, the dam body damping ratios in the calculation model will be enlarged to 15, 8, and 8% individually. The initial analysis holds that the actually measured damping values in the vibration tests have contained the dam body vibration energy to escape "radiation damping" effects to the boundless foundation in the far region. For this reason, the cause that led to the calculation result becoming large is likely that nonquality foundation assumes the even seismic dynamic input along the dam foundation. Actually, the seismic dynamic uneven input along the dam foundation has had some "averaged" function so as to make the foundation seismic input decrease.

In addition, the Deji double-curvature arch dam with a height of 181 m, in the Titi earthquake, with a dam site epicenter distance being 60 km, could not provide a strong seismic record of the dam foundation when in a major earthquake. However, the peak acceleration of 0.87g was recorded only in the overflow hole bottom on the dam crest. It is estimated that the dam site land surface peak value acceleration is about 0.3%. In the same year, on September 23, the dam bottom horizontal acceleration peak value of 0.139g was measured in an aftershock. The dam bottom horizontal and vertical peak value accelerations of 0.102g and 0.39g were measured respectively on September 27 after the shock.

1.2.2.2 Analysis and enlightenment of high arch dam seismic cases

Although there have been few seismic cases of high arch dams being subjected to strong earthquakes hitherto, as far as the evaluation of the seismic safety of high arch dams is concerned, in the mentioned few seismic cases, suffering from serious seismic disasters twice in particular, have had the completed strong seismic observation and measurement records and the data to describe the seismic case of the Pacoima arch dam in detail, which contained a great bulk of rare and valuable information. It is regrettable that though much progress has been made in the research on safety evaluation of seismic resistance of arch dams in recent years, it is rather difficult to carry out the quantitative tests of the feedback analysis via the actually measured dam site seismic dynamic inputs in the seismic cases at present. Not only can the question lie in the existing arch dam mathematic model, which is difficult to appropriately reflect the complicated physical and mechanical behaviors and damage mechanism in the real dam rock masses, but also the seismic resistance stability of the rock mass of the arch pier of an arch dam is the key to the seismic resistance safety of arch dams. What is even more important is that it is difficult to grasp the actual seismic dynamic input at that time. In general, the dam site seismic dynamic real measurement records that serve as the seismic dynamic inputs can include the dynamic interactive effects between the dam body and the foundation. Whereas the so-called "freedom filed" real measurement records of the downstream bedrock can be the seismic response to one point in the "scattering field" in the river valley before the dam construction and still include the geological condition effects of the river valley landform and near dam site field, but the overall seismic dynamic "scattering field" along the river valley uneven inputs cannot be given out. For this reason, when the earthquake is deduced in reverse from a few actually measured records of dam foundation and dam body downstream, the seismic dynamic inputs transmitted from the deep part of the crust in considering dam body and near regional foundation system bottom cannot be realized at present. This kind of inverse analysis belongs to the dynamic identification inverse problem without the existing uniqueness to the solution so that the difficult to seek solution is increased. Nevertheless, even if it is so, the consensus revealed in the newest scientific achievements can be used to carry out the deep thinking in such a way that many valuable and important enlightenments from the few seismic cases can be provided for qualification, tests, improvement, even breakthrough traditional and current concepts, thinking ways, methods, and so on, in evaluating the seismic safety of high concrete arch dams. They are primarily summarized next.

High arch dams relatively belong to the complicated special structures and seismic types are dense. The influences of high-order seismic types in seismic responses cannot be ignored. The dynamic enlargement effects of the dam body under the seismic action are rather obvious, and even if in the case of the larger structure damping, the dam crest dynamic magnification folds of the Pacoima arch dam increase as high as five times or so.

It can be seen through the comparison of actually measured acceleration time duration of the two banks of the Pacoima arch dam that even in the case of rather symmetric condition of river valley cross-sections, there are obvious differences in amplitude and phase of seismic motion of the left and right bank dam foundation due to the differences of mountain body topography and geological conditions in both banks. As a result, it is necessary to take the effects of uneven inputs along the dam foundation upon the arch dam seismic response into account. Proulx used the dam body downstream bedrock and actually measured acceleration time duration as earthquake motion input and adopted the frequently used nonquality foundation model to calculate the dam crest responses to three arch dams in Switzerland. The calculated results obtained are found to be much larger than the actually measured values. In that literature, these differences are simply finalized owing to the fact that the "homogenization" functions have resulted in a decrease in total inputs along the dam foundation uneven inputs when an earthquake takes place, and the literature holds that in the *in situ* measuring seismic tests, the measured dam body damping ratio in the case of microseismic situations has included the effects of the distance foundation radiation damping, being adaptable to seismic conditions. Accordingly, this analytical conclusion is worth discussing. In fact, the larger difference of dam crest response between the calculation and actual measurement are likely to reflect that the vibration energy of the dam body and neighboring foundation can escape the radiation damping impact to the unlimited distance foundation.

It should be pointed out that the land surface environment vibration (or pulse motion) excitation is used to carry out the dam body *in situ* vibration measurement, whose vibration amplitude value is in general only 1–2g, being much less than seismic motion input amplitude value when in the earthquake. When a mechanical eccentric device is used to produce the excited vibration force to carry out the suppressed vibration tests, the dam body vibration amplitude is extremely weak for such a huge dam body. Accordingly, the dam body weak vibration energy of the *in situ* measured vibration tests appears in the near regional foundation of the neighboring dam body so that there is no more energy that can escape to the far regional foundation. Under the seismic action, the excited dam body and the near regional foundation vibration energy are much larger than that of the *in situ* measurement. There is a rather part of outward wave energy that can escape to the far regional foundation. The influence of radiation damping will be obviously diminished in the seismic response to the dam body. The calculation results of more engineering indicate that in a strong earthquake, the radiation damping diminishing dam response can reach 20–40%, and the larger the seismic intensity is, the stronger the input of radiation damping is, being an important factor that cannot be neglected in analyzing arch dam seismic response. Of course, it is well known that although the damping of the dam body and foundation material in itself is characterized by nonlinear growth with an increase in seismic intensity, its value variation will not become large.

In addition, the actual seismic motion input "homogenization" along the dam foundation uneven inputs can only make the quantitative values reflect that the dam body total response acceleration and displacement have some reduction. As far as the dam body stress condition is concerned, the so-called "simulation static model" caused by the uneven displacement of seismic motion of the two bank dam foundations can lead to the obvious increase in local stress of the dam body due to the extreme sensitivity of relative displacement of the arch dam to the arch pier. In the present-day frequently used arch dam analysis model with nonquality foundation, in which the uneven inputs of the dam foundation cannot be taken into account, there is no method to account for the impact of radiation damping. It is considered that the stress growth caused by the uneven input may be offset to each other by the decrease in radiation damping to general response. It is just considered that the function of foundation quality inertia may be the same as the function of radiation damping being offset to each other. All these are the rough assumptions in concept confusion, being short of full evidence. The mentioned seismic cases can illustrate that the present-day frequently used arch dam dynamic analysis model with nonquality foundation can be difficult in coincidence with the dam body practical state as revealed by the seismic cases when an earthquake takes place.

The most important key to the safety of arch dams is the stability of the dam body arch piers. Under the seismic action, the arch piers of both banks on the arch dam upper parts are the weak locations of seismic resistance. The damages in the Pacoima arch dam in the two strong earthquakes were caused by the poor rock mass on the left bank. The severely damaged locations in the two strong earthquakes are basic repetition, whereby demonstrating its lawful nature. The serious consolidation treatment of the dam shoulder rock mass on the two banks of the Shapai arch dam is likely to become the important guarantee for the better seismic resistance.

The performance differences of both mentioned dams in a strong earthquake have fully shown that the stability of the dam shoulder rock mass has an important effect upon the safety of seismic resistance of arch dams. The seismic disasters of the Pacoima arch dam can also fully indicate that the dynamic deformation coupling between the dam body and arch seat foundation rock mass in a strong earthquake is extremely apparent. Under the action of a strong earthquake, the rock blocks under the gravity pier of the left dam shoulder of this arch dam are ruptured into A and B blocks, as shown in Fig. 1.7. The lower A rock block had once surpassed the limited balance state and lost the stability to slip, with obvious residual displacement. But owing to the seismic reciprocating function, rock blocks can maintain stability after an earthquake. The upper B rock blocks deformed obviously, leading to the cracks of the joint seam between the gravity pier and dam body on its upper part. The crack width on the dam crest reached 50 mm, but the crack width was diminished to 6 mm at 3 m above the pier bottom. The microcracks on the pier bottom indicate that B rock blocks cannot be out of stability, so that the rock mass block on the left dam shoulder still maintained stability. However, coupling deformation resulted in the cracks of the joint seam between the dam body and gravity pier so as to cause the dam body crack to suffer from damage. If there were no gravity piers, arch seats would directly connect with the rock mass, and the dam body located in the arch pier would be bound to suffer from serious damage because of rock block deformation. The performance states of the A and B rock blocks in the earthquake on the left dam shoulder of the Pacoima arch dam cannot be tested and explained by the traditional rigid limited balance method. This is because the rigid limited balance method is based on the static concept, which can neither reflect the influence of dynamic deformation coupling between the dam body and dam shoulder rock masses nor consider the thrust of the arch seat to the latent sliding rock blocks, the magnitude and directions of rock mass inertia force in itself, and the continuous changes in dynamic magnification effects as well as the reciprocal performances of seismic motion in itself in the course of earthquake. In fact, the A rock block instantly reaching the limited balance state cannot be out of stability to slip and fall down finally, whereas the B rock block without reaching the limited balance state makes the gravity and neighboring dam body crack and be damaged seriously because of the over-deformation. Accordingly, the seismic disasters of the Pacoima arch dam can illustrate that the application of the traditional rigid limit equilibrium method is difficult to check the stability of the arch dam shoulder seismic resistance of arch dams.

In addition, it is worthwhile to point out that owing to the dam line being too near the river course bend, the left bank mountain head at the Pacoima arch dam has had two air faces, being single-thin plus poor geological conditions of rock masses, whereby leading to the strong intensity of seismic strong response in the mountain head. The seismic peak value acceleration of the earthquake motion in the mountain mass at the dam crest in the two earthquakes reached 1.25g and 1.39g, respectively. And hence, the magnification effects of the seismic dynamic in the single-thin mountain mass are so apparent that the response differences of the mountain mass on the left bank and the right bank will be enlarged. For this reason, the dam line near the river bend is likely to be favorable for the layout of the junction of engineering structure, but in the seismic area, the requirements for the seismic safety of the dam body in the mountain mass scale should be appropriately considered when the dam line is selected.

The horizontal seams on each section of the Pacoima arch dam have shown the obvious signs of opening and closing in the course of a strong earthquake. Seismic cases have tested and proved that the arch dams cannot be taken as a whole structure to carry out the analysis under the strong seismic actions, whereas the contact nonlinear influence of repeated opening and closing of the horizontal joint seams must be considered in the course of a strong earthquake. The Pacoima arch dam is taken as a whole structure to carry out the seismic response analysis, whose results show that there will be a huge arch direction motion stress, but actually in the two strong earthquakes, the left bank is subjected to the influence of dam abutment rock mass deformation. In the outside affected region, the beam direction cracks were not found. The main reason is that this cannot be subjected to the horizontal joint opening of tensile stress, whereby leading to the apparent decrease in arch direction stress so that the functions of arch dam are weakened. Particularly in the two earthquakes, the Pacoima arch dam was in winter low water level operation, and water depth in front of the dam was only 2/3 of the design value, so that the horizontal joints were easy to open. The opening degrees of the arch dam horizontal joints were wide in the upper part and narrow in the lower part, and they disappeared gradually, plus the keyways set in the horizontal joints. Therefore, arch dams are taken as an ultra-high-static structure arch dams, and in the case of horizontal joints opening and closing, the dam body is of a certain complexity, whereas the stress distribution is reregulated. The dam section in the horizontal joint opening cannot like the overhanging beam to render the downstream dam facing beam direction stress to increase much more. The seismic disaster practical example seems not to indicate the necessity to lay out reinforcement in the beam direction for consolidation.

It has been suggested that as to arch dams in the seismic areas, it is necessary to check the dam section with the transverse joint open. In assuming that a certain distance (for example, 20 m) is below the dam crest, and in the case of the upper and lower through fractures along the horizontal construction joint, the tilt and slip safety degrees for the breakage dam blocks on the top part must be checked under the seismic actions. In the 1994 strong earthquake, the dam section on the left arch seat of the Pacoima arch dam had a slight disturbance at 15 m construction joint below the dam crest, being subjected to the influence of gravity pier deformation, but the amount of disturbance was only 10–12 mm, which could not threaten the safety of the dam body, and tilt and slip-falling could not take place. Accordingly, the seismic example indicates that the necessity of check of this kind seems not important.

When the Pacoima arch dam was designed, the dam concrete tests obtained indicate that the average compressive strength of concrete is 17.93 MPa within 28 days. After 40 years of operation, a drilling hole for taking core tests was conducted in the years of 1967–1968, and the test results indicate that the concrete average compressive strength reaches 33.78 MPa, whose qualities and strength are not worsened because of their long-term loading. Forty-three years after the San Fernando earthquake, a drilling hole for taking core samples in the dam body was conducted to carry out compression tests and photographic observation and exams within holes, whereby proving that the dam body concrete was consolidated and concrete qualities were good. This indicates that the dam concrete seems to have no decrease in its strength under the long-term loading actions and to be subjected to having no influence by the seismic disturbance, since an earthquake is likely to take place after the dams have been put into operation for many years. The data for the concrete being subjected to long-term loading to affect its qualities are of important reference significance for the evaluation of seismic safety of high arch dams.

After the earthquake in 1971, drilling holes for taking cores in the dam body to carry out compressive tests and photographic observation and exams within holes had detected no relative displacement in the contact face between the dam body and bedrocks. The further surveys indicate that the new cracks in the foundation rock mass and the reopening of the existing cracks are found. After the earthquake, an amount of seepage has obviously increased and the prevention seepage curtain has suffered from damage, whereby showing that the foundation is subjected to disturbance in a strong earthquake. However, in the present analysis of dam body seismic response, it is only considered that the concentration of dam heel tensile stress will lead to the cracks on the dam foundation contact face in assuming that dam foundation will not break. Actually, though the consolidation grouting is carried out in the dam foundation rock mass, there are still certain cracks, and the tensile ability of rock mass cracks is very poor, being much lower than the tensile ability of the dam foundation contact face. Therefore, when the earthquake took place and under the action of larger tensile force of the dam head, the rock mass cracks in the neighboring dam bottom should first be pulled open, whereby rendering the stresses to release in concentration and not making the dam foundation contact face have cracks. This kind of realization seems to be confirmed from the investigated results of the Pacoima arch dam after the earthquake. However, how to test through analysis and calculation is rather difficult at present.

Some important enlightenment calling for deep thought can be obtained from the analysis of only a few of the existing concrete arch dam seismic cases. This is of irreplaceable significance for deepening the understanding and realization of actual performance behaviors of concrete dams under strong seismic actions, whereby providing the orientation and basis for making the breakthrough in traditional concepts, thinking ways and methods, and making innovation and improvement so as to seek for realistic practices in the seismic resistance design and research. For this reason, this is a great and important function to play in guaranteeing the seismic safety of a large number of high concrete arch dams under construction in the seismic areas.

Based on the existing engineering practice and seismic disaster cases at home and abroad, as long as great attention is paid to the safety of seismic resistance, elaborate design, and construction, it is entirely possible to construct high concrete arch dams in strong seismic areas. Nevertheless, the seismic function in itself and seismic resistance of high concrete arch dams are very complicated; the existing real cases of seismic disasters of high concrete arch dams are only a few after all, and there have been no seismic cases of high arch dams of over 200 m suffering from a strong earthquake. The seismic resistance performances are likely to have the changes in nature from 100-m-grade high arch dams to 300-m-grade high arch dams. Accordingly, a series of high arch dams have been built in the strong seismic areas at present in western China. In how to prevent the likely maximum earthquake in the dam sites, there will be no severe disaster variations to occur, and particularly, the destruction of high dams and big reservoirs will result in unthinkable catastrophes, which are the serious challenges to be confronted. There can be no try-one's-luck mentality to relax one's vigilance.

To sum up the mentioned descriptions, the seismic safety of hydraulic concrete structures in China can be finalized in the following consensuses:

1. Although there are more severe seismic situations and serious seismic catastrophes in China, which is regarded as one of the multiple seismic countries, the construction of high dams in a scientific and orderly way, including in the west seismic areas, is still the requirement of social and economic development being suitable for the national situations in China.
2. Seismic safety problem is an unavoidable confrontational challenge in the construction of high concrete dams in the western China.

3. The engineering practice of high dam construction in countries with more earthquakes and the existing strong seismic cases preliminarily demonstrate that the high concrete dams with serious seismic resistance design and elaborate construction are of rather seismic resistance capacity. The high dam construction engineering in the seismic areas in western China can satisfy the seismic safety requirements basically.
4. Based on the seismic uncertainty and the complexity of high dam seismic resistance problems, intensive efforts must be made to deepen research on seismic resistance of the 300-m-grade of special high dams, so that the strategic priority of research is to prevent the serious seismic disaster variations of high dams and big reservoirs.

SEISMIC INPUTS AT SITE OF HIGH ARCH DAM

2   OUTLINE OF BASES OF SEISMIC FORTIFICATION AND SEISMIC HAZARD ANALYSIS AT DAM SITE. 27

3   DETERMINATION OF CORRELATION DESIGN SEISMIC MOTION PARAMETERS ON DAM SITE 53

4   DESIGN ACCELERATION TIME PROCESS 83

5   DAM SITE SEISMIC MOTION INPUT MECHANISM. 97

In dam safety evaluation, there are three elements: (1) seismic inputs; (2) structure seismic responses; and (3) structure resistance force. Seismic motion input is the important prerequisite as well as a weak link in dam engineering seismic resistance. The dam engineering seismic motion input involves many concepts and problems in the growing interconnection of different disciplines like engineering seismology. But the professional structural engineers of earthquake engineering who are not familiar with it are mainly dependent on the seismic department, while the seismic department is mainly focused on solving more generalized engineering problems in a macroscopic level. And it is unlikely that the seismic department is required to have more understanding of dam engineering characters and the concrete requirements of engineering technologies in realignment with it. At present, in the case of a lack of exchange and communication of the interconnection of different disciplines, when high arch dams are built in strong seismic areas, there is a certain lack of systematic and rational consensus in the technical way of solving the seismic motion input problem related to the dam site, thus resulting in the fact that there exist some concept confusions and theoretical puzzlements on the basis of seismic design prevention, seismic motion parameters, and seismic motion input mechanism, and so on, when the dam seismic resistance design and safety evaluations are carried out. For this reason, some regulations, parameters, and their adaptable conditions as well as the existing problems, etc. indicated by the seismic department cannot be correctly understood and employed. This can not only affect the correct evaluation of the seismic safety of high dam engineering structures but is also unfavorable to the development of engineering seismology. The contents in this section mainly deal with the discussion of some basic concepts and common problems in seismic design and practice of dam construction based on the specific features and requirements of hydraulic structures.

The contents in this section are Chapters 2–5, with four chapters in total.

Chapter 2 introduces the basis of seismic fortification and the seismic hazard analysis within the project area, including the dividing potential seismic source zones and determining seismic activity parameters and their probabilities of transcendence. Also, the seismic fortification levels and their corresponding performance objectives of other countries are introduced, and the guidelines for formulating the framework of seismic fortification levels are presented.

Chapter 2 discusses the genesis and scale of reservoir earthquakes as well as some commonly accepted viewpoints in the dam community, and also suggests that reservoir earthquake may not be included in the seismic fortification framework of dams.

Chapter 3 mainly introduces the peak value acceleration and the effective peak value acceleration concepts, discusses, and presents the rational value fetch method, introduces the response spectrum behavior and its attenuation relations and standard response spectrum, the behaviors of the probability-consistent and site-specific response spectra, points out the limitations of the standard design response spectrum, and suggests the use of the site-specific response spectrum based on the earthquake scenario related to the seismo-geological conditions of the dam site.

Chapter 4 presents a new method of generating accelerograms nonstationary in amplitude with frequency and consistent with target evolutionary spectrum according to the given seismic magnitude and epicenter distance, and introduces the solution process to the method in detail and finally the basic idea and concept of the stochastic finite fault method. The procedure of directly generating accelerograms of the dam site by using this method is included.

In Chapter 5, the basic concept of seismic input of design accelerations and the input mechanism of ground motion at the dam site with a free field pattern is presented. Also, the methods to solve the wave propagation problem with artificial transmitting boundaries or viscous damping boundaries are introduced. Finally, some problems existing at present in this aspect for seismic safety evaluation are discussed and corresponding suggestions are provided.

OUTLINE OF BASES OF SEISMIC FORTIFICATION AND SEISMIC HAZARD ANALYSIS AT DAM SITE

2.1 BASES OF SEISMIC FORTIFICATION DESIGN AND PREVENTION AND GENERAL SEISMIC DANGER ANALYSIS OF ENGINEERING WORKSITE

2.1.1 BASES FOR DESIGN AND PREVENTION

An earthquake is the long-term accumulated deformation energy due to the earth's crust formation movement, and is a result of the instant abruption of the rock body to release energy to be converted into dynamic energy, of which a part of the energy is radiated by the seismic waves formed in the abruption process. On the assumption that an earthquake is taken as a point source, seismic occurrence at a certain depth of the earth's crust is called seismic source, and the distance of the seismic source vertical to the land surface is called seismic source depth. The projection of seismic source on the land surface is known as the epicenter. The distance of the site away from the epicenter is called epicenter distance. The distance of the site away from the seismic source is called seismic source distance. The vertical distance from the site to the fault is known as the fault distance. The nearest distance from the site to the land surface projection of the seismic fault face is known as the fault projection distance. Seismic waves transmitted within the earth's crust media include the compression wave (frequently called the longitudinal wave) transmitted within the earth's crust body and the shear wave (frequently called the horizontal wave) of two types, as well as the Rayleigh wave and Lepu wave of two types, which are the secondary waves reflected many times by the body waves via the stratum boundary surfaces, being only limited to transmit in the earth's crust near the land surface layer. In seismic waves, the longitudinal waves arrive at the site earliest because of their high speed, and then the horizontal waves arrive and the secondary surface waves arrive finally.

Total energy released from the seismic source includes rupture energy, seismic wave energy, and heat energy produced by frictions. Seismic intensity depends on the part of energy radiated through seismic waves. This is mainly based on a certain amplitude recorded by a seismograph in an earthquake to determine the relative magnitude of the earthquake in the case of set conditions, on the basis of which the earthquake magnitude is classified. Therefore, the earthquake magnitude is the quantity to measure the scale of an earthquake itself. In some related literatures concerned with the seismic inputs of engineering seismic resistance, we frequently meet with several kinds of different definitions of earthquake magnitude so that the engineering personal should understand the differences and relations among them. The earthquake magnitude suggested in the earliest initial stage is based on the specified seismograph adopted in the Southern California region in the United States. This specified seismograph with a certain distance away from the seismic source records the displacement peak values within a cycle of 0.1~0.5 s, which is used to carry out the quantitative standardization known as the local earthquake magnitude M_L. In order to adapt the needs of earthquake magnitude determined by the global seismic stations, and based on the recorded seismic waves in the shallow seismic source being over 200 km of epicenter distance, the surface wave earthquake magnitude M_{LS} is defined mainly in terms of the level component displacement values between the cycles of 3~20 s. Later on, with an aim at the deep seismic source, the scale of the earthquake is measured with 1 s or so of seismic body wave earthquake magnitude M_b. Accordingly, there exists the following empiric formula among the mentioned three kinds of earthquake magnitudes defined through measuring the amplitudes in seismic waves as follows:

$$M_S = 1.13M_L - 1.08 = 1.59_{mb} - 4 \tag{2.1}$$

There are the so-called saturation phenomena of seismic magnitudes in the three kinds of earthquake magnitudes mentioned. The cause is that the earth's crust has a limited strength. Each time, the energy released by a small part of the earth's crust rupture is limited, and the seismic wave includes many kinds of frequency components; however, the definition of earthquake magnitude of various kinds is standardized only in terms of the energy of component with a certain frequency in a seismic wave, but this energy cannot increase with a growth in the rupture length. High-frequency component influence in the displacement frequency spectrum of seismic waves will decrease in earthquake magnitude. For this reason, it is again suggested that the so-called distance earthquake magnitude M_W be defined by seismic distance M_O. The distance earthquake magnitude is an earthquake magnitude calculated by the basic physical parameters so that there would be no saturated phenomena in earthquake saturated phenomena, and it describes the sizes of sliding motion on the seismic rupture surface, which can be obtained by means of the wave-form back calculation.

When $M_W = 3 \sim 7$, $M_W = M_L$; when $M_W = 7 \sim 7.5$, $M_W = M_S$; when $M_W > 7.5$, M_W is over M_S and M_{LO}.

In addition, when $M_L < 7$, M_L is over M_S, whose difference value is 0.7 ~ 0.2, which will decrease with an increase in earthquake magnitude. When $M_L = 7.5$, $M_L = M_S$; only when $M_L > 7$ can M_L be slightly smaller than M_S.

Earthquake magnitude M confirmed in the "Specifications for Seismic Magnitude of Earthquake" (GB17740-1999) in China is the surface wave earthquake magnitude.

The concerns of engineering seismic safety evaluation are likely to be the occurrence of earthquakes that affect the degrees of a site. This depends not only on the distance from the seismic source of earthquake occurrence to the engineering site, because the seismic waves produced from the seismic source are gradually attenuated in the process of several reflection transmissions. Land surface vibrations and their strong and weak effect degrees are usually characterized by the seismic intensity.

Seismic intensity roughly describes the measurement of seismic effect degrees through such microphenomenon qualifications as the person's feelings, instrument responses, house and building damages, and changes in natural phenomena, and the seismic intensity table to describe this kind of microphenomena can be used to make the earthquake magnitudes, so as to take the average features of comprehensive assessment of seismic effect in a specific region within a certain range as the symbols. Owing to the randomization of ground movement and the complexity of engineering structure, the evaluation accuracy of earthquake intensity scale is limited so that the earthquake intensity can only be taken as an integer in accordance with the international practice. The China Current Earthquake Regulations Table GB/T17742-2008 is shown in Table 2.1. This table adopts the principle of intensity of

Table 2.1 China's Seismic Intensity

Seismic Intensity (Degrees)	Person's Feelings	Building Seismic Disasters			Other Seismic Disaster Phenomena	Horizontal Seismic Motion Parameters	
		Types	Seismic Disaster Degrees	Mean Seismic Disaster Indexes		Peak Value Acceleration (m/s²)	Peak Value Special (m/s²)
I	No feelings.						
II	Some individuals in static states have feelings in the room.						
III	A few persons in static states have feelings in the room.		Doors and windows sound slightly.		Hangings move slightly.		
IV	Most of the persons in the room and a few persons outside the room have felt it; a few persons dreaming are awakened from sleep.		Doors and windows sound suddenly.		Hangings sway obviously and household utensils sound suddenly.		
V	Over a majority of persons in the room and most of persons outside the room have felt it. Most of the persons dreaming are awakened from sleep.		Doors and windows, roofs, and roof truss vibrate with sounds. Lime soil falls down. The plastering in individual rooms appears to have fine cracks.		Hangings shake greatly, unstable household utensils shake, or are upside down.	0.31 (0.22–0.44)	
VI	Most of the persons do not stand steadily. A few persons flee for their lives to the outdoors in astonishment.	A	A few medium houses are damaged but most of them suffer from microdamages, or they remain basically perfect.	0.00–0.11	Furniture and goods are moved. There are cracks appearing in the riverbanks and soft soil layers. There are sand boils appearing in the saturated sand layers.	0.63 (0.45–0.89)	0.06 (0.05–0.09)
		B	Only a few medium houses are destroyed, a few houses suffer from microdamages, and most houses remain basically perfect.				

(Continued)

Table 2.1 China's Seismic Intensity (cont.)

Seismic Intensity (Degrees)	Person's Feelings	Building Seismic Disasters			Other Seismic Disaster Phenomena	Horizontal Seismic Motion Parameters	
		Types	Seismic Disaster Degrees	Mean Seismic Disaster Indexes		Peak Value Acceleration (m/s²)	Peak Value Special (m/s²)
	Most persons do not stand steadily. A few persons flee for their lives to the outside in astonishment.	C	A few houses suffer from microdamages, and most houses remain basically perfect.	0.00–0.08	Furniture and goods are moved. There are cracks appearing in the riverbanks and soft soil layers. There are sand boils appearing in the saturated sand layers.	0.63 (0.45–0.89)	0.06 (0.05–0.09)
VII	Most persons flee for their lives to the outside in astonishment. Those who ride bicycles feel it; drivers and passengers who are in the buses are feeling it.	A	Only a few buildings are seriously destroyed or damaged. Most medium-sized buildings suffer from slight damages.	0.09–0.31	Materials fall down from the frames, collapse, and appear on the riverbanks. Sand boils are often seen in the saturated sand layers. There are more cracks in the loose and soft soil layers.	1.25 (0.90–1.77)	0.13 (0.10–0.18)
		B	Only a few medium-sized buildings are damaged; most buildings suffer from microdamage or remain basically perfect.				
		C	Only a few medium-sized buildings suffer from microdamages or remain basically perfect.	0.07–0.22			
VIII	Most of the persons are shaking and deranging and have difficulty in walking.	A	Only a few buildings are seriously destroyed; most buildings are severely or immediately damaged.	0.29–0.51	Dry and hard soil layers appear to have cracks. Most of the saturated sand layers appear to have sand boils.	2.50 (1.78–3.53)	0.25 (0.19–0.35)
		B	Individual buildings are completely destroyed; a few buildings are severely destroyed. Most of them are immediately or slightly damaged.				

Table 2.1 China's Seismic Intensity (cont.)

Seismic Intensity (Degrees)	Person's Feelings	Building Seismic Disasters			Other Seismic Disaster Phenomena	Horizontal Seismic Motion Parameters	
		Types	Seismic Disaster Degrees	Mean Seismic Disaster Indexes		Peak Value Acceleration (m/s²)	Peak Value Special (m/s²)
		C	Only a few buildings suffer from severe or immediate damage; most of buildings suffer from the microdamages.	0.20–0.40			
IX	Persons who are walking fall down.	A	Most engineering structures are seriously damaged or severely destroyed.		Dry and hard soil layers appear to have cracks. Cracks in the bedrock, disturbances, landslides, and collapse are frequently seen in seismic areas.	5.00 (3.54–7.07)	0.50 (0.36–0.71)
		B	Only a few buildings are destroyed; most buildings suffer from seriously or immediate damages.				
		C	A few buildings are seriously destroyed or severely damaged. Most buildings suffer from medium or slight damages.				
X	Those who ride bicycles fall down. Persons in shaking states fall down, away from their original places, with feelings of throwing up.	A	Majority of engineering structures are completely destroyed.	0.69–0.91	Mountain collapse and seismic cracks come out; arch bridges on the bedrock collapse.	10.00 (7.08–14.14)	1.00 (0.72–1.41)
		B	Most of them are completely destroyed.	0.58–0.80			
		C	Most buildings are destroyed.				
XI		A	Majority of engineering structures are completely destroyed.	0.89–1.00	Seismic faults greatly extend; a large number of landslides take place.		
		B					
		C		0.78–1.00			
XII		A	All the engineering structures are completely destroyed.		There are sharp and violent changes on the land surface, and mountains and rivers are changed.		

VII classification, corresponding to the majority of the international seismic intensity table, but only the Japanese seismic intensity table adopts the principle of intensity of VII classification. In China, when earthquake intensity is evaluated, the intensity of I–V is mainly based on a person's feelings and other seismic damage phenomena on the ground and in the first floor of houses and buildings. The intensity of VI–X is mainly based on the house or building seismic disasters, and with the references to other seismic disasters. The intensity of XI and XII should be based on the comprehensive house and building seismic catastrophes combined with the land surface earthquake disaster phenomena.

Earthquake magnitude and seismic intensity are entirely two different concepts, but in the engineering community, they are often confused with each other. When an earthquake with a certain earthquake magnitude takes place, the seismic intensity in the engineering site is related not only to the epicenter distance, but also to the site geologic and topographic conditions. Accordingly, only to say that a certain area or an engineering can resist an earthquake with several earthquake magnitudes without noticing its epicenter distance is, in general, not scientific and accurate, and even if the seismic source is right below the site bottom, the so-called "straight-down earthquake," the site seismic intensity (i.e., epicenter intensity I_0) is related to both earthquake magnitude and seismic source depth. The seismic source depth of the earthquake within the continental plate in China is mostly within the range of 10–30 km deep. In considering the large error in determining the depth of the seismic source, at present, seismic source can be approximately assumed as the point source, and seismic depth can be taken as the constant, in such a way that the statistic relationship between the epicenter intensity and earthquake magnitude can be given out. As a result, the Chinese seismologist Li Shanbang has developed the relation formula as follows:

$$M = 0.58I_0 + 1.5 \qquad (2.2)$$

Seismic intensity can indirectly reflect the strong and weak degrees of seismic motion itself through seismic motion effect degrees. For this reason, in the past engineering of a seismic design, earthquake basic intensity values of the site region provided in the "China Earthquake Intensity Zoning Map" by the Seismic Department in the whole country are, in general, used as the basis to determine the dam body to bear seismic functions. Seismic intensity given by this zoning map can be the intermediate- and long-term prediction for the likely damages from the largest seismic effect in the future in the region. Largely owing to seismic uncertainty, what the zoning map gives out is the 10% of estimated values being transcendental probability within 50 years.

Seismic intensity only indirectly characterizes the qualitative symbol of assessment of seismic functional strength. What the engineering design requires is accurate quantitative parameters. Up to now, the seismic engineering community has still extensively adopted seismic motion peak value acceleration to characterize the main parameters of seismic motion. For this reason, the seismic intensity table issued and implemented in 1999 has used the seismic motion strength parameters in the 1980 seismic intensity, whereby giving out the scope and average values of the corresponding level peak value acceleration and peak value speed for the seismic intensity of V–X scales. In addition, being the design and prevention standards source, the engineering community in China has for long made reference to the corresponding relations between the seismic intensity and seismic motion peak value acceleration derived from the statistic mean sense, which are prescribed in a seismic design norm issued by the department of each undertaking. For seismic intensity to be of practical significance for engineering earthquake resistance can be the intensity of VI, VII, VIII, and IX, corresponding to seismic motion level peak value accelerations of 0.05, 0.10, 0.20, and 0.40g, respectively ($g = 9.8$ m/s^2 is the gravity acceleration). The corresponding relations were once published by the authorities concerned with construction in the country and extensively adopted by each undertaking. These and seismic intensity corresponding to peak value acceleration values are all smaller than the mean values set in the current seismic intensity table. Since the relations between seismic intensity and seismic motion parameters are very complicated, the corresponding discreteness derived from the statistic mean sense has been rather large. What the macroseismic intensity characterizes can be the consequences of average seismic effect within the rather range, whereas the actually measured seismic motion parameters are the data of the individual measuring point. Accordingly, both cannot be obviously well in correspondence with each other.

As a result, the "China Earthquake Motion Parameter Zoning Map" (GB18306-2001) was published in 2001. This zoning map has been compiled in terms of the principle of seismic response spectrum of acceleration designed in accordance with the double parameter standardized criteria. Seismic motion acceleration response spectrum is of the single-freedom linear elastic system with multidifferent self-vibration cycle T; of the same damping ratio ξ. Under the action of given seismic motion acceleration intervals $a_g(t)$, the maximum value in each absolute acceleration interval, and relative to the corresponding self-vibration cycle can form the curve $S_a(T, \xi)$. $S_a(T, \xi)$, and is divided by the peak value in $a_g(t)$ to obtain the nondimensional value $\beta(T, \xi)$, which is called the regularized acceleration response spectrum. Figure 2.1 indicates the standard design acceleration response spectrum given out in the aseismic design specifications.

The statistic mean value of acceleration response spectrum is obtained through the smoothing finalization from the selected several actually measured seismic motion acceleration intervals on the given site of various types.

a_m, β_m, T_0, T_g, and r indicate the design seismic motion peak value acceleration, the platform value of regularized acceleration response spectrum, the first inflection point cycle value, the second inflection point value, and the descending speed controlling value in the descending section following $(T_g/T)^r$ rules, respectively. The zoning map corresponds to the design and prevention level being transcendental probability of 10% within 50 years. The effective peak acceleration (EPA) and the effective peak velocity (EPV) are given out in two maps. Taking $T_g = 2\pi EPV/EPA$, of which the EPA and EPV taking the values are as follows: in the acceleration (velocity) response spectrum $S_a(T, \xi)$, $S_v(T, \xi)$ given out by the selected seismic motion acceleration intervals $a_g(t)$, the mean value of different cycle spectrum

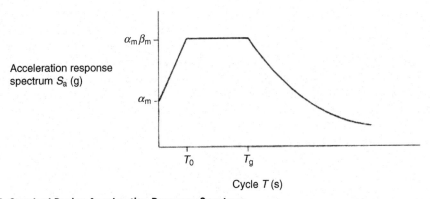

FIGURE 2.1 Standard Design Acceleration Response Spectrum

values at the two ends of the platform is divided by the superior value of 2.5 of the enlarged coefficient of the platform section. The zoning map is adaptable to the general engineering construction. As for the important big dam engineering likely to have an occurrence of secondary disasters, its aseismic design and prevention requirements can only be decided after a special study is made of the analysis of seismic danger on the engineering site.

The "Seismic Design Code of Hydraulic Structures" (DL5073-2000) in China stipulates that under the general conditions, seismic design must be based on the design peak acceleration determined by the zoning map, but the regularized acceleration response spectrum of the design peak acceleration given out in the code must be adopted in the design. It is necessary to point out that what the zoning map gives out are the values of the general site soil foundation of types, being inclined to the safety of arch dams constructed on the rock bed. As for the large-scale engineering projects of dams of over 200 m in height in the regions with basic seismic intensity of VI and above, or the reservoir storage capacity of over 10 billion m^3 and dams of over 150 m in height in the regions with basic seismic intensity of VII and above, their design peak acceleration specified in the code should be decided in accordance with the analytical results of seismic danger in special studies of engineering sites.

2.1.2 GENERAL DESCRIPTION OF EARTHQUAKE HAZARD EVALUATION IN ENGINEERING SITE

2.1.2.1 Basic concept of earthquake hazard evaluation

Seismic occurrence is random in time, space, and strength, and seismic wave transmission is also a complicated process with many uncertain factors. For this reason, it is necessary to base on probability theory when carrying out earthquake hazard evaluation. Earthquake hazard evaluation involves many problems in the fields of seismic and geologic specialty. The contents in this part are only to make a brief introduction to the basic concept and procedures of earthquake hazard evaluation as viewed from the angle of the ordinary engineering needs of aseismic work. When there is a deeper understanding of aseismic work, it is necessary for the engineers to locate the information from the special literatures of engineering seismology.

In carrying out earthquake hazard evaluation, to begin with, it is necessary to determine the special range of seismic occurrence events. It is generally considered that an earthquake that occurs can have an impact upon the site seismic motion in the region with the site as the center and a radius of no less than 150 km. The likely similar regions where the destructive earthquake may take place in the future within this range can be finalized, as each latent seismic source in each part within every seismic source area, to produce seismic motion upon the site can be taken into account. In this way, as viewed from the angle of probability statistics, and as far as every seismic source is concerned, it is necessary not only to analyze the probability of occurrence of each earthquake with different seismic magnitudes as well as the probability of different numbers of seismic occurrences but also to consider the probability of seismic occurrence in different locations in the latent seismic source area. The seismic waves are produced by the seismic source, whose energy in the transmission process can diffuse into the surroundings, but part of its energy in damping media is consumed and weakened gradually. Accordingly, seismic motion caused by every seismic event in the given site is closely related to the distance, seismic magnitude, and attenuation laws among them. After the comprehensive considerations are given to the probability of different seismic motions caused in the site by the likely seismic events within the range, the results of site seismic hazard evaluation are given out with the transcendental probability $P[A \geq a]$

curve form. In terms of required design probability level, the design seismic motion parameters for an earthquake likely to suffer from the nonexceeding design probability level in the future can be decided in the engineering site.

China's mainland interior earthquakes, with strong earthquakes in particular, almost take place within the active fracture zone, being subjected to earthquake formation. In the engineering seismic community, the so-called active faults refer to having active faults since the Late Quaternary Period, that is, the last active year from today has been the fault of less than 10,000 years. At present, the viscous and slip dislocation of active faults can generally serve as the seismic genesis mechanism, and the active faults being likely to generate destructive earthquakes are called the seismic active faults, usually judged as the genesis seismic faults. The identification of the faults of this type should, at the same time, satisfy the formation stress environment, the latest formation deformation, media formation of viscous and slip movement, and large enough conditions for fracture scales, and so on. As for large-scale regional faults, these conditions need to be identified in sections.

2.1.2.2 Classifications of latent seismic area
In China and neighboring regions, several seismic areas are classified in terms of the principle of geologic environment and similar seismic conditions. In each area and in terms of similarity of seismic activities, the seismic zones are concentrated in space, and also, enough historical seismic information and data can be used as the principle of statistic units of seismic activities so that a series of seismic zones are classified. In each seismic zone, many latent seismic source zones are classified in terms of already-occurred seismic formation conditions as the key link and with already-occurred destructive earthquake geologic formations and the analogy principle of seismic activity background.

The determination of latent seismic source length should take the occurred earthquake formation scales and sectional lengths into full account; these can serve as the main bases to determine the upper limits of earthquake magnitude of the likely occurred earthquake in these latent seismic source zones.

The width of latent seismic source zone is related to the types of faults. As far as the strike–slip fault is concerned under the general conditions, 5 km width of the two ends is taken. As to the apparent dip normal fault, the consequent dip site 10 km in width, including the surface fault belt can be adopted. If the fault dip is ambiguous, 10 km in width at the two sides will be taken. As far as the reverse scouring fault is concerned, the projection width on the surface of the deep slope section concerned with earthquake occurrence must be taken. It is necessary to point out that the latent seismic source area with high-appearing limits of seismic magnitude is unsuitable for the over-wide classification, for assuming that the present earthquake in the space inside the latent seismic source area distributes in terms of equal probability, and if the latent seismic source area is classified over-wide, the contributions made to the site seismic motion parameters will be "diluted."

Many latent seismic source areas are divided in the hinterland and the neighboring areas in China by the National Seismic Department, which are the bases for the complication of the "China Earthquake Motion Parameter Zoning Map" and the implementation of earthquake hazard evaluation. Figure 2.2 shows the schematic diagram of the dam site latent seismic source area division of the Xiaowan arch dam.

2.1.2.3 Determination of seismic activity parameters in latent seismic source areas
Seismic activity parameters in the latent seismic source area are first dependent on the seismic activity parameters falling into the earthquake belt, including the possible maximum and minimum earthquake

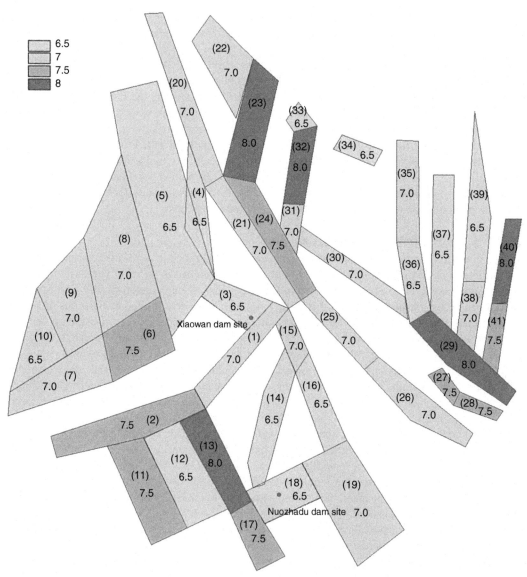

FIGURE 2.2 Schematic Diagram of Dam Site Latent Seismic Source Area Division of Xiaowan Arch Dam

magnitude scope within the belt or zone. The minimum earthquake magnitude in each earthquake belt in China has taken 4.0; the values of occurred frequency (N) of each earthquake magnitude (M) are determined in terms of Gudengbao and Richter logarithm expression of the relation $\lg N = a - bm$, where a is related to total number of earthquakes, and b expresses the activity relations of large and small earthquakes.

At present, adopting the probability of numbers of assumed events occurrence and Poisson's model without any relations to the numbers of already-occurred events, can be used to characterize the probability of seismic occurrence numbers n within the time section t, with $P(n) = (vt)^n e - vt/n$, where v is the yearly average occurrence rate. The probability of occurrence once at least is $1 - e^{-vt}$. In the latent seismic source areas, seismic activity time and special regularity, credibility, and other factors should be considered in the middle- and long-term research on seismic prediction. With the special distribution function $f_i m_j$, the seismic yearly average occurrence rate v in the seismic belt is weighted and distributed into each earthquake magnitude archive (m_j) of each latent seismic source area (i), which is used as the seismic yearly average occurrence rate.

2.1.2.4 Determination of transcendental probability of dam site seismic motion parameters

The procedures of Dam Site Earthquake Hazard Evaluation can be generalized as follows:

1. Based on the deepening research on geologic structure and seismic activity conditions in each latent seismic area (i), it is necessary to further determine the adopted latent seismic source area and the upper limit (M_s) of earthquake magnitude and initial calculation of earthquake magnitude (M_O). As described previously, at present, M_O is taken as 4.0 scales.
2. The yearly average seismic occurrence rate in each latent seismic source area is determined in accordance with the seismic yearly average occurrence rate v in the regional seismic belt and the special distribution function $f_{i,m}$ of each earthquake magnitude archive (m_j) in each latent seismic source area (j).
3. Density function $f_i(m)$ of seismic occurrence probability with different earthquake magnitudes is derived in terms of earthquake magnitude–frequency relations.
4. The approximate point source zonation (l) can be made within each latent seismic source area, and assuming that their occurrence probabilities are the same, some special distribution probability density function $f_i(R)$ becomes the value of the ratio between each zonation area and total area of the latent seismic source area.
5. Based on the record of actually measured seismic motion acceleration on the bedrock surface and in considering the saturated phenomena of a new field of strong earthquakes, the relevant statistic mean attenuation relations and their standard differences among the seismic motion peak acceleration and seismic magnitude (M) and seismic source distance (R) can be obtained, that is:

$$\lg A = C_1 + C_2 M + C_3 M^2 + C_4 \lg[R + C_5 \exp(C_6 M)] \tag{2.3}$$

6. The transcendental probability of the occurrence peak acceleration a at the dam site in each latent seismic source area can be obtained as shown in the formula (2.4), that is:

$$P[A \geq a] = \frac{1}{a} \sum_i \sum_l \sum_{j=M_O}^{M_u} P_i[A \geq a | M, R] f_i(R) \mathrm{d}M \mathrm{d}R \tag{2.4}$$

7. Largely owing to adaptation of statistic mean attenuation relations and under the assumption of obtained $\lg A$ transcendental probability subordinate normal distribution, the uncertainty is rectified with total probability within $\pm 3\sigma$ range and then, the transcendental probability $P[A \geq a]$ curves of the final seismic motion peak acceleration are given out (Fig. 2.3).

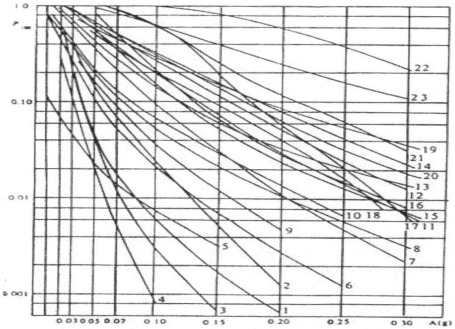

FIGURE 2.3 The Transcendental Probability Curves of Dam Site Seismic Peak Acceleration of Several Engineering Structures in China

2.2 ASEISMIC GRADATION DESIGN AND PREVENTION LEVEL AND DETERMINATION OF CORRESPONDING PERFORMANCE OBJECTIVE OF HIGH ARCH DAMS

2.2.1 ASEISMIC DESIGN AND PREVENTION LEVEL AND CORRESPONDING PERFORMANCE OBJECTIVE OF FOREIGN DAMS

At present, in different countries, aseismic design and prevention gradation, design and prevention level, and the corresponding performance objective are the same, so their general description is as follows:

1. The International Commission on Large Dams (ICOLD) issued No. 72 Communiqué, "Guidelines for selecting seismic Parameters for Large Dams" in which two grades are suggested for design and prevention for large dams, that is, maximum design earthquake (MDE) and operating basic earthquake (OBE). In addition, maximum credible earthquake (MCE) is defined, based on the actual possible occurrence of maximum earthquakes in terms of reservoir and dam earthquakes and geologic conditions. It is likely to cause the severe secondary disasters to the big dams (the majority of dams in this type of dam). Therefore, MDE takes for MCE. MCE is confirmed by means of conformation analysis or in terms of earthquake hazard evaluation. It takes a long time for obtaining the transcendental probability of a 50% earthquake, whose performance objective is not to cause dam failure but to make reservoir water discharge out of control, whereas OBE takes

the transcendental probability of 50% earthquake within 100 years (recurrence period is 145 years), whose performance objective is to allow to yield the repairable local damages. Dam sites can be divided into A groups in terms of earthquake hazard evaluations, and large dams can also be divided into four groups in terms of their importance and degrees of accident hazards, which are based on the confirmation of adopted dam body seismic responsive analysis methods and the involvement of seismic motion parameters. It is necessary to point out that the suggestions in the guidelines are that the peak acceleration corresponding to aseismic design and prevention levels should take the average value of an 84% guarantee rate plus a variance. In addition, the guidelines also introduce the reservoir-induced earthquake (RIE) design and prevention, having been revised and known as reservoir-triggered earthquake (RTE) in the recently revised guidelines, and they hold that it is impossible to surpass the earthquake magnitude of triggered seismic occurrence faults so that this should not be listed in the framework of large dam design and prevention.

2. Although the "Guidelines for Selecting Seismic Parameters for Large Dams" by ICOLD mainly refers to the relevant guidelines of the United States Committee of Large Dams (USCOLD), there are no unified national universal large dam aseismic design codes or guidelines in the United States so that the USCOLD and the majority of units concerned have taken the recurrence period of large dam design earthquakes of 1,000–10,000 years.

3. The dam safety guidelines issued by Canada in 1995 stipulate that as far as severe consequences of dam failure are concerned, the MDE must take the recurrence period of 10,000 years MCE, and their corresponding performance objectives cannot cause dam failure but cause the reservoir discharge to be out of control. As to the larger consequences of dam failure after an earthquake, MCE should be selected between 50% and 100% based on the actual conditions, with the recurrence period of 1,000–10,000 years. As to light consequences of dam failure after an earthquake, the recurrence period of 100–1000 years can be taken for design and prevention level. In 2005, the new stipulations in seismic hazard considerations in the "Dam Safety Analysis in Dam Safety Guidelines: Practices and Procedures T201," drafted by the Canadian Dam Association, to be checked are the earthquake design ground motion (EDGM), determined as grade I design and prevention, based on the engineering site earthquake hazard evaluation, is the similar MCE limit earthquake as the performance objective without dam failure, whose recurrence period can be divided into three groups of 10,000, 2,500, and 1,000 years, based on the consequences of engineering structure damages. It is considered that satisfying application requirements is not the control element for large dam aseismic safety control so that the employers can make the free choices as viewed from economic considerations.

4. The earthquake guide to seismic risk to dams in the United Kingdom was issued by the United Kingdom in 1998, in which although two-grade aseismic design and prevention levels are defined, the detailed stipulations are made for the performance objectives of safety evaluation earthquakes (SEE), without occurrence of dam failure catastrophes. The large dams are divided into four groups in accordance with ICOLD guidelines, whose SEE design and prevention levels corresponding to the seismic recurrence periods are 1,000, 3,000, 10,000, and 30,000 years by year orders. As to the most important group of large dams, design and prevention levels can adopt the MCE defined by ICOLD. As to the engineering without having a special seismic hazard analysis of the dam site, the seismic design and prevention levels can be determined in terms of a zoning map. The whole country is divided into A, B, and C zones by the zoning map. In the A zone with the strongest seismic activity, the corresponding recurrence periods are 30,000,

10,000, and 1,000 years, whose seismic motion peak ground accelerations are 0.375, 0.25, and 0.10g, respectively. There are no stipulations made for OBE. In fact, this is only the regulation for classification design and prevention and not multigrade design and prevention.

5. Italy issued the Aseismic Design and Recheck Guidelines for the newly built and ready-built large dams in 1959 and 2001, respectively, which are similar to the guidelines issued by the United Kingdom, but the four groups of large dams with SEE corresponding to recurrence periods take the larger year than 2,500, 2,500, 1,000, and 500 years, respectively. In the national zonations, there are four zones. In the zones with the strongest and the weakest seismic activities, SEE peak ground accelerations corresponding to the recurrence periods of 2500 years are 0.6 and 0.2g, respectively. As for OBE, it is stipulated that the recurrence period in each zone is 1/2 of peak acceleration of 500 years.

6. Large dams are divided into five groups from I to V, by the order in terms of their importance and again divided into four groups from A to D by the order in terms of seismic hazards upon the infrastructure facilities in accordance with specifications set in the "Code for Design and Safety Assessment of Dams and Hydraulic Structures," published by Romania in 2002. SEE or OBE are adopted for large dams with different grades. As to SEE of I/A and II/B groups of large dams, their peak ground accelerations should be selected in the recurrence periods of 457–800 years in terms of the results of special seismic hazard evaluation of dam sites. As to the large dams of groups III, IV, V, and C/D, their peak ground accelerations can be determined in terms of the OBE zoning map with the recurrence period of 100 years, whose values groups of SEE, OBE aseismic design and prevention of large dam engineering, grade I aseismic design and prevention level is adopted for the given grade concrete engineering. For this reason, this is only the "classification design and prevention," and not the "multigrade design and prevention."

7. In the "Earthquake Analysis of Dams" issued by Austria in 1996, seismic design and prevention levels are divided into MCE and OBE of two grades, of which MCE corresponding to the performance objective is the damages restricted in a certain range, while OBE is just a slight damage. It is stipulated that the aseismic analysis and check must be made of dams with the height of over 15 m or dams with the reservoir storage capacity of over 500,000 m^3 in accordance with MCE and OBE, and the rest of the conditions can be checked only in terms of OBE. OBE can be determined in terms of zoning map, whose peak accelerations are between 0.06g and 0.14g. The peak accelerations of MCE zoning map are between 0.11g and 0.3g, but it is suggested that MCE is determined by the earthquake hazard special evaluation of the dam site. It is specially stipulated that when the analog static method is adopted dynamic strength cannot be taken into account. In addition, a check is made of the requirements to satisfy dam body stability and deformation in dynamic analysis.

8. Aseismic design and prevention levels are divided into two grades in the Russian "Guidelines for Design and Construction and Stipulations for Hydraulic Structures in Seismic Areas." The recurrence period of MDE adopts 5,000–10,000 years, with the corresponding performance objective of having no dam failure. Another grade is in correspondence with OBE strength level earthquake (SLE), whose recurrence period is 100–500 years, and the corresponding performance objective is to allow the structures to have slight damage, but they are still under normal operation. In the stipulations, hydraulic structures are divided into four groups. Backwater structures of Grade I and II need to be checked in accordance with design and prevention levels of two grades, and the rest are checked only in terms of SLE. In Russia,

the seismic intensity zoning map for the recurrence period of 500, 1000, and 5000 years has been published. Russia has adopted the MSK-64 intensity table with 12 scales, being similar to the international universally used and revised Mercalli (MM) intensity table. Peak ground acceleration values and intensity corresponding to the table are given out for three kinds of foundation soils from the minimum $0.06g$ of intensity VI in grades II and III foundation soils, to the maximum $0.48g$ of intensity X in grade I foundation soils, in the stipulations. Large dam seismic response analysis method corresponding to different design and prevention levels and damping ratios of dam bodies and foundation of various types are stipulated.

9. In New Zealand, the New Zealand Society of Large Dams (NZSOLD) published the "Dams Safety Guidelines" in 2000, in which aseismic design and prevention levels are divided into SEE and OBE. It is worth pointing out that SEE recurrence period to the high, middle, and low areas with seismic activities adopt 10,000, 2,500, and 500 years, respectively. Usually in high or low seismic activities in different areas and corresponding to the sizes of seismic strength, the design and prevention levels (recurrence periods) are decided by engineering structure grades and seismic performance objects, and are embodied by corresponding to different seismic motion peak ground accelerations. SEE corresponding performance objectives are that there should be no dam failure for the completed dam, and there should be no serious damages to the newly built dams. In the case of the same aseismic design and prevention levels, the different aseismic performance objectives should be adopted for the ready-completed and newly built large dams, and this is another characteristic. The OBE corresponding to performance objective with the recurrence period taking 150 years requires having slight damage. MCE is known as controlling maximum earthquake (CME). CME adopts SEE in general unless the CME recurrence period is very long. It is worth noticing that the guidelines particularly point out that CME corresponding to the peak ground acceleration is not to adopt the average values but 84% of the quantile of guarantee rate, and its recurrence period is not over 10,000 years. Usually, the peak ground acceleration value corresponding to different transcendental probabilities given out by earthquake hazard r = evaluation are not the average value, but the values after being checked so that the values are, in general, larger than the average value.

10. In Switzerland, the federal government issued the "Swiss Guidelines on the Assessment of the Earthquake Behavior of Dams" in 2000, in which only one aseismic design and prevention level SEE is adopted, whose performance objective is to make reservoir water discharge not be out of control, and the ancillary structures and components related to safety cannot have nonrepairable damages. That is why the OBE level is no longer adopted to do two-grade design and pretension. It is considered that the structures that are able to meet the needs of SEE performance objectives cannot be likely to affect the operation after being subjected to SEE. Large dams are divided into three grades in terms of dam height and reservoir storage capacity, whose SEE corresponding recurrence periods are 10,000, 5,000, and 1,000 years. Peak ground acceleration corresponding to these are decided in terms of the corresponding intensity zoning map and the conversion on relationship between intensity and peak ground acceleration.

11. In Japan, large dam aseismic design has basically adopted the seismic coefficient method without probability meaning. However, after the 1995 Kobe earthquake took place, "The Branch Commission of Seismic Engineering and Aseismic Design Norm" under the Civil Works Association has worked out the suggestion of a two-grade aseismic design and prevention

standard. The first-grade design and prevention standard is corresponding to the presently adopted standard, whose performance objective is that large dams cannot have structure destructions. The second-grade design and prevention standard is the standard of the occurrence and earthquake of 6.5 scales near the dam site. As far as concrete dams are concerned, there should be no occurrence of inclined falling-down, sliding motion, throughout cracks in the dam body and foundation, and dam body destruction. As to earth-rock dams, there should be no occurrence of throughout cracks in the seepage control area, overtopping, and dam body destruction. Japan is a country with a frequent occurrence of strong earthquakes. In general, the recurrence cycle of occurrence of the plate earthquake of VIII is 100–200 years, while the recurrence cycle of occurrence of earthquake of VI–VII within the plate inner active faults is 1000 years.

Comprehensive analysis of foreign large dam seismic design and prevention levels or standards and performance objectives can be finalized as follows:

1. Different design and prevention levels or standards are adopted for the large dams of different grades. For this reason, each country has basically determined the aseismic design and prevention levels or standards in terms of recurrence periods through probability analysis. The final achievements of dam-site earthquake hazard evaluation can express the probability curve with yearly transcendental probability as the axis of ordinates and with the peak ground acceleration as the axis of abscissas. Yearly transcendental probability reciprocal is the recurrence period, but it is not the recurrence cycle on seismic occurrence faults corresponding to a certain earthquake magnitude.
2. As far as large dams with different grades are concerned, different design and prevention levels or standards are adopted. Although MDE (or SEE) and OBE (or SLE) of two-stage aseismic design and prevention levels or standards are adopted in the codes or guidelines for large dam aseismic design by the majority of countries, Canada, the United Kingdom, Switzerland, and so on, carry out large dam aseismic design only in terms of MDE in practice. In the aseismic design of important large dams, the recurrence period is 100–200 years OBE, being lower than the design and prevention probability level stipulated by the seismic zoning map in China so that it cannot play the control role. Some countries adopt the MDE as OBE for the low-grade dams.
3. As to the important large dam, most countries adopt the MCE to carry out the MDE, and MCE recurrence period is 500–10,000 years, or are obtained by the qualification method, whose performance objectives cannot make reservoir water discharge out of control, not because of dam failures. OBE performance objectives are able to repair light damages and are mainly decided by the engineering practical experience.

2.2.2 THE GUIDING THINKING FOR WORKING OUT ASEISMIC DESIGN PREVENTION LEVEL OR STANDARD FRAMEWORK

It can be seen from making a comprehensive survey of the large dam aseismic design code of each country all over the world that the engineering aseismic design and prevention classifications are usually determined in terms of engineering importance and seismic disaster consequences, as well as the dam site earthquake hazard evaluations. Aseismic design and prevention levels or standards and corresponding performance objectives, as well as the adaptation of the structure-response analysis method, are specified in terms of aseismic design and prevention classifications.

In determining aseismic design and prevention levels or standards and the corresponding performance objectives, and in the 1990s, the United States suggested the so-called concept of performance-based seismic design, "performance-based" is translated into "function," but it is mostly called as "research frontier," "development orientation," and "new concept" both at home and abroad, while they have become the hot-point problems in engineering earthquake resistance. Accordingly, there is the advent of the aseismic performance behavior and various kinds of definitions such as "adopting risk analysis," "stressing aseismic measures," and "attention paid to comprehensive aseismic abilities," stressing structural general behavior to the individual character. Expanding the individual analysis to the structural system analysis, "shifting the linear analysis to the nonlinear analysis," "with main check strength transferred into check deformation," and "with reliability design to replace confirmation design" from satisfying a single performance objective to satisfying multiperformance objectives.

Actually, aseismic performance design can be briefly generalized under the different specialized seismic design and prevention levels. It is necessary to make the structures satisfy the corresponding aseismic performance required aseismic design, whose connotation should include the following: (1) aseismic design and prevention levels and performance objects must correspond to each other; (2) the grade design and prevention and multiperformance objectives must be adopted; and (3) the performance objectives must be concretely quantified. In order to make clear the guiding thinking for working out large dam aseismic designs and a prevention-level framework in the country, first of all, it is necessary to carry out a profile analysis of real conditions in combination with large dam behaviors, and to current hydraulic structures.

2.2.2.1 Aseismic design and prevention levels corresponding to performance objectives

Any kind of aseismic design, whether a general engineering design is designed in terms of specifications, or an important engineering design needing special study beyond the range of the application of specifications, should have the aseismic design and prevention levels corresponding to a certain performance objective. Accordingly, the aseismic design and prevention levels without the performance objectives are meaningless. This must, first of all, be defined clearly in the general introduction to all aseismic design specifications and guidelines, and so on, at home and abroad.

Under the prerequisite of satisfying the function requirements, any design should embody the relative balance between safety and economy. The aseismic specifications in all the departments have stipulated the corresponding aseismic design and prevention levels and the performance objectives for the different grade engineering structures, thus embodying this principle. As to the important engineering structure, a large dam in particular is an engineering structure with strong social and public interests and a far-reaching social impact after an earthquake so that safety guarantee is first and foremost. Of course, what the specifications stipulate are usually of common characteristics and safety that must satisfy the lowest requirements. Under the market economic conditions, this is not contradictory to the higher requirements for design and prevention, as well as performance objectives by the employer in accordance with the economic conditions and the special individual behavior of the concrete engineering structures.

As to the economy, it is determined to a great extent by the economic and social environment and national situation of each country in the world. For this reason, regulations in the specifications of each country are of strong policy and society, continuing to develop and change and being of stages. In addition, there can be no doubt that adopting "investment-benefit guidelines" as the principle for balancing the project cost and performance objective requirement is the rational development orientation.

There are many analytical theories and methods, but the problems lie in the operation and reliability of basic data and information for the benefit estimation of concrete structure seismic safety investments and reduction of seismic disaster losses. This principle used in practice and serving as the basis in the specifications, particularly for large dams involving extensively complex structures, have had rather the biggest difficulties at present. The secondary disasters in particular caused by large dam severe catastrophes are likely to be greater than the losses caused by the structure itself, of which there exist many controversies on how to estimate and embody the losses of life values and social impacts. In addition, the estimation of losses also involves the prediction of social and economic development downstream while a big dam is in operation, and this can increase the complexity of the problem. Therefore, as far as the flood disasters over the occurrence, frequency, and disaster cases of seismic disasters are concerned, the risk evaluation methods are under discussion.

It is necessary to point out that the design specification, including the aseismic specifications in our country, has the trend of gradually changing from compelling specifications to recommending specifications. At the same time, great efforts are being made to strengthen the requirement and management of the compelling articles concerning the public security and social impact or influences. For this reason, aseismic design and prevention levels or standards, at least for large and important dam engineering, serve as the compelling articles stipulated so that these articles are not lowered at random in terms of requirement of investment benefits guidelines by the employers.

Aseismic design and prevention levels corresponding to the performance objective should be the basic principle for aseismic design. At present, structural aseismic design and its specifications neither pay attention to the structural safety but to disregard economy, nor to stress common behavior basic requirement but to restrict the higher individual behavior requirement. Therefore, it is not suitable to consider that the stipulation guidelines and performance guidelines are different in essence but not to set them against each other.

2.2.2.2 Adoption of staging design and prevention and multiperformance objectives

Different design and prevention levels or standards are adopted for the different stages of dams. This kind of principle of classification design and prevention has come to the common consensus in each country. The classification design and prevention for different types of large dams and the adoption of multistage design and prevention for the same dam engineering are two kinds of concepts. The classification design and prevention for the different stages of engineering structures are suggested for the different types of requirements of aseismic design and prevention levels or standards based on their importance, scale, and possible disaster degrees caused by seismic catastrophes. The multistage design and prevention for the same engineering structure with an aim of different performance objectives requirements are suggested for the demands of aseismic design and prevention levels. As far as the hydraulic structures are concerned, our country has worked out the related specifications for the engineering stage classification and the corresponding flood control design standards, in which apart from the normal operation conditions, there are the special operation conditions with the aim of checking floods. As far as engineering in seismic areas are concerned, the accidental operation conditions must be recorded in seismic functions. Based on the accidental operation conditions, seismic design and prevention levels and the corresponding performance objectives require the higher standards, so that the specifications can no longer require the staging design and pretension in aseismic design.

Staging aseismic design and prevention requirements are, first of all, suggested in the aseismic design of nuclear power stations. The nuclear power stations via aseismic design must stop the reactors for checks or examinations when the basic earthquake operation takes place. After conforming nonsurpassing design function requirements are able to maintain normal operation, restarting can be made in the conditions operation. When a safe shut down earthquake (SSE) occurs, the nuclear power stations must be ensured to stop the reactors safely within the specified time and without the occurrence of a nuclear leakage accident. Meanwhile, when the structures or requirement dynamic responses are checked in aseismic designs, the different damping ratios to be adopted are stipulated with an aim at the two groups of aseismic design and prevention levels. Since the structures in SSE operation conditions or damping ratios of the equipment are higher than OBE operation conditions, OBE operation conditions are likely to be controllable.

In addition, the staging aseismic design and prevention principles are presented in the early 1973 edition of specifications by the Structural Engineers Association of California in the United States and the 1982 CEB-FIP (European Concrete Commission-International Prestress Association) concrete structure design specifications. At the end of the 1980s, the two-scale design and prevention standards were used to carry out aseismic design in terms of the requirement of earthquake-resistant design performance objectives of "there being no damages with micro-earthquake, meso-earthquake with damages to be repaired, and macro-earthquake without dam failure" stipulated in the "Code for Architectural Aseismic Design" in China. The concrete contents are as follows: the linear elastic analysis is adopted to check the element strength in terms of 63% frequent occurrence of earthquake of transcendental probability within 50 years, and to check the maximum elastic–plastic displacement values in between layers of the overall structure in controlling over collapse by the framework structure in terms of 2–3% of a rare meeting with earthquakes of transcendental probability. The mentioned standards and their corresponding performance objectives are definitive and quantitative; however, at present, the repairable performance objectives in the meso-earthquakes are still based on the engineering experiences, for the structures of various types are still hard to be quantified.

It must be pointed out that the more detailed the design and prevention levels or standards, the more difficult the corresponding performance objectives are differentiated in accurate quantification. There has long existed one aseismic design and prevention level or standard corresponding to multiperformance objectives early in the aseismic engineering practice. For instance, concrete dams have for a long time needed checking for their different performance objectives such as compressive resistance, tensile strength, and stability against sliding. Of course, with the engineering itself and design methods to keep pace with times, the performance objectives may be changed or increased. Particularly, as far as the important large dams are concerned, making a dam body earthquake deform, cracking degrees, and seismic function effects serve as the quantitative performance objectives, not corresponding to aseismic design and prevention levels and risk degrees difficult to form the consensus, at least in recent times. As to ancillary structures on the large dams or equipment and nonstructural performance objectives, their corresponding quantitative performance objectives are checked on the basis of the large dam structure seismic responses, and the corresponding aseismic measures are adopted.

Accordingly, stating design and prevention and multiperformance objectives adopted in aseismic behavior design cannot be the new concept of aseismic designs.

2.2.2.3 Concrete quantification of performance objectives

In aseismic design, the performance objective corresponding to aseismic design and prevention levels are only restricted to be macro-quantified. They are nonoperable, whose concrete quantification can be embodied by bearing force or a normally-applied limit state equation, of which physical quantities selected to characterize seismic function effects, resisting force to determine materials, and their structure and safety criterions are included. These must be realigned with the analysis method of structural earthquake response, resisting force, and the value taking of safety abundance. The design method, including all the elements, is always being improved with the changes in national situations, accumulation of engineering experiences, and progress made in scientific and research work, so as to make them be in concert with the developing national situations and reflect the practical experience and mature scientific research achievements of engineering and seismic disaster examples. Also, design codes or specifications are bound to be revised in adaptation to the changes in time. For instance, much work should be done from the analysis of structure members to the whole system, from the check of strength to the deformation, from the analysis of linear elasticity to the nonlinear elasticity–plasticity, from the analysis of certainty to the reliability and from the analysis of member reliability to the system reliability, and so on. However, the design specifications or codes are worked out in terms of summarizing the existing engineering practical experiences and reflecting the mature scientific research achievements. Some frontier research achievements that are not fully tested through engineering practices cannot be included in the specifications or codes to serve as the design basis. For this reason, the specifications or codes can mainly give out the common regulations within the adaptable scope. As far as the important or specially required engineering is concerned, the specifications require carrying out the special research and evaluation, of which model experiment tests must be carried out. Largely owing to promotion by engineering construction development, the concrete engineering design concept and method are worth further improving with some breakthroughs made, so as to render the performance objectives of aseismic design and prevention to become more rational and more practical in such a way that more accurate quantity and better operations can be achieved. These are the inherent requirements for aseismic design and codes, and they are not the special symbols for the performance design.

To summarize these descriptions, it seems that as far as large dams are concerned, at least, there are no differences in the basic concept and connotation in comparison with the current aseismic performance design with the present current hydraulic engineering aseismic design practice.

2.2.3 SUGGESTIONS ON REVISION OF ASEISMIC DESIGN AND PREVENTION LEVEL FRAMEWORK IN CHINA

To summarize these descriptions, the basic principles for the establishment of large dam aseismic design and prevention level frameworks are as follows:

1. The aseismic design and prevention level framework must be in agreement with the national situation in China.
2. The aseismic design and prevention level framework must be based on seismic disaster real examples, engineering practice, and mature scientific research achievements.
3. It is necessary to differentiate the different grades and types of large dams in different treatments.
4. Design and prevention levels must be in correspondence with the performance objectives.
5. The methods to determine design and prevention levels and the performance objectives must continue to be improved.
6. The performance objectives must be quantified and must be operable.

The newly revised "Aseismic Design Code for Architecture Structure" stipulates that as to architecture structure with special requirements for use functions or other aspects, there are always more concrete or higher aseismic design and prevention objectives when aseismic performance design is adopted. For this reason, one section of the "architecture aseismic performance design" is increased and shown in the appendix, and the quantification aseismic performance design objectives and design principles concerning meso-earthquake and macro-earthquake suggested in terms of seismic disaster quantified classifications are increased in "Classification Standards of Architecture Seismic Damage Grades," issued by the Ministry of Construction in 1990.

As far as hydraulic engineering structures are concerned, it is difficult to classify more design and prevention levels or standards and the corresponding performance-controlling objectives and give out the quantification standards of common consensus and able-operations, largely owing to the differences of site landforms and geological conditions of engineering structures of various types. Therefore, at present, it is impractical to classify the quantification standards for seismic damage grades of different hydraulic structures, and there are no necessities. In fact, in the codes or specification stipulations for the safety of engineering structure earthquake resistance, there is no limit to raise the higher requirements over the stipulations set in the specifications for some engineering structures or some part of seismic performances by the relevant departments in charge or the owners' willingness under the conditions, through the special studies. However, it seems unnecessary, and it is hard to work out the concrete regulations in the codes or specifications.

In China, there is a special code to classify their grades in terms of the importance of large dams and in case of disasters from dam failure. Grade One design and prevention levels of MDE is adopted in the current "Code for Hydraulic Structure Earthquake Resistance Design" (DL5073-2000), whose corresponding objective is that if there are local damages, dams after being treated in general can still be put into normal operation. The concrete contents of this performance objective correspond to the design codes of large dams of various types, being recorded in the special loading combination operational requirement of seismic functions, for they are in coordination with the basic design specifications of large dams of various types, being of quantified operations. The types of aseismic design and prevention of hydraulic structures are determined in terms of their magnitudes and site aseismic intensity. Under the general cases, their peak ground acceleration design is determined in terms of recurrence period of 475 years (corresponding to the transcendental probability of 10% within 50 years) of the national zoning map. As to the dam height of over 200 m in the regions with basic seismic intensity of VI or over VI or the large-scale engineering with reservoir storage capacity 10 billion m^3, the design and prevention bases should be determined in terms of bedrock provided by the special site earthquake hazard evaluation. The large dam aseismic design and prevention classifications are all grade A, whose MDE recurrent period is taken as 4950 years, corresponding to the transcendental probability within 100 years of 2%. It should be pointed out that the aseismic design and prevention levels or standards of the important large dam MDE in China are close to the MCE levels of most of the foreign countries in the world so that their performances objectives are near OBE requirements. As to the aseismic design and prevention classifications lower than those of A classification of large dams, their recurrent periods of design and prevention levels or standards are required to be higher than those of OBE in general. For this reason, the earthquake resistance specifications or codes for hydraulic structures in China have stipulated the strict requirements for seismic design and prevention levels or standards to the important large dams. This is because China is a large country with more earthquakes as well as the largest dams built in the world, as of China's national situation. In considering the determination of aseismic design and prevention levels or standards, this, in fact, belongs to the seismic middle- and long-term predictions under exploration, both at home and abroad. Therefore, there exists a very large uncertainty in such a way that once the high dam and big reservoir have suffered from complete failures, the secondary catastrophes they cause can bring about losses to society that are much more severe than the consequence to the engineering structures in themselves. Accordingly, earthquake resistance safety is extremely important so that the higher requirements for the large dam aseismic safety are strictly stipulated. In the 2008 Wenchuan earthquake, there were many dams in the strong seismic zone that suffered from the seismic disaster to a certain extent, but there were no occurrences of disaster changes of dam failures, where there are four dams of different types located in the strong seismic zone, which are the high dams with a height of over 100 m, designed in terms of code requirements and constructed in guaranteeing the qualities and experienced through the strong seismic tests. Also, the codes or specifications for hydraulic structure earthquake resistance are identified, whereby illustrating that the existing engineering structures, designed in terms of higher MDE design and prevention levels, constructed in guaranteeing qualities, and managed in operational procedures are ensured with their whole aseismic safety. In addition, adopting increased OBE two-stage design and prevention levels is of no practical significance, because it is difficult to think that large dams that do not require satisfying OBE performance objective demands can satisfy MDE performance objective requirements. Accordingly, it can be considered that as far as the past existing engineering structures are concerned, the current large dam earthquake resistance design codes or specifications have a prevention level framework that is basically rational and feasible.

In addition, China has constructed a series of high arch dams with a height of 300 m at the designed peak ground acceleration in the western regions with high seismic intensities in recent years. Therefore, aseismic operational conditions have become the controlling factors in design. For instance, the Dagangshan arch dam with a height of 210 m is on the Dadu River, whose design peak ground acceleration reaches 0.56g. Up to now, there has been a certain lack of ensuring seismic safety of precedents and engineering experience, and even no seismic disaster cases for references in building such super large types of arch dams, in the strong seismic regions at home and abroad. Once these high dams and big reservoir suffer from strong earthquakes and sustain serious damage, thus resulting in unthinkable secondary disasters, the earthquake resistance safety becomes the priority to be highly concerned about by society, owners, and the engineering and technical personnel. As far as these types of important large dam engineering structures are concerned, prevention of damage from the maximum possible limit earthquake exceeding design prediction and occurrence of out of control reservoir water discharge can lead to disaster change in dam failure by the serious secondary catastrophes. This should be the strategic priority for the current research on aseismic safety of important high dams in China. For this reason, aseismic design and prevention standards or levels for the important high dams and big reservoirs need to increase the check stipulations in the performance objective without dam failure within the limit earthquake.

At present, the key technologies to be urgently solved in the evaluation of prevention of major seismic disaster changes are as follows:

1. It is rational to determine the MCE and seismic motion parameters related to sites.
2. It is necessary to determine the quantification guidelines and methods for seeking solutions for the limit stages of large dam failures of various types.

On the basis of which, the following initial suggestions are given for the amendment of large dam design and prevention levels or standards in China:

1. Single-state large dam aseismic design and prevention levels or standards determined in terms of MDE in current application in China are rational and practical.
2. As far as the design peak ground acceleration exceeding $0.3g$, a high dam of 200 m or reservoir storage capacity of over 10 billion m^3 are concerned, special research and checks should be made on the performance objectives without dam failures in the case of MCE.
3. Intensive efforts should be made to deepen research on the rational determination of MCE and on the qualification of guidelines for large dam failures of various types.

At present, it is still difficult to come to the common consensus to give out concrete quantification guidelines for technical approaches and methods on the performance objectives without dam failures of various types in the case of a MCE. Nevertheless, taking the general deformation of large dam systems as the core but not taking local stress over standard cracking in the dam body as the concept to evaluate the overall seismic safety of dam engineering structures, have gradually come to the consensus in the engineering community. The application of practical arch dams to the existing important engineering structures is relatively rather mature. No matter how important large dam engineering structures are, they should pass through special studies, whose principle of performance objective without dam failure checked in the case of MCE should be listed in the codes or specifications and also put on the agenda.

2.3 **RESERVOIR EARTHQUAKE**

Should reservoir earthquakes be listed in large dam aseismic design and prevention levels or standards? The "Guidelines for Large Dam Seismic Motion Parameters Selection" in the Past International Dam Commission had listed the RIE in the framework of dam aseismic design and prevention levels or standards. The United States seismo-geology scholar Allen suggested in 1982 that when dams of any reservoirs are over 80–100 m high, their designed earthquake should be based on the possible occurrence of earthquakes of 6.5 magnitude. When the World Bank consultants carried out the inquiry of Xiaolangdi engineering, in considering that there exist active faults in the reservoir region, they suggested an average earthquake magnitude of occurrence of reservoir seismic engineering structures of 6.0 scales in the world hitherto be taken as the design earthquake magnitude for Xiaolangdi engineering.

Reservoir earthquake refers to the fact that an impounding reservoir has resulted in changes in environmental physical states, whereby causing the occurrence of seismic phenomena within the inundated range of the reservoir dam region by the normal water storage level. Since 1931, after the 60-m-high Marathon dam in Greece began to impound water, a series of earthquakes have occurred. Since 1935, after the 220-m-high Hoover dam in Lake Mead in the United States began to impound water with the occurrence of earthquakes, over 30 countries in the world have reported over 100 cases of occurrences of earthquakes related to reservoir water impounding, but whether some cases are related to reservoir water impounding are still under controversy, of which less than half are widely confirmed to be related to reservoir water impounding. The communiqué, "Reservoir Induced Earthquake-Knowledge: General Description," issued by ICOLD, has listed 39 seismic cases of reservoir earthquakes. No matter how the statistics are made, the cases of reservoir earthquakes caused by reservoir water impounding account for an extremely small percentage among the numerous global reservoirs. Although the percentage of high dams is slightly high, only a few are confirmed. Among all the seismic cases, there are only few reaching over 5 seismic magnitudes, being of great significance for earthquake resistance. Up to now, there have been four cases in which the earthquake magnitudes are over six scales, that is, the 103-m-high Koyna dam in India ($M = 6.5$), the 120-m-high Kremasta dam in Greece ($M = 6.3$), the 105-m-high Xinfengji-ang dam in China ($M = 6.1$), and the 122-m-high Kariba dam in Zambia boundary ($M = 6.0$). And the highest earthquake magnitude is under 6.5 scales.

The genesis mechanism of reservoir earthquakes is very complicated and remains unsolved up to now. The main causes are not understood fully enough to realize the rock mass behavior state in the deep part of the rock strata where the reservoir earthquake source locates and water mass movement laws are present. For this reason, it is difficult to establish a physical model to describe this complicated process, so only the statistic and analog methods are used to probe into their essential laws. Reservoir earthquake genesis can be roughly divided into two groups. One group is related to the non-structural-types of shallow stratum microearthquake relevant with karsts, mine tunnels, land surface rock mass stress adjustment, and so on. In general, the occurrence of reservoir earthquakes is soon after reservoir impounding water and a great change in water level, being apparent in the positive correlation with water level, whose maximum earthquake magnitude cannot exceed 3–4 scales, so that no more damages will be caused to the engineering structures and reservoir areas. All the seismic cases of reservoir earthquakes all over the world have fallen into this group. Another group is reservoir water triggered to pass through, or the neighboring reservoir regions are under critical states of reservoir earthquake formation types in the seismic occurrence faults. A continuous belt correlated with the seismic occurrence formation often appears gradually in the microearthquake after the reservoir impounds water so that the intensity of earthquake magnitude is increasing till the main earthquake takes place, whose rear rock strata stress aftershocks will last for a period of time. Since the seepage process of reservoir water into the deep part of the rock strata is rather slow, the time for the main earthquake occurrence always lags behind the reservoir water impounding to the highest water level or the obvious changes that occurred in time. The maximum magnitude of the main earthquake is likely to be higher, being mainly concerned about by the engineering earthquake resistance. Up to now, the four cases of RIEs with seismic magnitudes of over 6 grades all fall into this group of foreshock–main shock–aftershock types. Accordingly, what the society and engineering community are concerned about can be mainly the structure type of reservoir-induced earthquake.

Reservoir earthquakes have been in the past known as RIEs. Since most reservoir earthquakes fall into the nonstructure types, the reservoir-induced earthquakes called in the past are still called the same nowadays. However, the structure type of reservoir earthquake, about which the dam engineering community is concerned, may be likely to damage the engineering and reservoir region, and has been called RTE by ICOLD, rationally. The tectonic stress of the seismic energy of this type originates from the seismic occurrence structure itself, and it is the latent seismic source without a reservoir built. In comparison with the vast energy released when an earthquake happens, reservoir water impounding has had a very weak effect upon the fault strain energy. As a result, it is only before the reservoir is built that the accumulation of strain energy along the seismic occurrence fault face be close to the critical state, in such a way that reservoir-impounded water can play the trigger role.

The function of reservoir water in reservoir earthquakes still needs to be further studied and evaluated, but the present-day extensive common consensus is that reservoir water seepage and its weight are the main factors.

Much research has been done on the calculation of the self-weight of reservoir water in the structure type of reservoir earthquake, with many research achievements obtained. The computation models to calculate the selfweight of water mass in itself to the crust rock mass stress field are mostly rough that it is difficult to understand the value-taking of rock parameters in deep understanding and changes in fault morphology in such a way that accurate quantification is hard to make. In qualification, this research indicates that even if a self-weight of over 10 km under land surface are taken as the effect of rock mass stress field near the seismic source of the initial crack points, they are insignificant and negligible in comparison with the effect of the self-weight of the rock mass.

It can be considered that the important trigger function of reservoir water to the structure type of reservoir earthquake is to make pore pressure infiltrate into the deeper part, so as to reduce normal stress on the fault face and weaken its shear strength, thus leading to lowering of overall shear strength but causing the fault structure face to be instable. Although the high temperature and high pressure of water mass seepage and rock mass mechanics behavior located at a depth of over 10,000 m underground are still not understood, this concept at present has been widely accepted by the large dam engineering community.

Based on the fact that reservoir water seepage is the important functional concept that causes the structure type of reservoir earthquake, a series of recognitions of the structure type of reservoir earthquake can be further discussed:

1. After the reservoir impounds water, the effect of reservoir water upon the additional impact of rock strata stress state of seismic source in comparison with seismic occurrence fault stress field is so very weak that it can only play a trigger role. For this reason, the structure type of reservoir earthquake and its triggered natural earthquake have had no differences in their nature, whose main earthquake magnitude cannot exceed the upper limit of a natural earthquake magnitude triggered by the possible occurrence of faults.

2. As viewed from seismic geologic environment, the structure type of reservoir earthquake, triggered by reservoir water needs to have a certain background condition of seismic geology and hydrologic geology; that is, there must be seismic occurrence structures within the reservoir region and the neighboring areas. This seismic occurrence structure is close to the critical state prior to the reservoir construction; there exist the hydrogeological conditions for reservoir water to infiltrate from the reservoir basin into the rock strata in the deeper part of the reservoir region after the reservoir has impounded water. It is obvious that there are only a few of high dams and large reservoirs with these background conditions so that the seismic case occurrence of the strong structure type of reservoir earthquakes are very few among the many completed high dams and large reservoirs of each country in the world.

3. As viewed from the time relation, when a reservoir impounds water, or in the early stage when there is a large change in water level, the frequency and strength of microseismic activities have had a remarkable increase, as compared with native or autochthonous earthquakes before the reservoir was built in this region, and is also in correlation with water level changes. As far as the structure type of earthquake is concerned, the occurrence time of the main earthquake always lags behind water level changes. When a reservoir impounds water to the highest water level or after the main earthquake takes place, the seismic activities in the reservoir region can gradually recover to the background state in this region, with a gradual regulation in rock mass stress of the reservoir region.

4. As viewed from the spatial correlation, the influencing range or scope of reservoir water infiltration cannot exceed the first watershed divide in general, owing to being subject to the

limitation of the infiltration scope of reservoir water along the fault belt, whereby confining that reservoir earthquake falls into a special range. Under the condition of there being no regional active faults in the reservoir region, the reservoir earthquake is, in general, in the range of 5–10 km from the reservoir bank.

5. As far as the fault behaviors are concerned, in terms of Mohr–Coulomb theory and as viewed from the analysis of the effect of reservoir water upon the fault face stress state, gravity loading in the reservoir region is helpless to the relative disturbance between the two sides of the fault of various types. On the other hand, the strike–slip fault and normal inclined fault, under the action of reservoir water seepage pressure, are triggered more easily than the reverse scouring fault. Up to now, there have been four structure types of reservoir earthquakes of over six scales over the world, whose seismic occurrence faults are normal fault or strike–slip faults, but not nonreverse scouring faults.

6. As viewed from seismic features by themselves, and in comparison with a natural earthquake, the seismic source to trigger reservoir earthquakes is shallow, being 5 km deep or so in general, whose seismic motion attenuates faster with the distance from the epicenter. The structure type of reservoir earthquake follows the foreshock–main shock–aftershock type in general. Up to now, there have been four structure types of reservoir earthquakes of over six scales over the world falling into this type. The ratio between main shock and maximum aftershock is high. In seismic magnitude–frequency formula $\lg N = a - bm$, b is also higher. The main frequency, as well as vertical and horizontal component ratios in seismic motion acceleration, are all higher.

As far as reservoir earthquake safety evaluation of large dam engineering is concerned, first of all, it is necessary to base on the principal influencing factors, hydrogeological conditions of reservoir basins infiltrating into the deep parts of the rock mass, dam heights, reservoir storage capacity rock behaviors, structure faces, and karst development degrees in the reservoir region to determine the reservoir section where the reservoir is likely to trigger earthquake cases and engineering analog principle, both at home and abroad at present. In terms of the important influencing factors, the deterministic method is adopted to carry out the comprehensive assessment or various kinds of cluster, and fuzzy analysis methods are adopted to give out the occurrence probability of seismic magnitude sections.

As far as the occurred earthquakes in the reservoir region where the completed reservoirs are concerned, whether they belong to the judgment of reservoir earthquake should be mainly based on the monitoring and contrast of seismic activity before and after reservoir impounding water in the reservoir region. For this reason, the current "Seismic Design Code of Hydraulic Structures" in China requires that, as to dam heights of over 110 m and reservoirs with storage capacity of over 500 million m^3, and if possible, the reservoir may induce an earthquake with the intensity of over 6, a special and highly accurate seismic monitoring station and network be established before the reservoir impounds water, in order to carry out the restage monitoring of earthquakes so that the basic seismic activity laws can be obtained before building the reservoir for impounding water in this region, with a view to make a comparison with the reservoir impounding water.

In summarizing these descriptions, it can be considered that an earthquake triggered by the structure type of reservoir is impossible to exceed the seismic magnitude of occurrence of seismic faults triggered in the reservoir region and being subjected to the confines of reservoir water seepage time and spatial conditions so that it cannot, in general, exceed the MDE. Therefore, it should not list them into the framework of large dam seismic design and prevention levels or standards.

DETERMINATION OF CORRELATION DESIGN SEISMIC MOTION PARAMETERS ON DAM SITE

In engineering aseismic design, seismic wave amplitude value, frequency components and sustained time are the three main factors for seismic function. Seismic waves include longitudinal and transversal body waves, and Rayleigh and Love face waves of two groups, as well as the refracted waves and reflected waves within the nonhomogeneous medium in the earth's crust. The attention and arriving land surface times of the wave velocities of these waves are uncertain, the same as in the process of transmission in such a way that they have become irregular and complicated wave shapes with more frequency components. The seismic response spectrum reflects the parameters of seismic motion input wave frequency behaviors. Seismic response spectrum $s(T, \cdot \xi)$ is free vibration cycle T, damping ratio ξ single mass point system response maximum value under the horizontal seismic motion action, which is a function changing with cycle T. The single mass point system response may be the displacement, velocity, and acceleration, which corresponds to displacement, velocity, and acceleration response spectrums, respectively. Response spectrum frequently applied in engineering is the regularized nondimensional acceleration response spectrum by horizontal seismic motion acceleration a_g, namely, $\beta(\xi) = Sa(T \cdot \xi)\bar{a}_g$, that is, free-vibration cycle $T \cdot$ damping ratio ξ single-mass point system dynamic enlarged coefficient of horizontal seismic motion peak ground acceleration (PGA) response. At present, the seismic engineering community at home and abroad has extensively used the maximum horizontal seismic motion PGA a_g and response spectrum β to characterize two main parameters for designing seismic motion inputs. Also, there is only $Sa(T, \cdot \xi)$, called the seismic influence coefficient, to be used as the main parameter. Hence, PGA is the rigid structure response spectrum value with zero cycle. It is necessary to consider such factors as the opening and closing influences of arch dam transverse cracks and dam foundation under the seismic action or local cracks and sliding influence in rock mass near the dam foundation and the nonlinear behaviors of the dam body and foundation materials, and so on, in analyzing the nonlinear problems so that seismic motion acceleration time process is an important parameter of seismic motion acceleration time process that still needs to be derived from the fitting-given response spectrum.

3.1 DESIGN OF PEAK GROUND ACCELERATION

Peak ground acceleration usually refers to the maximum acceleration. There always appears to have some plus types of high frequency peaks in the practically measured seismic motion acceleration time process so that the maximum peak can decide PGA. The calculated results indicate that if a few maximum peaks in seismic motion acceleration time process are cut off artificially, despite

Seismic Safety of High Arch Dams

53

PGA being much lower, the influence on acceleration response spectrum is still small. In a large dam aseismic design, one practically measures seismic motion acceleration time process near the Pacoima arch dam in the 1994 Beijing earthquake in the United States has been frequently taken as an example. If the originally recorded PGA of 0.424g is reduced to 0.260g, that is, PGA is reduced by 38% artificially, the calculated response spectrums of both are the same, and the maximum value of the latter is only decreased by 8% (see Fig. 3.1), and the time process of these pulse types of high-frequency PGA is much shorter than the basic free-vibration cycle of general ground architecture, thus giving out a small initial velocity, but attenuation is very fast in seismic wave transmission so that the damaged consequence caused by them is not serious. This has also been tested in some practical examples of strong seismic disasters. Although the plus peak of high-frequency acceleration in some practical measurements of ground surface seismic acceleration is large, the destructed degrees of the surrounding architecture structures are very slight. As far as structure seismic safety is concerned, only the parameters responding to structure effect significantly are important. Serving as the main parameters in aseismic design, the traditionally-used seismic motion PGA is not to reflect the realistic aseismic design parameters for seismic functional intensity. Particularly when high dams are constructed in strong seismic areas, designing seismic motion PGA is mostly controlled from the principal earthquake to the nearby earthquake, whose basic frequency is much lower than the high frequency of seismic motion PGA and also is not sensitive to the plus-type peak. For this reason, the "China Earthquake Motion Parameter Zonation Map" has adopted the seismic motion acceleration response spectrum corresponding to effective peak acceleration (EPA) to characterize the main aseismic design parameters as seismic function strength, being more rational. It is necessary to point out that the "China Earthquake Motion Parameter Zonation Map" and "China Earthquake Motion Response Spectrum Behavior Cycle Zonation Map" are aimed at shear wave velocity of 250–500 m/s, belonging to the common medium-hard sites of Grade II, and at the same time, the corresponding regulation values are also stipulated for seismic motion response spectrum behavior cycle values in the rest of the site types. In fact, as for the bedrock sites, the PGA given in Fig. 3.1 adopted is inclined to be safe.

The most important keys to hydraulic structure aseismic design are the bedrock PGA and response spectrum, serving as the main parameters for seismic motion inputs. For this reason, an analysis is made of 154 horizontal strong earthquake records of earthquake magnitudes (M_s) being over 4.5 grades, recorded on the bedrocks in the western United States in the years of 1933–1994. In order to compare the influence between earthquake magnitude M and epicenter distance R, these records should be classified into four groups in terms of $M_s < 6.5$, $R \leq 30$ m (or >30 km) or $M_s \geq 6.5$, $R \leq 30$ km (or >30 km), whose record numbers can be 48, 27, 27, and 52 clauses, respectively. Figure 3.2 gives out the average value of response spectrum $\beta(T, 5\%)$ of each group with damping ratio of 5%, of which the bold line is the regularized $\beta(T, 5\%)$ average value, and the fine line is the horizontal acceleration response average value versus average value response spectrum regularized value and both are basically in concert or agreement. It can be seen from the Fig. 3.2 that as to the different earthquake response spectrum peak cycle T is 0.2 s or so, while the average $\beta(T, 5\%)$ maximum value is 2.5 or so. For this reason, it is suggested that effective peak β acceleration should be defined as follows:

$$\text{EPA} = \frac{Sa(0.2, 5\%)}{2.5} \tag{3.1}$$

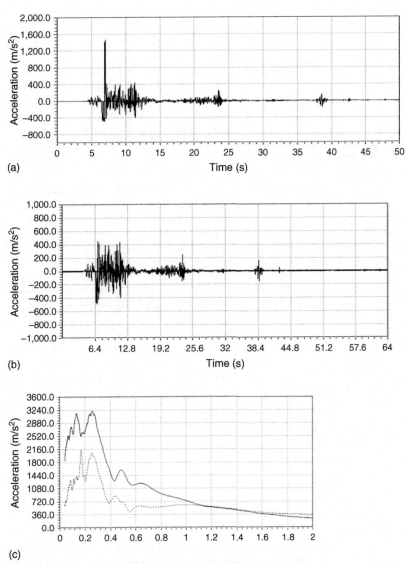

FIGURE 3.1 Response Spectrum Comparison Before and After Reducing PGA

(a) Peak ground acceleration is 0.42g of original time process; Maximum value = 1468.30 m/g^2 (time: 7 s). Sampling frequency = 50.00 Hz, (b) Peak ground acceleration reduced to 0.260g of time process. Maximum value = 484.75 m/g^2 (time: 6.64 s). Sampling frequency = 50.00 Hz acceleration (m/s^2) (c) Response spectrum comparison. Maximum value = 3211.914 m/g^2 (cycle = 0.258 s). Sampling frequency = 50.00 Hz acceleration (m/s^2).

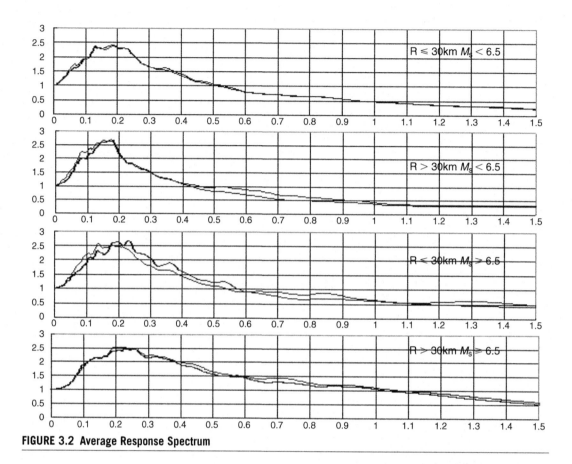

FIGURE 3.2 Average Response Spectrum

Figure 3.3 gives out the average value and standard difference of the ratios of each group, EPA and PGA. More important concerns about the important engineering works can be $M_s \geq 6.5$ and $R \leq 30$ km events with small probability and strong earthquakes. It can be seen from Fig. 3.3 that as far as statistic mean values are concerned, bedrock ground surface EPA is slightly smaller than PGA, but there is not much difference between the two. However, as far as the individual seismic event is concerned, the difference of both may reach ±40%. Accordingly, design acceleration time process should be selected from the existing strong earthquake records, or when its amplitude values are fitted in terms of given response spectrum to generate time process, it is likely to have a larger difference based on EPA or PGA, but adopting EPA can be more rational.

After bedrock EPA is decided, the weakening relation can replace the present-day adopted PGA weakened relation, thus obtaining that the results of earthquake motion hazard evaluation of the important engineering site are based on EPA serving as earthquake motion parameters. Although PGA was adopted, the EPA concept correlated with response spectrum in the earthquake motion parameter zonation map, the earthquake hazard evaluation of the large dam engineering site has been based on PGA concept hitherto, with the corresponding improvement made.

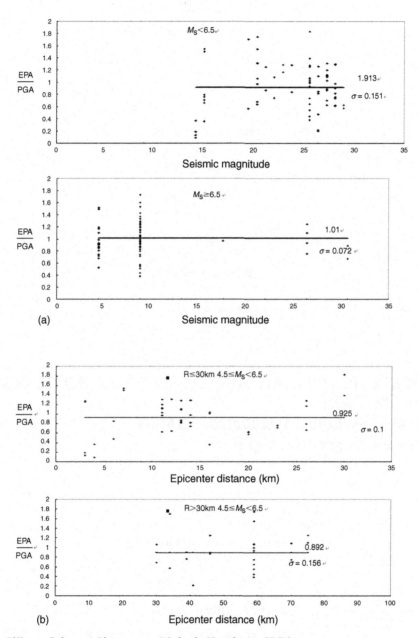

FIGURE 3.3 Different Epicenter Distances and Seismic Magnitudes EP/PC

(a) Relation map of bedrock site and seismic magnitude. (b) Relation map of bedrock site and seismic magnitude.

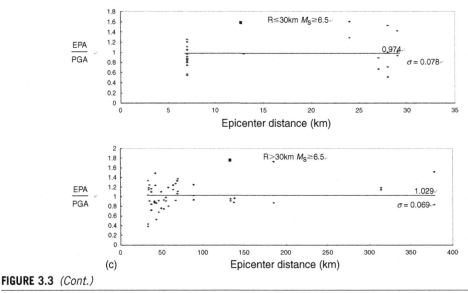

FIGURE 3.3 *(Cont.)*

(c) Relations map of bedrock site EP/PC and epicenter distance.

3.2 RESPONSE SPECTRUM ATTENUATION RELATIONS AND ITS DESIGN RESPONSE SPECTRUM

3.2.1 RESPONSE SPECTRUM ATTENUATION RELATION

3.2.1.1 Response spectrum characteristics

Response spectrum is the most important key parameter in seismic motion inputs. As described earlier, response spectrum $Sa(T,\cdot\xi)$ is free vibration cycle T. Damping ratio ξ single mass point system, being under the action of horizontal earthquake, the function of the response maximum value changes with the cycle T. Accordingly, the solution can be found in the following.

Single-degree freedom system under the action of horizontal earthquake motion acceleration $a_g(t)$ can be expressed by the vibration system consisting of quality M, rigidity K, and damping C based on rigidity foundation, whose dynamic equation is as follows:

$$M\ddot{x}(t) + C\dot{x}(t) + Kx(t) = -Ma_g(t) \tag{3.2}$$

The two sides of Eq. (3.2) are divided by M; after sorting, we have:

$$\ddot{x}(t) + 2\xi\omega\dot{x}(t) + \omega^2 x(t) = -xa_g(t) \tag{3.3}$$

$$\omega = \sqrt{K/M}$$

$$\xi = c/2\omega M$$

where, ω is the system inherent circular frequency; T is the system free vibration cycle; and ξ is the system damping ratio. The Eq. (3.2) general solution can be:

$$x(t) = e^{-\xi\omega t} (c_1 \cos\omega't + c_2 \sin\omega't) = Ae^{-\xi\omega t} \cos(\omega't + \phi) \qquad (3.4)$$

where, $\omega' = \omega\sqrt{1-\xi}$ is the inherent circular frequency of the damping system. Integral constants are c_1, c_2 or A, Φ are determined by the initial displacement x_0 and velocity x'_0 in the system. Substituting them into Eq. (3.4), we will have:

$$x(t) = e^{-\xi\omega t} \left(x_0 \cos\omega't + \frac{\dot{x}_0 + \xi\omega x_0}{\omega'} \sin\omega't \right) \qquad (3.5)$$

To solve, the Duhamel integral can be given, namely:

$$x(t) = \int_0^t \frac{-a_g(\tau)}{\omega'} e^{-\xi\omega(t-\tau)} \sin\omega'(t-\tau)d\tau \qquad (3.6)$$

Equation (3.6) can be twice differential and added with ground surface acceleration. In this way, the absolute acceleration response to the mass point can be obtained as follows:

$$\ddot{x}(t) = \int_0^t \omega e^{-\xi\omega(t-\tau)} \left[\frac{1-2\xi^2}{\sqrt{1-\xi^2}} \sin\omega'(t-\tau) + 2\xi\cos\omega'(t-\tau) \right] \ddot{x}_g(\tau)d\tau \qquad (3.7)$$

Since damping ratio is less than 10% in general when the actual structure vibrates, $\omega' = \omega$ can be approximately taken; thus, we can have:

$$\ddot{x}(t) = \int_0^t \omega e^{-\xi\omega(t-\tau)} \left[(1-2\xi^2)\sin\omega(t-\tau) + 2\xi\cos\omega(t-\tau) \right] \ddot{x}_g(\tau)d\tau \qquad (3.8)$$

In Eq. (3.8), the basic cycle is T_i, damping ratio is ξ single mass point system. The maximum value in absolute acceleration response time process is the spectrum value $S(T'_i, \xi)$ in the response curve corresponding to abscissa cycle T_i. Although seismic motion response spectrum is characterized through the maximum value of the single mass point system response to the different basic cycles, it can reflect the frequency component of seismic motion inputs, as well as one of the main basic behaviors of seismic motion. Owing to different basic cycle single mass point systems to constitute the response spectrum, their response maximum values happen in different times so that the response spectrum cannot reflect the phase difference influences among each frequency subcomponent in seismic motion acceleration waves, as well as the sustained time influence in seismic motion waves.

Any response spectrum curves of actually measured acceleration time process are all irregular so that integrating the response spectrum curves in the same types of ground foundation can obtain the smoothing mean value. For it to be convenient for the application to engineer earthquake resistance design, they still need to be further simplified. Usually, the response spectrum taking the amplitude values from $T = 0$ to the response spectrum peak β_{max}, corresponding to the part of performance cycle; T_g are β_{max} horizontal straight line section. The structure acceleration response mainly depends upon the high frequency component in seismic motion acceleration so that with an increase in T value, the acceleration response value decreases. In this way, the rear part, which is usually larger than feature cycle T_g, takes the curves in terms of exponent attenuations, that is, $\beta = \beta_{max} (T_g/T)^\gamma$. In order to avoid the too-low value of the response spectrum of the part extending to the long cycle, the response spectrum values, being over certain cycle values, should be limited and revised. For this reason, the design response spectrum adopted in aseismic design mainly depends upon such three parameters as β_{max}, T_g, and y.

Based on the basic cycle short of many medium- and small-sized dams and in theory, $T = 0$ rigid structures without dynamic amplification effect, the part of response spectrum values from $T = 0$ to $T = 0.1$ s in design response spectrum takes the inclined section from 1.0 to β_{max} set in the Code of Earthquake Resistance Design for Hydraulic Structures in China. At present, the design response spectrum has been used in the aseismic design codes by each department in the country.

The embodiment of seismic motion input frequency component and performance response spectrum are obtained through the actually measured earthquake acceleration records. The actually measured earthquake acceleration records are subject to the influences of the sizes of seismic magnitudes, seismic source performances, transmission media and site conditions, and other factors. The low-frequency component in seismic waves enlarges with an increase in earthquake magnitude, while the high-frequency component attenuates with an increase in distance. Therefore, seismic wave frequency constituents and performances are closely related with the earthquake magnitude and the distance from the seismic source. It can be seen that the response spectrum is bound to be related with the earthquake magnitude and the distance from the seismic source. The correlations among the response spectrum, earthquake magnitude M, and distance from seismic source R characterized by the attenuation relations are the first and foremost important to engineering aseismic design.

3.2.1.2 *Attenuation relation of response spectrum*

The transmission process in the earth's crust medium and seismic wave is extremely complicated. At present, it is hard to finalize and summarize the correlative and extensive regulations or laws among the response spectrum, earthquake magnitude M, and the distance from seismic source R in theory. Therefore, it can only be based on the actually measured records to derive from statistical significance or sense to reflect the correlative attenuation relations among the response spectrum and earthquake magnitude, as well as the distance from seismic source.

Up to now, there are only the western United States and Japan holding abundant seismic motion acceleration data actually measured from the strong earthquakes many times, which are enough to make statistic attenuation relations of seismic motion parameters. The strong earthquakes in Japan belong to the earthquakes between the plates, whose seismic motion behaviors are different from the earthquakes within the plates in the United States and China. At present, there have been extremely few actually measured strong earthquake acceleration recorded data but an abundance of historical earthquake records in China, from which the relation of intensity attenuation can be finalized. For this reason, in the earthquake hazard evaluation site in China, the attenuation relations of seismic motion parameters are obtained from the conversion of seismic intensity information, with the western United States falling into the same seismicity within the plates as the reference zones. By using the abundant seismic intensity information in our country, the conversion of seismic intensity and seismic motion parameters between two regions can be carried out in terms of isoseismic magnitudes, isodistances, or image method. Based on this basic assumption of difference in intensity attenuation relations and the attenuation laws of seismic motion parameters in different regions, we thus make the corresponding points to earthquake magnitudes and seismic intensity correspond to seismic motion PGA. However, the assumption on which this kind of conversion is based may have the following several points to be worth further discussing. To begin with, seismic intensity is a macro, comprehensive seismic disaster qualitative description and refers to the average seismic disaster states in a certain range of the region (e.g., 1 km or so). Only a part of seismic disasters depends on the foundation being out of effects and other factors that have no connection with seismic motion, such as fact disturbances and landslides. PGA in seismic motion parameters is only one of the major factors affecting seismic foundations. In addition, a strong seismic record is obtained from only one fixed place, while intensity corresponds to a certain range in the region, in which the variation in PGA in each point is very complex, and there exists a certain difference.

In the second place, taking the western United States as the reference region, the earthquake magnitude in the relation of seismic intensity attenuation is derived in terms of empirical formula conversion, with the embodiment of the isoseismic line of intensity attenuation taking the circular. Whereas in China, the seismic intensity model parameters are determined using the earthquake magnitude determined by instruments, intensity isoseismic lines take the oval with the differentiation of long and short axes.

Although there exist the above problems, at present, in the case of a lack of actually-measured strong earthquake records, this kind of intensity conversion method is not out of its expedient measure. However, if the assumption on which this conversion method is based can be extended to consider the difference of intensity attenuation relations in different regions and similar to the attenuation relation difference of each cycle component (T) in seismic motion response spectrum, it is hard or different to understand it and be accepted.

The Chinese scholar held that, in dealing with the Chinese hinterland seismic activity as viewed from seismic basic genesis environment, the occurrence of earthquakes in the Chinese mainland falls into the inner Eurasian plate (within plate) seismic type, that is, the mainland seismic structure system earthquake. The main body of the mainland seismic structure system is located in the northern latitude of $20\sim50°$, traversing the Eurasian mainland and North American mainland, whose eastern and western two half zones appear to have asymmetric seismic activities. The majority of earthquakes in the mainland structure system take place in the shallow source earthquakes within the earth's crust, showing the level or horizontal disturbance in the main. The Chinese mainland and North American mainland are similar to a certain extent in structure, earth's crust constituents, modern stress state, seismic genesis, and seismic activity performance, and so on, which is comparable. Accordingly, the interborrowing and interapplying of seismic records from two regions have certain structural bases. It can be seen from this that at present, in the case of a shortage of strong seismic records, in our country, the direct use of statistic mean attenuation relation of acceleration response spectrum fitting each cycle component $Sa(T)$ obtained from the western United States strong records can be more rational than that obtained from the response spectrum attenuation formula through the intensity conversion. The differences, based on the information and fitting method, will lead to different response spectrum attenuation relations so that rational selection of acceleration response spectrum attenuation is first and foremost important.

The existing seismic motion attenuation relations are based on the point seismic source assumption to describe the attenuation laws of seismic motion parameters with the distance from seismic source under the special conditions, without considering crack patterns of seismic occurrence faults and influence factors of site conditions upon seismic motion parameters. In fact, under the conditions of nearby sites, the cracking direction effects of seismic occurrence faults, hanging side effects, and deposit basin efforts, and the influence of nearby seismic performance sites upon seismic motion cannot be ignored. Being subject to the quantity of strong earthquake records and range restriction, the statistical analysis of early seismic motion attenuation relations can only be set up on the basis of the information of far site and mid-sized earthquakes. The lack of nearby fault seismic records makes it difficult to obtain the nearby site seismic motion attenuation performances by statistic method. More and more research indicates that the nearby site seismic motion attenuation laws are different from those of far site seismic motion attenuation. Using far site information to estimate the nearby site is easy to underestimate the seismic motion parameters in a small earthquake, but also to overestimate the seismic motion parameters in a large earthquake. In recent years, the strong-motion seismographs have been laid out and popularized over the world so that the United States, Japan, Turkey, and Taiwan of China have obtained a certain quantity of the nearby fault seismic records, so as to lay the data foundations for studying and predicting the nearby site seismic motion attenuation laws, and at the same time prove that there exists a rather large deviation in the seismic motion prediction model obtained for the nearby fault region through the deduction based on the mid- and far-site recording attenuation relations. New record accumulation is bound to promote the renewal of old attenuation relations, and the seismic information or data obtained in recent years should be stored in the data bank in such a way that the influence of nearby site seismic motion performances upon the response spectrum attenuation relations can be evaluated and treated more objectively.

Next generation attenuation (NGA) research is aided financially by the United States Pacific Seismic Engineering Research Center, which serves as an applied research project of the Lifeline System Aseismic Engineering Program, sponsored by the California Communication Department, California Energy Resource Commission, and the Pacific Gas and Electric Power Company. This research project is carried out mainly in the west of the United States, and on the basis of the world's main strong seismic real measurement of acceleration recorded information or data. All the data come from PEER NGA data bank (Flatfile 7.2 Edition) including 3,551 strong seismic records in 173 earthquakes, whose site conditions are average shear wave velocity $V_{s30} = 116 \sim (2,016$ m/s) free fields in the depth of 30 m deep (see Fig. 3.4). It can be seen from Fig. 3.4 that the free field records in the NGA data bank are mainly concentrated on the earthquake magnitudes of $5 \sim 7.7$ and within the distance range of $5 \sim 100$ km. Although a large number of records are supplemented, including the nearby site strong seismic records that have occurred for several times over the world in recent years, there are still very few of the nearby site strong seismic records of earthquake magnitudes of 6.5 within the distance of 10 km.

Research is responsible for by the well-known authority teams in the five research fields including Abrahamson and Silva (2007), Boore and Atkinson (2007), Campbell and Bozorgnia (2007), Chiou and Youngs (2007), and Idriss (2007), respectively. Research objectives are suggested by each research team individually for adapting the shallow seismic source condition in the western United States, and efforts are made to consider to a certain extent the new seismic motion attenuation relations of nearby site large earthquake influencing factors and cracks, fault depth, seismic source effects, deposit basin effects, and so on.

The NGA's five sets of attenuation relations aim at each different cycle component of acceleration response spectrum, of which $T = O$ component is corresponding to the PGA, while $T = 0.2$ component divided by 2.5 is corresponding to the previously described effective peak ground acceleration. The five sets of attenuation relations have resulted in a certain difference existing among these attenuation relations because of incomplete similarities of selections and recording quantity, specified bedrock conditions, and considered nearby site performances. Table 3.1 lists five sets of NGA attenuation relation

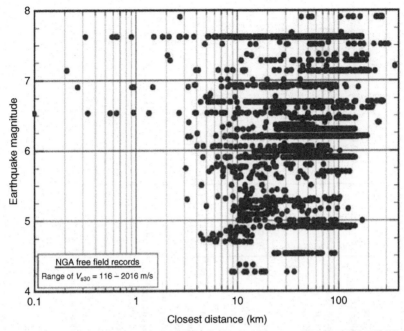

FIGURE 3.4 Earthquake Magnitude Ground Surface Seismic Record-Distance Distribution in NGA Data Bank

models. Different NGA attenuation relation models involve the basic parameters, which are not completely the same with the distance serving as an example; some can adopt epicenter distance, seismic source distance, fault distance, or fault projection distance. The differences of these distance definitions have had not much effect upon the far site engineering site earthquake motion, but rather much effect upon the nearby site earthquake motion.

It can be seen from the comparison of five sets of attenuation relations that:

1. Mathematical models are different from each other, but they involve more parameters that need regression determination and show much complexity. If they are not analyzed and simplified, they are difficult to be directly used in engineering practice.
2. Though strong seismic records used in regression originate from the same data bank, there are differences in the selection of earthquake magnitudes, distances, shear wave velocity range, and so on, with the data or information materials selected by Abrahamson–Silva, Chiou–Youngs being rich.
3. Earthquake magnitudes all take matrix seismic magnitude M_w. As to strong earthquakes with over 5 scales, matrix seismic magnitude M_w is basically the same as face wave earthquake magnitude M_s.
4. Except that Boore–Atkinson adopts the nearest distance (R_{jb}) from the site to the bottom end ground surface projection of fault fracture face, the rest all adopt fault distance.
5. The bedrock is defined with average wave velocity V_{S30} being over 550 m/s of site earth to define the bedrock, of whom Boore–Atkinson, Chiou–Youngs, and Idriss take V_{S30} being over 1300, 1500, 900 m/s of the site earth to define the bedrock, respectively.

Table 3.1 Five Sets of NGA Attenuation Relation Comparison

Attenuation Relations	Selected Records	Adaptable Conditions	Bedrock	Contents Considered	Distance	Basic Model
Abrahamson–Silva (AS07)	2,675 strong seismic records of 130 times of earthquakes	1. $M = 5\sim8.5$ (SS); 2. $M = 5\sim8$(RV'NML); 3. $0km \le R_{rup} \le 200$ km	$V_{S30} \ge 760$ m/s	1. Fracture patterns 2. Hanging side effects	Fault distance	$\ln S_a(g) = f_1(M, R_{rup}) + a_{12}F_{RV} + a_{13}F_N + f_5(PGA_{1100}, V_{S30}) + f_{10}(z_{1,0}, V_{S30}) + F_{hw}f_4(R_{jb}, R_{rup}, W, dip, Z_{top}, M) + F_{RV}f_6(Z_{top}) + (1-F_{RV})f_7(Z_{top}, R_{jb}, M) + f_8(R_{rup})$
Boore–Atkinson (BA07)	1,574 records of 58 times of main earthquakes	1. $M = 5\sim8$; 2. $R_{jb} < 200$ km; 3. $V_{S30} = 180\sim1300$ m/s	$V_{S30} \ge 760$ m/s	Fracture patterns	Joyner–Boore distance R_{rup}	$\ln Y = F_M(M) + F_D(R_{jb}, M) + F_S(V_{S30}, R_{jb}, M) + \varepsilon\sigma_T$
Campbell–Bozorgnia (CB07)	1,574 strong seismic records of 64 times of earthquakes	1. $M = 4\sim8$; 2. $0km \le R_{rup} \le 200$ km	≥ 760 m/s	1. Fracture patterns 2. Hanging side effects 3. Basin effects	Fault distance R_{rup}	$\overline{\ln Y} = f_{mag} + f_{dis} + f_{flt} + f_{hng} + f_{sed} + \varepsilon$
Chiou–Youngs (CY07)	3,297 strong seismic records of 131 times of earthquakes	1. $M = 4\sim8.5$ (SS); 2. $M = 4\sim8$(RV'NML); 3. $0km \le R_{rup} \le 200$ km; 4. $V_{S30} = 150\sim1500$ m/s $V_{S30} = 450\sim900$ m/s	≥ 760 m/s	1. Fracture patterns 2. Hanging side effects	Fault distance R_{rup}	$\ln * y_{surface} * = \ln * y_{1130} * + f_{site} + \sigma_{zj}$ $\ln * y_{1130} * = c_1 + f_{source} + f_{path} + f_{HW} + \tau z_i$
Idriss (IMI07)	987 strong seismic records of 74 times of earthquakes		≥ 550 m/s	Fracture patterns	Fault distance R_{rup}	$\ln * PAA(T)* = \alpha_1(T) + \alpha_2(T)M - *\beta_1(T) + \beta_2(T)M^* \times \ln(R_{rup} + 10) + \gamma(T)R_{rup} + \varphi(T)F$

6. Taking the nearby site earthquake magnitude saturated effect into consideration.

7. Taking the influence of fault types into consideration.

8. Only Abrahamson–Silva, Campbell–Bozorgnia, and Chiou–Youngs account for three sets of attenuation relations into the hanging side effects.

In contrast with the mentioned five kinds of attenuation relations, Abrahamson–Silva's attenuation relation is based on the rich strong seismic recording information, which is mainly taken from the western United States. The factors considered are also complete and the bedrock site attenuation relation forms are relatively simplified. The specifications for rock mass shear wave velocity are also adaptable to the rock II conditions for the dam bottom constructing foundation face, whose regularized response spectrum β is close to the fitting average response spectrum β, by taking bedrock nearby site, which really measured strong seismic records. Therefore, it can be considered that they can basically be used in high dam aseismic design in our country.

The results from the analysis and comparison of five sets of attenuation relation indicate:

1. There exists a rather large difference among each set of attenuation curves in the nearby site ($R_{rup} \leq 10$), the far site ($R_{rup} \geq 10$), and strong earthquake ($m \geq 7$inte); when $M_w = 8.0$ intensity, the occurring seismic fault type is the strike–slip type, and five sets of attenuation relation curves of bedrock response spectrum components with T values being 0, 0.2, 1.0, and 3.0 s, respectively, are shown in Fig. 3.5. In the case of the nearby site strong earthquake, the differences are rather apparent.

2. The five types of attenuation relations obviously show the "saturation" performances of the nearby site seismic magnitudes.

3. In Fig. 3.6, Abrahamson–Silva (ASOT) and Idriss (IMI$_0$), two kinds of attenuation relation models give out the Fault distance R_{rup} of 10 km, and earthquake magnitude M of 6.5, 7.0, 7.5, and 8.0, so that three kinds of different fault fracture types can affect the regularized response spectrum. The rest of attenuation relation models and five kinds of attenuation relation models appear to have the similar laws in the case of far distances. Therefore, it can be considered that fault fracture types have had not much influence upon the response spectrum attenuation relations. Particularly, there can be little impact upon the nearby site large earthquake short cycle component, which can be ignored.

4. It can be seen from Fig. 3.7 that Abrahamson–Silva (ASO7), Campbell–Bozorgnia (CBO7), and Chiou–Youngs (CYO7) consider the hanging side effect attenuation relations, in which when fault fracture types are the reverse fault and inclination of 45°, the growth rate of earthquake motion acceleration response spectrum, after the hanging side effect accounted into it, has had a greater difference with an existence of distance distribution pattern. In the Abrahamson–Silva attenuation relation, the hanging side effect influence is the most protruding. After the hanging side effect is accounted into, the growth rate of response spectrum, in corresponding to $T = 0$ and $T = 0.2$ s, is about 60% in the epicenter, being gradually increasing with an increase in distance till the distance is 10 km, the rate reaches 200% of the peak or so, and then, it reduces to zero rapidly with an increase in distance. Figure 3.8 shows that in terms of the Abrahamson–Silva attenuation relation, when in different earthquake magnitudes and distances, the different fault inclinations can influence the hanging side effect. As seen generally within the distance of 10 km or so near the site and I_s inner short cycle component, and when the fault inclination is no more than 70°, the hanging side effect can play an increasingly obvious role in enhancement of earthquake motion, but when the fault inclination is over 70°, the hanging side-effect is obviously diminished. In Campbell–Bozorgnia (CBO7) and Chiou–Youngs (CYO7) attenuation relation, the influence of the hanging side effect is not the same as the protruding of the Abrahamson and Silva attenuation relation. However, the three have shown the laws of large earthquake magnitude, near distance, occurrence seismic fault inclination less than 70°, and obvious

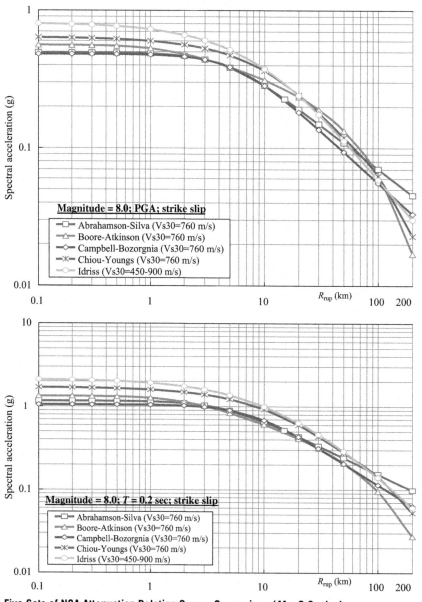

FIGURE 3.5 Five Sets of NGA Attenuation Relation Curves Comparison (*M* = 8 Scales)

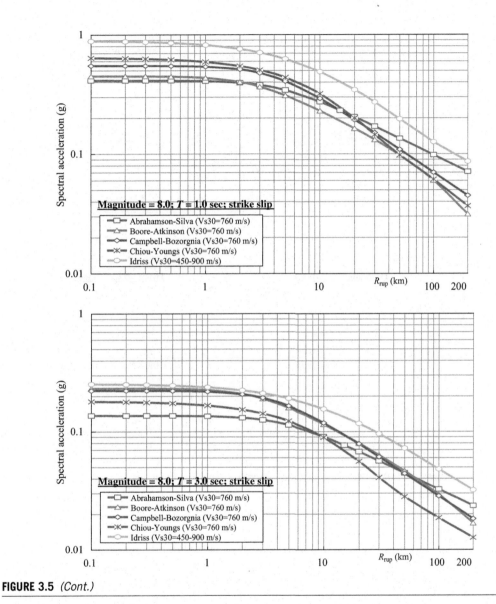

FIGURE 3.5 *(Cont.)*

influence of the hanging side effect of the response spectrum short cycle component are similar or approximate.

High dam small probability design and prevention levels are the most nearby large earthquake sites, whose basic cycle is within I_s. When the seismic occurrence fault inclination is less than 70° or so, while the right and light dam sites are located in the hanging sides, attention must be paid to the influence of the hanging side effect. The existing seismic disaster cases indicate that, when in strong earthquakes, the seismic disasters located in the upper hanging site area on the fault

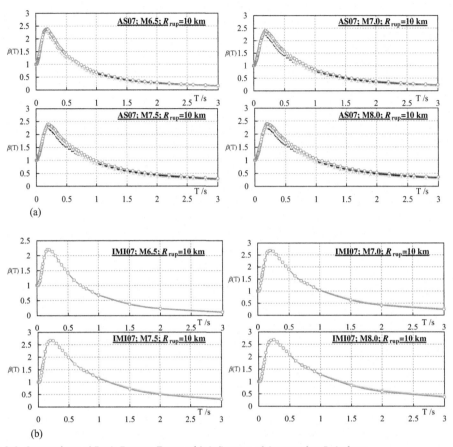

FIGURE 3.6 Comparison of Fault Facture Types with Influence of Attenuation Relations

(a) Comparison of fault fracture-type influences in Abrahamson–Silva attenuation relations model. (b) Comparison of fault fracture-type influence in Idriss attenuation relations model.

fracture face are more obviously strengthened than those in the lower hanging site regions. For instance, in 1999, the Chelungpu occurrence seismic fault in the Chi-Chi earthquake in Taiwan was on the reverse fault with the inclination of 25~45°, whose hanging site effect is very protruding, so the architectural structures located at the upper hanging site suffered from severe damages, while the architectural structures located at the lower hanging site, and even the architecture structures two to three m from faults were found to have no destruction. At present, in engineering site earthquake hazard evaluations, the attenuation relations adopted cannot account for the influence of the hanging site effect. It is likely to consider because there are more steep inclinations in the occurrence of seismic faults in our country. The United States architecture code (UBC97) stipulates that the nearby site effect should be taken into account for the strong seismic area with PGA of 0.4g set in the seismic motion parameter zonation map, whose seismic motion parameters need to be multiplied by the nearby site factors. When the seismic magnitude is not

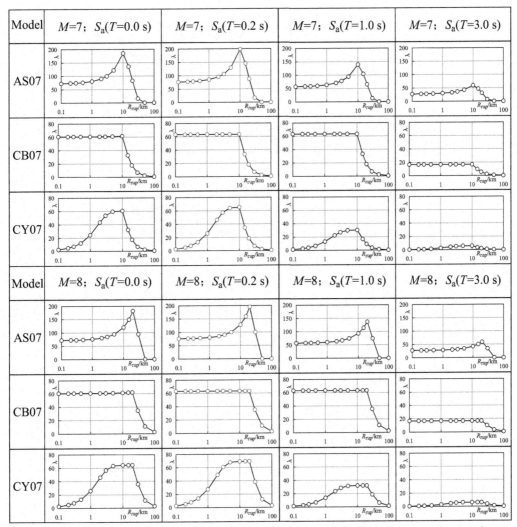

FIGURE 3.7 Increasing Rate of Seismic Motion Acceleration Response Spectrum After Accounting Hanging Side Effect

less than seven and fault distance is not over two and five km, seismic motion acceleration nearby site factors are 1.5 and 1.2, respectively. China's "Design Codes for Architectural Earthquake Resistance," under deliberation and approval, stipulates that the structures located in both sides of fracture of seismic occurrence within the range of 10 km must be accounted into the nearby site effect, within the range of 1.5, and outside five km, suitable to be multiplied by increasing the coefficient, being not less than 1.25. It is just here that the nearby effect seems mainly to consider the influence of long sustainable time pulse type of high energy inputs, but the influence of hanging site effect of the nearby site large earthquake will not be included.

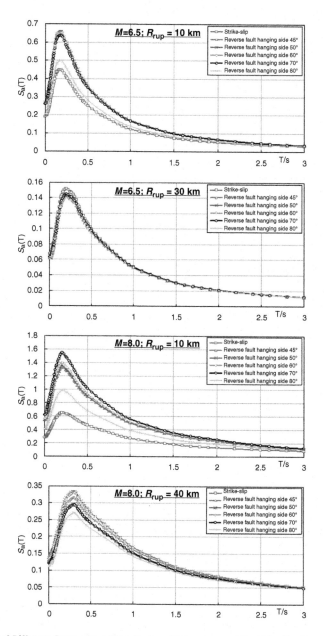

FIGURE 3.8 Influence of Different Earthquake Magnitudes, Distances, and Fault Inclinations Upon the Hanging Side Effect

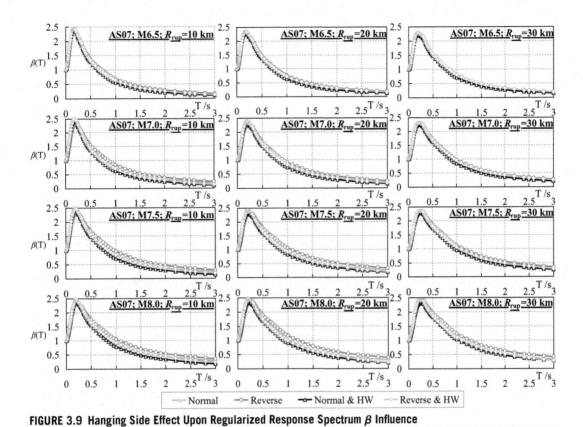

FIGURE 3.9 Hanging Side Effect Upon Regularized Response Spectrum β Influence

Figure 3.9 gives out the hanging site effect upon regularized response spectrum β influence in terms of Abrahamson–Silva (ASOT) attenuation relations, whereby illustrating that the hanging site effect has had not much influence.

5. The regularized bedrock response spectrum is 2.5 or so, having no apparent relation with earthquake magnitude and distance, and basic corresponding cycle $T = 0.2$ s component. When there are large earthquakes and far earthquakes, the seismic motion long cycle component increases, and the response spectrum peak corresponding to the cycle has had a slight increase. But after being larger than peak cycle, the attenuation exponent (r) of the response spectrum falling section has had an increasing trend with an increase in seismic magnitude and distance.

Engineering aseismic design is based on the criterion of bedrock seismic motion inputs adopted in earthquake hazard evaluation. The influence of the covering earth layer over the site bedrock involves the nonlinear dynamic behaviors of each layer covering the earth. At present, it is only in terms of really measured thickness of larger covering of the earth and mechanics parameters that can be served as a one-dimensional problem to seek the solution to land surface earthquake motion from rockbed seismic motion inputs.

As to bedrock, Abrahamson–Silva acceleration response spectrum attenuation relation can be simplified into the formula to be conveniently used in engineering structures in the case of only accounting the hanging side effect and not accounting the fault type differences, namely:

$$\ln Sa(g) = f_1(M, R_{rup}) + F_{HW} f_4(R_{jb}, R_{rup}, W, \text{dip}, Z_{top}, M) \tag{3.9}$$

Where, F_{HW} can take 1.0 when the site is located on the fault fracture face hanging side, or else take O. The basic attenuation relation of acceleration response spectrum, without accounting the hanging side effect, can be as follows:

$$f_1(M, R_{rup}) = \begin{cases} a_1 + a_4(M - c_1) + a_8(8.5 - M)^2 + \left[a_2 + a_3(M - c_1)\right]\ln(R) & M \leq c_1 \\ a_1 + a_5(M - c_1) + a_8(8.5 - M)^2 + \left[a_2 + a_3(M - c_1)\right]\ln(R) & M > c_1 \end{cases} \tag{3.10}$$

$$R = \sqrt{R_{rup} + c_4} \quad (c_4 \text{ takes } 4.5) \tag{3.11}$$

$$f_4(R_{jb}, R_{rup}, W, \text{dip}, Z_{top}, M) = a_{14} T_1(R_{jb}) T_2(R_x, W, \text{dip}) T_3(R_x, Z_{top}) T_4(M) T_5(\text{dip}) \tag{3.12}$$

$$T_1(R_{jb}) = \begin{cases} 1 - \dfrac{R_{jb}}{30} & R_{jb} < 30 \text{ km} \\ 0 & R_{jb} \geq 30 \text{ km} \end{cases} \tag{3.13}$$

$$T_2(R_x, W, \text{dip}) = \begin{cases} 0.5 + \dfrac{R_x}{2W \cos(\text{dip})} & R_x \leq W \cos(\text{dip}) \\ 1 & R_x > W \cos(\text{dip}) \text{ or dip} = 90 \end{cases} \tag{3.14}$$

$$T_3(R_x, Z_{top}) = \begin{cases} 1 & R_x \geq Z_{top} \\ \dfrac{R_x}{Z_{top}} & R_x < Z_{top} \end{cases} \tag{3.15}$$

$$T_4(M) = \begin{cases} 0 & M \leq 6 \\ M = 6 & 6 < M < 7 \\ 1 & M \geq 7 \end{cases} \tag{3.16}$$

$$T_5(\text{dip}) = \begin{cases} 1 - \dfrac{\text{dip} - 70}{20} & \text{dip} \geq 70 \\ 1 & \text{dip} < 70 \end{cases} \tag{3.17}$$

where M is matrix seismic magnitude; R_{rup} is fault distance, km; R_{job} is the nearest distance from the site to the fault fracture face bottom end land surface projection, km; R_x is the level distance from the site to the hanging side fracture top end, km; W is the width of fault downward fracture, km; dip is fault inclination; Z_{top} is the depth of land surface distance from fracture top end. Being corresponding to earthquake magnitudes of 5, 6, 7, and 8, they can take 7 km, 3 km, 0km, and 0 km, respectively; a_1, a_2, a_3, a_4, a_5, a_8, a_{14}, and c_1 are coefficients in close relation with cycle T.

3.2.2 DESIGN RESPONSE SPECTRUM
3.2.2.1 Standard response spectrum

Standard response spectrum is to refer to the designed acceleration response spectrum specified for the general architectural structures in aseismic design code. It is the acceleration response spectrum statistics mean value obtained on the basis of many real measure acceleration records of the same type of foundations. At present in our county, only in the particularly important nuclear power engineering design, can mean value plus one-fold standard difference be adopted as the designed standard response spectrum. Usually, nondimensional β value divided by PGA is adopted to express the standard response spectrum. The acceleration response spectrum directly expressed by seismic influence coefficient α (m/s) in "Architecture Aseismic Design Code" in China is adopted to determine seismic functions. The simplified standard response spectrum adopted in aseismic design mainly depends on such three parameters as $\beta_{max}(\alpha_{max})$, T_g, and r. As described before, the response spectrum is the correlation with seismic farness and nearness and magnitudes. The regularized response spectrum platform β_{max} is, in general, subject to less seismic farness and nearness and magnitudes, which can be neglected at present. For this reason, in aseismic design standard response spectrum, the performance cycle T_g can only be used to consider seismic farness and nearness, and magnitudes, as well as the influence of foundation types. Since large earthquakes, far earthquakes, and soft foundation response spectrum platforms tend to extend to long-term cycle direction, the special performance cycle T_g value is increased.

In large dam engineering aseismic design, great attention must be paid to the performance cycle T_g value, serving as seismic motion input base rock response spectrum. Concrete dams in China are mostly constructed on bedrocks; the arch dam foundation requires better rock-beds. The current "Aseismic Design Code of Hydraulic Structures" has considered that the transcendental probability of large dam design and prevention is so small that most of them come from the contributions of the occurrence of seismic faults near the dam sites. Accordingly, the performance cycle T_g value of rock-beds near the seismic response spectrum is taken as 0.2 s. This is in agreement with the rock-bed response spectrum performance cycle T_g value set in the Aseismic Design Code of Hydraulic Structures published in 1978 and in 2001. As described previously, the "China Earthquake Motion Parameter Zonation Map" and "China Earthquake Motion Response Spectrum Performance Cycle Zonation Map," issued in 2001, are aimed at the mid-hard Grade II of sites. As to hard Grade I of foundation, the nearby seismic performance cycle T_g value can be adjusted to be 0.25 s, which is coordinated with specifications adopted in the Architecture Aseismic Design code published in 1989, of which the defined Grade I foundation corresponding to shear wave velocity takes 500 m/s.

But the architecture aseismic design code was revised after 2009, in which Grade I of foundation is divided into two groups: I_0 and I_1. I_0 group is definitely defined as the rock site with the shear velocity being over 800 m/s, corresponding to nearby seismic performance cycle T_g value taken as 0.2 s. With regard to the classification of rock foundation shear wave velocity, it is defined as 700 m/s in the Nuclear Power Aseismic Design Code in our country, but this is defined as 760 m/s in the United States Architecture Aseismic Design Code, as well as the NGA as 800 m/s in the European Aseismic Design Code. In fact, taking the dam bottom construction foundation face requiring Grade II of rock mass, the shear wave velocity V_s is far greater than 800 m/s, but in general, $V_s = 2000$ m/s above. The mentioned

154 pieces of horizontal strong seismic magnitude M_s over 4.5-scale in the western Unites States during the years of 1933–1994, and the United States NGA giving out the results of bedrock nearby seismic response spectrum, all have shown that the response spectrum mean value is about 0.2 s. Accordingly, in the current aseismic design code of hydraulic structures, and in the case of considering nearby seismic situations, dam body rock mass response spectrum performance cycle T_g value taking 0.2 s is rational.

Although the seismic wave long cycle component in a large earthquake and far earthquake increases, the response spectrum performance cycle also has increased, thus making long cycle structure earthquake response increase. The statistic results of bedrock strong earthquake records in the western United States and the United States NGA relations indicate that the influence of earthquake magnitude upon bedrock response spectrum performance cycle T_g value is less obvious than the distance. In the newly complied "Architecture Aseismic Design Code," the site epicenter distance R is roughly divided into such three groups as $R \leq 10$ km, 10 km$<R<50$ km, and $R \geq 50$ km. Each group performance cycle value increases 0.05 s progressively in succession.

The long cycle structure is sensitive to the long cycle component in seismic motion. For this reason, as viewed from safety, the current seismic design code of hydraulic structures stipulates that as to a hydraulic structure with a basic cycle exceeding 1.0 s, its response spectrum performance cycle T_g value adopted in aseismic design should take 0.25 s. The newly compiled "Architecture Aseismic Design Code" requires that the performance cycle value should increase $0.05g$ when the seldom-occurring seismic actions of VII and IX magnitudes are calculated. This is because in considering the actions of seldom-occurring earthquakes of VIII and IX magnitudes, the structure has entered into the elastic–plastic stage, whose cycle can increase. For this reason, as viewed from a safety angle, the design response spectrum performance cycle value can increase.

It is necessary to point out that in the types of site foundations and earthquake magnitudes, not only can far and near distances in particular and other factors affect the response spectrum performance cycle values, but also the influence of exponent r value of the falling section of response spectrum curves cannot be disregarded, because it is directly related to the fastness and slowness of response spectrum curve falling after the peak cycle.

The basic cycle of hydraulic structures is not over I_s in general. For instance, the full reservoir cycle of the Jinping Grade I arch dam, the world's highest (305 m), is 0.67 s; the full reservoir basic cycle of the 295 m Xiaowan arch of located in the wide river valley is only I_s; and the full reservoir basic cycle of the 112 m Xiaolangdi water-intake tower assembly is 0.8 s. Therefore, as far as the aseismic design for the common hydraulic structures is concerned, the most important key is the bedrock standard response spectrum value in the cycle without exceeding I_s. As to the extremely few flexible higher structures, such as the TGP ship-lifting elevator tower column structure of 145-m high, its traverse basic cycle is over 2 s, and in general, they are the reinforcement concrete structures, whereby referring to design response spectrum specified in the architecture Aseismic Design Code.

As the previously simplified Abrahamson–Silva acceleration response spectrum attenuation relation formula (3.9), the effects of exponent r values of falling section of response spectrum curves upon the different earthquake magnitude M and the response spectrum cures from distance R are calculated. The calculated results indicate that if $\beta_{max} = 2.5$, $T_g = 0.2$ s, $\gamma = 0.6$, fitting response spectrum can be adopted in the case of R, without being over 20 km, the results are roughly close to those calculation in terms of Abrahamson–Silva acceleration response spectrum attenuation relation formula. When the cycle is less than 1.0 s, there is about 20% increase as against the arch dam standard spectrum value in terms of the current aseismic design code of hydraulic structures. Figure 3.10 gives out the comparisons among

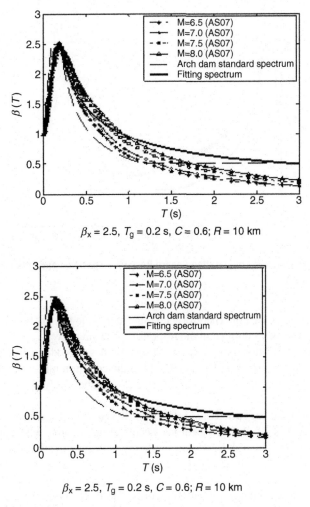

FIGURE 3.10 Abrahamson–Silva Response Spectrum and Fitting Response Spectrum in Terms of $\beta_{x\,max} = 2.5$, $T_g = 0.2$ s, $\gamma = 0.6$ and Comparison of Standard Spectrum $\beta_{x\,max} = 2.5$, $T_g = 0.2$ s, $\gamma = 0.9$ When the Damping Ratio of the Current Large Dam is 5%

the Abrahamson–Silva response spectrum and the fitting response spectrum in terms of $\beta_{x\,max} = 2.5$, $T_g = 0.2$ s, $\gamma = 0.6$, when the distance R is 10 and 20 km, and the standard spectrum $\beta_{x\,max} = 2.5$, $I_g = 0.2$ s, $\gamma = 0.9$, when the damping ratio of the current large dam is 5%. In the current seismic design code of hydraulic structures and Architecture Aseismic Design Code, the stipulations of both to the regularized standard response spectrum are identical in the case of $T<3$, so that they take $\gamma = 0.9$. In the newly compiled Architecture Aseismic Design Code, taking $\gamma = 0.9$ still remains. It is investigated that this is due to considering many architectures. Particularly, there is an increasing number of high-rise buildings,

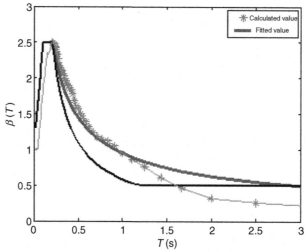

FIGURE 3.11 Real Measured Rocked Acceleration Record Response Spectrum and Fitted Response Spectrum
($\beta_x = 2.5$, $T_g = 0.2$ s, $\gamma = 0.6$)

whose basic cycle large values are big. Therefore, taking small value for γ is bound to lead to the over I_s long cycle part values to be over-enlarged in the response spectrum curves.

To summarize these descriptions, in the revised aseismic design code of hydraulic structures, it is suggested that the bedrock design response spectrum and the common mid-hard Grade II site standard response spectrum parameters take $T_g = 0.2$ s and $T_g = 0.25$ s, respectively, while the attenuation exponent y values all take 0.6.

In order to further test the suggested standard response spectrum for aseismic design code of hydraulic structures, 88 pieces of checked bedrock acceleration records with earthquake magnitude $M \geq 6.4$ scale and epicenter distance $R < 45$ km are selected from the strong seismic records occurring within the plate earthquakes in the range of the world prior to the year of 2007 to calculate their mean response spectrum. Figure 3.11 gives out the comparisons among the mean response spectrum calculated from the really measured acceleration records, and the response spectrum fitted in terms of $\beta_x = 2.5$, $T_g = 0.2$ s, $\gamma = 0.6$ and the current arch dam standard spectrum ($\beta_x = 2.5$, $T_g = 0.2$ s, $\gamma = 0.9$). The results indicate that the mean response spectrum curve within the cycle is less than I_s. It is very close to the standard response spectrum curve with $T_g = 0.2$ s, $\gamma = 0.6$, adopted by suggestions.

3.2.2.2 Uniform probability response spectrum

At present, the uniform probability response spectrums based on earthquake hazard evaluations have been adopted as the design response spectrum at home and abroad, and also called as the equal-hazard response spectrum or uniform hazard response spectrum. That is, the statistic attenuation laws and PGA corresponding to each cycle T response spectrum β are obtained in terms of strong seismic records. Similarly, the family of transcendental probability curves for each cycle $\beta(T)$ can be obtained, and then, the response spectrum is constituted by each point on the family of $\beta(T)$ transcendental probability level in uniform probability. Figure 3.12 is the family of yearly transcendental probability curves of each cycle component in bedrock response spectrum of the dam site of Xilo aqueduct arch dam.

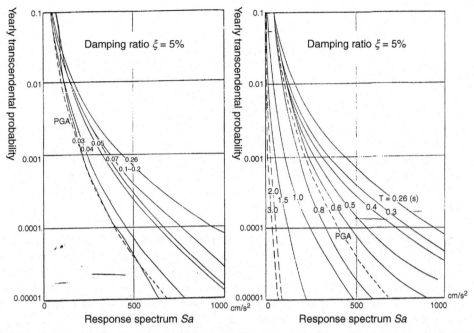

FIGURE 3.12 The Family of Yearly Transcendental Probability Curves of Different Cycle Components in Bedrock Response Spectrum of Dam-Site of Xilo Aqueduct Arch Dam

This type of so-called uniform probability response spectrum is in fact not extended in the wide application to the large dam engineering, whose main problems are as follows:

1. Acceleration response spectrum $S(T_i,\xi)$ characterizes the PGA a_g of seismic motion amplitude value and the regularized frequency spectrum $\beta(T_i,\xi)$ product. Both are the random variables of earthquake magnitude M and epicenter distance R. Supposing that both are the independent random variables, their probability must be the product of both. If taking $S(T_i,\xi) = a_g\beta(T_i,\xi)$ probability and a_g are uniform, from the probability sense, this means that $\beta(T_i,\xi)$ is the determinate variable, which is obviously not in agreement with the facts.

2. The results of the probability method are the comprehensive influences in different distances and earthquakes with different seismic magnitudes within many latent seismic source regions. The short cycle elements of the uniform probability response spectrum are always controlled by the nearby seismic small earthquake groups, and the long cycle elements are controlled by far seismic strong earthquake groups, thus making them have an envelope curve nature, whereby leading to the response peak zone being widened and performance cycle value enlarged, so as to render the response spectrum value at the mid-long cycle of important influence upon the high dam earthquake response to be larger than the design response spectrum values adopted in the current aseismic code. Therefore, it is difficult to make designers understand and accept. In addition, owing to the lack of a definite concept of earthquake magnitude and epicenter distance, it is difficult to evaluate the parameters of seismic motion sustainable time, and so on, which are the

first and foremost important to high dams as the nonlinear system seismic response and artificially generated random seismic motion time process.

Obviously, the uniform probability response spectrum cannot reflect the site's actual and possible damage from the strong earthquake itself with inherent frequency spectrum performance. For this reason, it has had no relation with the sites.

3.2.2.3 Site correlation design response spectrum

In seismic design code, the standard design spectrum based on the acceleration response spectrum statistic mean values of the existing strong acceleration records cannot embody the correlativity of specified engineering site seismogeological conditions. As to the important high dam engineering dam site seismic response spectrum, it should be aimed at the actually possible occurrence of seismic events in the case of a site area with seismogeological conditions. Therefore, under the prerequisite of given design PGA design and prevention standard based on the satisfaction of the site earthquake hazard evaluation, it is necessary to seek the solution of design response spectrum correlated with the engineering sites. In order to seek for more rational solving ways correlated with the engineering site's actual seismogeological conditions, it is suggested that the probability method be combined with the determinative method to design the earthquake to determine the ways to design the seismic response spectrum. This design response spectrum method, based on the scenario earthquake to determine the specified engineering site seismogeological conditions correlated, has been used in some practical large dam engineering aseismic design.

The method and procedures from a scenario earthquake to determine the design response spectrum are as follows:

1. The earthquake PGA corresponding to design and prevention probability is selected, based on site earthquake hazard evaluation to determine the contributed latent seismic source. As to the important engineering small capability (for example, $P_{100} = 0.02$) design and prevention standard earthquake, usually, there are one to two latent seismic source zones that can produce the corresponding earthquake PGA values to the dam site.

2. The maximum latent seismic source is selected for the site given PGA value contribution, along with the latent seismic source trunk fracture. The nearest section from the important engineering site can, in general, be taken as the epicenter distance R_{se} for the set earthquake. In the site, many possible epicenters can be produced for the given design acceleration values. The earthquake magnitude being the nearest from the site is minimum. Based on seismic magnitude–frequency formulation $\ln/N = a - bm$, earthquakes with the minimum seismic magnitude have had the maximum occurrence probability. If the latent seismic source has had a certain grade of seismic magnitude spatial distribution function f_i, m_j with a protruding role, the maximum seismic magnitude of occurrence probability after being adjusted in terms of f_i, m_j can be taken.

3. After determining the set earthquake location or earthquake magnitude, the earthquake magnitude M_{se} and epicenter distance R_{se} of the set earthquake can be obtained in terms of given site PGA and the corresponding attenuation relations. It is necessary to point out that in the present engineering aseismic design practice, the design seismic motion peak acceleration values determined in terms of design and prevention levels or standards have been checked in uncertainty. For this reason, the design response spectrum has taken its mean value in general so that it is unnecessary to carry out the uncertainty revision of the response spectrum attenuation

relation. It should be necessary to adopt the corresponding level probability and the site PGA values without uncertainty check to substitute into the attenuation relation formula.

4. After the known set earthquake magnitude and epicenter distance, the appropriate response spectrum attenuation relation should be selected to obtain the design response spectrum in correlation with site seismogeological conditions, which should be regularized in terms of PGA values.

The suggested method to determine the design response spectrum by the set earthquake is characterized by the following:

1. The results of earthquake hazard probability evaluation can serve as the prerequisites. The set earthquake prerequisite is determined in terms of the results of the earthquake hazard probability evaluation accepted by the engineering community and the design and prevention standard level and the corresponding PGA approved for engineering structures.

2. Being of definite physical sense. Generally speaking, the earthquake hazard probability evaluation of a certain site point can obtain the given design seismic motion probability standard (that is, design standard or norm, for instance, $P_{100} = 0.02$, etc.) and the corresponding seismic motion PGA values. It is important to find out one or several contributive latent seismic source zones in which some site points can be detected in terms of a certain principle to produce the concrete earthquake events of this seismic motion PGA values so as to derive earthquake magnitude M and epicenter distance R of this earthquake. It is also necessary to further estimate the design response spectrum and seismic motion acceleration time process at the site points. Setting earthquake connects the seismic motion strength value of the earthquake hazard probability results with a certain concrete seismic event with a definite physical sense, whereby being of probability significance for the determined earthquake.

3. Correlating with the occurrence of earthquake structure are the latent seismic source zones. The occurrence of most strong earthquakes in mainland China is related with the fracture. Accordingly, setting determination of earthquakes should take the correlation of seismic occurrence structures within the latent seismic source zone into full account. Since the important engineering site can produce the seismic motion parameters for the given probability level, the most contributive latent seismic source zones are usually very near the sites. Therefore, the seismic occurrence structures or the main trunk fractures within the latent seismic source zones clearly exist or have been understood definitely. Setting seismic selection must be based on the seismogeological conditions within the latent seismic source zones, being closely connected with the seismic occurrence structures or the main trunk fracture occasions. In addition, it can be based on the types of seismic occurrence faults, inclinations and the dam-site relative positions to determine whether they can be accounted in the hanging side effects.

4. It is in agreement with the principle of maximum occurrence of probability. Setting earthquake is an affirmation event. The selective prerequisite is that it must yield the design seismic motion PGA corresponding to stipulated design and prevention probability levels in the engineering sites. Since the important engineering design seismic motion is a small probability event, only a few latent seismic sources can usually satisfy this prerequisite condition. In connection with these few latent seismic sources, corresponding to the area and within the upper limit range of earthquake magnitude, there may be several seismic sources to trigger design seismic PGA in the sites. In terms of selected attenuation laws, each seismic source has had the different earthquake

magnitudes M and distances R to the sites, and their occurrence probabilities are different with each other. Each earthquake is able to satisfy this prerequisite condition of given design seismic motion PGA values in the sites. Setting earthquake magnitude M_{se} minimum at the nearest R_{se} away from the engineering sites is, in general, along with the main trunk fracture belt, so that the probability occurrence is maximum. Of course, if each seismic grade spatial distribution function f_i, m_i role in seismic source is taken into account, it is possible, in the individual case, to make the probability distribution of a certain seismic grade be particular protruding, thus leading to the fact that the possible setting earthquake with the maximum occurrence probability is not the nearest distance from the sites. It is necessary to point out that there are no longer many satisfied prerequisite conditions with different occurrence probability earthquakes to determine the setting earthquake in terms of the principle of mathematical expectation values. This is because in such a manner as to obtain the set earthquake expectation seismic magnitude and expectation epicenter distance in the sites, yielding a seismic motion PGA value that is no longer likely to satisfy the site-designed seismic motion PGA values determined in terms of design and prevention probability standards.

5. The response spectrum attenuation relation converted by the intensity is not used, but the response spectrum attenuation relation regressed by the strong earthquake records of the western United States can be directly adopted. This is because the western United States has a certain similarity and comparability or comparativeness in many aspects with China such as geological structure, earth crust components, modern stress state, seismic genesis, and seismic active behaviors.

6. Being of definite probability implication. Setting earthquake gives out the design response spectrum corresponding to the definite transcendental probability value, whose determination method is as follows: the seismic epicenter distance R_{max} of the site yielding given seismic PGA values and corresponding to the upper limit of earthquake magnitude m_u can be obtained along with the selected latent seismic source main trunk fracture. The intervals are divided as Δm several seismic magnitudes between the minimum setting earthquake M_u, of which, values are M_j, $M_{min} \leq M_j \leq M_u$. Each earthquake magnitude grade of the I latent seismic source along the main trunk fracture occupies a certain length L_i and the corresponding latent seismic source area A_i. Earthquake hazard evaluation usually assumes that the earthquake within the seismic zone evenly distributes in space, while the setting seismic determination principle takes the existence of earthquake occurrence structures into full consideration. When the seismic occurrence probability is decided in terms of area, the fault influence is always ignored, so it is used to replace A_i. Accordingly, the calculated results by L_i to select the setting earthquakes are more in agreement with seismogeological structural conditions. Accordingly, this is more rational. The occurrence number N of earthquake magnitude M in the seismic belt distributes in terms of Gudenbao–Liket seismic magnitude–frequency relation formula $InN = a - bM$.

In considering seismic motion PGA and the normalization design response spectrum serves as the independent random variable, the site acceleration response spectrum value yearly transcendental probability yielded by the setting earthquake is expressed by the formula, whose estimation formula can be expressed as follows:

$$P(M_{se}, R_{se}) = v \frac{L_{se} \times f_{l,m_i} \times e^{-b(M_{se}-M_0)}}{\sum_{i=1}^{n} L_i \times f_{l,m_i} \times e^{-b(M_i-M_0)}} \tag{3.18}$$

where v is the maximum contribute latent seismic source yearly transcendental probability value to yield the given seismic motion PGA values in the site. f_i, w_i is the spatial distribution foundation, that is, to reflect the time-space uneven seismic year average occurrence rate weighted coefficient of different earthquake magnitudes in each latent seismic source zone. m_0 is the starting calculation of earthquake magnitude; the seismic activity weakened regions usually take the earthquake magnitude of 4.0, while the seismic activity strong regions take the earthquake magnitude of 5.0 in earthquake hazard evaluation. L_i is the I grade earthquake, corresponding to seismic occurrence location, to occupy the length of the main trunk fracture in the latent source zone. L_{sc} is the set seismic magnitude grade of earthquake corresponding to the seismic occurrence location to occupy the length of the main trunk fracture in the latent source zone.

The seismic motion acceleration response spectrum $S(T_i,\xi)$ is the product of seismic motion PGA a_g, and the normalized design response spectrum $\beta(T_i,\xi)$, nearby, namely, $S(T_i,\xi) = a_g\beta(T_i,\xi)$. If a_g and $\beta(T_i,\xi)$ are viewed as statistic independent random variables, in terms of total probability formula $P(\beta/a_g) = P(SE)/P(a_g)$, where, $P(SE)$ is the response spectrum probability for the computer setting earthquake; $P(a_g)$ is the design and prevention probability standard of the PGA a_g; condition probability $P(\beta/a_g)$ is the probability of the normalization response spectrum $\beta(T_i,\xi)$. Obviously, in many possible earthquakes in the site given seismic motion PGA values, though the setting seismic probability $P(SE)$ is the maximum, the integrated many possible seismic contributive a_g design and prevention probability $P(a_g)$ can be small.

Figure 3.13 gives out the arch dam standard response spectrum in the current aseismic design code of hydraulic structures and uniform probability response spectrum and the site correlation response spectrum of the setting earthquake. It can be seen from Fig. 3.13 that the morphology and sizes of the uniform probability response spectrum are apparently irrational, whereas the site correlation response spectrum of the setting earthquakes and the arch dam standard response spectrum in the current aseismic design code of hydraulic structures are rather close to each other.

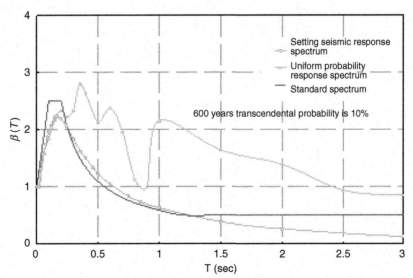

FIGURE 3.13 The Code Response Spectrum, Uniform Probability Response Spectrum, and Setting Earthquake Site Correlation Response Spectrum Used in Xiaowan Arch Dam Engineering

DESIGN ACCELERATION TIME PROCESS

4.1 AMPLITUDE AND FREQUENCY NONSMOOTH ACCELERATION TIME PROCESS

The large dam structure is complex. Under the action of a strong earthquake, dams appear to have strong nonlinear effects. At present, the important dam engineering seismic responses must have a nonlinear dynamic analysis. For this reason, the response spectrum method is no longer adaptable so that the analysis method of inputting seismic motion acceleration time process is adopted. In view of the randomization of seismic motion, the designing response spectrum is usually adopted as the objective spectrum to artificially fit the seismic motion acceleration time process. The traditional synthesis method is to use the even distribution random phase angle between trigonometric series and 0–2π to construct the smooth random process. Then, after the strength envelope function "changing with time" is introduced, the objective response spectrum is fitted and the peak ground acceleration is designed in terms of power spectrum and response spectrum approximate formula and through the iterative adjustment amplitude. The phase remains unchanged in the course of amplitude iteration, whereby leading to the iterative convergence to be unrealistic. The practical seismic motion acceleration time process amplitude and frequency component are of nonsmooth behavior as the time changes. However, the traditional artificial fitting random seismic motion acceleration time process method does not take the seismic motion frequency nonsmooth behaviors into account, while as to having strong nonlinear structures, the different frequency components affecting structure response in the seismic motion time process is obvious. The large dam structure has gradually suffered from damage in the process of a strong earthquake. Its resistance force continues to become worsened and the inherent cycle can be lengthened so that the long cycle component in the later-coming seismic motion is more sensitive. Accordingly, in the analysis of the dam engineering seismic response, it is necessary to take the amplitude of seismic motion acceleration time process inputs and frequency nonsmooth influence into account. In recent years, each country in the world has done much research on it, of which the key is to analyze the frequency spectrum of the seismic motion acceleration time process, varying with times and to seek for the regularity correlated with earthquake magnitude and distance.

Priestley (1965) suggested the concept of evolutionary power spectrum. The evolutionary power spectrum can reflect the energy distribution behaviors with changing time of each frequency component in the seismic course, being of a clear physical concept. The solution to the seismic motion evolutionary power spectrum is rather complex. Kameda et al., 1980 made the statistical regression to fit the empiric model of the evolutionary power spectrum based on the Japanese real measured seismic motion acceleration records. Nakayama et al. (1994) suggested the effective method to seek the solution to the evolutionary power spectrum through the narrow belt and low-pass filter.

Based on these researches, Zhang et al., 2007 suggested a method of nonsmooth random seismic motion acceleration time process to generate fitting objective evolutionary power spectrum amplitude and frequency in terms of given earthquake magnitude M and epicenter distance R.

4.1.1 SEEKING A SOLUTION TO THE EVOLUTIONARY POWER SPECTRUM

The seismic motion acceleration time process serves as a weak-smooth double random process $\{X(T)\}$, to satisfy all t, $E[x(t)] = 0$, $E\left[\left|x(t)\right|^2\right] < \infty$ conditions. It can be known from the Wiener–Khintchine theory $X(t) = \int_{-\infty}^{\infty} e^{it} \omega \mathrm{d}z(\omega)$ that if the oscillating function family $\phi(t,\omega) = A(t,\omega)e^{i\omega t}$ is used to replace $\{e^{i\omega t}\}$ [where, $A(t,\omega)$ is the complex amplitude function], there will be:

$$X(t) \int_{-\infty}^{\infty} A(t,\omega)e^{t\omega} \mathrm{d}z(\omega) \tag{4.1}$$

where $z(\omega)$ is the orthogonal increment process; letting $\mathrm{d}u(\omega) = E\left[\left|\mathrm{d}z(\omega)\right|^2\right]$

$$\mathrm{Var}\{x(t)\} = R(t,t) = \int_{-\infty}^{\infty} \left|A(t,\omega)\right|^2 \mathrm{d}\mu(\omega) \tag{4.2}$$

$\mathrm{d}G(t,\omega) = \left|A(t,\omega)\right|^2 \mathrm{d}\mu(\omega)$ is defined as the evolutionary power spectrum at t time, whose key lies in the complex amplitude function $A(t,\omega)$ solution.

Letting $\mathrm{d}F_1(t,\omega) = A(t,\omega)\mathrm{d}z(\omega)$, and $\mathrm{d}F_1(t,\omega) = \left|\mathrm{d}F_1(t,\omega)\right| = \mathrm{d}F_1(t,\omega)e^{i\phi(t,\omega)}$, thus, from (4.1), we will have:

$$x(t) = \int_{-\infty}^{\infty} \left|\mathrm{d}F_1(t,\omega)\right| e^{i[\omega t + \phi(t,\omega)]} \tag{4.3}$$

where, $\phi(t,\omega)$ is time angle of amplitude of variation, whose variation depends on the $\mathrm{d}F_1(t,\omega)$ real part and imaginary part:

$$\phi(t,\omega) = \mathrm{arctg}\left\{\frac{\mathrm{d}F_{1\zeta}(t,\omega)}{\mathrm{d}F_{1R}(t,\omega)}\right\} \tag{4.4}$$

Formula (4.3) integration is approximately expressed with a narrow-band filter:

$$x_i(t) = 2\left|\mathrm{d}F_1(t,\omega_i)\right| \cos\left(\omega_i t + \phi(t,\omega_i)\right) \tag{4.5}$$

The two ends of formula (4.3) are multiplied by $\cos(\omega_i, t)$ and $\sin(\omega_i, t)$, and the low-pass filter is used to remove the $2\omega_i$ frequency component:

$$\begin{cases} F\{x_i(t)\cos\omega_i t\} = \left|\mathrm{d}F_1(t,\omega_i)\right| \cos\phi(t,\omega_i) \\ F\{x_i(t)\sin\omega_i t\} = -\left|\mathrm{d}F_1(t,\omega_i)\right| \sin\phi(t,\omega_i) \end{cases} \tag{4.6}$$

Phase angle changing with times in ω_i value can be:

$$\phi(t,\omega) = \mathrm{arctg}\left[\frac{-F_1\{x_i(t)\sin\omega_i t\}}{F_1\{x_i(t)\cos\omega_i t\}}\right] \tag{4.7}$$

From formula (4.4) and formula (4.7), we can obtain:

$$\begin{cases} \mathrm{d}F_R(t,\omega_i) = F_1\{x_i(t)\cos\omega_i t\} \\ \mathrm{d}F_1(t,\omega_i) = -F_1\{x_i(t)\sin\omega_i t\} \end{cases} \tag{4.8}$$

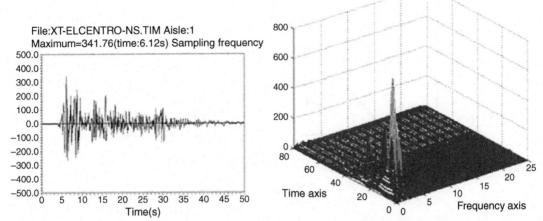

FIGURE 4.1 E1 Centro-N Seismic Motion Acceleration Time Process and Evolutionary Power Spectrum

The two ends of the formula $dF_1(t,\omega) = A(t_i,\omega_i)dz(\omega)$ are substituted by each item real part or imaginary part. We can obtain:

$$\left\{ \begin{array}{c} A_R(t,\omega) \\ A_I(t,\omega) \end{array} \right\} = \left[\begin{array}{cc} dz_R(\omega_i) & -dz_1(\omega_i) \\ dz_I(\omega_i) & dz_R(\omega_i) \end{array} \right]^{-1} \left\{ \begin{array}{c} dF_R(t,\omega_i) \\ dF_1(t,\omega_i) \end{array} \right\} \qquad (4.9)$$

After complex amplitude function $A(t,\omega)$ is derived, $x(t)$ evolutionary power spectrum $dG(t,\omega)$ is obtained through the definition of evolutionary spectrum.

Figure 4.1 is the El Centro-N seismic motion acceleration time process of the Imperial Valley earthquake in May 19, 1940, and the evolutionary power spectrum sketches obtained by the earlier methods.

4.1.2 OBJECTIVE EVOLUTIONARY POWER SPECTRUM EMPIRIC MODEL FITTING

In terms of real measured seismic motion acceleration records and with the reference to seismic motion acceleration time process strength envelop forms, the evolutionary power spectrum statistic regression model of each frequency component can adopt the exponential function consisting of such three parameters as the initial time $t_s(f)$, time difference $t_p(f)$ of power spectrum peak time-up, time-start $t_p(f)$, and power spectrum peak root $Qm(t)$ (Fig. 4.2):

$$dG(t,2\pi f) = a_m^2(f)\left[\frac{t-t_s(f)}{t_p(f)}\right]^2 \exp\left\{2\left[1-\frac{t-t_B(f)}{t_p(f)}\right]\right\} \qquad (4.10)$$

The three parameters can be made of statistical regression analysis in terms of seismic motion attenuation relation formula. We will have the following:

$$\left\{ \begin{array}{l} \lg a_m(f) = a_1(f) + a_2(f)M - a_4(f)\lg(R + a_5(f)e^{a_6(f)M}) \\ \lg t_p(f) = p_1(f) + p_2(f)M + p_4(f)\lg(R + p_5(f)e^{p_6(f)M}) \\ t_s^1(f) = t_s(f) - t_m = s_0(f) + s_1(f)R \end{array} \right. \qquad (4.11)$$

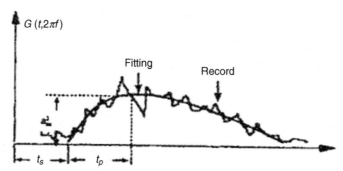

FIGURE 4.2 Time Variation Power Spectrum Empirical Model Parameters

where, M is earthquake magnitude, R is epicenter distance, km; t_m is average value in considering frequency scope t, of which $t_s^1(f)$ is mainly subject to epicenter distance R influence.

These parameters in regression-fitting empirical model have selected 80 pieces of bedrock records with earthquake magnitude ≥ 6.5, and epicenter distance ≤ 45 km in the western United States, of which two level components in the same station are considered as two pieces of independent seismic records. The results of regression fitting are as follows:

$$\lg a_m(f) = a_1(f) + a_2(f)M - a_4(f)\lg\left(R + 0.2347e^{0.6297m}\right) \tag{4.12}$$

$$\begin{cases} a_1(f) = 0.3342 + 1.4894 \times \lg f - 4.9727 \times (\lg f)^2 \\ a_2(f) = 0.4470 - 0.1573 \times \lg f + 0.5651 \times (\lg f)^2 \\ a_4(f) = 1.4708 + 0.0105 \times \lg f - 0.0720 \times (\lg f)^2 \end{cases} \tag{4.13}$$

$$\lg \mathrm{tpf}(f) = p_1(f) + p_2(f)M + p_4(f)\lg(R + 1.8733e^{0.3386M}) \tag{4.14}$$

$$\begin{cases} p_1(f) = -0.2051 - 1.5745 \times \lg f + 1.4641 \times (\lg f)^2 \\ p_2(f) = 0.1058 + 0.2008 \times \lg f - 0.2324 \times (\lg f)^2 \\ p_3(f) = 0.1479 + 0.1868 \times \lg f \end{cases} \tag{4.15}$$

$$t_s^1(f) = t_s(f) - t_m = s_0(f) + s_1(f)R \tag{4.16}$$

$$\begin{cases} s_0(f) = 0.0688 + 1.6678 \times \lg f - 1.5073 \times (\lg f)^2 \\ s_1(f) = 0.0529 - 0.0549 \times \lg f \end{cases} \tag{4.17}$$

Figure 4.3 is the evolutionary power spectrum sketch example for different earthquake magnitude M and epicenter distance R fitting.

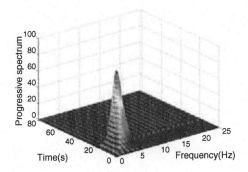 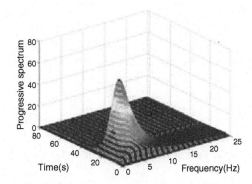

FIGURE 4.3 Evolutionary Power Spectrum Sketch Example for Different Earthquake Magnitudes M and Epicenter Distance R Fitting

(a) $M = 6.5$, $R = 15$ km. (b) $M = 7.3$, $R = 45$ km.

4.1.3 ARTIFICIAL FITTING AMPLITUDE AND FREQUENCY NONSMOOTH SEISMIC MOTION ACCELERATION TIME PROCESS GENESIS

After the objective evolutionary power spectrum is determined in terms of setting earthquake magnitude M and epicenter distance R, the amplitude and frequency nonsmooth random earthquake motion acceleration time process can be generated in terms of the following model:

$$a(t) = 2\sum_{i=1}^{N} \sqrt{dG^{T}(t,\omega)} \cos\left|\omega_1 + \phi(t,\omega_1)\right| \tag{4.18}$$

where, $dG^{T}(t,\omega)$ is the objective evolutionary spectrum; $\phi(t,\omega_i)$ is time variation amplitude angle correlated with time and frequency.

The concrete procedures for the realization are as follows:

1. It is necessary to determine objective evolutionary spectrum $dG^{T}(t,\omega)$ for each fixed point frequency point ω_i and to calculate $dG^{T}(t,\omega_i)$ along the time axis initial time $t_s^i\omega_i$, time difference $t_p^i\omega_i$ and peak $a_m^T(\omega)$ between the time-up and initial time.
2. Taking the even random distributed phase ϕ and objective evolutionary spectrum $dG^{T}(t,\omega)$ within $[0,2\pi]$ range, to substitute them into formula (4.18) so as to obtain the initial time process $a^0(t)$.
3. The initial time process $a^0(t)$ is treated with narrow-band filter and low-pass filter so as to obtain $dG^{T}(t,\omega_i)$ and $\phi(t,\omega_i)$.
4. It is necessary by order for each frequency component point ω_i to calculate $dG(t,\omega_i)$ initial time $t_s(\omega_i)$ peak time-up and the parameters of the initial time difference $t_p(\omega_i)$ and peak $a_m(\omega_i)$, and so on.
5. It is important to check the three parameters of each frequency component point and the objective evolutionary power spectrum corresponding parameter relative errors. If the requirement of less than allowable error values cannot be satisfied at the same time, the phase angle $\phi(t,\omega_i)$ is regulated in terms of parameter $t_p(\omega_i)$ error algebraic values and then, the objective evolutionary spectrum and the regulated phase angle is substituted into formula (4.18), time process $a^1(t)$ can be recalculated.

6. Repeating steps three to five, and iterating them into all the required frequency component point to satisfy the setting error requirements. Taking the 47,379 stations zero direction recorded the evolutionary spectrums of the 1989 Loma Prieta, CA earthquake as the initial objective spectra to serve as the demonstration examples, and they are smoothly treated in time and frequency direction. Time region smooth treatment is as follows:

$$G(t,\omega) = \frac{1}{T_a} \int_{1-(T_\varphi/2)}^{1+(t_\varphi/2)} G(t,\omega) dt \tag{4.19}$$

where, Ta is decided by the required smooth degree; frequency region treatment is as follows:

$$G(t,\omega) = \int_{-\infty}^{\infty} G(t,\omega)W(\omega-\Omega) d\Omega$$

$$W(\omega) = \frac{3}{4}u \left[\frac{\sin\frac{u\omega}{4}}{\frac{u\omega}{4}} \right]^4 \tag{4.20}$$

$$u = \frac{280}{151b}$$

where, $W(\omega)$ is Parzen window function; b is Parzen window bandwidth (the larger the bandwidth is, the smoother the spectrum value will be). In this case, Ta takes 3.0, b value takes 1.4.

Figure 4.4 is the initial recorded acceleration time process of the Loma Prieta, CA earthquake and three artificial fitting random process diagrammatic sketches with different initial phases.

The correlative coefficients of artificial fitting random time process with three different initial random phases in Table 4.1 are very small, whereby illustrating that their interrelations are not correlative with each other.

In comparison with the currently used multiple wave filter genesis amplitude and frequency nonsmooth random seismic motion acceleration time process method, the suggested method is characterized by the following:

1. In the multiple wave filter method, evolutionary power spectrum is obtained through the single mass point system stable response sum approximate relation. For this reason, the damping ratio needs to be enlarged so as to lower the instance response effect to satisfy the relatively slow assumption of the evolutionary power spectrum varying with times; however, the filter wave requires to reduce the single mass point system damping ratio so as to guarantee the accuracy requirement of the filter wave. Both contradictions can affect the determination of damping ratio and fitting accuracy. Accordingly, the suggested method is to directly seek for a solution to the evolutionary power spectrum through the narrow-band and low-pass filter, without affecting the damping ratio problem.
2. The amplitude and phase angle of the evolutionary power spectrum are interaffecting. In fitting objective evolutionary power spectrum to generate amplitude and frequency nonsmooth artificial simulation random seismic motion acceleration time process, the random selection of evolutionary power spectrum phase angle is bound to affect the amplitude variation. For this reason, it is

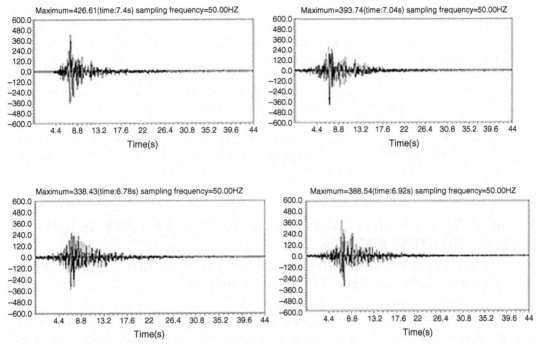

FIGURE 4.4 Fitting Time Process Curves in Terms of Real Measured Seismic Evolutionary Power Spectrum as the Objective Spectrum

(a) Initial seismic motion time (b) Initial random phase 1 genesis time (c) Initial random phase 2 genesis time process (d) Initial random phase 3 genesis time process.

Table 4.1 Zero Direction Seismic Motion Time Process Correlative Coefficients of 47,379 Stations and Time Process Synthesized Based on Objective Evolutionary Spectrum and Random Phases

Wave Types	Real Seismic Wave	Random Phase 1 Synthesized Wave	Random Phase 2 Synthesized Wave	Random Phase 3 Synthesized Wave
Random phase 1 Genesis wave	$-6.4505346E-04$	–	–	–
Random phase 2 Genesis wave	-0.1705572	$-6.1712906E-02$	–	–
Random phase 3 Genesis wave	-0.2312825	$1.6099496E-02$	$9.495770-02$	–

necessary only through repeating the iteration of amplitude and phase angle that the accurate requirement of approximation objective evolutionary power spectrum can be satisfied at the same time. The iterated results are particularly sensitive to the variation of phase angles so that in the process of iteration of only regulating the amplitude of the evolutionary power spectrum, the convergence is not always realistic. In the suggested method, the iteration in every time calculates the evolutionary power spectrum amplitude and phase angle, and at the same time, the evolutionary power spectrum phase angle and amplitude must be regulated, and also, the direction of phase angle regulation can be determined in terms of the positive and negative errors directly, whereby effectively accelerating the iterative process.

3. In the suggested method, the parameters of objective evolutionary power spectrum model are fitted in terms of selection of strong seismic acceleration records of the western United States. Such behaviors as geologic structure, crust components, modem stress states, and seismic genesis and seismic activities in the western United States are in similarity with those in our country, so that in the selected strong seismic acceleration records, the earthquake magnitude and epicenter distance range are approximately corresponding to the important large dam engineering, and they are more adaptable to the safety check of large dam earthquake resistance.

4.2 ADOPTING RANDOM FINITE FAULT METHOD TO DIRECTLY GENERATE ACCELERATION TIME PROCESS

4.2.1 ENGINEERING BACKGROUND OF ADOPTING RANDOM FINITE FAULT METHOD TO DIRECTLY GENERATE ACCELERETION TIME PROCESS

As previously described, in recent years, China has built a series of important hydropower works with high arch dams of 300 m high in strong seismic areas. Up till now, there have been shortages of precedents and engineering experiences for ensuring seismic safety at home and abroad, let alone real seismic disaster examples or cases as references in such over-large-sized arch dams in strong seismic areas. Once these high dams and large reservoirs suffer from strong earthquakes and serious destructions happen, unthinkable secondary disasters will be caused. In order to prevent these large and important seismic catastrophes from occurring, the engineering designers, proprietors, and societies are bound to show great concern for the evaluation of important disaster changes. Considering that seismic uncertainty and seismic forecast is the world's problem that remains unsolved, the strategic priority is given to earthquake resistance and disaster prevention to prevent the occurrence of reservoir water from discharging out of control, whereby causing serious secondary disasters of reservoir failure variation when the maximum possible extreme earthquake happens, which is out of the design predictions. For this reason, it is necessary to decide the possible occurrence of earthquake limit at the dam-site rationally, that is, the maximum believable earthquake and quantitative judgment codes serve as the dam failure limit states corresponding to performance objectives. Accordingly, these two problems are the main obstacles and the frontier research projects to be urgently solved in the current seismic safety evaluation of large dams.

At present, there are the probability method and determinate method in determining the maximum reliable earthquakes. The probability method is usually adopted, and the corresponding transcendental probability in earthquake hazard evaluation taken is 10^{-4}, that is, the earthquake with a recurrence period of 10,000 years can serve as the maximum reliable earthquake. After the Wenchuan earthquake, the Chinese National Development and Reform Committee issued a document to require that the check of some important engineering projects should be put under a seismic safety of limit seismicity and to point out that the recurrence period of limit seismicity can take 10,000 years. This is likely to be an expedient measure for the emergency assessment of the requirements of important waterworks and hydropower engineering structures earthquake resistance safety. In fact, a 10,000 year seismic frequency that can serve as the design and prevention standard for the maximum reliable earthquake is unsuitable; the main reasons are as follows:

1. An earthquake with a recurrence period of 10,000 years will not be the limit seismicity.
2. Earthquake hazard evaluation, mainly based on the mid-strong seismic statistic sample books, has given out the transcendental probability curves extension to the 100,000 years' frequency small probability event, whose reliability and accuracy are very poor.
3. The limit seismicity is mostly the near-fracture large earthquake, whose seismic motion inputs are sensitive to the distance uncertainty in attenuation relations.
4. It is difficult to take many behaviors of the near-site large earthquakes into account in conventional earthquake hazard evaluations, based on the point source model assumption. For instance, these are such influences as model and process of seismic occurrence fault fractures, and hanging side effects, and so on.

The Dagang Mountain high arch dam engineering in the west of China, with a design peak ground acceleration of 0.557g, can serve as the Moxi fracture in the latent seismic source, whose earthquake magnitude upper limitation is 8 scales and only 4.5 km away from the dam site and the maximum believable earthquake belongs to the near-fault large earthquake. Taking the 2008 occurrence of the Wenchuan earthquake as an example, its disturbance of seismic occurrence fault began from the epicenter in Wenchuan, and fractures developed at the velocity of about 3.1 km/s along the northeast direction gradually to Qinchuan, extending for over 300 km and lasting more than 100 s and releasing the maximum energy in the nonepicenter of North Beichuan. Such a fracture process for Zipingpu engineering structures not far from Wenchuan fell into the near-fault strong earthquake, whose dam-site seismic motion was obviously different from the results given out by the conventional earthquake hazard evaluation based on the present point source model assumption.

For this reason, the suggested improvement is to adopt the random finite fault method based on seismicity theory or half-experience statistic relation to directly generate the maximum convincible earthquake (MCE) seismic motion acceleration time process. This is the frontier research project in the current engineering seismicity discipline.

4.2.2 RANDOM FINITE FAULT METHOD BEHAVIORS, BASIC THINKING WAYS AND EFFECTS

The random finite fault method can be generalized in the following points:

1. *Behaviors*: The random finite fault method is mainly based on the seismicity physical model but still needs to determine the semitheory and semiexperience method of some relevant parameters in empirical statistics, and it is not based on the determinate method of probability theory of earthquake hazard evaluation. Consequently, this method does not involve many uncertainty factors in the empiric attenuation relations.
2. *Basic thinking ways*: The main trunk fault in the latent seismic source can be divided into a series of subfaults to serve as the point sources, whose fractures are of a certain model state and time series. Sequence iteration of the effects of each point source upon the dam site can give out the near-site strong seismic dam-site earthquake motion.
3. *Advantages*: It can reflect the near-abruption large earthquake motion behaviors without any way to consider in the conventional earthquake hazard evaluation based on the point source model assumption, such as finite seismic source body influence including seismic source dimensions, fault abruption patterns, transmission direction, as well as embodying the hanging-wall effect seismic source-site special relative position, and so on.

4. *Effects*: It is mainly important to consider the effect of seismic energy released by the nearby fracture upon the dam site, whereby improving the large seismic motion deviation problem caused by taking the whole fracture face a_0, a point source assumption. Accordingly, the larger the earthquake magnitude is, the more apparent the improvement will be.
5. *Problems*: At present, it is difficult to completely forecast some existing uncertainties in the fault fracture model and its course so that it is only likely to make the safety selection in working out the possible scheme for the engineering structures.

4.2.3 STEPS OF RANDOM FINITE FAULT TO GENERATE SITE SEISMIC MOTION TIME PROCESS

The steps of random finite fault to generate the site seismic motion time process can be generated as follows.

4.2.3.1 Determination of fault face geometric characteristics

In terms of seismic occurrence fault position coordinate, strike, trend, inclination, and other parameters as well as earthquake magnitude, it is decided to take the rectangle fracture face as the length width dimensions of the finite three-dimensional special face source serving as the macrogeometric characteristics based on empirical statistics.

In terms of matrix earthquake magnitude M_ω, Wells and Coppersmith (1994) statistic model is adopted to determine the rectangle fault area A and fault length L:

$$\lg A = -3.49 + 0.91 M_\omega \quad (4.8 \leq M_\omega \leq 7.9) \tag{4.21}$$

$$\lg L = -2.44 + 0.59 M_\omega \quad (4.8 \leq M_\omega \leq 8.1) \tag{4.22}$$

Fault buried depth is difficult to predict, but strong earthquakes often disturb the land surface.

4.2.3.2 Classification of subfault face

It is necessary to classify fault fracture face into a series of subfault fracture faces serving as a point source.

Seismic energy can be characterized as $E_0 = \Delta_a M_a / 2G$, of which Δ_a is the motion stress drop of the seismic occurrence faults expressed in terms of disk fracture model with fracture radius a:

$$\Delta_a = \frac{7m_0}{16a^3} \tag{4.23}$$

where, seismic matrix M_a equals to GDA; and D, G are the fracture face shear dislocation or malposition and the relations between shear module and matrix M given by Hanks and Kanamori (1979):

$$M_\omega = \frac{2}{3} \lg M_0 - 10.7 \tag{4.24}$$

Guoxin (2001) gave out the subsource length ΔL, which can be derived by the following formula:

$$\lg \Delta L = \lg L - 0.5t \left(M_\omega - M_2 \right) \tag{4.25}$$

where, earthquake magnitude M_2 of the subearthquake is $5\sim5.5$ scale to be suitable in general. Since seismic fracture and sliding course are uneven along the fault face, each subsource seismic matrix needs to have a weighted distribution, but their total sum should equal to total seismic matrix of the fault face. The weight of subsource earthquake magnitude can be embodied through the disturbance quantity of momentum D and fracture area A. There exists the protruding asperity of sliding quantity of momentum on the fault face. The inversion analysis statistic results from the representative seismic source by Someville et al. (1999) indicated that every earthquake contains and occupies one to three asperities of 22% of total fracture area on the average, of which the largest asperity occupies 16% of total fracture area or so. The sliding quantity of momentum of asperity is about twice as much as the average value.

4.2.3.3 Seismic motion time process generated by subsource fracture in dam sites

The physical model based on seismology is adopted to give out shear dislocation, sliding time function method in time domain for empiric green function, and so on. Taking as its path function is difficult or hard to reflect the complexity of transmission media so that the random methods of seismic source spectrum and attenuation function have been mostly adopted at present in the frequency field.

The dam-site seismic motion spectrum displacement $y(M_0,R,f)$ yielded by the subsource fracture of the point source include the seismic spectrum $E(m_0,f)$, path function $P(R,F)$, site function $G(f)$, and seismic motion parameter type $I(f)$:

$$y(m_0,R,f)=E(m_0,f)P(R,f)G(f)I(f) \tag{4.26}$$

here, f is frequency.

As to the rock-bed, the site function $G(f)$ is not accounted for.

The seismic motion acceleration parameter type $I(f)$ takes $(2\pi f)^2$.

Seismic source spectrum $E(m_0,f)$ can be expressed as follows:

$$E(m_0,f)=\mathrm{cmos}(m_0,f) \tag{4.27}$$

At present, the Brune (1983) ω^2 model is widely adopted to express the seismic source displacement Fus spectrum:

$$S(m_0,f)=\cfrac{1}{\left[1+\left(\dfrac{f}{f_0}\right)^2\right]} \tag{4.28}$$

$$f_0=4.9\times10^6\beta_s\left(\frac{\Delta_a}{m_0}\right)^{1/3} \tag{4.29}$$

where, f_0 is inclined angle frequency, Hz; β_s is the medium shear wave velocity at seismic source, km/s; Δ_a is stress drop, bars; and m_0 is seismic matrix, dyne-cm.

In formula (4.27), the constant C can be expressed as follows:

$$C=\frac{(R_{\theta\phi})VF_1}{(4\pi\beta_s\beta_s^3R_0)} \tag{4.30}$$

where, $R_{\theta\phi}$ is reflecting seismic radiation model and seismo-station orientation effect, usually taking the mean value; V is the level component coefficient of whole shear wave energy, usually taking $1/\sqrt{2}$; F_1 is freedom land surface amplification effect, usually taking 2; P_s is the seismic nearby medium density; R_0 is reference distance, usually taking 1, km.

Formula (4.29) indicates that the inclined angle frequency f_0 depends on seismic matrix M_0, while M_0 is correlative with seismic source fracture area. For instance, assuming that the sizes of earthquakes are all in coincidence with $M_0 f_0^3 = \text{constant}$ deciding norm laws, the inclined angle frequency f_0 depends on the magnitudes and quantities of subsource classifications, which is obviously irrational. For this reason, it is suitable to adopt the dynamic inclined angle frequency concept suggested by Motazedian and Atkinson (2005), that is, M_0 in formula (4.29) corresponding to each fracture subsource area of the fault face, whereby making inclined angle frequency f_0 change in the fracture course with time, and converge to low frequency.

In terms of setting the initial fracture point, fracture patterns and direction as well as relative position of the dam-site space, and after the superimposition of each point source in time sequence can produce the seismic motion time process, the near-site strong earthquake dam-site seismic motion time process can be obtained.

In the case of near-site conditions, R is less than 70 km, the simplified path effects can be expressed as the geometric diffusion function $(1/R)$, and the nonlinear attenuation function product related to transmission medium quality factor Q:

$$P(R,f) = \left(\frac{1}{R}\right)\exp\left[\frac{-\pi f_R}{Q(f)\beta_s}\right] \tag{4.31}$$

In terms of the dam site earthquake motion displacement spectrum $y(M_0, R, f)$ synthesis acceleration time process, the sustainable time includes the obtainable reciprocal inclined angle frequency f_0 of seismic source spectrum of seismic source sustained time and $0.05\,R$ (R in km) path sustained time. Usually, the determined window function for seismic source sustained time is set up in addition.

4.2.3.4 Site earthquake motion time process synthesis

After each subsource seismic occurrence yielding earthquake motion acceleration time process in the site is obtained through the point source model calculation, the acceleration time process $a(t)$ of the dam-site earthquake motion input can be synthesized by the superimposition of the following formulas:

$$a(t) = \sum_{i=1}^{n_1}\sum_{j=1}^{n_m} a_g\left(t - t_{ij}\right) \tag{4.32}$$

$$t_{ij} = \frac{|\xi_{ij} - \xi_0|}{v_R} + \frac{R_{ij} - R_0}{\beta_s} \tag{4.33}$$

where, n_1, n_m is the number of subsources classified along the fault length, width directions; $a_{ij}\left(t - t_{ij}\right)$ is subsource (i, j) synthesis acceleration time process; t_{ij} is time delay of each subsource seismic occurrence synthesis acceleration time process, including the fracture from the initial fracture point developing to time delay caused by subsources, and in the wave transmission course, each subsource to the site with different distances can cause the time delay; ξ_{ij} is subsource (i, j) coordinate; ξ_0 is fault initial fracture point coordinate; v_R is fault fracture velocity, taking average $0.8\,\beta_s$; R_{ij} is subsource (i, j) distance to the site; R_0 is the distance of initial fracture point to the site; β_s is shear wave velocity.

4.2.3.5 Real examples of engineering applications

The designed seismic acceleration of the Dagangshan hydropower engineering dam site reaches as high as $0.557g$, which is very rare in world dam design. The dam-site seismic motion input is mainly determined by Moxi fracture of the Xianshui River fracture zone in the nearest distance of 4.5 km away from the dam site. The upper limit of this seismic magnitude of this seismic source is 8.0 scales and also a typical nearby fault large earthquake, so that this large dam can serve as the engineering application real example of random finite fault method. In calculating Moxi fault occurrence of earthquake of 7.5 scale and 8.0 scale, 4.5 km or so of the engineering site is likely to record seismic motion acceleration time process. It has been known from the evaluation report of seismic safety of engineering site earthquake that the Moxi fault strike is 335–345°, inclined SW, inclination of 60–80°, with a total length of about 150 km.

To begin with, the fault fracture face and seismic source basic parameters should be calculated by the empirical statistic formula, as shown in Table 4.2.

Based on the site and relative position of the Moxi fault and with reference to the near surrounding record of the year of 1786, the strongest earthquake with 7 3/4 magnitudes between Kangding and Luding Moxi, the dam-site epicenter distance was 56 km. For this reason, the edge rupture method is adopted to compare the influence of the subsource dimensions, so that two schemes have taken the subsource lengths of 6.0 and 10 km, respectively, whose corresponding subseismic magnitudes are 5.0 scale and 5.4 scale, respectively. The fault areas are 186 km × 30 km and 190 km × 30 km, respectively, the epicenter distances are 49.022 km and 46.089 km, and the azimuth angles are 153.28° and 152.52°, respectively. Each subsource classification and their weight distributions on the fault face are seen in Fig. 4.5.

The motion inflection point frequency is adopted to synthesize the two schemes with each having 30 pieces of acceleration time processes, whose peak ground acceleration (PGA) mean values are 0.552 and $0.580g$, respectively, whereby indicating that the subsource classification has not much influence, being less than 5%. In terms of the results of special earthquake hazard evaluation of this engineering dam-site, it is shown that the transcendental probability is 1% of PGA of $0.662g$ within 100 years. The maximum convincible earthquake value given out by the random fault method is at least less than 12% of the results by the probability method. The random finite fault method is used to generate two schemes, each having two dam-site seismic motion acceleration time processes, as shown in Fig. 4.6.

Table 4.2 Fault Fracture Face and Seismic Source Basic Parameters

Parameter Names	Parameter Value	Parameter Names	Parameter Value
Fault strike/incline	345°/70°(90°)	Shear wave velocity (km/s)	3.47
Fault dimension (length × width, km)	About 190.55 × 32.36	Fracture velocity (m/s)	0.8 × shear wave velocity
Fault depth (km)	15	Medium density (g/cm³)	2.8
Stress drop (Pa)	5 × 106	Quality factor Q	250

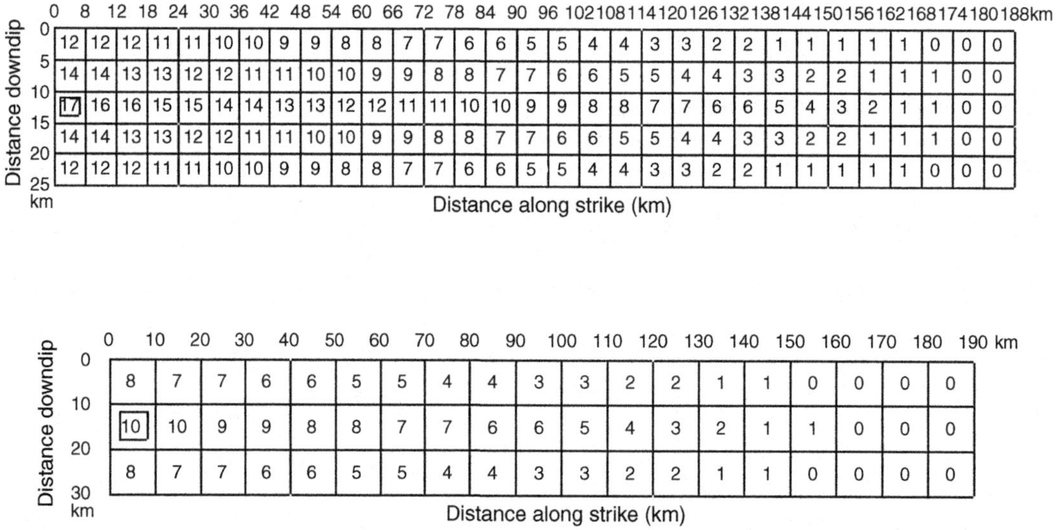

Distance downdip (km) — Distance along strike (km), scale: 0 8 12 18 24 30 36 42 48 54 60 66 72 78 84 90 96 102 108 114 120 126 132 138 144 150 156 162 168 174 180 188 km

0	8	12	18	24	30	36	42	48	54	60	66	72	78	84	90	96	102	108	114	120	126	132	138	144	150	156	162	168	174	180
12	12	12	11	11	10	10	9	9	8	8	7	7	6	6	5	5	4	4	3	3	2	2	1	1	1	1	1	0	0	0
14	14	13	13	12	12	11	11	10	10	9	9	8	8	7	7	6	6	5	5	4	4	3	3	2	2	1	1	1	0	0
17	16	16	15	15	14	14	13	13	12	12	11	11	10	10	9	9	8	8	7	7	6	6	5	4	3	2	1	1	0	0
14	14	13	13	12	12	11	11	10	10	9	9	8	8	7	7	6	6	5	5	4	4	3	3	2	2	1	1	1	0	0
12	12	12	11	11	10	10	9	9	8	8	7	7	6	6	5	5	4	4	3	3	2	2	1	1	1	1	1	0	0	0

Distance downdip (km) 0–30 — Distance along strike (km), scale: 0 10 20 30 40 50 60 70 80 90 100 110 120 130 140 150 160 170 180 190 km

0	10	20	30	40	50	60	70	80	90	100	110	120	130	140	150	160	170	180
8	7	7	6	6	5	5	4	4	3	3	2	2	1	1	0	0	0	0
10	10	9	9	8	8	7	7	6	6	5	4	3	2	1	1	0	0	0
8	7	7	6	6	5	5	4	4	3	3	2	2	1	1	0	0	0	0

FIGURE 4.5 Subsource Classification on Fault Face and Weight Distribution

(a) Scheme 1. (b) Scheme 2.

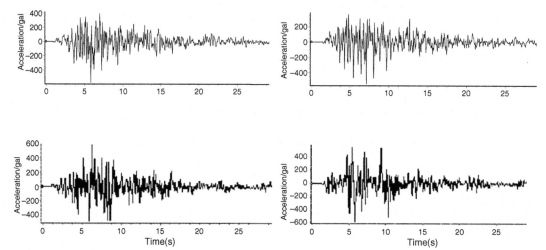

FIGURE 4.6 Dam-Site Seismic Motion Acceleration Time Process Generated by Random Finite Fault Method

(a) Scheme 1(1) (b) Scheme 1(2) (c) Scheme 2(1) (d) Scheme 2(2).

The analysis of the mentioned real engineering examples indicates that the maximum convincible earthquake at the near-site strong earthquake engineering site determined by adopting the random finite fault method is more rational in theory as well as feasible in practical application. However, at present, there are still many problems that are worth further studying and solving. They are mainly the determination of the stress drop in the fault face as well as of asperity distribution and weight so that the progress can be gradually obtained through the in-depth researches by the earthquake departments and in combination with the accumulation of engineering practice experiences.

DAM SITE SEISMIC MOTION INPUT MECHANISM

The seismic motion input mechanism includes the basic concept of site design seismic motion peak ground acceleration and the dam-site seismic motion input mode of two aspects. At present, in some important large dam aseismic designs, there have always existed different understandings of the basic parameters for design seismic motion acceleration to characterize the dam-site earthquake function strength. Whereas the aseismic analysis results with different methods provided by the different units in the same engineering project are difficult to carry out the intercomparison and check based on the same seismic motion inputs so that the results of the engineering seismic safety evaluation are directly affected. For this reason, as viewed from the engineering aseismic angle, it is necessary to carry out further discussion on the basic concept concerning the design of seismic peak ground acceleration in designing seismic motion input mode, as well as some concept confusions existing in the present application so as to come to the common consensus.

5.1 BASIC CONCEPT OF SITE DESIGN SEISMIC MOTION PEAK GROUND ACCELERATION INPUT

At present, in designing large dam engineering earthquake resistance, the main parameter seismic motion peak ground acceleration used to characterize the dam-site seismic function strength is either determined in terms of the "China Earthquake Motion Parameter Zonation Map" or provided by the transcendental probability curves in the dam-site-specific seismic hazard evaluation, based on the achievements of seismic hazard evaluation by the probability method. The "China Earthquake Motion Parameter Zonation Map" only aims at more macroanalysis on the national scale, while the dam-site-specific seismic hazard evaluation aims at the engineering site zone on a certain scale. In engineering site earthquake hazard evaluation, the attention relation of real measured seismic acceleration statistic recorded on bedrock is adopted, and it is based on the real measured strong earthquake acceleration recording sample group from the seismo-stations set up on the bedrock surface to obtain the attention relation among the bedrock earthquake motion peak ground acceleration, earthquake magnitude, and epicenter distance through statistics. The existing strong seismic records can only provide the information concerning soil types of land surface sites where seismo-stations are set up, but there is no way to have a better understanding of the geological conditions and the surrounding topographic conditions under their deeper stratum foundations. What is more important for a certain range of regions is, where there are not many real measured strong seismic records that can be used as a statistical sample book. Even if they are the records obtained from the seismo-stations set up on the bedrocks, the foundation geological conditions below each recording seismo-station are not possibly the same. Therefore, it can only be assumed that what the records represent is the flat land surface seismic motion of homogeneous rock mass along the horizontal finite extension. That is, the plane fixed type wave in the realistic elastic

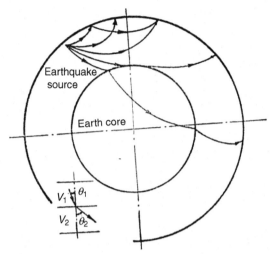

FIGURE 5.1 Seismic Wave Radiation Paths

medium to satisfy the standard wave motion equation, whereas the earthquake hazard evaluation in the engineering site area gives out the maximum horizontal seismic motion peak ground acceleration of the flat freedom land surface of semifinite space homogeneous rock mass in the engineering site areas. Accordingly, the so-called freedom site here is a general concept that can neither take the real land form conditions of engineering site and rock mass concrete geological conditions into full account, nor involve the engineering structure types to be built in the site and engineering aseismic design indexes of the earthquake foundation strength in engineering site areas.

The seismic waves starting from the seismic source through the transmission process in the crust medium and several refractions and reflections are difficult to determine their incidence directions (see Fig. 5.1).

Seismic wave amplitude, frequency spectrum components, and transmission and vibration direction in the whole transmission process are on the continuous changes or variations, whose transmission velocity of each frequency component is different. Seismic waves, including the longitudinal wave, transverse wave, and Love and Reynolds waves constituting two types of wave surfaces, are changing with the different arrival of times. At present, it is difficult to distinguish each wave shape from the component in seismic motion acceleration wave shapes from land surface records, but it is only known that the longitudinal wave has the maximum wave as the last, for these waves are overlapped with each other. However, seismic motion acceleration at the land surface on a certain point in the site is always decomposed into the interorthogonal three components, and two horizontal directions along the land surface in vertical direction with the land surface are usually taken. In fact, the percentages of the three orthogonal components continue to change so that the synthesis shock and transmission mode are also on continuous change, and they cannot maintain a certain wave shape in space and transmit along a fixed direction. As viewed in general, the crust medium density increases from the land surface down to the stratum deep depth, but in physics, the refraction and reflection laws of wave transmission in different media are from the crust's deep part to the land surface transmission seismic waves, whose inclination direction will be gradually close to the vertical level land surface. Therefore, what the earthquake hazard evaluation in engineering site zone gives out is the maximum component peak ground acceleration in the near land surface half infinite space homogeneous rock mass transmitted from the crustal depth along the vertical level to the land surface. In terms of the results obtained from statistical average of real measured strong earthquake records, it can be considered from the statistical sense that the two level component peak ground accelerations of seismic motion and the acceleration response spectrums are roughly the same. However, since the peak values of these level component seismic motion peak ground acceleration time process are impossible to appear at the same time, they are not simply replaced by one peak value multiplied by 1.4-fold and synthesis level seismic acceleration time process, so that the two interorthogonal random level acceleration time processes should be taken or adopted. As far as the ratio between the vertical seismic component and the level seismic component is concerned, it is changing with the earthquake magnitude and the distance to the engineering site. At present, the seismic community has extensively held that vertical component peak value is roughly equal to two-thirds of the level component as viewed from the statistical sense. Accordingly, these three interorthogonal random seismic motion acceleration components are not correlative in statistics.

This basic concept concerning the design of seismic motion peak ground acceleration in the engineering site zone is widely accepted in the seismic engineering community, at present, and this can be also known as the freedom field design seismic motion in the engineering site zone. Also, it is necessary to indicate that seismic motion in a real earthquake is very complicated, and the existing real measured strong earthquake records demonstrate that the land surface earthquake motion away from several 10 m may have the obvious differences. This is likely that the near land surface rock mass is not of realistic homogeneous behaviors, and that owing to the existing face wave, the seismic wave transmitting from the deep to the land surface cannot be completely vertical to the land surface. The freedom field seismic motion behaviors in designing seismic motion of the above engineering site zone are in concert with the basic assumptions and bases in earthquake hazard evaluation widely adopted at present.

5.2 DAM SITE SEISMIC MOTION INPUT MODE

Based on the basic concept of designing seismic motion in the engineering site zone, the large dam-site motion input and mode, as well as the analysis of dam body structure system seismic response are directly in close correlation with the mathematical model. In engineering earthquake resistance, the seismic response analysis of an engineering structure system starts from the dynamic equation, that is:

$$[m]|\ddot{u}| + [c]|\dot{u}| + [k]|u| = |F_1|$$ (5.1)

where [m], [c], [k] are structure system quality, damping, and rigid matrix, respectively; $\ddot{u}$, $\dot{u}$, u are acceleration, velocity, and displacement response, respectively; and F_1 is external force.

As far as seismic motion is concerned, the equation solution has mainly two kinds of ways to take as the vibration problem in an enclosed system and the wave motion problem in an opened system.

As the vibration problem to seek the solution, the interaction between the structure and foundation is not, in general, counted on. The quality, damping, and wave motion matrix in the equation do not include the foundation; the structure seismic inertia $F_1 = -Ma_g$ is taken as the external force, of which a_g is the design ground surface seismic motion peak ground acceleration in the site zone, and its function is on the foundation face of the so-called rigid pan body. The structure acceleration, velocity, and displacement response are all relative values corresponding to land surface movement, which are adaptable to the relatively small structure rigidity, low input wave frequency, and structure dimensions, which should be considered to be far smaller situations of the seismic shortest wave length of the earthquake wave. At present, the kinds of modes are mostly adopted by the common industries and civil architectures and structures.

As to the important large dam engineering with large dimensions and heavy weight, the importance of dam body structure and the interaction of foundation dynamic state should show the protruding role in analyzing the seismic response. This dynamic interaction includes the influence of the foundation upon the structure upon seismic motion input, of which the main seismic wave energy is dispersing to the far region foundation. For this reason, it is necessary to take the dam structure and foundation as a whole system to analyze the seismic response. In the arch dam in particular, deformation of the abutment arch support rock mass and stability performances are the key for the evaluation of arch dam aseismic safety. The land forms and faults, fault surfaces, joints, soft weak zones neighboring dam body foundation, and the physical and mechanical properties of different rock masses and geological formation conditions have an apparent impact upon structure foundation system responses. In this system, the foundation itself should be considered as a system with masses, damp, and stiffness or rigidity. The masses, damp, and rigid matrix in dynamic equation are included in the foundation; seeking for the solution to structure system acceleration velocity and displacement response is an absolute value included in the land surface movement.

This kind of dam body structure system is, in general, taken as the wave motion problem to seek for a solution in the opened system, the right end of the equation depends on the foundation model and the boundary condition selection, but the key is to determine the given free-field table for engineering site zones to design seismic motion peak ground acceleration corresponding to the free-field input seismic wave transmitted from the crust's deeper depth. The given free-field table to design seismic motion acceleration is the free-field table response to free-field incident seismic waves. At present, the assumption extensively accepted by the seismic engineering community and the engineering seismic community is to consider that when the incident seismic wave infinite semispace homogeneous rock mass transmits to the land surface, it will overlap with the reflected wave produced by the free boundary conditions so that the bedrock free-field takes the given design seismic motion amplitude value to half of the design seismic motion amplitude value.

5.3 FREE-FIELD INCIDENT SEISMIC MOTION INPUT MECHANISM

The structure and foundation dynamic interaction problem is an important problem in hydraulic structure aseismic design. Under the seismic action, dam body foundation dynamic interaction includes two aspects, that is, foundation simulation and seismic motion input. Being relative to the dam body, the foundation can be viewed as the semi-infinite body with masses. The infinite regional dam foundation can be classified into the near regional foundation of neighboring regional dam body and the outer far regional foundation. The near regional foundation is counted into the dam foundation two-bank landform and geological structure conditions of various types, including each latent sliding rock blocks on the two-bank dam abutments. Dam body structure seismic wave input by the crust as well as the outer traveling dispersion waves produced because of existing river valley foundation and dam body. The outer traveling wave transmits toward the mountain mass in the course of which its energy is gradually escaping dispersedly to geometrical dispersion and damping energy consumption within the inner part of the foundation, whereby leading to the vibration reduction in the dam body structure, whose effects are the same as damping so that this is known as radiation damping, whose physical concept is quite obvious. Radiation damping effects are the main contents of interaction of the dam body structure and foundation dynamics. The so-called radiation damping handling way close correlation for vibration energy escaping dispersedly to the far regional foundation should be fully considered in the dynamic analysis mathematical model for free-field incident seismic motion input mechanism and dam body structure system.

The dam body structure analysis model can only include the finite range of near regional foundation. Under the strong seismic action of the large dam structure, its escaping dispersion effects of the acceleration system outward traveling energy toward the far regional foundation may also obtain enough gains from the near regional foundation. The theory requires that the near regional foundation boundary away from dam foundation distance should be $L \geq CT / 2$, where, C is the maximum wave velocity in the infinite regional foundation medium and T is the earthquake sustained time. The mesh dimensions in dynamic calculation and analysis are restricted by the minimum wave length in such a way that it is bound to make the calculation and analysis require very large storage quantities and calculation work, but it is, in fact, difficult to realize. Nowadays, the near regional foundation range is mostly restricted to be one to two times the dam height, being the distance extending from the dam foundation to every direction, while the far regional foundation can be replaced by the artificial boundary set in the near regional foundation surroundings. If the artificial boundary adopted the usual fixed or free boundary, the outward traveling wave energy consumes a rather part dispersedly in the outer-side artificial boundary foundation, which will reflect back the dam body and the near regional foundation from the artificial boundary, whereby obviously affecting the seismic response results.

However, there is another viewpoint that the *in situ* real measured dam body damping values have included the radiation damping influence, for it is unnecessary to consider it in the calculation and analysis. In fact, damping behaviors of the dam body and its strong nonlinear behaviors will increase with an increase in excitation force. When the *in situ* real measured dam body damping value is carried out, and in order to guarantee dam body safety, either the machinery type of the excitation device or the blasting with small quantity of explosives is adopted, the energy of its excitation response is very small, and the dam body shock response is small as well, and the scattering wave transmitting range in the foundation is small.

These can by no means compare with the huge energy and their influence when a strong earthquake takes place. For this reason, in the aseismic design of engineering structures, the material damping values are larger than those of real measurement, and the damping values should adopt very large ones in the occurrence of a stronger earthquake than the weak one. For instance, in designing earthquake resistance of nuclear power stations, the damping ratio of safety limit earthquake (SL-2 or SSE) is much larger than the operational safety seismicity (SL-1 or OBE). The inner friction energy consumption structure and the near regional foundation material damping have a much smaller effect upon the seismic response than that of the far regional foundation radiation damping. Hence, as to taking smaller near regional foundation conditions usually, and in the case of a strong earthquake, when the large dam seismic response dynamics is analyzed, the radiation damping influence cannot be disregarded. At present, there have been many practical examples of aseismic calculation and analysis of high arch dam engineering projects, thus, indicating that radiation damping effects upon seismic response influence can reach 20–40% and grow with an enlargement in dam body volume and a decrease in the foundation deformation modulus and cannot be ignored, but should be taken as the additional safety factor to be considered.

The artificial boundary has always been a focus point problem that has been studied at home and abroad. The correlative research work may be divided into two research orientations or directions, namely, the global artificial boundary treatment method and the local artificial boundary treatment method.

The global artificial boundary condition simulating outer traveling waves can pass through the whole artificial boundary of the near regional foundation and enter the far regional and semi-infinite foundation, whereby guaranteeing that the whole artificial boundary outer traveling waves can satisfy all the site equations and physical boundary conditions in the infinite regions, so as to form the integration type of artificial boundary. This kind of boundary condition through Fourier changes is first established in the frequency region, and then converted into the time region through Fourier reversion changes. Such methods as the boundary element method (Dominguez et al., 1992; Chuhan et al., 1995), thin sheet method (Smith, 1975), method based on similarity (Dasgupta, 1982), wave function developed method, and the method based on Huygens' principle all belong to the range of the global artificial boundary conditions.

The local artificial boundary condition simulating outer traveling waves can pass through each point of the artificial boundary and enter the semi-infinite foundation individually, whereby guaranteeing that the outer traveling waves that radiate any point on the artificial boundary can pass through the boundary so as to form the micro-type of the artificial boundary. The local artificial boundary conditions are, in general, established in terms of the single-orientation wave transmission theory, Clayton–Engquist artificial boundary conditions, Higdon–Keys artificial boundary condition, infinite element method, viscosity boundary condition, viscoelastic boundary condition, and artificial homological boundary condition all belong to the artificial boundary condition of this type.

In considering many existing radiation damping and its corresponding input mechanism manners for large dam earthquake resistance analysis, the following two kinds of methods are mainly adopted at present.

5.3.1 ARTIFICIAL HOMOLOGICAL BOUNDARY METHOD

This method sets near regional foundation boundary as the local artificial boundary to satisfy the single outer traveling wave $f(x-ct)$ condition, of which x is the boundary outer normal direction. At present, in the important large dam engineering aseismic design, the artificial homological boundary method to the general infinite regional model with the universal adaptability developed by academician Liao Zhangpen is extensively adopted in our country. The artificial homological boundary theory is based on ensuring that the direct simulation is the outer scattering wave along the vertical boundary direction with artificial wave velocity in single-direction transmission. The transmission nature at the interface of the artificial homological boundary is in agreement with that in the original continuous media, that is, waves passing through the artificial boundary interface have no reflection effects but occur to have a complete homology. In this way, it can make the interior point motion equation of finite element combine with extrapolated method at the artificial boundary point in such a manner as to use the finite calculation model to simulate the wave motion process in the infinite medium. After the incident wave is subtracted from the total obtained response to get the other traveling wave, the direct simulation of outer traveling wave on the boundary can be carried out from the inner part of the finite model passing through the artificial boundary to the outer homological process.

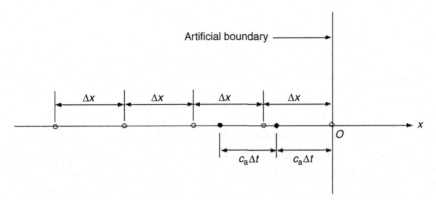

FIGURE 5.2 Relation Between Computation Point and Finite Element Dispersion Nodal Point

This method is evaluated strictly in theory. In seeking the solution method, the artificial boundary inner part in the space region can be dispersed by concentrating the quality finite element method. In the time region, the center difference method is used to carry out the dispersion. After the time-space coupling is solved, the stepwise integration is used to obtain the explicit solution in the time region, that is, the current nodal point motion equation is used to deduce the next time nodal point motion. It is not necessary to carry out the whole integration of the unit rigidity, unit quality, and unit damping matrix, whose right end term formation only needs to accumulate the contributions of effective loading vector in terms of each unit on the grade one level in such a way that the whole calculation is carried out basically in the grade one level of the units. Therefore, the very small high-speed storage zone is needed only, with high computation efficiency. Particularly when a series of the unit rigidity matrix, quality matrix, and damping matrix are similar, it need not carry out the repeated computation, with higher efficiency and being adaptable to the nonlinear problems of various types. The homology accuracy of the outer traveling wave on the boundary can continue to be improved in theory with an increase in the number of stages in homologies for many times, but in the practical engineering application, the second-order accuracy is usually employed. Also, the detailed research work has been done on the requirement for the gradual integration of numerical stability and the way of realization. This method has, at present, extensively been used in many high arch dam engineering earthquake resistance designs in our country.

The artificial homological boundary seismic motion input is based on the assumption of the overlay of the artificial wave velocity transmission in horizontal direction and vertical plane waves. The way of direct input of free-field incident seismic displacement wave from the boundary is adopted.

As shown in Fig. 5.2, let one incident wave be the artificial wave velocity c_a moving along x axis from the left side to the artificial boundary point O, and count point j at p time displacement expression formula $u_j^p = u(p\Delta t, j\Delta x)$, where, j and p are the integers, Δt is time step distance, Δx is space step distance. If $\Delta t = \Delta x / c_a$, u_0^{p+1} is determined directly by using the inner nodal point displacement u_j^{p+1-j}, being vertical to the artificial boundary, namely:

$$u_0^{p+1} = \sum_{j=1}^{N}(-1)^{j+1} c_j^N u_j^{p+1-j}$$

$$C_j^N = \frac{N!}{(N-j)!\,j!}\,(N=1,2\ldots) \tag{5.2}$$

Equation (5.2) is N order many times homological formula. This is the one-dimensional homological boundary principle, which can be conveniently expanded as two-dimensional and three-dimensional problems.

5.3.2 SUBSTRUCTURE METHOD BASED ON DYNAMICS

This kind of method takes the dam body and near regional foundation as the substructure to isolate from the infinite regional foundation so as to satisfy the radiation condition dynamic resistance and free-field incident seismic wave response to characterize the far regional foundation effects. The boundary condition and seismic motion input mechanisms are determined through the interaction inner force balance of the substructure and the far regional foundation contact face and the displacement continuous conditions. The substructure dynamic equation may be expressed by the dynamic resistance form of the subzone between the inner part and the boundary nodal points as follows:

$$\begin{bmatrix} S_{ss} & S_{sb} \\ S_{bs} & S_{bb} \end{bmatrix} \begin{Bmatrix} u_s \\ u_b \end{Bmatrix} = \begin{Bmatrix} 0 \\ p_b \end{Bmatrix} \tag{5.3}$$

where, S is substructure dynamic resistance; U is displacement response; labels s and b are the inner part and boundary zone.

The interaction force acting on the substructure boundary is $\{p_b\} = [S_b^g]\{u_b^g - u_b\}$, where, S_b^g, u_b^g are the far regional foundation free boundary dynamic resistance and the scattering displacement field under the action of incident free-field input wave. Substituting p_b into Eq. (5.3), we can have the following:

$$\begin{bmatrix} S_{ss} & S_{sb} \\ S_{bs} & S_{bb} + S_B^G \end{bmatrix} \begin{Bmatrix} u_s \\ u_b \end{Bmatrix} = \begin{Bmatrix} 0 \\ S_b^g u_b^g \end{Bmatrix} \tag{5.4}$$

It is easy to prove $[S_b^g]\{u_b^g\} = [S_b^f]\{u_b^f\}$, where, S_b^f, u_b^f are the free-site foundation substructure boundary dynamic resistance and the displacement under the action of incident free-field input wave, respectively. Dynamic resistance matrix S_b^g, S_b^f obtain solutions in the frequency region, coupled in the space region, and also are the full matrix for the whole near regional foundation boundary freedom. As far as the analysis of large-sized complex structure seismic response is concerned, the computation is very complicated. Particularly when the dam body and near regional foundation are of nonlinearity, it is necessary to seek for the corresponding global solution to time region through Fuchs counter variance, for it is difficult to use in practical engineering. Even if the approximate method is adopted in a certain frequency range, each main item of the dynamic resistance matrix in the frequency region must be fitted through the quadratic curve in such a way that quality, damping, and each nodal point integrated parameters of rigidity can be derived, whose computation is rather complicated and rough.

At present, the frequently used method to derive the dynamic resistance matrix S_b^g integrated parameters in the time region of far regional foundation free boundary is that the outer traveling scattering waves within the substructure be taken as the wave source problem. Starting from the standard wave

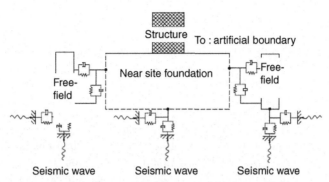

FIGURE 5.3 Dam Body and Near-Regional Foundation Substructure Boundary Model

motion equation of the plane, column surface or spherical surface waves in the realistic medium, the relation equations among normal and tangential stresses, displacements, and velocity on the substructure boundary are used to derive the damping of dynamic resistance matrix and the integrated parameters of rigidity with the time region, corresponding to setting up of the damper on the boundary and the viscoelastic boundary of the spring, that is the so-called viscoelastic boundary (see Fig. 5.3).

Though this method is of certain accuracy, physical sense is clear, for this method is also extensively used in the analysis of large dam earthquake resistance, and at the same time, in the case of incident free-field input action, this method is adopted to seek solutions to free-field stress σ_b^f displacement u_b^f and velocity $\dot{u}_b^f$ and to replace the far region foundation scattering displacement u_b^g, whereby making the computation further simplified.

In this method, the far regional foundation free boundary dynamic resistance S_b^g can be characterized through the nodal point spring rigidity K_b and damper damping coefficient C_b, corresponding to free-field longitudinal and transverse incident wave expression formula as follows:

$$K_{bp} = A_b \frac{\alpha_N E}{R} \qquad C_{bp} = \rho c_p A_b \text{ (Longitudinal)} \qquad (5.5)$$

$$K_{bs} = A_b \frac{\alpha_T G}{R} \qquad C_{bs} = \rho c_s A_b \text{ (Transverse)} \qquad (5.6)$$

where, A_b is the affected area vector of boundary mesh nodal point; E, G, ρ are far regional foundation medium elasticity modulus, shear modulus, and mass density, respectively; R is distance from the scattering wave source to near regional foundation boundary, since the wave source of the scattering site in substructure is not the nodal point source, taking value is of certain randomness; α_N and α_T are boundary normal and tangential revised coefficients are related to the plane surface or spherical surface wave shapes adopted in wave source problem, respectively. As to the incident seismic wave from the bottom boundary of the near regional foundation, when assuming that the vertical and horizontal components take the plane surface waves transmitted by longitudinal and transverse wave velocities, respectively, their values take one-half in all.

Seismic motion input corresponding to this model can take the near regional foundation as the substructure of semi-infinite space free-field foundation, being given out by the boundary interaction force P_b^f under the action of incident seismic wave, namely:

$$\left\{ P_b^f \right\} = \left[K_b \right] \left\{ u_b^f \right\} + \left[C_b \right] \left\{ \dot{u}_b^f \right\} + \left[\sigma_b^f \right] \left[n \right] \left\{ A_b \right\} \qquad (5.7)$$

where, [n] is boundary normal direction cosine vector; $[K_b]$, $[C_b]$ are viscoelastic boundary rigidity matrix and damping matrix.

The boundary input in the near regional foundation bottom side serves as the free-field incident seismic wave of the plane surface wave.

Viscous boundary may be considered as the special form of viscoelastic boundary. Viscous boundary was suggested by Lysmer and Kuhlemeyer (1969), in the earliest time, whose basic thinking was to set up a damper on the artificial cut-off boundary to absorb the outward transmitted wave energy and was unable to simulate the elastic recovery ability of semi-infinite foundation. This kind of treatment method is simple, which has been realized in many large-scale commercial software such as ABQAUS, FLAC, LS-DYNA, and ADINA. Viscous boundary disadvantages lie in the fact that it is unable to absorb the large angle incident outer transmitted wave energy, and the existing low-frequency drift unstable problems.

5.4 SEVERAL PROBLEMS NEED TO BE CLARIFIED AND FURTHER DISCUSSED IN SEISMIC MOTION INPUT MODE

At present, in large dam earthquake resistance design, there have been some confusion in seismic motion input serving as the prerequisite of evaluating aseismic safety, having a direct effect on the evaluation of engineering earthquake resistance safety. It is urgent for us to clarify them and to test and discuss their influence so as to come to a common consensus.

1. Design peak ground acceleration given out by earthquake hazard evaluation should be well understood. As previously described, what earthquake hazard evaluation gives out in the engineering site zone is the maximum horizontal component peak ground acceleration values that are transmitted from the crustal deep depth along the vertical to the horizontal land surface to the land surface vertically in the near land surface semi-influence space homogeneous rock mass. The incident seismic wave satisfies standard wave motion equation plane surface fixed wave in the realistic elastic medium. However, in many large dam engineering aseismic designs, it is always a mistake to directly connect the site zone design peak ground acceleration with the land surface of the real engineering site to assume the base level elevation of the free-field surface at random. For instance, the earth-rock dam takes the covering strata land surface, the gravity dam takes the dam foundation rock surface, and the arch dam takes the dam crest or the upper assumed level rock surface with a certain height. In terms of engineering geological concrete geological conditions and through counting in the foundation real mechanical behaviors and damping one-dimensional inversion analysis, the incident seismic wave of the foundation bottom boundary can be decided, whereby resulting in the fact that not only can the aseismic analysis results of different units in the same engineering be compared and checked with each other on the basis of the same seismic motion input, but also in comparing different dam types of schemes, the incident free-field seismic wave amplitude values in the same engineering site zone foundation bottom rock mass boundary have great differences. Particularly when there are covering strata or the foundation rock masses are rather weak, this will lead to a very great change in the incident free-site seismic wave behavior.

2. The quality-free foundation assumption is usually used to directly input design seismic motion peak ground acceleration. At present, the quality-free foundation is still employed in some engineering aseismic design. The quality-free foundation assumption only takes the effects of near regional foundation elasticity upon the rigidity into full account, but disregards the effect of foundation inertia and damping, and it is only adaptable to counting in foundation rigidity impact when the dam body dynamic behaviors are determined, but it is not suitable to be extended in application of dam body seismic response analysis. Particularly taking the near regional foundation boundary as rigidity has neglected the radiation damping influence as the major factor in structure–foundation interaction, which cannot reflect the effect of structure upon the dam foundation seismic motion and the uneven distribution impact or influence of both banks of the river valley dam foundation seismic motion. Taking land surface design seismic motion peak ground foundation to extend about one time of dam height can be served as the foundation rigid boundary even seismic motion input, being obviously not in consistence with actual conditions. The analysis of real engineering indicates that this kind of assumption clearly renders the results to become large. Switzerland has made the systematic strong seismic observation and measurement of such three high dams as the 250-m-high Mauvosin dam, 180-m-high Emosson dam, and the 130-m-high Punt-dal-Gall dam (Proulx et al., 2004). The results obtained from the observation and measurement indicate that taking dam foundation real measurement acceleration as the input, the dam response computed in terms of quality-free foundation model is much larger than the corresponding real measured value, whereby testing the real situation, thus illustrating that the foundation radiation damping and the foundation shock uneven distribution influence cannot be ignored.

3. As to the near regional foundation bottom side, the rigid boundary input can be taken to design seismic motion peak ground acceleration. The initial importance of quality-free foundation is that within the range of foundation taken in the depth of one time of dam height, the foundation quality, rigidity, and material damping influence must be taken into full account, but energy absorption boundary is set up on the two sides of the foundation to count in radiation damping influence, while the foundation bottom sides can still serve as the rigid boundary to be input into designing seismic motion peak ground acceleration. This kind of seismic motion input mechanism was once extensively recommended, but its neglect of foundation vertical resonance to affect the results has lacked serious discussion hitherto. Particularly when foundation depth reaches near $CT/4$ (of which C and T are the wave velocity and remarkable cycles of seismic wave, respectively), the influence can be more obvious. Also, there had been some doings to set up the dampers on each boundary including the bottom foundation, but the nodal point quality of the foundation bottom boundary can increase much artificially, so as to guarantee that the bottom boundary seismic acceleration can be half of the land surface design seismic motion peak ground acceleration. It is still worth discussing whether this kind of input mode can reflect the interactive force of the finite foundation bottom boundary in practical action as well as its outer traveling scattering wave radiation damping effects.

4. Land surface design seismic motion peak ground acceleration can replace the scattering wave along the river valley before the dam is built. In our country, research on large dam engineering earthquake resistance has provided the research achievement. Also, after the foundation dynamic resistance matrix along the dam foundation face is obtained, the integrated parameters have

approximately been fitted in the time region, but when the seismic motion is input into the dam foundation surface, the uniform land surface design seismic motion has replaced the scattering wave site adopted along the river valley before the dam was built, which lacks theoretical bases, obviously.

5. Substructure boundary is adopted to set up dampers and the spring element model but to neglect free-field stress. The previous description is based on the dynamic substructure. The second group of seismic motion input mode in the near regional foundation boundary input interactive force P_b has been adopted by some foreign large-sized commercial software, and also often used in large dam earthquake resistance design in our country. However, some can only set up the dampers in the near regional foundation boundary, but neglect spring elements or the free-field stress functions on two sides of the boundaries cannot be counted in seismic motion inputs. This neglect of result effect has lacked comparison and assessment hitherto.

5.5 SUGGESTIONS

At present, our country is encountering the severe challenges to construct a series of precedent high dam engineering works in strong earthquake areas. The seismic motion input is the important prerequisite of large dam aseismic safety evaluation, to which great attention must be paid. For this reason, and in terms of the problems and existing conditions in large dam-site seismic motion input mechanism at present in our country, we have suggested that the priority concerns and in-depth discussion be given in the following aspects:

1. As viewed from the engineering point of view, large dam seismic motion input mechanism involves the determination and seismic functions and the model and method of large dam body structure system analysis. The two aspects are very complicated and have uncertainty. The engineering community has lacked the deep understanding of this. In the evaluation of large dam aseismic safety, seismic input accuracy must match the seismic response analysis of the dam body structure system and the accuracy determined by the dynamic structure of dam building materials. As viewed from the engineering point of view, it is impossible to require that seismic motion input be of very high accuracy, and it is unnecessary to seek for the over accuracy. However, it is essential to combine with practical engineering to make the sensitivity analysis of uncertainty of the main influencing factors in input mechanism, whereby making the designers understand influencing degrees so as to simplify the less influencing factors upon the results as viewed from the engineering angle.

2. Many various kinds of methods are used to carry out comparisons and checks. Owing to the complexity of problems plus the different methods of seismic motion input, with a certain limitation and assumption, it is very necessary to combine with practical engineering to carry out the intercomparison and check of the case of same conditions so as to make it convenient for the engineering designers to make the integrated consideration and decision-making.

3. The common consensus and specifications should be formed as early as possible. Based on points 1 and 2, work and summarization of engineering practical experience, and the close coordination with the earthquake departments, the consensus of the basic concept and method of seismic motion input should be obtained as possible from the present-day recognition level at home and

abroad. In conforming to the distinctive physical concept, intuition, simple and easy method, the engineering designers should understand and grasp the principles so as to make the dam-site seismic motion input specialization as early as possible.

4. Efforts are made to strengthen practice examination or tests. Close attention should be paid to a few rare seismic cases. Practical efforts must be made to strengthen the strong seismographic stations and networks in the reservoir and dam zone and the indoor and outdoor test studies and also, much more efforts should be made to deepen the understanding of the dam-site seismic motion input mechanism in order to carry out tests and improvement.

HIGH ARCH DAM BODY – RESERVOIR WATER – FOUNDATION SYSTEM SEISMIC RESPONSE ANALYSIS AND SEISMIC SAFETY EVALUATION

6 HIGH ARCH DAM BODY–FOUNDATION SYSTEM THREE-DIMENSIONAL CONTACT NONLINEAR DYNAMIC
   ANALYSIS METHOD 115

7 DAM ABUTMENT AND ARCH SUPPORT ROCK BLOCK STABILITY AND SEISMIC SAFETY EVALUATION
   OF HIGH ARCH DAM. 137

8 RESEARCH ON PARALLEL COMPUTATION OF HIGH ARCH DAM STRUCTURE SEISMIC MOTION RESPONSE 155

9 ENGINEERING REAL EXAMPLE ANALYSIS OF PARALLEL COMPUTATION 207

The high arch dam system includes the dam body, dam foundation, reservoir water, and interaction integration system, whose seismic response analysis is the core of seismic safety evaluation. In recent years, a series of 300 m high arch dam engineering projects have been completed in the high seismic region in the western part of China, encountering the severe challenges of many important key technical problems. For instance, the Dagang mountain arch dam is only 4.5 km away from the Moxi seismic occurrence fault to the upper limitation of earthquake 8 magnitudes, and it is necessary to resist a few rare seismic functions of near fracture large earthquakes with the designed seismic acceleration being as high as $0.56g$ in the world. There have been many technologically difficult problems such as large flood discharge flow, complex geologic and topographic conditions, high steep-side slopes, and large-sized underground cave stability, having become the unprecedented engineering of the world's highest arch dam – the Jinping arch dam. In these rare precedents of the world's first-rate engineering works, considering seismic function has always become a controlling factor over the engineering conditions in design. The traditional aseismic design concept, method, and technical way based on the existing engineering experiences have been difficult to adapt to the rapid development of engineering construction requirement since there is a lack of real examples of high dam engineering works experiencing strong earthquakes. The high dam engineering works aseismic safety located in the earthquake-frequent west regions has attracted the extensive attention from every community in society. Particularly guaranteeing a high dam and big reservoir without the occurrence of dam failure and disaster variation has become a strategic priority in the current water resources and hydroelectric power engineering earthquake resistance, of which high arch dam body–reservoir water–foundation system seismic response analysis and aseismic safety assessment are the key to seismic disaster quantitative judgment. For this reason, this section introduces the innovative research achievements obtained closely around this strategic priority to carry out scientific researches, they are as follows: research and development can exactly reflect the nonlinear seismic response analysis model and seeking a solution method of high arch dam system of engineering real conditions when earthquakes take place; taking high dam system displacement response mutation to serve as the whole loss efficiency to judge quantitative standard new concept and operable method is suggested; the hardware platform construction and software application research and development of high performance and parallel computation technology in practical high dam engineering aseismic applications, and so on, are integrated.

This section is mainly based on the mentioned research achievements to introduce high arch dam body–reservoir water–foundation system seismic response analysis and aseismic safety evaluation method, whose contents are in Chapters 6–9.

Chapter 6 first introduces that China has faced a series of difficult subjects in evaluating seismic safety when many super-high arch dams of over 300 m have been constructed in strong earthquake areas in the western part of the country, discusses the main factors that should be considered in the seismic safety evaluation of high arch dam systems, summarizes the simulation of artificial boundary treatment methods and principles of infinite regional foundation, and introduces the contact nonlinear static–dynamic combination computation model for arch dam–foundation system. Finally, Chapter 6 introduces the solution method for wave motion problems in time region, that is, the finite element inner point motion equation combined with the extrapolation method of artificial boundary nodal points; and the finite computation model is used to simulate the wave motion process in the infinite medium.

Chapter 7 introduces the performances of aseismic stability analysis of arch dam abutments and revelation of correlative seismic disasters, points out the problem of the current arch dam aseismic stability analysis method and rigidity limit balance method localization, while the basic connotation of losing stability safety coefficient of arch dam abutment rock blocks is analyzed and suggests arch dam aseismic stability based on the deformation as the core and the aseismic safety evaluation idea. Finally, Chapter 7 illustrates the feasibility of taking deformation as the core to carry out the abutment aseismic stability analysis thinking method through the application of the method in large arch dam real engineering.

Chapter 8 covers the main contents of high arch dam structure seismic motion response high performances and parallel computation research work. It mainly introduces the high performances parallel computation development situation in hydraulic structure fields in our country concretely, analyzes the important significance of conducting high dam structure seismic motion response parallel computation research, gives out the explicit difference scheme in time region of considering motion contact nonlinear problem dynamic equation, and discusses the factors affecting the explicit integral computation mode of numerical stability, motion contact problem of Lagrange's method of multiplier and the nondiagonal quality matrix handling method for explicit integral computation, introduces the implementing steps of artificial homological boundary, deals with the viscoelastic boundary model physical significance, gives out the detailed viscoelastic boundary input formulation and establishes the virtual pattern of the dynamic equation including artificial viscoelastic boundary condition.

Finally, Chapter 8 develops, based on the principal finite element parallel program generating system (PFEPG), a high arch dam structure seismic motion response high performance parallel computer software using artificial homological boundary and viscoelastic boundary.

The contents in Chapter 9 are the practical application examples of the high arch dam structure seismic motion response high performance parallel computer software. Of which the parallel program developed is used to carry out seismic safety evaluation of the Xiawan and Xiluodu high arch dam engineering works. At the same time, in the process of engineering application, Chapter 9 probes into the homological boundary method to the sensitivity of net grid dimensions. Finally, Chapter 9 carries out the rock mass sliding block stability of the Beiheitai left abutment side slope, with the obvious efficiency of parallel computation indicated.

HIGH ARCH DAM BODY–FOUNDATION SYSTEM THREE-DIMENSIONAL CONTACT NONLINEAR DYNAMIC ANALYSIS METHOD

6.1 PERFORMANCE AND ENGINEERING BACKGROUND OF HIGH ARCH DAM ENGINEERING SEISMIC RESPONSE ANALYSIS IN STRONG SEISMIC AREAS

At present, the arch dam design specifications in our country are mainly based on the arched-beam divide loading method of structural mechanism to carry out the dam body structural linear elastic analysis under the action of static and dynamic loading. In this method, the dam body is taken as a three-dimensional (3-D) whole structure. Only with Fulget coefficient, the effect of elastic rigidity of the homogeneous dam foundation rock mass upon dam body structure function is taken into consideration; in the case of separation of supporting dam abutment rock blocks from the dam body, the rigidity limit balance method is used to carry out quasistatic stability analysis; when in the earthquake operating conditions, the even seismic motion input should be made on the dam foundation surface. In design, although the finite unit element method is adopted to carry out the check, the results can only be used for reference, whereas the dam body is still used as the whole structure in the finite unit element method adopted in the earthquake operating conditions. If the dam foundation is assumed as the quality-free rock mass, it can only be counted into elastic rigidity of the dam foundation rock mass, for it is only to replace the Fulget coefficient method that the unevenness of the dam foundation rock mass elastic rigidity can be considered. Owing to adoption of the quality-free foundation, its results can lead to the fact that we can only make the even seismic motion input on the dam foundation surface. Accordingly, the structural model and seismic motion input mechanism and arched-beam divide loading method have had no differences in nature, and they all belong to the closed shock system.

Although the finite element method is more adaptable to the complicated dam body detailed structure and dam foundation topographic and geological conditions, the linear elasticity finite element is at the key dam heel position and the stress concentration phenomena caused by the corner effect is very sensitive to mesh dimensions, whereby making the determination of safety standards in design have a great uncertainty so that it cannot become the major method to be based on in large dam design up to now.

With the hydropower development in western China where a large number of super-high arch dams of over 300 m will be built in the strong earthquake areas, seismic safety has often become the controlling operational condition in the design, whose seismic performance dynamics compared with the conventional arch dams appears to exist with great changes in nature, whereby encountering a series of challenges in key technical difficult subjects in the following:

1. High arch dam can serve as a large volume concrete structure. The dam body foundation reservoir water dynamic interaction under the seismic action affecting seismic response has been very important. Quality-free foundation assumption does not coincide with objective reality, and foundation inertia function cannot be neglected. The near regional foundation rock mass neighboring dam body, its uneven rock behaviors, and various kinds of geological structures, such as the unavoidable small-scale fractures and joints, and so on, can affect the dam body performance dynamics when an earthquake happens. Particularly, the previously described shock energy escaping dispersion to the far regional foundation radiation damping may have an apparant effect upon the seismic response to high arch dam body. Accordingly, the dam body–foundation-reservoir water should be considered as the whole opening wave motion system.

2. The dam body transverse interstice opening and closing need to be considered under the reciprocating action of a strong earthquake. In order to consider the influence of temperature function, the arch dam must be poured in stages or sections with the width of 20 m or so in the construction process, until the dam body concrete cools to a stable temperature, and transverse interstices between dam sections or monoliths distributed along the radial direction must be grouted. Under the action of static loading or under the action of a weaker earthquake of low arch dam, the dam body transverse interstices are, in general, compacted by their bearing of the upstream reservoir water, thus, forming a complete whole structure. The super-high arch dam under the reciprocating action of strong earthquake can be taken as a whole structure, and its upper part arched-direction tensile stress may reach as high as 5~6 MPa. Accordingly, at present, it is difficult to pour such large volumes of concrete with high resistance to tensile strength so that in terms of this computation result, there can be no way to construct super-high arch dams of 300-m high in the strong earthquake areas in the west part of our country. In fact, the grouted transverse interstice in the dam body can only transfer the compression stress and have almost no tensile strength. For this reason that in the process of reciprocating seismic action, the interstices are bound to open and close, and the arched direction tensile stress must be released to result in stress redistribution so that the high tensile stress obtained from the whole structure computation does not exist actually. As a result, a high arch dam under the action of strong earthquake no longer exists in the whole structure. Accordingly, it is essential to take the dynamic boundary nonlinear contact effect of transverse interstice opening and closing into full account in the seismic response analysis.

3. Arch dam engineering safety mainly depends on the rock block stability of dam abutment arch supports, and this basic concept has particularly embodied in the aseismic safety. The current quasi-static stability analysis result based on dam abutment arch support rock blocks by the rigidity limit balance method cannot consider the dynamic coupling functions between dam body and its support arch abutment rock blocks and seismic reciprocating performances in the earthquake process so that it is difficult to reflect the real conditions when an earthquake is under way. Therefore, deformation should be the core in the aseismic stability analysis of high arch dam supporting abutment rock blocks. It is necessary to consider dam body–foundation–reservoir water as the dynamic coupling impact of the whole system.

4. The existing monitoring and measuring data of the arch dam body earthquake indicate that in a real earthquake, seismic motion of each position of the construction foundation surface of the arch dam, whether their amplitudes or phase positions are in significant differences, with high arch dam protruding in particular. As the high secondary super static structure arch dam, it is very sensitive to the difference deformation of both bank arch supports so that it is necessary to consider the dam foundation face seismic motion input unevenness in analyzing the high arch dam seismic response.

It is particularly necessary to point out that many performances of seismic response analysis of the said super-high arch dams in strong earthquake areas are intercoupling and interaffecting so that they must be comprehensively considered in a unified analysis model, and this can increase the difficulty.

To sum up these descriptions, a breakthrough must be made to the current tradition from the concept to the method in order to solve the problem of seismic safety of constructing the world's rare super-high arch dams of 300 m high in the west part of our country where 80% of water energy resources are concentrated at present. Therefore, it is necessary to carry out the comprehensive consideration of the mentioned seismic response analysis model and seek a solution method of each item performances so as to solve the urgent, basis, and key technical problems of high arch dam engineering earthquake resistance in the west part of our country.

6.2 **CONTACT PROBLEM HANDLING OR SOLVING METHOD**

Contact problem is a widely existing mechanical problem in engineering. In high arch dam seismic response analysis, there exist contact problems in dam body transverse interstices, foundation, and dam abutment possible sliding blocks, whose handling method mainly lies in the seeking solution process of dynamics equations, and the contact constraint conditions should be adopted for the contact boundary.

Contact conditions are usually embodied in two forms, that is, two kinds of introducing additional conditions to form a revised basic equation pan function method: one is to use the penalty function to make the additional conditions as the product form be introduced into the pan function extreme value problem penalty function method and then with contact face to set up normal and tangential spring element embodiment, whose advantages are to use the penalty function to seek the solution to the universal function condition stagnation value problem without increasing unknown numbers of quantities. However, there exist the objective randomizations in these spring rigidity taking values and in theory they cannot satisfy the nonpenetrability of contact conditions. The other kind is Lagrange multipliers. This method expresses the contact surface normal and tangential contact force with Lagrange multipliers so as to satisfy contact constraint conditions, thus introducing the basic universal function and then seeking solutions to the revised condition universal function. Although Lagrange's method of multiplier increases the equation to seek for the unknown quantities, the function method's weak points can be avoided and has been extensively used in the engineering community.

The contact problem should conform to the following guidance: (1) the contact objects should not interintrude into each other; (2) the contact force normal component must be the positive pressure; and (3) tangential contact friction conditions. These conditions are different from the general constraint conditions, whose performances are the single side inequality constraint with the strong nonlinearity, belonging to the so-called contact nonlinear problem. The contact problem possible contact states have had the isolation state, viscosity state, and sliding state of three kinds.

In terms of classification of contact forces, the contact problem's contact conditions are mainly divided into tangential and normal contact conditions.

6.2.1 NORMAL CONTACT CONDITION

The normal contact condition judges whether it enters into contact and has entered into the due observational conditions. This condition includes the kinematics conditions and dynamics condition of two aspects.

1. *Nonpenetrability*: This condition is the kinematics condition among the contact surfaces. Nonpenetrability refers to object A and object B position shapes, which are not allowed to penetrate into each other (intruded or covered) in the movement process. In the computation of contract process, the fundamental question is how to satisfy the nonpenetrating conditions.

   The nonpenetrating condition can be described by means of a definition distance function ξ_n on the contact surface. Let some one point-to-middle point corresponding to objects A and B coordinate values u_A, u_B, n be the point corresponding to normal direction:

$$\xi_n = (u_A - u_B)_n \tag{6.1}$$

   $\xi_n > 0$ expresses separation; $\xi_n = 0$ denotes intercontact; and $\xi_n < 0$ illustrates the interintrusion, that is, object A and object B have completely penetrated with each other. Because Eq. (6.1) fits any point on the contact face so that the penetration cannot be required, it can be generally expressed as follows:

$$\xi_n = (u_A - u_B)_n \geq 0 \tag{6.2}$$

2. *Normal direction contact force is the pressure*: This condition is the dynamics condition among the contact faces. Without considering the adhesion conditions, the normal direction contact force among them can only be the pressure stress.

6.2.2 TANGENTIAL CONTACT CONDITIONS

Tangential contact condition judges the concrete contact states of the contact surface of two entered contact objects and their own obeyed conditions:

1. *Frictionless model*: If the contact face of two objects is absolutely smooth, or the friction between them is ignored, the frictionless model can be used to carry out the analysis; that is, the tangential friction force between the contact face is zero. The two objects on the tangential contact face can relatively slide freely.
2. *Friction model*: If the friction between the contact faces must be taken into account, the friction model should be adopted. In engineering analysis, the Mohr–Coulomb friction model is widely used because of its simplicity and applicability. The Mohr–Coulomb model holds that there is no relative sliding among the objects in static friction states, when friction force increases in the case

of static friction constraint condition is unable to satisfy the requirement, the objects begin to have relative slides. The sliding friction stress λ_s is only related with sliding friction coefficient μ and normal direction positive pressure λ_n so as to satisfy the following:

$$|\lambda_s| \leq \mu \lambda_n \tag{6.3}$$

when $|\lambda_s| \leq \mu \lambda_n$, it indicates that there is no relative sliding; when $|\lambda_s| = \mu \lambda_n$, it indicates that the relative sliding takes place between the contact face.

6.3 HIGH ARCH DAM BODY–FOUNDATION SYSTEM SEISMIC RESPONSE ANALYSIS METHOD

6.3.1 HIGH ARCH DAM BODY–RESERVOIR WATER–FOUNDATION SYSTEM NUMERICAL MODEL

The high arch dam system includes the dam body foundation, reservoir water, and interactive comprehensive system. Seismic response analysis mainly deals with arch dam deformation and stress distribution under the seismic action, interaction between the dam body and foundation, interaction between dam body and reservoir water, and other problems. The existing research indicates that since reservoir bottom absorbs energy, particularly being very remarkable to the sediment-laden river, the foundation frequency of high arch dams being smaller than that of compressible reservoir water body, but both frequencies being smaller than that of bed rock foundation, it is impossible for the compressible reservoir water to have the occurrence of resonance phenomena so that reservoir water hydrodynamic pressure under the seismic action can be used as the additional quality to be considered, whereby simplifying the problem of liquid and solid coupling and mainly taking the structure–foundation interacting problem into full consideration.

To begin with, the foundation can be divided into the near regional and far regional foundations. The dam foundation can be extended 1–1.5 times the dam height to two horizontal directions and the deep depths to form the limited foundation range serving as the near regional foundation, of which considerations must be given to the two banks' land forms and uneven rock behaviors of rock mass as well as the small-scale faults, joints, and weak rock mass and various geological structures affecting the dam body, including dam abutment arch support two banks latent sliding rock blocks cut off by this kind of structure. The outer part foundation beyond the near regional foundation can serve as the far regional foundation, because the far regional foundation is far away from the dam body, which can be viewed as the homogeneous linear elasticity infinite continuous media mainly in considering its energy escaped in dispersion as the radiation damping effect. As previously described, at present in the analysis of large dam visco-elasticity boundary based on the substructure method boundary, realizing the wave motion accurate simulation in the original continuous media must guarantee the agreement of transfer performances at the artificial boundary face with that in the original continuous media. That is, there is no reflect effect when the wave passes through the artificial boundary face, whereby making the outward traveling wave or be absorbed by visco-elasticity boundary.

The arch dam is a kind of high secondary super-static structure, whose dam type is, in general, determined by an arch ring control equation that can be divided into three-center circular arch upstream and downstream face parabolic arch, logarithmic spiral arch, elliptical arch, double curved

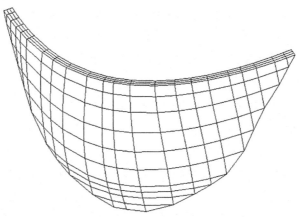

FIGURE 6.1 Finite Element Model of Arch Dam Body

arch, arch ring central parabolic arch, and so on. As far as arch dam finite element classification is concerned, the octal nodal point 3-D element is used, in general. First of all, it is necessary to determine some curved surface as the reference curved surface. In general, the vertical curved surface passing through the crown upstream side can be served as the reference curved surface of the dam body. When the design provides the geometric performance data for each design elevation arch rings, geometric shapes of the dam body can be determined. When the elevation of finite element mesh level cut-off surface is different from the design elevation, the interpolation method can be adopted to compute mesh elevation arch ring geometric parameters. At present, largely owing to the limitation of computation scale, the element dimension in arch dam 3-D finite element model can be 20–30 m in general. In the thickness direction of the dam body, the 3–5 layer elements are designed. Figure 6.1 is a finite element mesh schematic diagram of a practical arch dam body. When the arch dam is poured in blocks, it is required that the consideration be given to the existence of transverse interstice in computation. When the dam body finite element model is established, the double nodal point simulation is adopted for these interstice surfaces. If the arch dam finite element model cannot simulate all the transverse interstices, it is, generally, suggested that the real spacing be used as possible to simulate transverse interstices at the arch crown and one-quarter or so of arch ring near locations because the largest openings of the transverse interstices in the process of earthquakes emerge or appear in these locations.

As far as the division of finite element of near regional foundation is concerned, first of all, it is required that the dam foundation interface mesh be in agreement with the dam body, and then it is necessary to consider the division of dam body materials and the position and strike. The near regional foundation range taken is, in general, the left and right banks of the dam body, the upstream and downstream sides, the left and right cross-river direction, along-river direction and vertically extending 1.5–2 times the dam height below the dam foundation, respectively. As to the arch dam, the thrust of the arch acts downstream and the range of extending upstream at the dam body upstream side is slightly small. In addition, as far as the dangerous block bodies on the left and right banks related to the abutment stability in bed rock are concerned, the sliding fracture face needs setting up double nodal points to carry out simulation in the computation as when dam foundation interface fractures are considered,

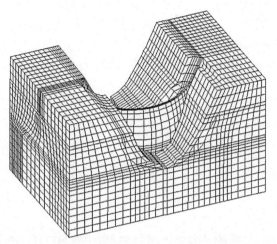

FIGURE 6.2 Some Arch Dam Body-Near Regional Foundation Finite Element Mesh Model

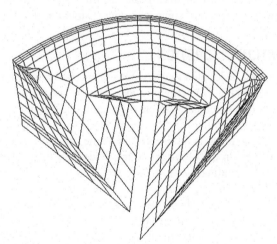

FIGURE 6.3 Relative Position Schematic Diagram of Considering Sliding Fracture Block Body and Dam Body in Foundation

the double nodal points need to be set up on the dam foundation interface. Figure 6.2 is the arch dam body–near regional foundation finite element mesh. Figure 6.3 is a schematic diagram for considering bedrock sliding blocks and dam body relative position.

The computation program for this chapter adopts the artificial homology boundary method (Zheng-peng, 2002) to consider far regional site radiation damping. In order to adopt the artificial homology boundary method, the effect of foundation infinite region upon large dam seismic response must be counted so that the artificial homology boundary zones consisting of 3–5 layers of elements should be set up. The connection of the homology boundary zone with the near regional foundation finite element mesh must be completely overlapped while the thickness of each outward extending layer element is

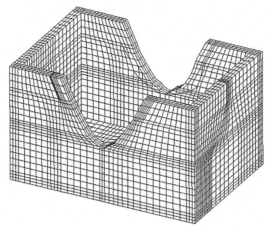

FIGURE 6.4 Homology Boundary Zone Mesh Diagram of Some Arch Dam System Finite Element Model

the same so as to use the homology formula to carry out the time–space extrapolation. Figure 6.4 is the homology boundary zone mesh diagram of some arch dam system finite element model.

6.3.2 DYNAMIC EQUATION DISCRETION

When the finite element method is adopted to carry out seismic wave motion response analysis of the near site zone, it can be based on the following physical concept: wave transmission velocity is finite, from which it can deduce, in finite element discrete mesh, any nodal point motion in the current time and the previous several times within the neighboring regions. As far as the inner nodal point motion equation is concerned, its natural discrete modes should be the time and space decoupling. In order to realize the wet-type solution within the time region, it is necessary to adopt the concentrated quality assumption of system quality including the upstream dam face reservoir water concentrated quality. The concentrated quality wet-type finite element nature deduces the next time nodal point motion from the current nodal point motion equation. Therefore, it is unnecessary to carry out the general assembly of rigidity, quality, and damping matrix, whose right end item formation only needs cumulating the effective loading vector contribution of each element on the element grade in such a way that the whole computation is carried out on the element grade I. Therefore, it only needs a very small high velocity register zone, with high computation efficiency, particularly when a series of element rigidity matrix, quality matrix, and damping matrix are the same, repeating computations can be avoided so as to make efficiency much higher.

When the time discrete is made of the dynamic problem, the central difference method in the direct integration is adopted, being suitable for the wave transmission problem to obtain the solution. It has been found through the analysis of recursion formula in central difference method that when [M] and [C] are the diagonal matrix, computation type is explicit formulation. For instance, the given nodal points have the initial disturbances, and after a step size Δt, the nodal points related with them will enter into motion, that is, these nodal points corresponding to the components in [u] will become nonzero quantity. With the time advancing, the rest of the nodal points will enter into motion in terms of this law by order. Therefore, this behavior is in agreement with wave transmission behavior.

Adopting the central difference method obtaining decoupling inner nodal point motion recursion type condition is quality matrix [M] and damping matrix [C], which are the diagonal matrix, that is, it is required to adopt the finite element model of the concentrated quality and concentrated damping. As to the quality matrix, the accuracy of wave motion simulation in the wave motion numerical simulation requires the finite element dimensions and wave length to be small enough so that when the dynamic balance equation is established, the space variation of inertia volume force within each finite element can be neglected. It can be seen from this that when the inner nodal point motion equation is established, the adoption of concentrated quality model is of rationality. The quality matrix [M] mentioned in this book will adopt the concentrated [C]; it is difficult to express with the diagonal matrix in most conditions. If Rayleigh's damping conditions are adopted, $[C] = a[M] + b[K] \times [C]$ are bound to be the nondiagonal matrix; for this reason, under this condition, the wet-type integration mode is established.

The motion balance equation for t time is established as follows:

$$[M][\ddot{u}]_t + [C][\dot{u}]u_t = |Q| \tag{6.4}$$

where [M] is quality matrix; [C] is damping matrix; [K] is rigidity matrix; $\{\ddot{u}\}_t$ and $\{\dot{u}\}_t$ are medium motion acceleration and velocity at t time element nodal point, respectively; [Q] is outer loading at t time.

In the time discretion, time step size Δt is adopted; as to the motion Eq. (6.4), the central difference method can be used for discrete velocity and acceleration into the following:

$$\{\dot{u}\}_t = \frac{1}{2\Delta t}\left(\{u\}_{t+\Delta t} - \{u\}_{t-\Delta t}\right) \tag{6.5}$$

$$\{\ddot{u}\}_t = \frac{1}{\Delta t^2}\left(\{u\}_{t+\Delta t} - 2\{u\}_{t-\Delta t}\right) \tag{6.6}$$

We can obtain the following:

$$\{\ddot{u}\}_t = \frac{2}{\Delta t^2}\left(\{u\}_{t+\Delta t} - 2\{u\}_t\right) - \frac{2}{\Delta t}\{\dot{u}\}_t \tag{6.7}$$

Substituting Eq. (6.7) into Eq. (6.4), we can derive the following:

$$\{u\}_{t+\Delta t} = \frac{1}{2}\Delta t^2 [M]^{-1}\left(\{Q\}_t - \{F\}_t\right) + \{u\}_t + \left(\Delta t[I] - \frac{1}{2}\Delta t^2 [M]^{-1}[C]\right)\{\dot{u}\}_t \tag{6.8}$$

where [I] and [K] are the same order element square matrix, $\{F\}_t = [K]\{u\}_t$ are the recovery force item. In order to obtain computation $\{\dot{u}\}_{t+\Delta t}$ explicit integration format, Xiaojun et al. (1992) adopted Newmark frequent mean acceleration mode to obtain the following mean approximate formula:

$$\frac{\{\ddot{u}\}_{t+\Delta t} + \{\ddot{u}\}_t}{2} = \frac{\{\dot{u}\}_{t+\Delta t} + \{\dot{u}\}_t}{\Delta t} \tag{6.9}$$

$$\frac{\{\dot{u}\}_{t+\Delta t} + \{\dot{u}\}_t}{2} = \frac{\{u\}_{t+\Delta t} + \{u\}_t}{\Delta t} \tag{6.10}$$

Parallel to seeking for a solution to t and $t + \Delta t$ time motion equation, we will obtain the following:

$$\Delta t\{\ddot{u}\}_{t+\Delta t} = \Delta t\{\ddot{u}\}_{t} + \frac{1}{2}\Delta t^2[M]^{-1}\left(\{Q\}_{t+\overline{\Delta}}\{F\}_{t+\Delta t} + \{Q\}_{t} - \{F\}_{t}\right) + \Delta t[M]^{-1}[C]\left(\{u\}_{t} - \{u\}_{t+\Delta t}\right) \quad (6.11)$$

Paralleling Eq. (6.8) with Eq. (6.11) to seek a solution, it will become the first format in damping matrix [C] as the nondiagonal matrix for the combination of the central difference with the mean approximation. Since this format shows a slight complexity in computation, it can be simplified, and hence the nodal point velocity recursion formula can be directly derived from the mean approximate Eq. (6.10) as follows:

$$\{\dot{u}\}_{t+\Delta t} = -\{\dot{u}\}_{t} + \frac{2}{\Delta t}\left(\{u\}_{t+\Delta t} - \{u\}_{t}\right) \quad (6.12)$$

Equations (6.8) and (6.12) can compose a self-starting wet-type difference format to seek a solution to the finite freedom and damping system dynamic Eq. (6.4). This is the second format in combination with the central difference with the mean approximation being more economical in calculation than the first format. The response acceleration can be determined by Eq. (6.9) to deduce the following equation:

$$\{\ddot{u}\}_{t+\Delta t} = -\{\ddot{u}\}_{t} + \frac{2}{\Delta t}\left(\{\dot{u}\}_{t+\Delta t} - \{\dot{u}\}_{t}\right) \quad (6.13)$$

In the explicit integration format, the central difference format is adaptable to damping matrix [C] being diagonal matrix situations, whereby proving that the cut-off errors of the central difference format recursion formula is of quantity grade $O(\Delta t3)$ so that this format is of binary grade accuracy. The central difference and the mean approximation combined format takes displacement and velocity as the variable's double-variable format. Whether it is the first format or the second format, the cut-off error quantity grade of displacement recursion equation is $O(\Delta t3)$ while the cut-off error quantity grade of velocity recursion equation is $O(\Delta t2)$. Accordingly, the accuracy of this kind of format is between grade I and grade II.

6.4 ARTIFICIAL HOMOLOGY BOUNDARY IMPLEMENTATION METHOD
6.4.1 HOMOLOGY FORMULA BASED ON FINITE ELEMENT MESH NODAL POINT

In the finite element analysis, the research objective (for instance, dam body and the finite range of foundation) is to carry out discretization. Then, nodal points in the discretization model can be divided into the interior points and the artificial boundary nodal points. The concentrated quality finite element model can be adopted to establish the nodal point motion equation of interior points. The artificial homology boundary can combine the finite element interior point motion equation with the artificial boundary nodal point extrapolated method. The finite computation model is used to simulate the wave motion process in the infinite medium.

With a single wave motion of one point on the transcendental artificial boundary as an example, let x-axis on the considered boundary point vertical to the artificial boundary and pointing to the exterior infinite region, the single wave along the sight transmission common expression is as follows:

$$u(t,x) = \sum_{j} f_j(c_{x_j}t - x) \quad (6.14)$$

where displacement $u(t,x)$ is the function of t,x, consisting of any single subwave motions $f_j(c_{x_j} t - x)$ (of which $j = 1,2,....$). These single subwave motion sight velocities are different in general; f_j is the unknown random function.

Equation (6.14) is of general characteristic, because these plane waves along the x-axis transmission sight velocity c_{x_j} and wave shape $f_j(*)$ are not restricted, and they are all unknown; what is known are that all these plane waves are the outer traveling waves, and after they are overlaid, the total wave field formed can satisfy the motion differential equation and physical boundary conditions, while there are no restrictions made to these differential equations and physical conditions. Since the total wave field displacement is known in wave motion numerical simulation, the components of each group are unknown, and it is just at this time that the direct simulation Eq. (6.14) should be passed through, and the displacement field u near the boundary point is used to form local artificial boundary conditions, and a common artificial wave velocity can be introduced so as to replace all wave motion components different sight wave velocity in their implementation.

Considering only one component in Eq. (6.14), we will have the following:

$$u_j(t,x) = f_j\left(c_{x_j} t - x\right) \tag{6.15}$$

It can be known from Eq. (6.15) that u_j is the wave motion independent variable $(c_{x_j} - x)$ function, if Δt is the assumed time step distance, in $t + \Delta t$ time, there will be the following:

$$u_j(t+\Delta t,x) = f_j\left(c_{x_j}(t+\Delta t) - x\right) = f_j\left(c_{x_j} t + c_{x_j}\Delta t - x\right)$$
$$= f_j\left(c_{x_j} t - \left(x - c_{x_j}\Delta t\right)\right) = u_j\left(t, x - c_{x_j}\Delta t\right) \tag{6.16}$$

Equation (6.16) indicates that the incidence wave along the x-axis sight transmission may be replaced through using point $x - c_{x_j}\Delta t$ at the t time and simulated by point x at $t + \Delta t$ time. But c_{x_j} is unknown so that it cannot give out the results, if j subwave physical sight velocity c_{x_j} is the artificial velocity c_a is replaced, and then, there will be the following:

$$u_j(t+\Delta t,x) = u_j\left(t, x - c_a\Delta t\right) + \Delta u_j(t+\Delta t,x) \tag{6.17}$$

As to j subwave, of which the error item will be:

$$\Delta u_j(t+\Delta t,x) = u_j(t+\Delta t,x) - u_j\left(t, x - c_a\Delta t\right)$$
$$= u_j\left(t, x - c_{x_j}\Delta t\right) - u_j\left(t, x - c_a\Delta t\right)$$
$$= \Delta u_j\left(t, x - c_{x_j}\Delta t\right) \tag{6.18}$$

Comparing Eq. (6.18) with Eq. (6.16) indicates that this error item has the same transmission velocity and transmission direction as the original single subwave motion, being the transmission wave motion with the wave velocity c_{x_j} along the x-axis. Therefore, the same Eq. (6.17) expresses the error as follows:

$$\Delta u_j(t+\Delta t,x) = \Delta u_j\left(t, x - c_a\Delta t\right) + \Delta^2 u_j(t+\Delta t,x) \tag{6.19}$$

If t and x in Eq. (6.18) are replaced by $t - \Delta t$ and $x - c_a\Delta t$ respectively, we will obtain the following:

$$\Delta u_j\left(t, x - c_a\Delta t\right) = u_j\left(t, x - c_a\Delta t\right) - u_j\left(t - \Delta t, x - 2c_a\Delta t\right) \tag{6.20}$$

Substituting Eq. (6.19) and Eq. (6.20) into Eq. (6.17), we will obtain the following:

$$u_j(t+\Delta t,x) = 2u_j(t,x-c_a\Delta t) - u_j(t-\Delta t,x-2c_a\Delta t) + \Delta^2 u_j(t+\Delta t,x) \qquad (6.21)$$

It is easy to prove that the second-order error item $\Delta^2 u_j(t+\Delta t,x)$ transmits with the wave velocity c_{x_j} in the same way along the x-axis; this reason is made by analogy until N order error item is introduced:

$$u_j(t+\Delta t,x) = \sum_{n=1}^{N} (-1)^{n+1} C_n^N u_j \left[t-(n-1)\Delta t, x-nc_a\Delta t \right]$$
$$+\Delta^N u_j(t+\Delta t,x) \qquad (6.22)$$

where $C_n^N = \dfrac{N!}{(N-n)!n!}$

After omitting the error item, there will be

$$u_j(t+\Delta t,x) = \sum_{n=1}^{N} (-1)^{n+1} C_n^N u_j \left[t-(n-1)\Delta t, x-nc_a\Delta t \right] \qquad (6.23)$$

Equation (6.23) is the j subwave N order multihomology formula, while the multi-sub-wave overlaying conform the general incident wave motion, for selecting the common artificial wave velocity can simulate every single subwave, having no concrete details with each subwave so that it is applicable to the general incident wave motion, there will be the following formula:

$$u(t+\Delta t,x) = \sum_{n=1}^{N} (-1)^{n+1} C_n^N u \left[t-(n-1)\Delta t, x-nc_a\Delta t \right] \qquad (6.24)$$

It can be seen from the deducing process that the multihomology formula is characterized by the following.

Multihomology formula is completely derived from the single direction wave motion expression equations, having no special performances with the infinite regional control difference equations and physical boundary condition so that it is of universal significance, that is, the same formula can be used in scalar quantity and vector wave. It can also be used in 2-D and 3-D problems, and it can be used not only in isotropy medium but also in anisotropy; it can be used in mono-phase medium and in multiphase medium; it can be used in artificial boundary nodal points either in uniform medium or in division interface of different medium; it can be used in the homology formula being not only displacement but also other wave motion physical quantities such as velocity, stress, and so on.

It can be characterized with the time and space decoupling explicit one-dimensional form, being very conveniently realized on computers.

The higher the number of orders is, the smaller the error is, and the higher the accuracy is. The multihomology formula accuracy is controllable. Selecting the appropriate homology order N can make the accuracy of the homology formula keep in agreement with that of motion equation of interior nodal points.

6.4.2 COMPUTATION METHOD OF BOUNDARY NODAL POINT DISPLACEMENT

The near field wave site consists of free field u_f and scattering field u_s:

$$u = u_f + u_s \tag{6.25}$$

where u_f is the solution to the uniform elasticity semispace problem, including the contribution of the incident wave and the uniform semispace free face reflection wave; u_s is an uneven medium of structure existence or local uneven medium, the scattering wave field caused by irregular landforms.

In the artificial boundary zone, u_s is the exterior transmission wave through the artificial boundary to the infinitely far region. Before its use, the introduced artificial nodal point current scattering wave value $u_s^{t+\Delta t}$ can be used, and the boundary nodal point neighboring several points of known scattering wave values prior to several times can be derived through the extrapolation method, with the second-order homology formula as an example to express as follows:

$$
\begin{aligned}
u_s\left(x_b, y_b, t+\Delta t\right) = {} & 2t_1 u_s\left(x_b, y_b, t\right) + 2t_2 u_s\left(x_b - \Delta x, y_b, t\right) + 2t_3 u_s\left(x_b - 2\Delta x, y_b, t\right) \\
& - t_1 u_s\left(x_b - J\Delta x, y_b, t-\Delta t\right) - t_2 u_s\left[x_b - (J+1)\Delta x, y_b, t-\Delta t\right] - t_3 u_s \\
& \times \left[x_b - (J+2)\Delta x, y_b, t-\Delta t\right] + t_3 u_s\left[x_b - (J+2)\Delta x, y_b, t-\Delta t\right]
\end{aligned} \tag{6.26}
$$

where

$$t_1 = \frac{(1-S)(2-S)}{2}, \quad t_2 = S(2-S), \quad t_3 = \frac{S(S-1)}{2}, \quad S = \frac{c_a \Delta t}{\Delta x}.$$

J takes its value as 1, 2, 3, thus making $x_b - (J+2)\Delta x$ be larger than the minimum value of $x_b - 2c_a \Delta t$.

Therefore, $t+\Delta t$ time artificial boundary nodal point total displacement field $u^{t+\Delta t}$ can be obtained from the free field displacement $u_f^{t+\Delta t}$ and Eq. (6.26) calculated scattering wave displacement sum.

6.4.3 CONCRETE COMPUTATION STEPS OF ARTIFICIAL BOUNDARY NODAL POINT DISPLACEMENT WHEN SEISMIC WAVE IS INCIDENT

1. Displacement response of each nodal point $t+\Delta t$ time within the zone is computed by the dynamic finite equation.
2. Each point free field displacement in $t+\Delta t$ moment artificial boundary zone is computed.
3. As for the artificial boundary zone, $t+\Delta t$ moment and related previous several moments total displacement fields minus the corresponding free field displacement can obtain the multihomology formula (here, the second-order homology formula is adopted), and then, the previous several moments scattering displacement fields of artificial boundary nodal points are needed by the extrapolated $t+\Delta t$ moments scattering wave displacement.
4. The homology formula is used to compute $t+\Delta t$ moment artificial boundary nodal point homology displacement.
5. The artificial boundary nodal point $t+\Delta t$ moment scattering field displacement plus the free field displacement can obtain $t+\Delta t$ movement artificial boundary nodal point total displacement.
6. Return to Step 1.

6.4.4 NUMERICAL SIMULATION OF ARCH DAM CONTACT NONLINEAR PROBLEM

The opening and closing and slip of transverse cracks within concrete dams and various crack faces in the foundations under the seismic actions have fallen into the boundary nonlinear contact problems, being necessary to satisfy the displacement on the special boundary contact face and the contact condition of contact force so that they can be reviewed as the foundational extreme value problems with constraints.

6.4.4.1 Boundary constraint conditions and current states of seeking a solution method to arch dam system contact problems

Under the seismic action, the control equation of the motion contact problems of the crack or interstice interface within the arch dam foundation system basically belongs to the low-speed collision and linear-elastic small deformation range, being without counting pseudo-static contact with speed changes before and after the collisions caused by the impulsion. In the contact conditions, the normal direction relative displacement conditions among the contact faces can guarantee that there will be no occurrence of intermosaic phenomena between the contact faces. As to tangential contact force, Mohr–Coulomb's law is used to describe the static motion sliding friction states before and after sliding motion.

In the process of dynamic load, the contact dynamic boundary is changing in such states as isolation, viscosity, and sliding motion, whereas different contact states are corresponding to different constraint conditions, and this constitutes the nonlinear dynamic contact problem. In addition, the static contact problem itself is also a stepwise problem so that the method of handling the contact problem is, at the same time, to adopt the static and dynamic analysis.

At present, the contact element method is extensively used in research and engineering application of transverse cracks or interstices of the arch dam body. The fictitious nonlinear contact element is set up to simulate the interstice face contact phenomena, whose constitution relation depends upon the contact states judged in terms of the relative positions among the contact faces. The contact elements need assuming moral and tangential rigidities, whose natures should be corresponding to the penalty function method in the universal extreme value problem. It is necessary to make the normal direction rigidity of the contract element be big enough in order to render the contact faces to have no insertions in the contact faces, but the contact force to produce oscillation with high frequency so as to affect the stability of computation results. In considering that there are no keyways set in between arch dam transverse interstices, the very large tangential rigidity of the contact elements has been taken. In fact, since the contact face keyways are inserted with inclines. When the transverse cracks open, there exists tangential sliding space in keyways. Therefore, there must be a certain normal direction relative displacement, under which there has been an assumption to decrease the tangential rigidity by a large margin. It can be seen from this that there exists a certain randomization in the contact element taking values. In theory, it is entirely impossible to simulate the contact situation accurately.

At present, the substructure method is adopted to seek the solution in the method of contact element to simulate arch dam transverse interstices. It is necessary to aggregate the finite element nodal points on the dam body contact faces afterward and the concealed nonlinear interactive method is employed again. In fact, a high arch dam has more transverse interstices, but in the past, the effect of transverse interstices in the dam body model upon the stress distribution of the dam body was considered, holding that when the number of transverse interstice simulation is over 3–5 cracks, the influences are not serious. However, in engineering practice and under the action of strong earthquake, the opening degree of overlarge transverse interstice is likely to cause damage to transverse crack water sealing structures. If there are only a few transverse cracks or interstices, they will result in opening concentration but larger than the real ones. When more transverse interstices are considered in the arch dam body model, the nodal points will increase on the contact face, whereby causing an increase in working quantity to concerted-type nonlinear interactive seeking solution. In addition, there have been the methods to adopt contact elements to simulate arch dam transverse interstices, in which the quality-free foundation assumption is still adopted, and the adoption of substructure concealed type method to seek the solution is not convenient to change the transverse crack position and quantity at random in the design scheme.

For this reason, the motion contact force model has been introduced into the research on seismic response to arch dam body transverse interstices and bedrock slipping crack faces, whose basic idea is directly to consider the motion contact force among the contact faces. The equation to seek the solution to motion contact force can be directly derived through the introduction of geometric and physical conditions on the motion contact faces, being in nature, corresponding to Lagrange's method of multipliers in the universal function extreme value problem. This method in theory may accurately satisfy the contact formulation to seek a solution method in the time region of the whole system.

6.4.4.2 Arch dam transverse interstice motion contact face force model
In the process of carrying out dynamic computation stepwise integration with the explicit formulation method, the variation inequality principle for the contact constraint to directly seek a solution can structure the linear complementary equation. In each step, computation of time regional integration through the decomposition of contact point displacement, the normal direction contact force and its caused nodal point displacement, and the tangential friction force and its caused displacement, are calculated respectively so as to obtain the contact positive stress on the contact face dynamic response among the contact faces in considering interface friction effects, whose concrete steps for seeking solutions are as follows:

6.4.4.2.1 Decomposition of contact point displacement
As far as the interstice face contact problem is concerned, as shown in Fig. 6.5, letting arbitrary interstice face S exist in the medium, the interfaces at S on two sides are known as S^+ and S^-. In the initial time, S^+, S^- are overlaid. If interstice faces vary smoothly, the normal vector $\bar{N}$ and tangential vector $\vec{\tau}$ exist (of which $\bar{N}$ directs N^+ direction). When the structure finite element discretion is carried out, the nodal point distribution of interstice's two ends are required to be the same in such a way that the stress caused by motion contact is able to emerge as the form of interaction force of nodal point pairs to match with the explicit finite element method. In the process of the inner point dynamic computation stepwise integration, the boundary contact condition can be directly applied to the point pairs on the interstice faces, whereby obtaining the contact force between two points as well as real displacement of the contact points.

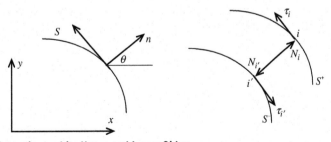

FIGURE 6.5 Contact Interstice and its Upper and Lower Sides

In explicit integration, the second format of the center difference method combined with average approximation method introduced in Section 6.3.2 can be adopted. Letting the contact point pairs i and i', of which i is located at S^+ side and i' is located at S^- side. First of all, as to any contact point i, it is necessary to let its upper subject to normal contact force N_i and tangential contact force τ_i actions so as to list the expression formula to seek for the solution to displacement with an aim at i nodal point j freedom, there will be the following:

$$u_{ij}^{t+\Delta t} = \frac{\Delta t^2}{2m_i}\left[\left(Q_{ij}^t + N_{ij}^t + \tau_{ij}^t\right) - \sum_{l=1}^{n_e}\sum_{k=1}^{n} K_{ijlk}\, u_{lk}^t - \sum_{l=1}^{n_e}\sum_{k=1}^{n} C_{ijlk}\,\dot{u}_{lk}^t\right] + u_{ij}^t + \Delta t\dot{u}_{ij}^t \tag{6.27}$$

where $N_{ij}, \tau_{ij} - N_i, \tau_i$ is the component on j direction, N_{ij}^t, τ_{ij}^t are in the motion state function, which are related not only to t and t previous moment motion state, but also to $t + \Delta t$ moment motion state so that there is no way for Eq. (6.27) to seek a solution to $u_{ij}^{t+\Delta t}$ directly.

Hence, $u_{ij}^{t+\Delta t}$ can be divided into three parts:

$$u_{ij}^{t+\Delta t} = \bar{u}_{ij}^{t+\Delta t} + \Delta u_{ij}^{t+\Delta t} + \Delta v_{ij}^{t+\Delta t} \tag{6.28}$$

$$\bar{u}_{ij}^{t+\Delta t} = \frac{\Delta t^2}{2m_i}\left[Q_{ij}^t - \sum_{l=1}^{n_e}\sum_{k=1}^{n} K_{ijlk}\,u_{lk}^t - \sum_{l=1}^{n_e}\sum_{k=1}^{n} C_{ijlk}\,\dot{u}_{lk}^t\right] + u_{ij}^t + \Delta t\dot{u}_{ij}^t \tag{6.29}$$

It may directly obtain the above from the previous motion state.

Letting $M_i = \dfrac{2m_i}{\Delta t^2}$, there will be the following:

$$\Delta u_{ij}^{t+\Delta t} = \frac{\Delta t^2}{2m_i}N_{ij}^t = \frac{N_{ij}^t}{M_i} \tag{6.30}$$

$$\Delta v_{ij}^{t+\Delta t} = \frac{\Delta t^2}{2m_i}\tau_{ij}^t = \frac{\tau_{ij}^t}{M_i} \tag{6.31}$$

which will be determined by the joint motion contact state.

In seeking a solution to the contact problem, and with the nodal point i pair and i as the research object, all the i and i' serve as the below labeled variables representing the column vectors consisted of components of i and i' points on the freedom in the following text (except for M_i and M_i')

6.4.4.2.2 Computation of normal contact force and caused nodal point displacement

To begin with, the normal contact constraint conditions are detailed in the following.

When i and i' happen to contact with each other, i and i' contact forces are equal to magnitude but opposite in direction:

$$N_i^t = -N_{i'}^t \tag{6.32}$$

$$\tau_i^t = -\tau_{i'}^t \tag{6.33}$$

At the same time, i and i' displacements satisfy the displacement coordination conditions when in contact, that is, normal requirement of non-inter-intrusion, expressing it as follows:

$$\bar{u}_i^T \left(u_{i'}^{t+\Delta t} - u_i^{t+\Delta t} \right) = 0 \tag{6.34}$$

When i and i' are separated, there will be as follows:

$$N_i^t = N_{i'}^t = 0 \tag{6.35}$$

$$\tau_i^t = \tau_{i'}^t \tag{6.36}$$

$$u_i^{t+\Delta t} = \bar{u}_i^{t+\Delta t} \tag{6.37}$$

$$u_{i'}^{t+\Delta t} = \bar{u}_{i'}^{t+\Delta t} \tag{6.38}$$

In the computation process, $\bar{u}_i^{t+\Delta t}$, $\bar{u}_{i'}^{t+\Delta t}$ are obtained from Eq. (6.29), and the following formula judges whether contact is to happen or not:

$$\bar{u}_i^T \left(\bar{u}_{i'}^{t+\Delta t} - \bar{u}_i^{t+\Delta t} \right) \geq 0 \tag{6.39}$$

When Eq. (6.29) is not satisfied, i and i' cannot happen to contact, normal contact force and tangential friction force is zero, $\bar{u}_i^{t+\Delta t}$, $\bar{u}_{i'}^{t+\Delta t}$ are the total displacement. When Eq. (6.39) is established, i and i' happen to contact with nonintrusion of normal conditions, there will be the following:

$$\bar{n}_i^T \left(\bar{u}_{i'}^{t+\Delta t} + \Delta u_{i'}^{t+\Delta t} + \Delta v_{i'}^{t+\Delta t} - \bar{u}_i^{t+\Delta t} - \Delta u_i^{t+\Delta t} - \Delta v_i^{t+\Delta t} \right) \geq 0 \tag{6.40}$$

where $\bar{n}_i^T \Delta v_{i'}^{t+\Delta t} = \bar{n}_i^T \Delta v_i^{t+\Delta t} = 0$, and let $\Delta_{1i} = \bar{n}_i^T \left(\bar{u}_{i'}^{t+\Delta t} - \bar{u}_i^{t+\Delta t} \right)$ therefore

$$\bar{n}_i^T \left(\Delta u_i^{t+\Delta t} - \Delta u_{i'}^{t+\Delta t} \right) = \Delta_{1i} \tag{6.41}$$

Substituting Eqs. (6.30) and (6.32) into Eq. (6.41), there will be the following:

$$N_i^t \frac{M_i M_{i'}}{M_i + M_{i'}} \bar{n}_i \Delta_{1i} \tag{6.42}$$

And then, Eqs (6.30) can be used to compute $\bar{u}_i^{t+\Delta t}$, $\bar{u}_{i'}^{t+\Delta t}$.

6.4.4.2.3 Computation of tangential friction force and its caused displacement

When the judgment i and i' happen to contact in normal calculation, the tangential friction force need calculating. The following expression equations are listed in detail.

When i and i' are in static friction states, there are no relative slips between two points:

$$\bar{t}_i^T \left(u_i^{t+\Delta t} - u_{i'}^{t+\Delta t} \right) = \bar{t}_i^T \left(u_i^t - u_{i'}^t \right) \tag{6.43}$$

At the same time, static friction force can satisfy Coulomb's law:

$$\left| \tau_i^t \right| \leq \mu_s \left| N_i^t \right| \tag{6.44}$$

where μ_s is static friction coefficient.

When i and i' are in dynamic friction state, there will be the following based on friction law:

$$\left| \tau_i^t \right| \leq \mu \left| N_i^t \right| \tag{6.45}$$

where μ is dynamic friction coefficient.

In the calculation process, first of all, assuming that i and i' friction states are the same as the previous time step, and if they are static friction states, i and i' are in tangential nonrelative displacement, can substitute Eq. (6.28) into Eq. (6.43), we will have the formula as follows:

$$\overline{t}_i^T\left(\overline{u}_{i'}^{t+\Delta t}+\Delta u_{i'}^{t+\Delta t}+\Delta v_{i'}^{t+\Delta t}-\overline{u}_i^{t+\Delta t}-\Delta u_i^{t+\Delta t}-\Delta v_i^{t+\Delta t}\right)=\overline{t}_i^T\left(u_{i'}^t-u_i^t\right) \tag{6.46}$$

where $\overline{t}_i^T\Delta u_{i'}^{t+\Delta t}=\overline{t}_i^T\Delta u_i^{t+\Delta t}=0$, and let $\Delta_{2i}=\overline{t}_i^T\left[\left(\overline{u}_{i'}^{t+\Delta t}-\overline{u}_i^{t+\Delta t}\right)-\left(u_{i'}^t-u_i^t\right)\right]$:

$$\overline{t}_i^T\left(\Delta v_{i'}^{t+\Delta t}-\Delta v_i^{t+\Delta t}\right)=\Delta_{2i} \tag{6.47}$$

Substituting Eqs (6.31) and (6.33) into Eq. (6.47), there will be the following:

$$\tau_i^t\frac{M_iM_{i'}}{M_i+M_{i'}}\overline{t}_i\Delta_{2i} \tag{6.48}$$

At the same time, Eq. (6.44) is needed to judge whether the static friction force value exceeds the allowed value or not. If the value exceeds the allowed one, it indicates that i and i' in between have transferred into the motion friction state. When the calculation is carried out in terms of i and i' being in dynamic friction state, Eq. (6.45) is used to compute dynamic friction force τ_i^t, whose sign is determined by Δ_{2i}. After τ_i^t is obtained, Eqs (6.30) and (6.31) are used to calculate $\Delta v_i^{t+\Delta t}$, $\Delta v_{i'}^{t+\Delta t}$.

In terms of i and i' being in dynamic friction state to carry out the calculation, i and i' need to be tested to see whether they coincide with the allowed conditions:

$$\overline{t}_i^T\left(u_{i'}^{t+\Delta t}-u_i^{t+\Delta t}\right)>\overline{t}_i^T\left(u_{i'}^t-u_i^t\right)[\text{when}\,\text{sgn}(\Delta_{2i})=1\text{h}] \tag{6.49}$$

$$\overline{t}_i^T\left(u_{i'}^{t+\Delta t}-u_i^{t+\Delta t}\right)<\overline{t}_i^T\left(u_{i'}^t-u_i^t\right)[\text{when}\,\text{sgn}(\Delta_{2i})=-1\text{h}] \tag{6.50}$$

If these formulas are not established, the nodal point pairs enter into frictions state, Eq. (6.48) will be used to recalculate τ_i^t.

From this calculation, $\overline{u}_i^{t+\Delta t}$, $\Delta u_i^{t+\Delta t}$, $\Delta v_i^{t+\Delta t}$ can be obtained, whereby calculating total displacement $u_i^{t+\Delta t}$ of contact points. This is the dynamic contact force model computation format, matching with the simulation and there exist no problems of artificial selection of contact rigidity and also the interintrusion phenomena in the contact faces will not happen, holding great superiority.

6.4.4.2.4 Realization of 3-D motion contact problem numerical computation

There is no difference between 3-D problem to normal contact force and seeking a solution to the caused displacement process and 2-D problem, but its tangential frictional force solution seeking needs to be especially treated because of the unknown tangential quantity. Figure 6.6 indicates the tangential behaviors of the 3-D contact problem.

In letting $\Delta=\left(\overline{u}_{i'}^{t+\Delta t}-\overline{u}_i^{t+\Delta t}\right)-\left(u_{i'}^t-u_i^t\right)$ be unthinkable $t+\Delta t$ moment contact force function, contact point pairs are the difference and t moment displacement difference. This difference value sum vector with its projection on the contact face point tangential plane is the possible occurrence of tangential friction direction. Let $\overline{n}$ represent the contact face normal line direction cosine vector $\{l_1,m_1,n_1\}^T$, Δ directional cosine vector Δ is $\{l_2,m_2,n_2\}^T$ and it is required that the tangential cosine vector of friction force direction be $\overline{t}=\{l_3,m_3,n_3\}^T$.

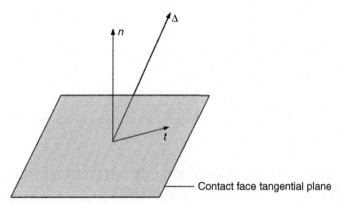

FIGURE 6.6 Tangential Sliding Direction of 3-D Contact Problem

Since $\bar{n}$ and $\bar{\Delta}$ may determine the plane, its plane equation can be:

$$\begin{vmatrix} x & y & z \\ l_1 & m_1 & n_1 \\ l_2 & m_2 & n_2 \end{vmatrix} = 0 \tag{6.51}$$

Equation (6.51) is developed to obtain $\bar{n}$ and $\bar{\Delta}$ determined plane equation:

$$(m_1n_2 - m_2n_1)x + (l_2n_1 - l_1n_2)y + (l_1m_2 - l_2m_1)z = 0 \tag{6.52}$$

Ordering $A = m_1n_2 - m_2n_1$, $B = l_2n_1 - l_1n_2$, $C = l_1m_2 - l_2m_1$, and then, $Ax + By + Cz = 0$, while $\bar{t}$ is in the plane determined by $\bar{n}$ and $\bar{\Delta}$, there will be the following:

$$l_3A + m_3B + n_3C = 0 \tag{6.53}$$

That is:

$$l_3(m_1n_2 - m_2n_1) + m_3(l_2n_1 - l_1n_2) + n_3(l_1m_2 - l_2m_1) = 0 \tag{6.54}$$

Again, $\bar{n}$ is vertical to $\bar{t}$ with each other, that is:

$$l_3l_1 + m_3m_1 + n_3n_1 = 0 \tag{6.55}$$

Again, it can be known from cosine vector that:

$$l_3^2 + m_3^2 + n_3^2 = 1 \tag{6.56}$$

Equations (6.54)–(6.56) may form the equation group to seek a solution to l_3, m_3, n_3, thus obtaining the following:

$$n_3 = \cfrac{1}{\left[1 + \left(\cfrac{l_2n_1^2 - l_1n_1n_2 - l_1m_1m_2 + l_2m_1^2}{m_1^2n_2 - m_1m_2n_1 - l_1l_2n_1 + l_1^2n_2}\right)^2 + \left(\cfrac{m_1n_1n_2 - m_2n_1^2 - m_2l_1^2 + l_1l_2m_1}{l_1l_2n_1 - n_2l_1^2 - n_2m_1^2 + m_1m_2n_1}\right)^2\right]^{\frac{1}{2}}} \tag{6.57}$$

$$l_3 = \frac{l_2 n_1^2 - l_1 n_1 n_2 - l_1 m_1 m_2 + l_2 m_1^2}{m_1^2 n_2 - m_1 m_2 n_1 - l_1 l_2 n_1 + l_1^2 n_2} n_3 \tag{6.58}$$

$$m_3 = \frac{m_1 n_1 n_2 - m_2 n_1^2 - m_2 l_1^2 + l_1 l_2 m_1}{l_1 l_2 n_1 - n_2 l_1^2 - n_2 m_1^2 + m_1 m_2 n_1} n_3 \tag{6.59}$$

Hence, obtaining $\bar{t} = \{l_3, m_3, n_3\}^T$, that is, the directional cosine of tangential friction force can be sought for its solution.

6.4.5 STATIC AND DYNAMIC COMBINATION COMPUTATION METHOD OF ARCH DAM–FOUNDATION SYSTEM CONTACT NONLINEARITY

When the dynamic analysis of the dam body system is carried out, attention should be paid to the weight itself, static water pressure, sediment pressure, and static loading functions, on the basis of which the dam body system will continue to bear the dynamic functions of earthquake motion. As to the linear structure, static and dynamic forces can be calculated separately, and then, the linear superposition can be carried out. However, as to the nonlinear situations existing in structure, since the superposition principle is no longer established, and after the static analysis must be completed, and on basis of static displacement and stress and strain fields, the consideration can be made of common functions of static loading and seismic wave inputs to carry out the calculation in the dynamic stage in such a way that the coupling function of static and dynamic loading can be considered under nonlinear conditions. Owing to arch dam seismic safety evaluation, the integrated static loading and seismic function effects are needed. In order to be able to combine the explicit finite element with local homology artificial boundary method, it is necessary to adopt static loading as the step function forms to apply to dam body-foundation system. After the displacement value is stable, the value is input into seismic waves by bedrock. Accordingly, the wave motion response analysis computation module for the system is carried out. This kind of simulation method is called as the static and dynamic combination computation method. The concrete process of static and dynamic combination computation method is as follows:

1. In response to computation structure under the static loading, these static loads include structure weight itself, temperature loading, seepage flow pressure, static water pressure, sediment pressure, and so on.
2. It is necessary to keep the static loading function of Step 1, and at the same time, to substitute the results of displacement response, contact information, and so on, for Step 1 and at this time to input the seismic motion displacement waves into seeking for the response to structure under the action of seismic loading. It is necessary to point out that just at this time, without application of seismic wave motion, this step response should be a straight line without changing in time.

   Seismic wave motion response analysis method and the dynamic contact force computation model carry out the contact problem seeking for a solution for the concrete large dam body–foundation system seismic numerical system foundation. It can be seen from this that this research project deals with and develops the high arch dam foundation system seismic response analysis method that can at the same time take a series of key factors intercorrelated in high arch dam systems into full consideration, including arch dam, dynamic interaction between foundation and reservoir water, foundation quality and energy escaping dispersion of radiation damping to the far regional foundation, reciprocal opening and closing between inner transverse interstices within dam body, the landforms of near region foundation of neighboring dam body including various geological structures and the latent sliding rock blocks, and so on. In combining with the previously described input mechanism of dam site seismic motion, the unevenness or nonuniformity of input space along dam foundation motion input should also be embodied, whereby making a breakthrough in a large number of being difficult in concert with practical limitations basically in the current arch dam aseismic design.

DAM ABUTMENT AND ARCH SUPPORT ROCK BLOCK STABILITY AND SEISMIC SAFETY EVALUATION OF HIGH ARCH DAM

7.1 ASEISMIC STABILITY ANALYSIS BEHAVIORS OF ARCH DAM ABUTMENT AND EARTHQUAKE DISASTER ENLIGHTENMENT

Dam abutment and arch support rock stability of high arch dams is the key to the whole engineering seismic safety and the main bases of evaluation. Aseismic stability analysis behaviors of arch dam abutment are as follows:

1. There can inevitably exist various types of faults, joints, cracks or fractures, and soft-weak structure planes of engineering geology in dam sites. Based on large-scale engineering geologic prospecting work on the dam site, the engineering geologic sci-tech personnel have classified the rock mass on both sides of two bank abutments of an arch dam into each latent sliding blocks cut-off from these structure planes. These are significantly different from the homogeneous soil body in which the most unfavorable sliding surface or face situations are sought for.
2. Dam abutment latent sliding blocks should, first of all, determine the bottom sliding plane and sliding surface, and assuming that a tension crack surface has been formed along the dam abutment upstream joint surface or soft and weak structure plane. The natural bank slope downstream is usually taken as the air face, or there exist many transversal river direction strike faults or soft and weak structure planes taken as transversal air faces, downstream, making the latent sliding blocks have the slip-movement geometric morphology.
3. The latent sliding blocks bearing the thrust transmitted by arch supports of the arch dam, the seepage pressure within the inner rock mass boundary face, self-weight of blocks themselves, and the seismic inertia force, as well as sliding resistance and reactive force produced by the rock mass shear strength of the side sliding face and the bottom sliding surface, whose sliding directions are along the intersecting line direction of the side sliding face and bottom sliding surface. When the normal stress on the side sliding face is the tensile stress, the sliding resistance disappears and the sliding direction is along the reluctant force acting on the rock blocks so that the sliding mode is changed into one-way sliding, whose sliding direction is decided by the shear stress reluctant direction on the bottom sliding surface.
4. Dam body and arch support thrust acting on sliding blocks are subject to the effect of rock deformation, so that it is necessary to take the deformation coupling influence between the dam body and abutment rock mass into full consideration. Though rock mass cannot lose stability, its

deformation is overly large, whereby resulting in dam body damages. This is an important feature that the abutment stability analysis of the arch dam is obviously different from the side slope stability analysis of mountain mass.

When an earthquake takes place, arch support trust, the amplitude value, and direction of seismic inertia force of abutment rock block are changing with time, whereby making the magnitude and direction of the resultant force acting on abutment rock block and its decides sliding mode and direction also are carrying with time. This is an important difference of dynamic and static stable analysis.

5. It is considered that an earthquake is a short-time reciprocating motion, reaching the limit balance leading to an instantaneous instability, which is called as limit stake. Although it is likely to affect the shear strength of sliding block mass, for the sliding movement is also in the reciprocating changes, the block mass is also likely to have a certain residual displacement after the earthquake. When resistance force produced by the block mass shear strength is still larger than the static active effect born by block mass, it can no longer continue to slide. As a result, local instantaneous instability can be uncertain to finally lead to the whole instability. This is another important difference of dynamic and static stability analysis.

6. The principal subject of dam abutment aseismic stability research is the dam body but not the latent sliding rock blocks on the dam abutment. This is the prerequisite for such problems as the aseismic stability definition, instability judgment evidence or bases, damage types, analysis method, and so on. For this reason, it is necessary to start with the influence of the dam body upon the aseismic safety, and to evaluate dam body seismic safety but not simply to discuss the dam abutment latent sliding rock mass stability from the mechanics stability in a theoretical concept.

Seismic disasters of the US Pacoima arch dam in the 1971 San Fernando and in the 1994 North Ridge strong earthquakes have fully embodied the mentioned performances or behaviors, whose main enlightenments are finalized as follows:

1. The arch dam is a higher-degree hyperstatic structure with a large regulation space of dam body local stress and better aseismic performances.

2. The first strong seismic disasters occurred mainly in the left abutment, and after the consolidation, in the second strong earthquake, the original place again suffered from the same seismic disasters, whereby fully demonstrating that the upper abutment of the arch dam is the weak location or part of earthquake resistance.

3. The dynamic amplification effect of abutment rock mass seismic motion in the two strong earthquakes is very apparent, with its peak ground acceleration exceeding $2.0g$, but there is no occurrence of the whole instability of dam body, and only the local fracture sliding move to produce deformation can lead to the dam body cracks.

4. In terms of making seismic response computations of the whole structure of the arch dam based on the real measurement of dam bottom acceleration, the dam body's largest arch direction tensile stress can reach as high as 4–5 MPa, but there are no cracks to be found in the dam body. This is because the expansion transverse cracks within the dam body are bound to be pulled open in a strong earthquake. Particularly in a strong earthquake, water depth is only two-thirds of the designed water depth so that the dam body arch direction tensile stress is greatly weakened. Under the seismic action, the opening and closing of dam body transverse cracks of the arch dam can

lead to the appearance of the largest tensile stress on the arch support dam foundation face part of two-bank abutments of the upper part arch dam.

Accordingly, the seismic cases indicate that it is different to account for the seismic disasters in a rational way with the traditional rigidity limit balance to check the abutment stability. In fact, it is likely that far before the abutment rock mass makes a whole rock mass slide, the dam abutment deformation caused by the local cracks and sliding moves has led to the serious cracks of the weak part of the arch support dam foundation face. The seismic disasters are mainly embodied in the dam body weak part because of insufficient strength to cause cracks. The usual application of safety coefficient for the dam abutment aseismic stability cannot reflect the real engineering aseismic ecology. It is only to base on the engineering experience as the relative analogy design to take the value standard.

7.2 BASIC CONCEPT OF DAM ABUTMENT ROCK BLOCK INSTABILITY SAFETY COEFFICIENT OF ARCH DAM

Consensus of instability safety coefficient is the prerequisite of abutment aseismic stability safety evaluation of arch dams. The structure bearing various kinds of functions, analysis method of deforming structure effects, and the resistance force of structure itself and safety judgment standards must be realigned.

The structure bearing capacity safety coefficient in design is an index to reflect the structure function effect, resistance force working conditions, and "the distance" among the limit states. There are two magnitudes to weigh or measure this distance: one is to increase the design function so as to make the design effect reach the limit states of design resistance state, that is, the so-called overloading safety coefficient; another is to decrease design resistance force so as to make the design effect reach the limit state, that is, the so-called decreasing strong safety coefficient. Strictly speaking, overloading or decreasing strong safety coefficient is only of practical and rich significance to linear elastic structure.

The problem of abutment aseismic stability involving abutment latent local fractures and sliding moves of arch dams is more complicated. Even if the material is itself of the dam body and rock mass is close to fragility damage to be regarded as the linear-elasticity basically, the transverse crack opening and closing of the dam body as well as the local cracks and sliding moves of abutment rock mass latent sliding block boundary in the whole system. As to the nonlinear system, even if an increase in each function and a decrease in resistance force strength parameter should be based on the same proportion, the structure working state should be changed, while the final effect amplitude value cannot be contracted and magnified in terms of similar proportion. For this reason, overloading or decreasing the strength safety coefficient can be served as distance scales or measures to assess working conditions and limit states. It is only under the condition that a certain function increase or some resistance force parameter decrease is likely to occur, can the evaluation of structure safety be of practical engineering significance. As to seismic function, the seismic occurrence of over design and prevention probability standard is practical and possible, for the dam-site design earthquake is, in fact, a well unsolvable middle- and long-term seismic predictable problem both at home and abroad. Therefore, there is a larger uncertainty so that the occurrence of strong earthquakes being far over the original designed basic intensity in Xintai, Tangshan, Wenchuan, and so on, are the good examples or cases. Accordingly, the overloading safety coefficient of seismic function can be of practical significance. In addition, the shear strength on the latent sliding block sliding surface is of large uncertainty. This is not only because the dam-site engineering geologic and hydrogeologic conditions are complicated and the internal conditions in the rock mass are unlikely to be predicted clearly, but also because the uncertainty is in the $f \cdot c$ value experiments and their data processing. In addition, factors such as the mountain mass disturbances in the process of construction excavation, the work mass in long-term water-soaking and seepage-flow influence after water impoundment, rock mass local damages and low cycle fatigue influence of reciprocating exchange function when in an earthquake, the gradual weakening of effects of consolidating measures adopted to improve $f \cdot c$ values, and the possible in effective factors, and so on, are all likely to lead to the fact that the actual resistance sliding strength of the abutment rock block sliding surface is lower than the values adopted in design. Therefore, it is considered that rock mass resistance force to decrease strength safety coefficient in a certain range is of practical significance.

But if one of the structure bearing functions is increased, the working state of function effect will be no longer the practical design working state similar magnification but completely change the original practical working state so that what is measured by this scale is not the original "distance" in significance. The conclusions do not correspond to the prerequisite, whereby causing the illogic in concept. Even if this overloading safety coefficient is taken as the relative scale to evaluate the safety of practical and rich significance among different engineering works, which cannot be taken, the enlarged function to the river valley landform and geologic conditions, dam types, dam height, and each engineering work with different material strength is, in general, not different in the function affectation and resistance force constituting the whole working conditions. For this reason, there is no comparison. For instance, there exists this kind of problem in the dam body serving as the main function water loading "overloading." For example, water level is heightened over the dam crest, and an overflowing dam is bound to occur, so that the dam working state will change greatly. Water level over the dam crest that cannot overflow water loading is unlikely to occur. Even the dam body that is taken as a linear system in strength check cannot be taken because of its disagreement with actual states and it is just the same with the nonlinear system of abutment rock mass aseismic stability. Accordingly, if water loading state is increased via enlarging water body bulk density, it will not actually take place, with which the overloading safety coefficient is obtained, being extensively used as a scale to evaluate the safety of arch dams, but its engineering practical significance is worth further discussing. The geomechanic model test based on the mentioned concept not only exists in the loading problem but also is used as the limit state symbol indicating that the dam body has too serious a fracture to continue loading. The dam body fracture or crack is caused by the tensile stress exceeding the material resistance tensile strength, while at present, model materials in this kind of experiment are far higher than that of what the similar laws require the resistance tensile strength values corresponding to large dam concretes. In addition, the seepage pressure, which is the most important to the stability of abutment of arch dams, cannot be ore-simulated in the tests. Therefore, the overloading safety coefficient obtained from this is different to reflect the safe and rich significance of practical arch dam engineering construction.

The sliding instability of any structure is the developing process of local deformation accumulation. Therefore, aseismic stability analysis of high arch dams must take the deformation as the core, in such a way that the dam body–dam foundation–reservoir water should be considered as the dynamic deformation coupling influence in the whole system.

7.3 PROBLEMS OF CURRENT ARCH DAM ASEISMIC STABILITY ANALYSIS METHOD

The current arch dam aseismic stability analysis adopts the traditional rigidity limit balance method. The rigidity limit balance method is the conventional design method adopted in the present analysis of various kinds of internal and external structures and side slope stability, thus forming the specifications of various kinds of designs, and they are familiarized and used by the broad masses of design personnel. But this method has had a larger localization either in concept or in seeking a solution method, and cannot reflect the practical situations and mechanism of structure and side slope instability. As far as the abutment aseismic stability of the arch dam is concerned, the whole rock blocks are still stable, and the local cracks and sliding moves of supporting abutment rock blocks are bound to lead to the dam-site to yield deformation coupled with the dam body, thus, making that part of dam body stress increase obviously so as to produce an allowed destruction in designs. On the other hand, in an earthquake, the seismic function is reciprocal and even the whole body of the dam abutment rock mass reaches the limit state instantaneously, being unlikely to lead to the final instability. Therefore, the rigidity limit balance method to check the results of dam abutment aseismic stability cannot reflect dam abutment rock mass real instability conditions and its effect upon dam body safety when an earthquake occurs.

Owing to introduce the contact face counter force along the normal action assumption and negation of force matrix, the rigidity limit balance method itself in seeking a solution method can result in large calculation result errors being inclined to unsafety. Particularly when the sliding rock block geometric morphology is complicated, more sliced divisions or slice block interaction simplified assumptions have to be introduced into seeking solutions, for there exist many uncertainties, either the lack of strict mechanic theory base or the difficulty to obtain the close quantitative results.

Some related aseismic specifications in current hydraulic engineering works are based on the traditional approaches in current engineering design:

1. After the latent sliding rock blocks are determined, the arch end largest thrust and direction can be determined in terms of dam body dynamic and static calculation of the most unfavorable results when an earthquake takes place.
2. When the latent sliding rock block inertia representative value is determined, its dynamic amplification effect cannot be counted and the assumed seismic inertia representative value must be in terms of unfavorable directional action, and its maximum value and arch end trust maximum value occur simultaneously.
3. Based on the geometric performances of the latent sliding rock blocks, the most unfavorable sliding mode changing without time is selected.
4. Seepage pressure within rock mass change will not be taken into account when an earthquake occurs.

These specifications are under the prerequisite of the separation of aseismic stability check of the latent sliding block mass of the abutment of an arch dam from the dam body aseismic strength check. The analog static method based on rigidity limit balance is adopted by neither considering abutment rock block action force and sliding mode changing with time nor counting on the dynamic deformation coupling between the dam body and rock mass and the seismic dynamic magnification effect of bank slope rock mass. This is in fact a static analysis method, which cannot completely reflect many behaviors in the arch dam abutment aseismic stability analysis. For this reason, this method has been improved; that is, the dam body and the near regional foundation, including two-bank latent sliding blocks, are taken as a structure system in such a way that a three-dimensional element model for the whole system is established with the artificial boundary counting into the radiation damping influence in the far regional foundation and based on the expansion cracks in the dam body of the motion theoretical simulation. The displacement time processing to satisfy design seismic motion input into the foundation bottom part can serve as the wave motion problem in time domain explicit seeking solution system static and dynamic integration stress response. Accordingly, the stress obtained from each boundary face of the latent sliding blocks at each time interval must be integrated, whereby obtaining arch support thrust acted on each block mass, the block mass weight and other seismic inertia force, and seepage pressure resultant force. In terms of the rigidity limit balance method, the abutment aseismic stability safety coefficient must be obtained at each time interval so that the minimum value in the whole time process can be taken as the design value. In this way, though aseismic stability safety coefficient changing in the seismic process has been considered, the limitation of rigidity limit balance method cannot be avoided.

7.4 ARCH DAM ASEISMIC STABILITY WITH DEFORMATION AS ITS CORE AND ASEISMIC SAFETY EVALUATION

As far as high arch dams 300-m high, where such an important engineering project is concerned, it is required that under the action of the "maximum credible earthquake," the uncontrollable reservoir water discharge that results in dam failure should be avoided. Dam failure taken as one performance objective in design is an inoperable and very fuzzy concept. As far as high arch dams with hyperstatic structure of higher degree are concerned, and the extreme conditions of the maximum credible earthquake, the local tensile stress exceeds the standard till the individual through cracks appear and only a few transverse tracks or fractures open large enough to cause the damages to the upper local sealing installations or structures. In fact, it is unlikely that there exist the dam heel tensile stress concentration or the phenomena of tensile stress surpassing seepage-proof curtain, all of which cannot characterize its whole structure failure. It is only when the abutment supporting arch dam to bear reservoir water pressure is out of stability that the whole engineering ineffective failure be characterized, whereas deformation is only the symbol variable to measure the arch dam stability so that it can also become the quantitative index to evaluate the aseismic safety of high arch dam engineering. In order to solve this key technical problem rationally and to probe the quantitative guidelines for evaluation indexes to assess aseismic safety of high arch dams, this research project has made a breakthrough in the localization of traditional rigidity limit balance method and suggested a new and more practical solving approach with deformation as the core.

In the research and development (R&D) project of high arch dam body–foundation–reservoir water systems, each sliding face of each possible sliding rock block determined in terms of two-bank abutments in the near regional foundation and dam foundation face as the aseismic weak location is treated with contact faces of Mohr–Coulomb performances. Possible sliding rock slip face and resistance sliding indexes on the dam foundation surface are determined in terms of the grade of dam body concrete. The possible advent of local damages and the system deformation in the process of topology in these locations are calculated under the action of strong earthquakes so as to realize the arch dam aseismic stability analysis with deformation as the core.

Since the foundation possible sliding rock block motion contact boundary is counted into the calculation, it must be considered that the initial stress states formed the terrestrial stress quarter among these boundary surfaces can affect the whole nonlinear dynamic analysis problem so that the whole calculation can be divided into three steps:

1. Since it is different to obtain the real terrestrial stress quarter data, approximation takes stress field as the terrestrial stress quarter field under the action of near regional foundation self-weight, but the stress on the possible sliding rock blocks motion contact boundary can only be kept with the foundation as the initial stress field. Therefore, the initial sliding move and opening degrees are set as zero, so that it is considered that before the dam is constructed, the joint surfaces have been filled with materials and compacted for long.
2. After the dams are constructed, it is necessary to consider the dam body self-weight, reservoir water pressure and reservoir sediment pressure, temperature and bedrock seepage pressure, and static response analysis under various actions of static loadings.
3. The seismic motion acceleration time processing of three components are input from the foundation bottom so as to carry out the seismic wave motion response analysis, and to seek for solutions to the displacement response to large dams and abutment rock mass in each location in the whole process of an earthquake.

When the deformation performance of the high arch dam system under the action of strong earthquakes is analyzed, the key problem is how to determine the quantitative guidelines for the symbol variables to measure arch dam aseismic stability and for the system deformation of quantitative indexes to evaluate high arch dam engineering aseismic safety. Since in each engineering dam-site landform and seismo geological conditions dam body types and sizes and materials are different, it is impossible to determine a unified deformation quantitative guideline. For this reason, the authors of this book have suggested the fundamental thinking that although the high arch dam system occurs to have damages and the damage developing process is very complicated, the whole system deformation process is a gradual growing process with an increase in the input seismic motion amplitude or a decrease in shear strength parameters of the latent sliding rock blocks. Deformation can be the absolute displacement of a key part or location of the high arch dam system or the residual relative displacement among the sliding faces. Once the gradually growing deformation becomes an obvious mutation with the seismic overloading folds or the bedrock sliding face shear index decrease folds, the mutation inflection point can symbolize that the internal behaviors of the system have developed from the quantitative changes to the qualitative changes, so as to take the deformation quantity corresponding to these behaviors as the symbol to make the judgment of high arch dam engineering becoming invalid and fails. The concrete implementation of this thinking can seek for the dam body controlling displacement or the carve inflection point as an objective in the variables correlated displacement with the change in overloading folds and the decrease in strength so as to carry out a lot of the dam–foundation system static and dynamic integrated response analysis. Accordingly, seismic motion input corresponding to influence point states and the maximum credible seismic ratio values can be used as the quantitative guidelines for the engineering failure-free dam under the action of maximum credible earthquake, and as a judgment to make sure of abundance magnitude of engineering latent overloading aseismic safety, or under the design action, the abundance magnitude of decreasing strength aseismic safety of high arch dam engineering system.

7.5 PRACTICAL EXAMPLES OF ENGINEERING APPLICATIONS

Based on the mentioned arch dam aseismic stability analysis concept with deformation as the core, and closely in combination with a series of practical high arch dam engineering works of grade I in Xiaowan, Xiluodu, Dagangshan, and Jinping in the strong seismic areas in west China, the calculation analysis of aseismic safety evaluation of high arch dam systems has been carried out to provide the references for the aseismic design of these large and important engineering works through the concrete applications to the engineering practice. In addition, summarizing the research achievements of aseismic safety analysis conducted in different high arch dam engineering works has further raised the realization of high arch dam body–foundation system aseismic safety performances so as to further improve and complete the aseismic safety analysis and evaluation method used in high arch dam systems.

Table 7.1 lists the basic data and information of such three grade I arch dam engineering works as Xiluodu, Dagangshan, and Jinping, as well as the calculation analysis of overloading and decreasing strength safety coefficient and a brief analysis of damage modes. Figures 7.1–7.3 are the finite element models and two-bank most unfavorable latent sliding rock mass shapes of the Xiluodu, Dagangshan, and Jinping grade I arch dam system.

Table 7.1 The Whole Aseismic Safety Analysis and Evaluation Comparison of Three Large-Scale High Arch Dam Engineering

Item Name of Engineering	Xiluodu	Dagangshan	Jinping Grade I
Dam height	278 m	210 m	305 m
Landform and geological conditions	River valley in dam-site zone basically appears to be U-shaped. Two-bank abutment rock mass consists of 4~12 stratum of E'mishan basalt, stratum occurrence is flat, slightly inclined downstream and deviation to the left bank, no faults or larger soft and weak faces are mainly in middle and internal disturbed belt and joining fractures.	River valley cross section appears to U-shaped, two-bank landform is basically symmetric. Rocks in dam-site zone are mainly acid granite in Chenjiang period, diabase-porphyrite, diabase-prophry, granodiorite-aplite, dioritite, felsite-porphyry, etc. Various kinds of vein rocks develop in weaving of granites, geological structure in dam-site zone is mainly with regional faults developing in the surrounding area, with the zone, the second vein rocks, small faults and joint fracture are the main features.	Dam-site landform is the typical deep-cut V-shaped valley so that engineering geological conditions are very complicated. The griotte strata distributed on the two banks have outcropped, sandstones spread at the altitudes of 1900~2300 above sea level. There exist the deep cracks and f5, f8 fault geological defects on the left bank. The right bank is marble rocks consequent slope, of which the inclination griotte strata have the discontinuous distribution of weak faces, and the dam site upper part on the left bank is consolidated with concrete.
Design seismic level peak ground acceleration	0.321*g*	0.5575*g*	0.197*g*

Table 7.1 The Whole Aseismic Safety Analysis and Evaluation Comparison of Three Large-Scale High Arch Dam Engineering *(Cont.)*

Item Name of Engineering	Xiluodu	Dagangshan	Jinping Grade I
Seismic overloading safety coefficient and damage mode	1.2: When overloading is 1.3-fold, a certain displacement difference appears on the right bank bottom sliding face two sides nodal point, and the displacement difference on the left bank bottom sliding face two sides has increased, but the opening crack sliding movement on the right and left banks sliding rock blocks is not obvious. The huge displacement difference value of the right bank dam foundation face two sides nodal point pair indicates that dam body slides downstream and losses stability, but rock mass displacement is not large. Damage process is not that dam foundation face passing through strength damage has led to an increase in dam body right side consequent river displacement and the rotating around the left makes the whole body loss the stability.	1.5: The left bank sliding block position is the weakest part in arch dam system. Under the design seismicity, there appears to have a certain sliding move between overloading folds, sliding move quantity increases fast, and the advent of residual sliding move location is gradually expanding. In the case of 1.5 overloading, the relative sliding move quantity reaches 8.41 cm as much. And the left bank sliding block relative to bedrock slides several tens of centimeters, being unfavorable for large dam bearing capacity and normal operation after the earthquake.	3.1: Dam foundation of the bank upper part is replaced by concrete consolidation so that it is not easy to have the foundation face damages. The sliding quantity of the left bank rock mass sliding block is large, but the right bank sliding block slides to bring about dam body consequent river sliding movement, whereby causing dam foundation interface to fracture and expansion. When in 3.2-fold of overloading operation, the left bank dam foundation interface is also pulled open so that sliding block and dam body consequent river sliding moves grow fast, resulting in damages.
Sliding fracture surface, safety coefficient, and damage mode	1.5: With f, c decrease, sliding quantity of the left bank dam foundation rock block is gradually increasing. When the strength coefficient decrease reaches 1.6, bedrock sliding increases enough to make the left local dam foundation interface open too enough to close to some extent. Though the bearing capacity cannot lose completely, the whole structure system subject to force working condition is apparently worsened, and local damage in structure system is shifted from the inner part of the left bedrock to dam foundation surface.	2.6: When the strength decrease coefficient cannot exceed 2.6, and under the design seismic action, the whole dam body-bedrock system working conditions cannot be worsened obviously. When the strength decrease coefficient reaches 2.7, and in the stage of static calculation, there can be no way to get the stable static solution because of the left bank sliding block losing stability, whereby illustrating that static strong store safety coefficient 2.6 can determine the system stability.	3.6: When the shear strength index of sliding fracture face decreases 3.7-fold, the sliding blocks on the left bank cannot mountain static stability, while the strength coefficient decrease will not exceed 3.6, and under the action of design seismicity, the whole dam body-bedrock system working condition will not be worsened.

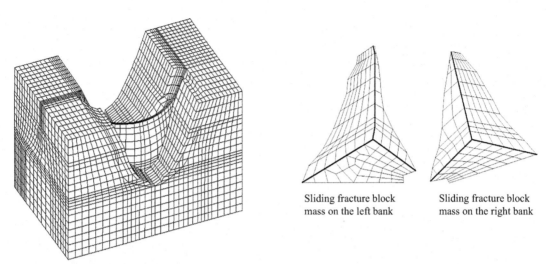

FIGURE 7.1 Xiluodu Arch Dam System Finite Element Model and the Most Unfavorable Latent Sliding Rock Mass Shape on the Bank

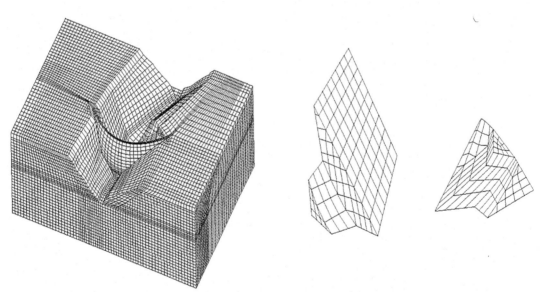

FIGURE 7.2 Dagangshan Arch Dam System Finite Element Model and the Most Unfavorable Latent Sliding Rock Mass Shaped on Both Banks

In the computation of the Xiluodu arch dam overloading safety coefficient, the nodal point displacement time process of the dam body main location is seen in Fig. 4.1. When seismic loading is magnified by 1.3 times, the dam body will appear to have the displacement mutation, and the displacement from the right arch end to the arch crown increases sharply, whereby indicating that the dam body has been out of stability and lost its bearing capacity. In order to study the destructive mechanism of the

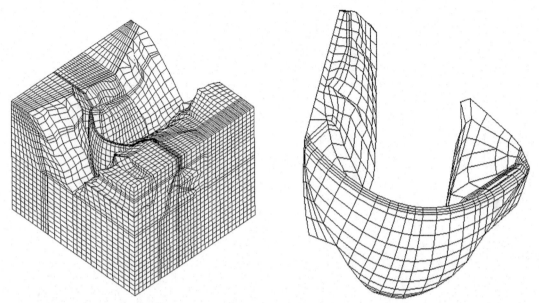

FIGURE 7.3 Jinping Grade I Arch Dam System Finite Element Model and the Most Unfavorable Latent Sliding Rock Mass Shape on Both Banks

dam body-bedrock system, it is necessary to make a comparison of the bedrock's inner part with dam body displacement conditions. When the overloading coefficient reaches 1.2 times, the relative sliding cannot happen in the right bank bottom two-side nodal points, while there is a certain displacement difference in the left bank bottom sliding face two-side nodal points, being the same as the design seismic operational conditions. The fracture faces within the bedrock cannot occur to have a large sliding move and local damages. When the overloading coefficient reaches 1.3 times, there appears to have a certain displacement difference on the right bank bottom sliding face two-side nodal points, and the displacement difference on the left bank bottom sliding two-sides can also have some increase, but generally speaking, the opening fracture sliding move of sliding rocks on the left and right banks is not significant. When the overloading coefficient reaches 1.3 times, the huge displacement difference of nodal point pairs on the dam foundation interface two sides on the right bank indicates that the dam body slides downstream and loses stability, but rock mass displacement is not large. It can be seen in combination with the cracking conditions that under the seismic overloading mode or pattern, the damage process of Xiluodu arch dam can be as follows: the damage of passing-through strength of the dam foundation interface has led to the whole dam body to lose stability, and when seismic loading magnified reaches a certain extent, dam body displacement is too big, whereby resulting in accelerating damages near the dam foundation interface because of stress increase, and a weakening dam foundation interface can further enlarge dam body displacement under state and dynamic loading in such a way that the dam body appears to have a losing balance or stability destruction. The analysis shows that the maximum main stress of the right bank of the dam body is obviously higher than that on the left bank because the bedrock is in uniformity. Accordingly, when seismic loading is magnified by 1.3 times, the cracking of the right bank of the dam foundation interface develops rapidly, whereby causing destruction (Fig. 7.4).

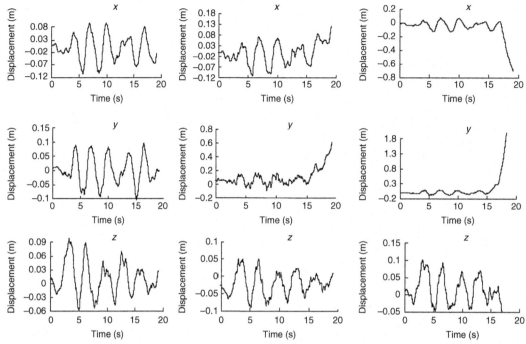

FIGURE 7.4 The Dam Body Main Part Nodal Point Displacement Time Process in the Case of Overloading Safety Coefficient of 1.3 of Xiluodu Arch Dam

The analysis is made of calculation results of the bedrock sliding crack face shear strength store safety coefficient of the Xiluodu arch dam. It can be seen from the nodal point pair displacement time process of the two-bank sliding rock block bottom sliding face that the local sliding move of the left bank sliding face grows obviously with an increase in strength-decreasing coefficient, while the dam foundation interface two-side nodal points of the top arch two arch ends have had very small displacement difference in the case of strength decreasing of 1.5 so that there is only an instantaneous opening between both or disturbance, and they are in the state of stickiness. When strength-decreasing coefficient is 1.6, there appears to have an apparent displacement difference on the left arch end dam foundation interface two-side nodal points, and there is no way for the two points to close after the earthquake finishes. This indicates that with a decrease in f, cc, the sliding quantity of the left dam foundation rock blocks is gradually increasing, and when the strength decreasing coefficient reaches 1.6, the bedrock sliding increases largely enough to make the left side local dam foundation interface open to a big enough degree that there is no way to have closure. In spite of having no completion of losing bearing capacity, the structure system subject to force working condition is obviously worsened so that the local destruction in the structural system has been shifted from the inner part of the left bank bedrock to the dam foundation interface. The residual displacement of the left arch end of the top arch changing relation with the strength decreasing coefficient is shown in Fig. 7.5. Obviously, before the strength decreasing coefficient reaches 1.5, the residual displacement coefficient of the left arch end of the top arch is basically zero; after the strength decreasing coefficient reaches 1.5, the residual displacement appears to have the linear growth. Therefore, f, c decreasing coefficient to 1.5 is a turning point of the variation large dam system working state.

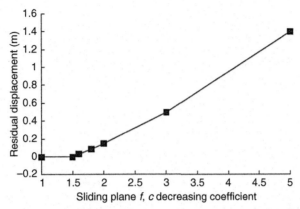

FIGURE 7.5 Dam Top Left Arch End Residual Displacement Varying Relation with Strength Decreasing Coefficient of Xiluodu Arch Dam

As to the Dagangshan arch dam with an increase in overloading coefficient, the residual sliding move quantity has had a nonlinear growth. It can be seen from the initial analysis of the damaged mechanism that the left bank sliding block position is the weakest location in the whole arch dam bedrock system. In the case of design seismicity, there appears to have a certain sliding move between the dam body and bedrock. With a fast increase, the location with the advent of residual sliding move is gradually expanding. When the overloading coefficient reaches 1.5, the sliding move quantity exceeding 1 cm location is expanding downward from the bank sliding block to the dam bottom at an elevation of 925 m above sea level, while when the overloading coefficient continues to increase, the whole large dam occurs to slide and move downstream, with an increase in the sliding quantity. It can be seen from Fig. 7.6 that overloading coefficient within the range of 1.5–1.7 is a turn of sliding move quantity

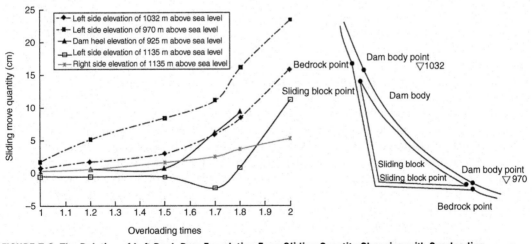

FIGURE 7.6 The Relation of Left Bank Dam Foundation Face Sliding Quantity Changing with Overloading Coefficient and the Position Resulting Among Dam Body, Sliding Blocks and Bedrocks of Dagangshan Arch Dam

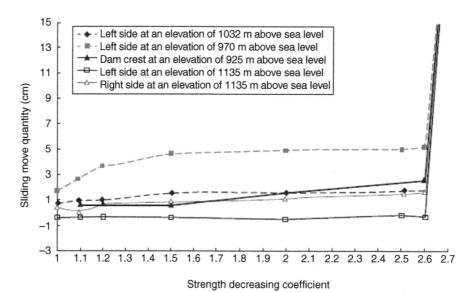

FIGURE 7.7 Variation Relation of Dam Foundation Interface Sliding Quantity with Strength Decreasing Coefficient of Dagangshan Arch Dam

increase from slowness to sharpness. In terms of this standard, the aseismic overloading safety coefficient of the Dagangshan arch dam can take 1.5, but under the operational condition of 1.5 overloading coefficient, the relative sliding and move quantity between the left dam body bedrock at an elevation of 970 m above sea level can reach 8.41 cm, and the sliding blocks relative to bedrocks on the left bank have slid several tens of centimeters so that the sliding move quantity is so relatively large that it is likely to bring about an unfavorable effect upon the large dam bearing capacity and normal operation after the earthquake.

In carrying out research on the bedrock sliding cracking face shear strength strong store safety coefficient of the Dagangshan arch dam, and in calculating the strength decreasing coefficient of 2.7, there is an advent of a larger displacement, and there is no way to obtain stable static solution in the stage of static calculation because of losing stability of the left bank sliding blocks, whereby illustrating that in terms of given sliding blocks and mechanics parameters, the static strong store safety coefficient of the Dagangshan arch dam should be 2.6 (see Fig. 7.7). Of course, in the case of static losing stability, it is not necessary to discuss the aseismic stability safety coefficient. When the strength-decreasing coefficient cannot exceed 2.6, as viewed from designing earthquake function of each calculation operation conditions, the strength-decreasing coefficient increases in the process from 1.1 to 2.6, the residual sliding move quantity after the earthquake between the dam body and bedrock only increases from near 3 cm to 5 cm or so, and there are no more changes in dam foundation interface damage conditions after the earthquake. Accordingly, it can be considered that under the action of designing seismic function, the whole dam body-bedrock system working conditions have no obvious deteriorations. Seismic motion is the reciprocating loading. In the case of a structural system being able to maintain the static stability, there occurs to have a certain local sliding move at an instantaneous moment under the action of seismic loading, but it will soon recover to the balance position because of seismic motion reciprocating functions without causing the whole system to lose stability. Therefore, as to the Dagangshan arch dam under the condition of strength decreasing coefficient being able to guarantee dam body bedrock system static stability, seismic loading cannot result in the whole system to close stability or working performances with an advent of obvious deterioration in such a manner that the great dam static and dynamic integration strong store safety coefficient can take 2.6 of static strong store safety coefficient.

As to the Jinping grade I arch dam, in the case of seismic overloading and through analysis, the sliding motion of the left bank sliding blocks and bringing about dam body to local deformation downstream can be the major factor to affect the large dam. One nodal point pair on the left bank sliding block bottom sliding face and the representative nodal point pair on the right bank sliding face at an altitude of 1900 m above sea level are adopted to investigate the sliding conditions of the left and right banks. Figures 7.8 and 7.9 are the relation curves of the consequent river sliding move quantity between the left bank sliding blocks and dam foundation face and the consequent river displacement of the dam crest left and right arch supports with the changes in overloading coefficient of 3.1. There appears to be inflection points on curves in two figures.

In the Jinping grade I arch dam system, when the bedrock sliding cracking face shear strength decreases to 3.6 times, in the stage of static calculation, there can be no way to obtain a stable static solution for the system has a large sliding move quantity of sliding block sliding cracking faces and dam foundation face representative position as well as the large consequence river displacement of the dam crest's left and right arch supports so that it can be considered by the static loading in the normal operation conditions.

It can be seen from the summarization of the mentioned engineering application of practical examples that in the analysis of finite element time process of the dam body foundation system, the dam

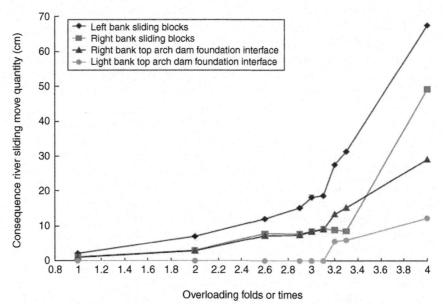

FIGURE 7.8 The Relative Curve of Consequence River Sliding Move Quantity with Changes in Overloading Coefficient Between the Left Sliding Blocks and Dam Foundation Face of Jinping Grade I Arch Dam

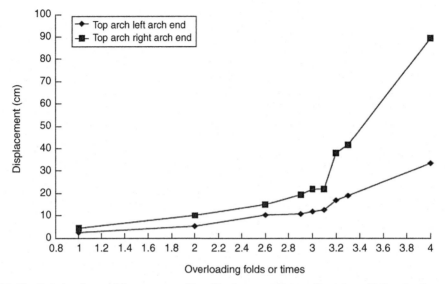

FIGURE 7.9 The Relation Curve of Consequence River Displacement Values Changing with Overloading Coefficient of Dam Crest Left and Right Arch Supports

body behavior displacement response mutation in a strong earthquake can serve as a guideline for making the judgment of the whole arch dam systems. However, in calculating the operational considerations of seismic overloading and sliding cracking face shear strength decreasing, the dam body displacement response to the mutation can be likely to change with different dam body shapes, dam-site landforms, geological conditions, and others. For this reason, it is necessary to carry out the concrete analysis of concrete engineering works. This displacement response appears to have some inflection points as the seismic loading or strength decreasing coefficient increases, which can be used as the basis to judge whether the whole arch dam system can lose its stability or not. When the concrete engineering is analyzed, it is necessary to combine the concrete behaviors displayed by the dam–foundation system seismic dynamic responses as well as a certain computation experience in such a manner as to determine the behavior displacement used to map displacement and overloading strength decreasing coefficient relation curves to look at the inflection points in mutation occurrence. At the same time, it must be noticed that static loading is different from seismic loading in different performances. Static loading is sustainable in action. When bedrock sliding cracking face shear strength decreases till there is no way to bear static loading, the sliding blocks and dam body will be pushed by this static loading to have the occurrence of sliding movement and losing stability, while the seismic motion is the reciprocating loading. In the case of its structural system being able to keep stability, a certain local sliding movement can take place at an instantaneous moment under the earthquake action, and the balance position will recover very soon because of seismic motion reciprocation function and then move in the opposite direction so that the whole instability will not be caused. For this reason, as to some arch dams, when the aseismic stability safety coefficient in carrying out sliding cracking face shear strength decrease is analyzed, and strength decreasing coefficient is able to ensure the dam body-bedrock system static stability conditions, the same seismic loadings are unlikely to lead to the whole system to lose stability, or for the working behavior to have an obvious deterioration. The strength-decreasing coefficient continues to be enlarged, and the large dam shows the static instability at first. It is just at this time that it is considered that the large dam static and dynamic integrated strong store safety coefficients are the same as those of static strong store safety.

"The thinking of taking deformation as the core for arch dam aseismic stability and aseismic safety evaluation" has been initially tested through the application in the important arch dam practical engineering works, whereby indicating that its achievements applied to practical engineering are also basically feasible and also that under the action of maximum credible seismicity, non-dam failure quantitative guidelines are provided for further in-depth study fundamentals in the future.

As far as the relative current arch dam aseismic stability analysis method is concerned, and based on the nonlinear seismic wave motion response analysis method, the dam body strength should be unified with the aseismic stability check to carry out the concept and method for the large dam–foundation system aseismic safety evaluation, and be really near to practical working state. As viewed from the present-day situations, it is still worth carrying out more in-depth studies on some aspects. Particularly in material fundamental models only restricted in linear elastic material model computation and analysis, although this research on the computation and analysis are able to meet the needs of the structure's normal operation states and nonlinearity, mainly in predesign contact boundary development situations, they are still unable to reflect the materials being gradually out of effectiveness and damaged regional development conditions after the structures enter into damaging and destroying stages. As for the large concrete dam, it is necessary to set up the concrete dynamic damage fundamental model combined with the dynamic contact models to simulate contact faces of various types. The nonlinear static and dynamic analysis system of the large dam–foundation system must be further perfected so as to further reveal the mechanism of large concrete dams suffering from seismic damages for carrying out the large dam aseismic safety evaluation. In addition, being subject to computation scale and computation condition restriction, only a few sliding block combinations are considered in the present-day high arch dam–foundation system whole aseismic safety evaluation method and the system, and it might be possible to refer to the rigidity limit balance method first for abutment stability analysis achievement, and at the same time, the cracking contact interface can be set up at the dam foundation interface, which still to belongs to a kind of assumption of structure weak face under the present-day computation conditions. For instance, a rather limiting consideration is given to the foundation rock mass tensile strength and dam foundation common boundary cracking possibly developing into the inner foundation. For this reason, these problems are worth further discussing and improving.

RESEARCH ON PARALLEL COMPUTATION OF HIGH ARCH DAM STRUCTURE SEISMIC MOTION RESPONSE

8.1 RESEARCH AND SIGNIFICANCE OF LARGE-SCALE STRUCTURE RESPONSE

8.1.1 DEVELOPING SITUATION OF HIGH PERFORMANCES PARALLEL COMPUTATION IN HYDRAULIC STRUCTURE FIELD IN OUR COUNTRY

Hydraulic engineering involves a wide range of items, and computation quantity is extremely large. High-performance parallel computation technology has been introduced into the analytical fields including reservoir optimum scheduling, watershed water and sediment numerical simulation, hydrological and water resources simulation, hydraulic structure design and aseismic check, research on large dam concrete three-dimensional micromechanics numerical model, and so on. Hydraulic engineering problems are characterized by the large-scale high parallel computation so that high-performance parallel computation has found an extremely wide application in the water resources field.

The problems that need to be solved in hydraulic structures include hydraulic structure design, arch dam abutment stability, high dam aseismic design, side slope resistance sliding stability, and so on. The mechanical analysis of all these problems is characterized by the obvious nonlinearity, more uncertain influencing factors, high safety requirements, long-term construction duration, and so on. These are many kinds of coupling problems and repeated calculation problems so that the fine simulation and high-speed calculation are required in carrying out numerical simulation and emulation analysis. In the research of Dai et al. (2006), the regional decomposition method was adopted for Lagrange's explicit difference serial program. Based on message passing interface (MPI), the dynamic serial–parallel explicit computation program for the geotechnical and hydraulic engineering has been developed, but this program cannot simulate the seepage temperature loading, and other factors required to be considered in hydraulic structures, whereby confining the range of this program application. Hohai University is the earliest scientific research institution in the country to have carried out hydraulic structure parallel computation research, thus the advanced and high-performance campus grid platform, on the basis of which, Hohai University has done much pioneering work on the introduction of high-performance and parallel computation into hydraulic structures (Lei et al., 2006, 2007). Chao et al. (2002) introduced the parallel computation into the dynamic optimal design research of high arch dams to solve the load balance problem in parallel computation, with good results obtained. If this program is used in the analysis of structure dynamic response, it is necessary to add the dynamic contact problem-handling, and this program adopts the quality-free foundation model. In order to simulate the foundation radiation damping influence, the boundary condition handling manner should be improved. Zhang Jianfei (2004) developed the parallel boundary element method applied to the analysis of hydraulic structure dynamic and static states. But it is not easy to obtain the basic solution to the requirement of complicated hydraulic structure boundary element method, whereby confining the application of this method.

In order to meet the needs of fast and highly efficient hydropower development in our country, it is necessary to improve the speed of dynamic analysis of hydraulic structures, and it is urgent to develop the parallel analysis program of all-round influencing factors including contact problems, seepage flow pressure, temperature loading, hydrodynamic pressure, material damage nonlinearity, and other influencing factors.

8.1.2 THE SIGNIFICANCE OF RESEARCH ON HIGH DAM STRUCTURE SEISMIC MOTION PARALLEL COMPUTATION

The important high arch dam of 300 m in height is a very complex special structure that has seismic response analysis needs. Meanwhile, counting into the dam body–foundation-reservoir water dynamic interaction, boundary nonlinearity of dam body expansion transverse crack open and close, shock energy escaping and dispersing in far regional dam foundation, the near regional complicated landforms and geological structure, the material nonlinearity of the dam body and foundation rock mass, the nonuniformity of seismic motion along the dam foundation, and other complicated problems of dam abutment latent sliding rock block dynamic stability under the coupling action of arch supports and foundation dynamics should necesitate to count in seismic response analysis. The finite element numerical computation of the whole system needs to seek solutions to thousands or even millions of freedom equation groups. The computation quantity of large dam responses under the dynamic loading appears to have a geometric series growth as the dam body freedom increases. With the nonlinear dynamic problem in particular, and in order to guarantee the convergence and stability of numerical computation, the time step length or size in time domain computation frequency needs to take several seconds of one ten-thousandth time, and even within one time step length, iteration should be conducted for several times, while the sustainable time for input seismic motion often reaches as long as tens of seconds. For instance, a standalone series program is used to carry out the aseismic computation of Xiluodu arch dam with 440,307 freedoms, and even if the present-day fastest PC computer is adopted, it will take 20 days to complete this calculation. Accordingly, in the time domain, it is to seek solutions to such a large-scale nonlinear dynamic problem that the computer speed has become the bottleneck for high arch dam seismic motion analysis and aseismic safety evaluation. At present, high performance parallel calculation has become one of the main means of solving the problem of large-scale scientific computation.

High performance parallel computation can adopt the supercomputer or high performance parallel computer group system. Application and maintenance cost of a supercomputer is extremely expensive, and its application cost is so high that it is not economically feasible. As a result, a high-performance parallel computer group system is able to solve the problem of a large-scale computation in hydraulic structure seismic response analysis, to raise the computation speed by hundreds or thousands of times and to make computing equipment have expansion capacity or spreading range. When the concrete problem computation requires to be raised, increasing computation nodal points can satisfy the growing requirements. Therefore, selection of a high-performance parallel computer group as the scheme to solve the complex hydraulic structure seismic response analytical problem is either technical consideration or economic feasibility as the starting point, also, being the inevitable choice involving super-large-scale dynamic calculation under the present conditions.

8.2 THE DEVELOPMENT AND EXISTING CONDITIONS OF FINITE ELEMENT PARALLEL COMPUTATION

At present, in the computation mechanics field, the finite element method around and based on the variation principle and the boundary element method based on the boundary integral equation in combination with various parallel computer application coming into being have formed a new scientific branch – the finite element parallel computation. This computation method is of very high efficiency, thus making the large-sized and complicated mechanical problems that cannot be solved or not well-solved by many present-day serial computers or serial computation methods obtain the satisfactory solutions so that its development speed is very surprising. The international upsurge has been stirred up to use the parallel computer to carry out the engineering analysis and research. During the two decades of 1975–1995, over 1000 papers or articles concerning the finite element method and corresponding to numbers of parallel computations have been published, and in the recent 15 years, research and development of the finite element parallel computations concerned have been more rapid so that the published research theses or papers have exceeded the aggregated sum of the previous 20 years.

The finite element parallel computation develops in two directions. One is to implement parallel-seeking various kinds of computation methods. Others are the parallel analysis methods including the finite element parallel computation method and the boundary element computation method. The former tends to become mature, while the latter has less research. With an aim at the research in this respect, it is to explore the finite element calculation in itself of the latent concurrency parallelism, and it is the fundamental problem in the finite element parallel computation.

With the ongoing improvement in the parallel computation hardware level at home and abroad, the parallel computer group system has continued to make a breakthrough in solving the large-scale computation latent energy. Nevertheless, bringing this huge latent energy into full play needs depend upon the ongoing progress made in parallel computation software. Therefore, it is only important to develop the adaptable advanced parallel computer group system parallel software that the parallel computation capacity can be brought into full play. In fact, as viewed from the current hardware price level, it is not necessary to need too much money to set up a parallel computer group system with high performance. It can be seen from this that hindrance to the overall application of parallel technology is not the hardware computation capacity and prices but lies in being adaptable to parallel system software.

8.2.1 BRIEF DESCRIPTION OF PARALLEL FINITE ELEMENT

Since A.K. Noor of the National Aeronautics and Space Administration (NASA) published the first article on finite element parallel computation, the finite element parallel handling technology and the parallel computer have developed synchronously. Based on the incomplete statistics by the year of 1992, over 400 papers concerning this aspect in the foreign countries have been published, of which the number of articles of the posterior to 5 years is the aggregate sum of the prior 12 years. The research contents have developed from the past computation methods to the computation method, research on the combination of software with hardware, and also, some practical large-sized structure analysis software for the computer types have been developed.

The finite element machine is specially used to operate the special finite element computation program computer. In the end of the 1970s, some people had published a paper concerning the finite element machine (FEM).

Some scholars of the Langley Research Center of NASA compiled the articles to introduce FEM designed for researches by this center in detail. This FEM consists of one processor array, one microcomputer serving as a controller, one parallel operating system, and some modularized general parallel computation programs. The users can use the other special functions of the text editor and the controller to establish the finite element computation model and to carry out analyses. For over a decade, some people have made the intensive efforts in this respect, but the special FEM development future is still uncertain.

Cardio matrix parallel machine (CPM) is, in the initial stage, used in parallel processing of signals and images. Owing to its matrix computation functions with high efficiency, in recent years, some people have used it in the finite element analysis, and also done some benefit trials.

The huge type of vector machine has exhibited its huge formidable force in the finite element analysis gradually. The US ENWEI company CM-2 is in the leading position. Many experts on structure analysis have employed this huge type of vector machine with 65,636 processors in the finite element computation. For instance, T. Belyschko et al. have adopted the explicit method to complete the shell nonlinear finite element computation with 32,768 units. Parallel efficiency is extremely high and its speed is nearly higher than one quantitative grade of the CRAY X-MP/14 parallel machine.

Parallel machine network and working station network are the developing direction in recent years. Shichuan Yuanji of Tokyo University has connected three sets of CRAY-MP machines to form the network by means of super-high-speed network to carry out the finite element analysis. Conjugate gradient method based on the regional splitting technique is adopted to seek a solution to the finite element equation. In seeking solution one million and the system average operational speed reaches as high as 1074 GFLOP. In addition, based on an engineering working station network, they have conducted a similar research in the case of parallel environments, and in 1990, the number of freedoms for seeking solution to the problem reached as high as 200,000. The parallel finite element has developed now, the number of freedoms for seeking solution to the finite element has reached as high as 10 million grades, and the speed for seeking solutions has been extremely improved.

8.2.2 DEVELOPMENT CONDITIONS OF DOMESTIC FINITE ELEMENT PARALLEL COMPUTATION

Parallel computation design and effective realization strongly depend on the parallel machine software and hardware environs. In the initial two decades, only a few domestic units have had parallel machines and their types are in disorder so that there are very few research personnel, starting late and being constrained only in a specific hardware environ. As viewed from the contents of finite element analysis method, several tens of published research essays (reports) have not shown a strong systematization. As viewed from the process of parallel finite element development in our counter, Zhu and Xin (1991) realized the finite element equation direct solving process by means of a transputer chip distribution pattern MIMD system and suggested the parallel computation method for pretreatment conjugate gradient method based on the substructure. Wang (1991) suggested the parallel direct solving process of variable bandwidth linear equation group in the finite element analysis based on the Transputer chip distribution pattern MIMD system, so as to complete a static analysis program. Zhou et al. (1993) realized the rigidity matrix computation, symmetric band matrix Cholesky decomposition, and parallel handling of the seeking solution to linear equation group on the YH-1 vector machine. With an aim at the problem of irregular structure engineering analysis, they adopted the variable bandwidth storage method and the seeking solution to variable bandwidth storage method and realized the rigidity matrix parallel computation and the seeking solution to variable bandwidth sparse linear equation group parallel direct process. Ruqing (1993) carried out research on the parallel computation method in the wide range by means of ELXSI-6400 shared storage type MIMD system, with the main research achievements including the following: (1) the substructure solving process parallel computation in static analysis and the parallel computation method of modal integration substructure method in dynamic analysis are suggested; (2) starting from wave-front method to seek solution to the large-scale structure to analyze problems is developed; (3) starting from Jacobi block iteration method and weighted residual method, the finite element equation parallel solving process and the finite element parallel basic format based on the asymmetric control finite element equation are derived; (4) the application of coloration theory in image theory has realized the rigidity matrix parallel computation; (5) SOR parallel iteration solving process based on the colored line subdivision has been achieved; (6) subspace iteration method, Lanczos method, and the application of multinomial secant line iteration method and vector iteration method to seek solution to structure inherent frequency and modal parallel computation method have been realized; (7) with an aim at the elasticity analysis, a kind of multiwave-front-substructure parallel computation method is suggested; (8) with an aim at the elastic–plastic contact problem, a kind of parallel solving process based on the parameter variation principle is presented; and (9) the one-step integration parallel handling is realized. Up till now, the domestic parallel finite element application framework has basically been established. In addition, the sci-tech personnel in Zhejiang University, the southwest computation center of the Chinese Academy of Sciences, the Northeast University and National Defense University of Science and Technology, and other research institutions have carried out the pioneering research work on the parallel computation of the finite element analysis, with rich research results obtained.

The results from the mentioned research indicate that the domestic parallel computation method research depends on the vector machine, distribution modal parallel machine, and shared storage parallel machine in hardware, and in the contents, research covering a wide range development is still far from the practical application and commercialization. Still, little has been done on the universal parallel computation method research without depending on parallel machine concrete environs. Similarly with an aim at structure, little has been done on the finite element analysis parallel computation hardware.

The factors constraining the high performance computation application level in our country lie in short of parallel computation application software, and mainly depend on the imports from abroad, being far from the requirements, while facing such a situation as the applied software development is almost in the blank state and cannot follow the development step of high-end computers. Nowadays, in the field of urgent need for application of computers with high performances, the existing problems are very complicated, most of which are the physical field problems and coupling seeking solutions, such situations are many. At present, software on the markets cannot satisfy the demands so that it is only to depend on the engineering personnel in the applied fields to compile the program to complete the computation, whereby leading to high cost for computation and great wastage in human resource, so that the superior capacity of the engineering personnel in their own fields cannot be brought into full play. At present, the engineering simulation in our country has been widely dependent upon the imported commercial application software so that this situation is very unfavorable for upgrading the sci-tech competitive force and national safety interests of our country. ANSYS, ABAQUS, and FLUENT Universal software are of a certain parallel function, but as to the specific problems, their parallel functions are low plus these software belong to foreign products. They have the restriction of numbers of nodal points to the users of our country so that all the functions of hardware cannot be brought into full play, whereby resulting in very low computation speed.

Professor Liang Gouping, an expert on numerical mathematics (Guoping, 1990), developed the finite element automatic program. The main thinking of the finite element automatic program is that the combination of element program design with the program automatic genesis technology must adopt the automatic genesis technology, element technology, and format tank technology. Based on the finite element method unified mathematics principle and the inner laws, and with similar mathematics formula deduction mode, the automatic genesis finite element program will be obtained from the finite element problem differential equation formula and the adaptation of seeking solution computation method. At the same time, based on the finite element automatic program, the parallel finite element program generator (PFEPG) has been developed. The PFEPG system's main functions are to let the computers automatically generate the parallel finite element computation sources program based on the regional decomposition method in terms of the users finite element expression formula. It can greatly reduce the requirement for organizational personnel. The users only input various kinds of expression formula needed by the finite element method, that is, the parallel finite element computation of the whole resource program can be automatically generated in such a way that a bulk of tedious programming labor can be eliminated to guarantee the correctness and uniformity of the program and that the program is easy to read and revise, whereby ensuring the reutilization and easy maintenance of software.

8.3 DYNAMIC EXPLICIT COMPUTATION FORMAT AND DYNAMIC CONTACT PROBLEM HANDLING METHOD

8.3.1 FINITE ELEMENT EXPLICIT COMPUTATION FORMAT IN STRUCTURE SEISMIC RESPONSE ANALYSIS

Finite element is a very important numerical simulation method in dynamic computation. In general, based on the problems with different seeking solution modes, finite element is divided into the explicit finite element and the implicit finite element. The implicit computation method can take the larger step length, but almost every increment step needs an iterating solution, and each iteration requires seeking a solution to the large-sized linear equation group. This process needs occupying a rather large quantity of computation resources, disk space, and internal memory; the explicit computation method has the fast computation speed. It is only when the time step length is small enough that there does not exist the convergence problem so that the internal memory needs less than the implicit computation method requires, and the numerical computation process can be easily carried out by the parallel calculation. Programming is relatively simple. Therefore, explicit discrete format is still adopted in the process of developing a parallel program, but the discrete modes for acceleration $\{\ddot{u}\}_t$, and velocity $\{\dot{u}\}_t$ are different from those discussed in Chapter 6.

It is here that the dynamic differential equation at t time can be written as the decreasing order form, that is:

$$\begin{aligned}
&\{v\}_t = \{\dot{u}\} \\
&[M]\{\dot{v}\}_t + [C]\{v\}_t + [K]\{u\}_t = \{Q\}_t
\end{aligned}$$

(8.1)

Difference ahead is adopted for acceleration:

$$\{\dot{v}\}_t = \frac{\left(\{v\}_{t+\Delta t} - \{v\}_t\right)}{\Delta t} \tag{8.2}$$

Substituting Eq. (8.2) into Eq. (8.1), there will be the following:

$$[M]\{v\}_{t+\Delta t} - [M]\{v\}_t + [C]\{v\}_t \,\Delta t + [K]\{u\}_t \,\Delta t = \{Q\}_t \,\Delta t \tag{8.3}$$

Difference afterward is adopted for velocity:

$$\{v\}_{t+\Delta t} = \frac{\left(\{u\}_{t+\Delta t} - \{u\}_t\right)}{\Delta t} \tag{8.4}$$

Substituting Eq. (8.4) into Eq. (8.3), there will be the following:

$$[M]\left(\{u\}_{t+\Delta t} - \{u\}_t\right) - [M]\{v\}_t \,\Delta t + [C]\{v\}_t \,\Delta t^2 + [K]\{u\}_t \,\Delta t^2 = \{Q\}_t \,\Delta t^2$$

Adopting Rayleigh damping $[C] = \alpha[M] + \beta[K]$, substituting into the earlier equation, we can obtain:

$$[M]\{u\}_{t+\Delta t} = \{Q\}_t \,\Delta t^2 - [K]\left(\{u\}_t + \beta\{v\}_t\right)\Delta t^2 + [M]\{u\}_t + [M]\{v\}_t \left(\Delta t - \alpha\Delta t^2\right) \tag{8.5}$$

Letting $[A] = [M]$

$$\{F\}_t = \{Q\}_t \,\Delta t^2 - [K]\left(\{u^t\} + \beta\{\ddot{u}^t\}\right)\Delta t^2 + [M]\{u^t\} + [M]\{\ddot{u}^t\}\left(\Delta t - \alpha\Delta t^2\right)$$

$$[M] = \begin{bmatrix} m_1 I & & & \\ & m_2 I & & \\ & & \ddots & \\ & & & m_k I \end{bmatrix} \text{ is the diagonal block matrix,}$$

And then Eq. (8.5) becomes the following:

$$[A]\{u\}_{t+\Delta t} = \{F\}_t \text{ or } [A]\{u\} = \{F\} \tag{8.6}$$

Therefore, Eq. (8.6) is the explicit computation recurrence formula when the finite element dynamic analysis is used to analyze structure.

The system consisting of the whole internal nodal points can be treated as a closed linear system with the finite freedoms. In terms of modal superposition concept, $\{v\}_t = (\{u\}_t - \{u\}_{t-\Delta t})/\Delta t$, corresponding to Eq. (8.6) discrete format single freedom form, it can be $u_{n+1} = (2 - 2\xi\omega\Delta t - \omega^2\Delta t^2)u_n + (2\xi\omega\Delta t - 1)u^{n-1}$, whose stable conditions can be $\omega\Delta t \leq 2\sqrt{1 + \xi^2} - 2\xi$, where, ξ is the damping ratio.

8.3.2 INFLUENCE FACTORS ON EXPLICIT INTEGRATED COMPUTATION FORMAT NUMERICAL STABILITY

At present, a batch of high dams of 300 m in height is being constructed in our country. The responses of these complicated structures under the seismic loading must be computed, and the freedom required for the solution increases remarkably so that the explicit computation method with the advantages of saving internal memory space becomes more and more apparent. The largest stable time step length in dynamic computation determines computation efficiency. Under the condition of ensuring calculation accuracy in step length can save computation time and raise computation efficiency, but time step length is subject to the constraint of stable conditions, whereas the discrete format stability is subject to many factors such as equation discrete method, mesh sizes, shapes, interpolation function, damping forms, sizes, and so on. Much research work has been done on the effects of various kinds of factors upon the numerical integration stability. Wang et al. (2002) carried out the analysis of numerical stability of Rayleigh damping medium finite element discrete model and discussed damping ratio and effects of discrete format upon the numerical stability of dynamic analysis. However, in the existing research, no research achievements have been found in considering unit inner integration mode and the unit nonminimum side length stabilizing time step length influence. The real calculation results indicate that apart from the above many factors affecting stabilizing time step length, the unit inner integration mode and the unit nonminimum side length affecting computing stable time step length need to make definitely clear these factors affecting stabilization mechanism so as to be more convenient to decide the best time step length and to improve computation efficiency in the computing process.

8.3.2.1 The unit inner integration mode affecting stability

Numerical instability originates in the process of model discreteness. Model discreteness includes computation equation discreteness (time discreteness) and computation objective entity discreteness (space discreteness). Computation equation discreteness, that is, what difference format is used to replace the original difference equation; computation objective entity discreteness is the essence of finite element, that is, the original complex and irregular geometric body is profiled into several geometric solids with small volume and regular shapes. Discreteness of entity model can take various kinds of modes. In order to be convenient for discussion, the eight-node cubic unit with representative is selected as the research objective, and the unit inner-interpolation function and the unit shape function take the same values, that is, the isoparameter unit to discuss the stability of numerical integration.

1. The eight-node cubic isoparameter shape function.
   The common eight-node cubic isoparameter unit shape function can take the following:

$$\phi_I = \frac{1}{8}\left(1+\xi_0\right)\left(1+\eta_0\right)\left(1+\zeta_0\right) \tag{8.7}$$

   where $\xi_0 = \xi_i\xi$ $\qquad \eta_0 = \eta_i\eta$ $\qquad \zeta_0 = \zeta_i\zeta$

2. In local coordinate, node (ξ,η,ζ) in the whole coordinate system can be as follows:

$$x_i = x_{iI}\phi_I(\xi,\eta,\zeta) \qquad i=1,\cdots,3, I=1,\cdots,8 \tag{8.8}$$

   Equation (8.8) has the implicit summation convention to the subscript (x_{1I}, x_{2I}, x_{3I}) of the I node coordinate value in the whole coordinate system.
   Meanwhile, the I node displacement is (u_{1I}, u_{2I}, u_{3I}), but in the local coordinate system, the coordinate is (ξ,η,ζ) and the node displacement coordinate is (u_1, u_2, u_3), thus expressing as follows:

$$u_i = \sum_{I=1}^{8}\phi_I\left(\xi,\eta,\zeta\right)u_{iI} \quad (i=1,\cdots,3) \tag{8.9}$$

Based on the unit node coordinative (x_I, y_I, z_I) and displacement (u_I, v_I, w_I), to derive any one point coordinative (x, y, z) and displacement (u, v, w) within the unit inner part, the same unit interpolation function $\{\phi_I(\xi, \eta, \zeta)\}_{I=1,\cdots,8}$, is adopted so that such a unit can be called as isoparameter unit.

3. The whole loading equivalence is node f_{xI}, f_{yI}, f_{zI}, the node displacement u_I, v_I, w_I, using virtual working principle (all the external force doing work on the virtual displacement equals to the internal stress doing work on the virtual strain), namely:

$$f_{il} u_{il} = \int_V \sigma_{ij} \varepsilon_{ij}\, dV \Rightarrow \sum_{I=1}^{8} f_{xI} u_I + \sum_{I=1}^{8} f_{yI} v_I + \sum_{I=1}^{8} f_{zI} w_I = \int_V \sigma_{ij} \varepsilon_{ij}\, dV \qquad (8.10)$$

The two ends of Eq. (8.10) versus time to derive the derivatives:

$$f_{il} \dot{u}_{il} = \int_V \sigma_{ij} \dot{\varepsilon}_{ij}\, dV \Rightarrow \sum_{I=1}^{8} f_{xI} \dot{u}_I + \sum_{I=1}^{8} f_{yI} \dot{v}_I + \sum_{I=1}^{8} f_{zI} \dot{w}_I = \int_V \sigma_{ij} \dot{\varepsilon}_{ij}\, dV \qquad (8.11)$$

Neglecting the nonlinear part distributed in the unit displacement field, Eq. (8.11) becomes the following:

$$\dot{u}_{il} f_{il} = V \bar{\sigma}_{ij} \dot{\bar{u}}_{i,j} \qquad (8.12)$$

Where $\bar{\sigma}_{ij}$ is the unit average stress and $\dot{\bar{u}}_{i,j}$ is the unit average speed gradient.

$$\dot{\bar{u}}_{i,j} = \frac{1}{V} \int_V \dot{u}_{i,j}\, dV\, (i = 1, \cdots, 8, j = 1, \cdots, 3) \qquad (8.13)$$

$$B_{i,j} = \int_V \frac{\partial \phi_i}{\partial x_j}\, dV\, (i = 1, \cdots, 8, j = 1, \cdots, 3) \qquad (8.14)$$

Based on the mentioned equation, Belyschko et al. (1981) obtained the maximum stable time step length formula under the center difference format as follows:

$$\Delta t \leq V \sqrt{\frac{\rho}{2(\lambda + 2\mu) B_{il} B_{il}}} \left(\sqrt{1 - \zeta^2} - \zeta\right)$$
$$\lambda = \frac{vE}{(1+v)(1-2v)} \qquad (8.15)$$

$$\mu = G = \frac{E}{2(1+v)} \qquad (8.16)$$

where ζ is the system Rayleigh damping ratio; E is elastic modulus; v is Poison's ratio; λ, μ the Lame coefficient; and subscript il is the summation and convention.

4. *Computation example*: Taking the cubic structure bearing its own weight under natural condition as an example, the side length of this cubic is 10-m long, its density is 2×10^3 kg/m^3, Poison's ratio is 0.3, Rayleigh damping coefficient takes $\alpha = 6.28$, $\beta = 0.0015$, and gravity acceleration is $g = 9.80$ m/s^2. Boundary condition can take bottom face vertical displacement limitation of zero.

Table 8.1 Result Comparisons Between Real Computation and Theoretic Prediction

Mesh Size Ratio	Max. Stable Time Step Length Ratio Value		Deviation Ratio (%)
	Calculated Value	Theoretic Value	
(1.25/1)1.25	1.55	1.5625	0.8
(1/0.5)2	4.02	4.0	0.5
(1.25/0.5)2.5	6.23	6.25	0.32
(1/0.625)1.6	2.58	2.56	0.78
(0.625/0.5)1.25	1.56	1.5625	0.26
(1.25/0.625)2	4	4	0

It can be known through calculation that in terms of the mentioned theoretical formulas, the derived 1 m × 1 m × 1 m mesh or grid maximum stable time step length is $\Delta t = 2.11 \times 10^{-4}$ s. The following is the obtained maximum stable time step length under different mesh sizes:

a. Gauss' integral mode is adopted within the internal unit:

1 m × 1 m × 1 m mesh maximum stable time step length is 4.5×10^{-5} s; 0.5 m × 0.5 m × 0.5 m mesh maximum stable time step length is 1.0×10^{-5} s; if mesh size is reduced by 1/2, the maximum stable time step length decrease is 1/4.

b. Vertex integral mode is adopted within the internal unit:

1 m × 1 m × 1 m mesh maximum stable time step length is 3.096×10^{-5} s; 0.5 m × 0.5 m × 0.5 m mesh maximum table time step length is 7.7×10^{-4} s; 1.25 m × 1.25 m × 1.25 m mesh maximum stable time step length is 4.8×10^{-5} s; 0.625 m × 0.625 m × 0.625 m mesh maximum stable time step length is 1.2×10^{-5} s. Table 8.1 lists the maximum stable time step length of real computations and theoretical postulation in the case of various kinds of mesh sizes. It can be seen from these that in eight-cubic unit, mesh size square ratio equals to the maximum stable time step length ratio.

At the same time, it can be seen that the integration within the unit is whether vertex integral mode or Gauss' integral mode is adopted, the maximum stable time step length obtained in terms of real computation is not larger than the maximum stable time step length ratio obtained in terms of Eq. (8.15). The advent of such case is likely that some factors affecting stability may be ignored by the equations, for instance, when the unit rigidity computation results to make discreteness become large.

5. The initial analysis of the unit internal integral mode affecting stable time step length causes. Vertex integral mode is of the first-order calculation accuracy, that is to say, if the integrated function is a linear function, the results obtained will be accurate. If the integrated function is the second or more orders, the greater errors will be produced, whereas Gauss' integral calculates the accuracy from the first-order to the higher-order in terms of selected different numbers of integral nodes (Wang and Min, 2003). Under the same conditions of numbers of selected integral nodes, Gauss' integral numerical accuracy is higher. The following example can show the accurate differences between Gauss' integral and vertex integral.

The computation integral formulas of vertex integral and Gauss' integral are used individually:

Accurate integral values $\int_0^1 \left(10^x + x^2 + x\right) dx = \left(\dfrac{1}{\ln 10} 10^x + \dfrac{x^3}{3} + \dfrac{x^2}{2}\right)\Bigg|_0^1 = 4.742$

1. **Vertex integral:**

Integral node position $x_1 = 0$, $x_2 = 1$; integral coefficient $D_1 = 0.5$, $D_2 = 0.5$; Integral domain is $[0,1]$; the integrated function value on the integral nodes $F(x_1) = 1.0$, $F(x_2) = 12$, there will be the following:

$$\int_0^1 \left(10^x + x^2 + x\right) dx = 1.0[D_1 F(x_1) + D_2 F(x_2)] = 6.5$$

Integral error $\varepsilon_d = 37.1\%$

2. **Two-node Gauss' integral:**

Integral node $g_1 = \dfrac{1}{2}\left(1 - \dfrac{1}{\sqrt{3}}\right) = 0.211325$, $g_2 = \dfrac{1}{2}\left(1 + \dfrac{1}{\sqrt{3}}\right) = 0.788765$, $F\left(g_1\right) = 1.883$, $F\left(g_2\right) = 7.558$

Integral coefficient $G_1 = 1/2$, $G_2 = 1/2$, there will be the following:

$\int_0^1 \left(10^x + x^2 + x\right) dx = \dfrac{1}{2}(1.883 + 7.558) = 4.72$, integral error $\varepsilon_d = 0.46\%$.

It can be seen from the example that in the case of the same integral nodes, Gauss' integral accuracy is much higher than that of vertex integral, for there exists this kind of difference; the unit rigidity on the wider data band or tape with accurate solution as the center can be obtained by adopting vertex integral computation so that self-shock frequency will be larger in accordance with large rigidity, and the unit self-shock frequency discreteness can be even larger, but the stable time step length derived from the unit self-shock frequency can be smaller.

In the example, the maximum stable time step length of common vertex integral in 1 m × 1 m × 1 m uniform mesh is 3.1×10^{-5} s, while the maximum stable time step length of two Gauss' node integral is 4.5×10^{-5} s. Therefore, in adopting Gauss' node integral, its maximum stable time step length can be raised by 45%. This shows that there exists important relation between the integral format stability and the adaptation of numerical integral mode within the inner unit and that Gauss' integral node high accuracy will have a favorable effect upon the numerical stability. There were no considerations given to the effects of integral mode upon the stability in researches by Flangan and Belytschko (1981) and Wang et al. (2002). The real computation results indicate that the effects of integral mode upon the stability cannot be neglected. Based on the analysis of the stability of explicit computation method, $\omega \Delta t \langle S(A)$ is known. It is here that $S(A)$ is the spectrum radius of recurrence matrix. $S(A)$ value is only decided by the recurrence formula. For this reason, in the case of no changes in discrete format, Δt maximum value is decided by the maximum value of unit rigidity K, while the unit rigidity K maximum value is dependent on the accuracy grade of the unit integral operation, whereby making the discreteness of the derived unit rigidity become large. The ω maximum value can decide Δt maximum value, of which the enlargement of unit rigidity maximum value has resulted in the enlargement of maximum value ω of self-shock cycle, whereby making the maximum stable time step length Δt become small.

Table 8.2 The Maximum Stable Time Step Length Obtained from Practical Calculation by Adopting Different Units

Unit Side Length (m)	Maximum Stable Time Step Length ($\times 10^{-5}$ s)
$1 \times 1 \times 1$	4.54
$1.25 \times 1 \times 1$	4.75
$1.25 \times 1.25 \times 1$	4.88
$2 \times 1 \times 1$	4.78
$2 \times 2 \times 1$	4.91
$2 \times 1.25 \times 1$	4.9
$1.25 \times 1.25 \times 1.25$	7.04
$2 \times 2 \times 2$	17.3

8.3.2.2 The nonminimum side length affecting stability in nonequilateral length unit

Still taking 10-m side length cubic structure bearing self-weight in the natural state for an example, the physical parameters used are the same ones in the earlier examples. Table 8.2 lists the maximum stable time step length obtained from practical calculation by adopting different units. It can be seen that the minimum side length decides the quantitative grade of the maximum stable step length, and at the same time, the unit other side lengths have some effect upon stable time step length but the effects are not serious. When the nonregular meshes are in general determined in their maximum stable time step length, the corresponding unit minimum side length can be decided through Eq. (8.15) to determine the initial quantitative grade, and then, there will be a corresponding slight raising in terms of other side lengths, and of course, this contains some experience factors.

It can be seen from the analysis that the numerical stability is very important when an explicit method is adopted to carry out the dynamic computation. This part analyzes in brief two factors affecting the numerical stability of the discrete format that can be often ignored by the researchers, that is, the inner unit integral mode and the unit nonminimum side length, and discusses the mechanism of both to affect numerical stability as well as the influence degrees upon stable time step length. As to these two factors, they are, not completely at present, introduced into the calculation formula of stable time step length, and it can only be based on the practical adaptation of the unit inner integral accuracy that the stable time step length be varied so as to obtain the optimal time step length. It is necessary to point out that when the integrated function in the inner unit is the linear function, the common vertex integral is able to obtain its accurate numerical solution, and at this time, it is to change to use the high-order calculation accuracy method that the time step length adopted in calculation cannot be improved in such a way that the flexible operation of this influencing factor can obtain the highest efficiency with the least working quantity.

8.3.3 LAGRANGE'S METHOD OF MULTIPLIERS OF MOTION CONTACT NONLINEARITY

Parallel program adopts an explicit algorithm to calculate the dam body to seismic loading dynamic response. Hydrodynamic pressure in analysis is considered as the attached quality attached to the dam body surface. The mountain mass and dam body are viewed as the linear elastic body, and dam body temperature loading is also considered. The concentration mass method is adopted in the process of equation discreteness, and the mountain mass and dam body serve as the linear elastic materials. After the discreteness, the dynamic basic equation adopts Eq. (8.6) form, that is, $AU = F$.

In arch dam seismic response analysis, there have frequently occurred many problems of the existing fault sliding move belts in the mountain mass, structure fractures, and temperature cracks in the dam body. In order to reflect fracture face contact, sliding move, and nonjoint process, Lagrange's method of multipliers' λ is introduced into the dynamic basic Eq. (8.6) to express the contact force, and the friction problem is converted into seeking solution to the following saddle point problem, whose energy universal function is as follows:

$$\begin{cases} \max_{\lambda} \min_{u} J(u,\lambda) = \frac{1}{2} U^T AU + U^T B\lambda + \lambda^T g - U^T F \\ \lambda_n^l + N \geq 0, \left| \lambda_\tau^l \right| \leq \mu\left(\lambda_n^l + N \right) + ca, \end{cases} \quad \text{of which} \left| \lambda_\tau^l \right| = \sqrt{\lambda_{\tau 1}^{l2} + \lambda_{\tau 2}^{l2}} \qquad (8.17)$$

where g is some node initial displacement constraint value vector in node pair; λ_n^l, λ_τ^l is normal multiplier, tangential multiplier under the local coordinate; μ is traction coefficient; N is the initial pressure on the contact surface; c is the cohesion; and a is node equal effect area.

As to every contact node pair $B_i = [I, -I], \lambda_i = [\lambda_n, \lambda_\tau]$; total Lagrange's method of multiplier number is the contact node pair number and nodal point freedom product; when the contact surface appears to have the sliding, $ca = 0$, from $\min_u J(u, \lambda)$, there will be the following:

$$AU + B\lambda - F = 0$$
$$U = A^{-1}(F - B\lambda) \qquad (8.18)$$

Matrix T is converted through the whole coordinate and the local coordinate, and the multiplier is converted under the local coordinate system, that is, $\lambda^l = T\lambda$, that is, the seeking solution Eq. (8.17) equals to seek solution to the following equation as follows:

$$\begin{cases} \min_{\lambda^l} \left(\frac{1}{2} \lambda^{lT} B^T A^{-1} B\lambda^l - l^{lT} T B^T A^{-1} F \right) \\ \lambda_n^l + N \geq 0, \left| \lambda_\tau^l \right| \leq \mu\left(\lambda_n^l + N \right) + ca \end{cases} \qquad (8.19)$$

And hence, it is first converted under the local coordinate system, and again to seek the solution to the multiplier and to first seek the solution to the multiplier under the coordinate system and again, the multiplier converted to the local coordinate is equal, whereby indicating that the explicit algorithm can directly seek solution under the whole coordinate, and then is converted under the local coordinate system to carry out the contact judgment and the corresponding modification is made. Therefore, for each nodal point pair, there will be the following:

$$B^T A^{-1} B = (I, -I) \begin{pmatrix} m_1^{-1} I & 0 \\ 0 & m_2^{-1} I \end{pmatrix} \begin{pmatrix} I \\ -I \end{pmatrix} = \frac{m_1 + m_2}{m_1 m_2} I$$

$$B^T A^{-1} F = (I, -I) \begin{pmatrix} m_1^{-1} I & 0 \\ 0 & m_2^{-1} I \end{pmatrix} \begin{pmatrix} f_1 \\ f_2 \end{pmatrix} = \frac{f_1}{m_1} - \frac{f_2}{m_2}$$

And so $\lambda = \dfrac{m_2 f_1 - m_1 f_2}{m_1 + m_2}$

$$\lambda^l = T \frac{m_2 f_1 - m_1 f_2}{m_1 + m_2}$$

After the modification with the following modes so as to make them satisfy the inequality constraint conditions:

1. if $\lambda_n^l + N \le 0, \lambda^l = 0$

2. if $\lambda_n^l + N > 0$: when $\left| \lambda_\tau^l \right| \le \mu \left(\lambda_n^l + N \right) + ca, \lambda_\tau^l = \lambda_\tau^l$; when $\left| \lambda_\tau^l \right| < \mu \left(\lambda_n^l + N \right)$

$$\lambda_\tau^l = \frac{\lambda_\tau^l \left[\mu \left(\lambda_n^l + N \right) + ca \right]}{\left| \lambda_\tau^l \right|}.$$

And then $\lambda = T^\tau \lambda^l$ is converted into the whole coordinate system, λ and μ value is derived from $AU = F - B\lambda$. In this way, seeking solution to Eq. (8.17) can obtain the nodal point displacement and the contact force of the contact face. Explicit algorithm format can make full use of special performances of concentrated mass matrix, whereby making every time step multiplier computation be directly obtained without its iterations.

8.3.4 POINT-BY-POINT MULTIPLIER METHOD FOR OFF-DIAGONAL ADDITIONAL MASS MATRIX

Hydrodynamic pressure and sediment dynamic loading functions should be considered in accordance with the Westergaard model and additional mass approximation. Owing to the existence of additional mass, mass matrix is no longer a diagonal matrix so that the earlier derived algorithm cannot obtain the nodal point contact force, but the nodal point without additional mass, mass matrix can still satisfy mass matrix diagonal conditions so that the previous method can be adopted to directly obtain the displacement and contact force. As to the nodal point with additional mass, the pointwise calculation method can be adopted to carry out seeking for solution.

Point-by-point or stepwise multiplier method (Tang, 2006) is for each point or node with additional mass to adopt the iteration method to derive the contact force by orders, whereby avoiding the inverse matrix of the large-scale matrix.

Equation (8.18) can be written into the following component form:

$$\sum_j a_{ij} u_j + \sum_k b_{ij} \lambda_j = f_i \Rightarrow a_{ii} u_i + b_{ii} \lambda_i = f_i - \sum_{j \ne i} a_{ij} u_j - \sum_{k \ne i} b_{ik} \lambda_k \qquad (8.20)$$

The subscript i expresses the i point pairs, in the case of three-dimensional situations, a_{ii} is 6×6 square matrix. The similar Gauss–Seidel iteration method is employed in Eq. (8.20) to seek the solution; there will be the following:

$$a_{ii} \overline{u}_i + b_{ii} \overline{\lambda}_i = f_i - \sum_j a_{ij} u_j - \sum_k b_{ik} \lambda_k + a_{ii} u_i + b_{ii} \lambda_i \qquad (8.21)$$

Letting $\sum_j a_{ij} u_j + \sum_k b_{ik} \lambda_k - f_i = r_i$, Eq. (8.21) can be simplified as follows:

$$a_{ii} \overline{u}_i + b_{ii} \overline{\lambda}_i = a_{ii} u_i + b_{ii} \lambda_i - r_i \tag{8.22}$$

Equation (8.22) will be combined with the friction contact constraint conditions of node pairs, the friction contact problem can be converted into the following solution equations:

$$\begin{cases} \min_{\overline{\lambda}_i^l} \left(\dfrac{1}{2} \overline{\lambda}_i^{lT} T_i B_{ii}^T A_{ii}^{-1} B_{ii} T_i^T \overline{\lambda}_i^l - \overline{\lambda}_i^{lT} T_i B_{ii}^T A_{ii}^{-1} f \right) \\ \overline{\lambda}_{in}^l + N \ge 0, \left| \overline{\lambda}_{is}^l \right| \le \mu_i \left(\overline{\lambda}_{in}^l + N \right) + ca \end{cases} \tag{8.23}$$

Gauss–Seidel iteration method can be used to derive the solution to node pair contact force $\overline{\lambda}_i^l$, and then kilograms per meter cube can be converted into the contact force $\overline{\lambda}_i$ under the whole coordinate system.

8.4 FEPG SYSTEM AND FINITE ELEMENT METHOD BASED ON FEPG

8.4.1 FEPG SYSTEM PROGRAM STRUCTURE PERFORMANCES

The FEPG system adopts the component program design method to decompose the finite element computation program generator into several element programs (Guoping, 1990). Each element program is a complete FORTRAN program that can carry out the compilation, connection, and operation independently, and programs communicate via magnetic disk file. The programs system consisting of element or component programs conducts the compilation and implementation by the batch process command of the command stream generator. The finite element computation program consists of START, BFT, SOLV, and E.V 5 element programs, of which the two E and U element programs are the program generators based on the expression formulas given by the users, and the three other programs are given by the system, changing not following the change in the expression formulas. The five element programs have the following functions:

1. *START program*: Gives each freedom of every node and the variables of algebraic equation group to be formed in the future (equation numbers) corresponding relationship (i.e., which node's which freedom will correspond to which variable of the equation group) and the initial solution values.
2. *BFT element program*: Gives out the boundary value to be solved at every moment, that is, to specify node displacement and loading as well as the time renew and the calculation results to be preserved.
3. *SOLV solver*: Used to iterate the formed general rigidity matrix and solution linear algebraic equation group.
4. *E element program*: Used to calculate single rigidity, single mass, unit loading, and so on, and to convert the expression of each freedom in its node into the variable expression by algebraic equation group, and at the same time, to deal with the boundary constraint conditions so as to form the right end items of the algebraic equation group.
5. *U element program*: Used to convert the variable displacement derived by the solver into each freedom displacement in node as well as other postprocessor calculations.

The operation of these element programs is completed through the batch command document of the finite element program. The finite element program batch command document is completed through CMD command stream document generator so as to operate this batch command to obtain the finite element solution.

The FEPG system to steady and dynamic states, linear and nonlinear finite element problem solution program is shown Fig. 8.1.

8.4.2 THE FINITE UNIT ELEMENT METHOD ON FEPG

The finite element method is a kind of numerical solution partial differential equation method. Its solution, in fact, is the weak solution of partial differential equation so that it is only necessary first to change the partial differential equation into the weak solution integral form that the finite element method can be used. In the FEPG system, the users can obtain the whole program far calculation needs with two kinds of ways, as shown in Fig. 8.2. One kind of the method is to use the formula tank or bank to generate the program. In such common physical problems as solid mechanics electric magnetic field, thermal conductivity, seepage flow, fluid mechanics, and so on, descriptive equations can be used using the finite element language and put into the formula bank in such a way that the users only need to have to point and click to the formula menu to generate all the finite element computation programs that they need. The users can employ the finite element language to write the control equation into VDE or PDE documents and to write the calculation method into GCN and GIO (MDI) documents based on their own research on the physical problems, and then use the FEPG system command (GIO command) to generate the whole finite element programs.

The following work must be, in general, done when an FEPG system is used to carry out the finite element programming and solution.

1. To compile PDE (VDE), FBC, GCN, and GIO (MDI) documents. If the network edition (IFEPG) is used, the document needs uploading.
2. Operation GIO (MDI) command can generate the computation program.
3. If material parameters are complicated, they need to be modified *.PRE parameter document can regenerate the program.
4. Operation preprocessing centers FEPG, GID to do preprocessing work.
5. To carry out the finite element computation.
6. To enter into postprocessing of computation results.

FEPG finite element documents and other generated FORTRAN sources documents logic corresponding relations are shown in Fig. 8.3.

8.5 ARTIFICIAL BOUNDARY REALIZATION IN FEPG

8.5.1 INPUT FORMULA OF SEISMIC WAVE IN ARTIFICIAL VISCOELASTIC BOUNDARY

In numerical computation, first of all, it is necessary to decide the semispace free wave field, and to compute the loadings needed to apply to the finite element nodes of all the artificial boundaries, that is,

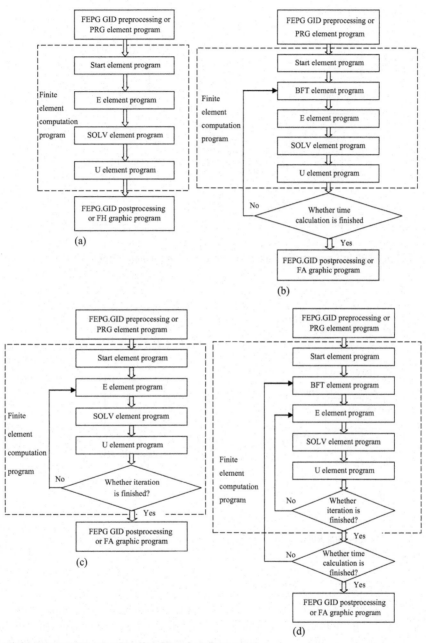

FIGURE 8.1 Finite Element Program Solution Flowchart Framework

(a) Linear steady problem solution flowchart framework. (b) Linear dynamic problem solution flowchart framework. (c) Nonlinear steady problem solution flowchart framework. (d) Nonlinear dynamic problem solution flowchart framework.

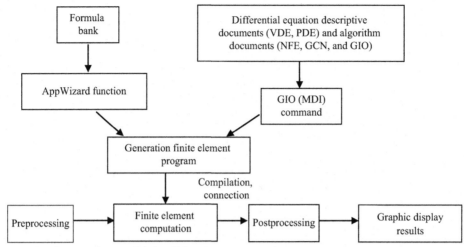

FIGURE 8.2 Applying FEPG in Carrying Out Finite Element Computing General Process

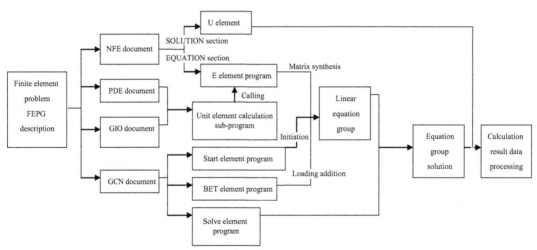

FIGURE 8.3 Logic Corresponding Relationship of FEPG Finite Element Document and Generated FORTRAN Sources Document (FEPG 2006)

the loadings $\{\bar{F}_b^f\} = \{\bar{X}_i^f\} + [c_b]\{\dot{u}^f\} + [k_b]\{u^f\}$ yielded by seismic waves are input into the artificial boundary, the detailed computation formulas for the loadings on each boundary surface in the near fields will be given out in the following: ground surface serves as an infinite great body surface, ground surface boundary condition can be used as the first group boundary and viewed as the face-free force surface, that is, $\bar{X} = 0; \bar{Y} = 0; \bar{Z} = 0$ while one assumed cut-off face parallel to the surface is taken as the first grade boundary, into which the seismic waves are input. Let seismic waves not change with the coordinate x and y, $\partial(*)/\partial x = \partial(*)/\partial y = 0$, that is, seismic waves are used as one-dimensional plane

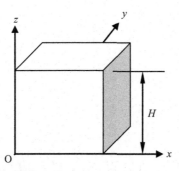

FIGURE 8.4 Physical Model of Viscoelastic Boundary

waves input from the bottom boundary as is shown in Fig. 8.4. Vertical input P wave, and SV waves are input along the directions of x-axis and y-axis individually.

8.5.1.1 Semi-infinite large body free field

The three free-field displacement components u, v, w can satisfy the one-dimensional wave motion equation:

$$\frac{\partial^2 u}{\partial z^2} = \frac{1}{c_s^2}\frac{\partial^2 u}{\partial t^2}; \quad \frac{\partial^2 v}{\partial z^2} = \frac{1}{c_s^2}\frac{\partial^2 v}{\partial t^2}; \quad \frac{\partial^2 w}{\partial z^2} = \frac{1}{c_p^2}\frac{\partial^2 w}{\partial t^2} \tag{8.24}$$

where, $c_p = \sqrt{\dfrac{\lambda+2\mu}{\rho}} = \sqrt{\dfrac{(1-v)E}{(1+v)(1-2v)\rho}}$; $c_s = \sqrt{\dfrac{\mu}{\rho}} = \sqrt{\dfrac{G}{\rho}} = \sqrt{\dfrac{E}{2(1+v)\rho}}$

$$c_s^2 = \frac{\mu}{\rho}; \ \rho c_s = \frac{\mu}{c_s} \quad c_p^2 = \frac{\lambda+2\mu}{\rho} \quad \rho c_p = \frac{\lambda+2\mu}{c_p} \tag{8.25}$$

Owing to $\partial(*)/\partial x = \partial(*)/\partial y = 0$, free-field strain can be simplified as follows:

$$\varepsilon_{xx} = \frac{\partial u}{\partial x} = 0; \varepsilon_{yy} = \frac{\partial v}{\partial y} = 0; \varepsilon_{xy} = \frac{\partial u}{\partial y} + \frac{\partial v}{\partial x} = 0$$

$$\varepsilon_{zz} = \frac{\partial w}{\partial z}; \varepsilon_{xz} = \frac{\partial u}{\partial z}; \varepsilon_{yz} = \frac{\partial v}{\partial z} \tag{8.26}$$

Obviously, the displacement field can have the following formal solutions:

$$\begin{cases} u^f = u_0\left(t - \dfrac{z}{c_s}\right) + u_0\left(t - \dfrac{2H-z}{c^s}\right) \\[3mm] v^f = v^0\left(t - \dfrac{z}{c_s}\right) + v_0\left(t - \dfrac{2H-z}{c_s}\right) \\[3mm] w^f = w_0\left(t - \dfrac{z}{c_p}\right) + w_0\left(t - \dfrac{2H-z}{c_p}\right) \end{cases} \tag{8.27}$$

$$\begin{cases}
\dot{u}^f = \dot{u}_0\left(t - \dfrac{z}{c_s}\right) + \dot{u}_0\left(t - \dfrac{2H - z}{c_s}\right) \\[2ex]
\dot{v}^f = \dot{v}_0\left(t - \dfrac{z}{c_s}\right) + \dot{v}_0\left(t - \dfrac{2H - z}{c_s}\right) \\[2ex]
\dot{w}^f = \dot{w}_0\left(t - \dfrac{z}{c_p}\right) + \dot{w}_0\left(t - \dfrac{2H - z}{c_p}\right)
\end{cases} \tag{8.28}$$

$$\begin{cases}
\dfrac{\partial u^f}{\partial z} = -\dfrac{1}{c_s}\left[\dot{u}_0\left(t - \dfrac{z}{c_s}\right) - \dot{u}_0\left(t - \dfrac{2H - z}{c_s}\right)\right] \\[2ex]
\dfrac{\partial v^f}{\partial z} = -\dfrac{1}{c_s}\left[\dot{v}_0\left(t - \dfrac{z}{c_s}\right) - v_0\left(t - \dfrac{2H - z}{c_s}\right)\right] \\[2ex]
\dfrac{\partial w^f}{\partial z} = -\dfrac{1}{c_p}\left[\dot{w}_0\left(t - \dfrac{z}{c_p}\right) - \dot{w}_0\left(t - \dfrac{2H - z}{c_p}\right)\right]
\end{cases} \tag{8.29}$$

The stress derived from the fundamental structure can be as follows:

$$\begin{Bmatrix}
\sigma_{xx} \\ \sigma_{yy} \\ \sigma_{zz} \\ \sigma_{yz} \\ \sigma_{xz} \\ \sigma_{xy}
\end{Bmatrix} = \begin{Bmatrix}
\lambda\varepsilon_{zz} \\ \lambda\varepsilon_{zz} \\ (\lambda + 2\mu)\varepsilon_{zz} \\ \mu\varepsilon_{yz} \\ \mu\varepsilon_{xz} \\ 0
\end{Bmatrix} = \begin{Bmatrix}
\lambda\dfrac{\partial w}{\partial z} \\[1.5ex]
\lambda\dfrac{\partial w}{\partial z} \\[1.5ex]
(\lambda + 2\mu)\dfrac{\partial w}{\partial z} \\[1.5ex]
\mu\dfrac{\partial v}{\partial z} \\[1.5ex]
\mu\dfrac{\partial u}{\partial z} \\[1.5ex]
0
\end{Bmatrix} \tag{8.30}$$

8.5.1.2 Artificial boundary surface loading

It can be known from Eq. (8.6) that the artificial viscoelastic boundary surface force can be
$\{\overline{F}_b^f\} = \{\overline{X}_i^f\} + [c_b]\{\dot{u}^f\} + [k_b]\{u^f\}$,

where

$$\{\overline{X}_i^f\} = \begin{Bmatrix}
\overline{X}_b^f \\ \overline{Y}_b^f \\ \overline{Z}_b^f
\end{Bmatrix} = \begin{bmatrix}
\sigma_{xx} & \sigma_{yx} & \sigma_{zx} \\
\sigma_{xy} & \sigma_{yy} & \sigma_{zy} \\
\sigma_{xz} & \sigma_{yz} & \sigma_{zz}
\end{bmatrix} \begin{Bmatrix}
l \\ m \\ n
\end{Bmatrix} = \begin{Bmatrix}
l\sigma_{xx} + n\sigma_{zx} \\
m\sigma_{yy} + n\sigma_{zy} \\
l\sigma_{xz} + m\sigma_{yz} + n\sigma_{zz}
\end{Bmatrix} \tag{8.31}$$

where l, m, and n are the exterior normal direction cosine on the surface.

In the following, seismic waves will be input into the near foundation bottom boundary as indicated in Fig. 8.4, thus giving out the bottom boundary and four side boundary surface force formulas.

1. In the bottom boundary ($z = 0$) incidence:
   Wave incidence in the bottom boundary, $u = u_0(t)$; $v = v_0(t)$; $w = w_0(t)$; $t \geq 0$, on the bottom boundary $z = 0$; $l = 0$; $m = 0$; $n = -1$, using $-z$ to express.

   Substituting Eq. (8.29) into Eq. (8.30), we obtain:

$$\sigma_{xz} = \mu \frac{\partial u}{\partial z} = -\rho c_s \left[\dot{u}_0 \left(t - \frac{z}{c_s} \right) - \dot{u}_0 \left(t - \frac{2H - z}{c_s} \right) \right]$$

$$\sigma_{yz} = \mu \frac{\partial v}{\partial z} = -\rho c_s \left[\dot{v}_0 \left(t - \frac{z}{c_s} \right) - \dot{v}_0 \left(t - \frac{2H - z}{c_s} \right) \right]$$

$$\sigma_{zz} = (\lambda + 2\mu) \frac{\partial w}{\partial z} = -\rho c_p \left[\dot{w}_0 \left(t - \frac{z}{c_p} \right) - \dot{w}_0 \left(t - \frac{2H - z}{c_p} \right) \right]$$

$$\left\{ \begin{array}{c} \overline{X}_b \\ \overline{Y}_b \\ \overline{Z}_b \end{array} \right\} = \left\{ \begin{array}{c} l\sigma_{xx} + n\sigma_{zx} \\ m\sigma_{yy} + n\sigma_{zy} \\ l\sigma_{xz} + m\sigma_{yz} + n\sigma_{zz} \end{array} \right\} = \left\{ \begin{array}{c} -\sigma_{zx} \\ -\sigma_{zy} \\ -\sigma_{zz} \end{array} \right\} = \left\{ \begin{array}{c} \rho c_s \left[\dot{u}_0 \left(t - \frac{z}{c_s} \right) - \dot{u}_0 \left(t - \frac{2H-z}{c_s} \right) \right] \\ \rho c_s \left[\dot{v}_0 \left(t - \frac{z}{c_s} \right) - \dot{v}_0 \left(t - \frac{2H-z}{c_s} \right) \right] \\ \rho c_p \left[\dot{w}_0 \left(t - \frac{z}{c_p} \right) - \dot{w}_0 \left(t - \frac{2H-z}{c_p} \right) \right] \end{array} \right\} \quad (8.32)$$

while

$$[k_b] = \begin{bmatrix} G/2r_b & 0 & 0 \\ 0 & G/2r_b & 0 \\ 0 & 0 & E/2r_b \end{bmatrix}; \quad [c_b] = \begin{bmatrix} \rho c_s & 0 & 0 \\ 0 & \rho c_s & 0 \\ 0 & 0 & \rho c_p \end{bmatrix} \quad (8.33)$$

   Substituting Eq. (8.32) and Eq. (8.31), we obtain the bottom boundary surface force formulas as follows:

$$\left\{ \begin{array}{c} \overline{F}_{bx}^{-z} \\ \overline{F}_{by}^{-z} \\ \overline{F}_{bz}^{-z} \end{array} \right\} = \begin{bmatrix} G/2r_b & 0 & 0 \\ 0 & G/2r_b & 0 \\ 0 & 0 & E/2r_b \end{bmatrix} \left\{ \begin{array}{c} u_0 \left(t - \frac{z}{c_s} \right) + u_0 \left(t - \frac{2H-z}{c_s} \right) \\ v_0 \left(t - \frac{z}{c_s} \right) + v_0 \left(t - \frac{2H-z}{c_s} \right) \\ w_0 \left(t - \frac{z}{c_p} \right) + w_0 \left(t - \frac{2H-z}{c_p} \right) \end{array} \right\} + \left\{ \begin{array}{c} 2\rho c_s \dot{u}_0 \left(t - \frac{z}{c_s} \right) \\ 2\rho c_s \dot{v}_0 \left(t - \frac{z}{c_s} \right) \\ 2\rho c_p [\dot{w}_0 \left(t - \frac{z}{c_p} \right) \end{array} \right\}$$

Attention, $z = 0$, in the preceding equation, there will be the following:

$$\begin{cases} \bar{F}_{bx}^{-z} = \dfrac{G}{2r_b}\left[u_0(t) + u_0\left(t - \dfrac{2H}{c_s} \right) \right] + 2\rho c_s \dot{u}_0(t) \\[3mm] \bar{F}_{by}^{-z} = \dfrac{G}{2r_b}\left[v_0(t) + v_0\left(t - \dfrac{2H}{c_s} \right) \right] + 2\rho c_s \dot{v}_0(t) \\[3mm] \bar{F}_{bz}^{-z} = \dfrac{E}{2r_b}\left[w_0(t) + w_0\left(t - \dfrac{2H}{c_p} \right) \right] + 2\rho c_p \dot{w}_0(t) \end{cases} \tag{8.34}$$

2. In vertical x axial side boundary incidence:
   a. In side boundary $x = 0$; $l = -1$; $m = 0$; $n = 0$ using $-x$ to express. Some node in the bottom boundary depends on the bottom boundary input waves, being the bottom boundary input wave and the ground surface reflection wave overlap in this node, there will be the following:

$$\begin{cases} v^f = v_0\left(t - \dfrac{z}{c_s} \right) + v_0\left(t - \dfrac{2H - z}{c_s} \right) \\[3mm] v^f = v_0\left(t - \dfrac{z}{c_s} \right) + v_0\left(t - \dfrac{2H - z}{c_s} \right) \\[3mm] w^f = w_0\left(t - \dfrac{z}{c_s} \right) + w_0\left(t - \dfrac{2H - z}{c_s} \right) \end{cases} \tag{8.35}$$

$$\sigma_{xx} = \lambda \frac{\partial w}{\partial z} = -\frac{\lambda}{c_p}\left[\dot{w}_0\left(t - \frac{z}{c_p} \right) - \dot{w}_0\left(t - \frac{2H - z}{c_p} \right) \right]$$

$$\sigma_{xz} = \mu \frac{\partial u}{\partial z} = -\frac{\mu}{c_s}\left[\dot{u}_0\left(t - \frac{z}{c_s} \right) - \dot{u}_0\left(t - \frac{2H - z}{c_s} \right) \right]$$

$$\begin{Bmatrix} X_b \\ Y_b \\ Z_b \end{Bmatrix} = \begin{Bmatrix} l\sigma_{xx} + n\sigma_{zx} \\ m\sigma_{yy} + n\sigma_{zy} \\ l\sigma_{xz} + m\sigma_{yz} + n\sigma_{zz} \end{Bmatrix} = \begin{Bmatrix} -\sigma_{xx} \\ -\sigma_{xy} \\ -\sigma_{xz} \end{Bmatrix} = \begin{Bmatrix} -\sigma_{xx} \\ 0 \\ -\sigma_{xz} \end{Bmatrix}$$

$$\begin{Bmatrix} \bar{F}_{bx}^{-x} \\ \bar{F}_{by}^{-x} \\ \bar{F}_{bz}^{-x} \end{Bmatrix} = \begin{bmatrix} E/2r_b & 0 & 0 \\ 0 & G/2r_b & 0 \\ 0 & 0 & G/2r_b \end{bmatrix} \begin{Bmatrix} u_0\left(t - \dfrac{z}{c_s} \right) + u_0\left(t - \dfrac{2H - z}{c_s} \right) \\[3mm] v_0\left(t - \dfrac{z}{c_s} \right) + v_0\left(t - \dfrac{2H - z}{c_s} \right) \\[3mm] w_0\left(t - \dfrac{z}{c_p} \right) + w_0\left(t - \dfrac{2H - z}{c_p} \right) \end{Bmatrix}$$

$$+\begin{bmatrix} \rho c_p & 0 & 0 \\ 0 & \rho c_s & 0 \\ 0 & 0 & \rho c_s \end{bmatrix}\begin{Bmatrix} \dot{u}_0\left(t-\dfrac{z}{c_s}\right)+\dot{u}_0\left(t-\dfrac{2H-z}{c_s}\right) \\ \dot{v}_0\left(t-\dfrac{z}{c_s}\right)+\dot{v}_0\left(t-\dfrac{2H-z}{c_s}\right) \\ \dot{w}_0\left(t-\dfrac{z}{c_p}\right)+\dot{w}_0\left(t-\dfrac{2H-z}{c_p}\right) \end{Bmatrix}+\begin{Bmatrix} \dfrac{\lambda}{c_p}\left[\dot{w}_0\left(t-\dfrac{z}{c_p}\right)-\dot{w}_0\left(t-\dfrac{2H-z}{c_p}\right)\right] \\ 0 \\ \dfrac{\mu}{c_s}\left[\dot{u}_0\left(t-\dfrac{z}{c_s}\right)-\dot{u}_0\left(t-\dfrac{2H-z}{c_s}\right)\right] \end{Bmatrix}$$

This equation can be expressed as follows:

$$\begin{Bmatrix} \bar{F}_{bx} \\ \bar{F}_{by} \\ \bar{F}_{bz} \end{Bmatrix}=\begin{bmatrix} E/2r_b & 0 & 0 \\ 0 & G/2r_b & 0 \\ 0 & 0 & G/2r_b \end{bmatrix}\begin{Bmatrix} u^f \\ v^f \\ w^f \end{Bmatrix}+\begin{bmatrix} \rho c_p & 0 & 0 \\ 0 & \rho c_s & 0 \\ 0 & 0 & \rho c_s \end{bmatrix}\begin{Bmatrix} \dot{u}^f \\ \dot{v}^f \\ \dot{w}^f \end{Bmatrix}+\begin{Bmatrix} \dfrac{\lambda}{c_p}\left[\dot{w}_0\left(t-\dfrac{z}{c_p}\right)-\dot{w}_0\left(t-\dfrac{2H-z}{c_p}\right)\right] \\ 0 \\ \dfrac{\mu}{c_s}\left[\dot{u}_0\left(t-\dfrac{z}{c_s}\right)-\dot{u}_0\left(t-\dfrac{2H-z}{c_s}\right)\right] \end{Bmatrix}$$

Namely

$$\begin{cases} \bar{F}_{bx}^{-x}=\dfrac{E}{2r_b}u^f+\rho c_p\dot{u}^f+\dfrac{\lambda}{c_p}\left[\dot{w}_0\left(t-\dfrac{z}{c_p}\right)-\dot{w}_0\left(t-\dfrac{2H-z}{c_p}\right)\right] \\[3mm] \bar{F}_{by}^{-x}=\dfrac{G}{2r_b}v^f+\rho c_s\dot{v}^f \\[3mm] \bar{F}_{bz}^{-x}=\dfrac{G}{2r_b}w^f+\rho c_s\dot{w}^f+\dfrac{\mu}{c_s}\left[\dot{u}_0\left(t-\dfrac{z}{c_s}\right)-\dot{u}_0\left(t-\dfrac{2H-z}{c_s}\right)\right] \end{cases} \tag{8.36}$$

b. In the side boundary $x=x_b$; $l=1$; $m=0$; $n=0$. Using $+x$ to express, the postulation process is similar to that of the side boundary $-x$; there will be the following:

$$\begin{cases} \bar{F}_{bx}^{+x}=\dfrac{E}{2r_b}u^f+\rho c_p\dot{u}^f-\dfrac{\lambda}{c_p}\left[\dot{w}_0\left(t-\dfrac{z}{c_p}\right)-\dot{w}_0\left(t-\dfrac{2H-z}{c_p}\right)\right] \\[3mm] \bar{F}_{by}^{-x}=\dfrac{G}{2r_b}v^f+\rho c_s\dot{v}^f \\[3mm] \bar{F}_{bz}^{+x}=\dfrac{G}{2r_b}w^f+\rho c_s\dot{w}^f-\dfrac{\mu}{c_s}\left[\dot{u}_0\left(t-\dfrac{z}{c_s}\right)-\dot{u}_0\left(t-\dfrac{2H-z}{c_s}\right)\right] \end{cases} \tag{8.37}$$

3. In the vertical y axis side boundary incidence:
   a. In the side boundary $y=0$; $l=0$, $n=0$. Using $-y$ to express, there will be the following:

$$\begin{Bmatrix} \bar{X}_b \\ \bar{Y}_b \\ \bar{Z}_b \end{Bmatrix}=\begin{Bmatrix} l\sigma_{xx}+n\sigma_{zx} \\ m\sigma_{yy}+n\sigma_{zy} \\ l\sigma_{xz}+m\sigma_{yz}+n\sigma_{zz} \end{Bmatrix}=\begin{Bmatrix} 0 \\ -\sigma_{yy} \\ -\sigma_{yz} \end{Bmatrix}$$

$$\sigma_{yz} = \mu \frac{\partial v}{\partial z} = -\rho c_s \left[\dot{v}_0 \left(t - \frac{z}{c_s} \right) - \dot{v}_0 \left(t - \frac{2H - z}{c_s} \right) \right]$$

$$\sigma_{yy} = \lambda \frac{\partial w}{\partial z} = -\frac{\lambda}{c_p} \left[\dot{w}_0 \left(t - \frac{z}{c_p} \right) - \dot{w}_0 \left(t - \frac{2H - z}{c_p} \right) \right]$$

$$\left\{ \begin{array}{c} \overline{F}_{bx} \\ \overline{F}_{by} \\ \overline{F}_{bz} \end{array} \right\} = \left[\begin{array}{ccc} G/2r_b & 0 & 0 \\ 0 & E/2r_b & 0 \\ 0 & 0 & G/2r_b \end{array} \right] \left\{ \begin{array}{c} u_0 \left(t - \dfrac{z}{c_s} \right) + u_0 \left(t - \dfrac{2H - z}{c_s} \right) \\ v_0 \left(t - \dfrac{z}{c_s} \right) + v_0 \left(t - \dfrac{2H - z}{c_s} \right) \\ w_0 \left(t - \dfrac{z}{c_p} \right) + w_0 \left(t - \dfrac{2H - z}{c_p} \right) \end{array} \right\}$$

$$+ \left[\begin{array}{ccc} \rho c_s & 0 & 0 \\ 0 & \rho c_p & 0 \\ 0 & 0 & \rho c_s \end{array} \right] \left\{ \begin{array}{c} \dot{u}_0 \left(t - \dfrac{z}{c_s} \right) + \dot{u}_0 \left(t - \dfrac{2H - z}{c_s} \right) \\ \dot{v}_0 \left(t - \dfrac{z}{c_s} \right) + \dot{v}_0 \left(t - \dfrac{2H - z}{c_s} \right) \\ \dot{w}_0 \left(t - \dfrac{z}{c_p} \right) + \dot{w}_0 \left(t - \dfrac{2H - z}{c_p} \right) \end{array} \right\} + \left\{ \begin{array}{c} 0 \\ \dfrac{\lambda}{c_p} \left[\dot{w}_0 \left(t - \dfrac{z}{c_p} \right) - \dot{w}_0 \left(t - \dfrac{2H - z}{c_p} \right) \right] \\ \dfrac{\mu}{c_s} \left[\dot{v}_0 \left(t - \dfrac{z}{c_s} \right) - \dot{v}_0 \left(t - \dfrac{2H - z}{c_s} \right) \right] \end{array} \right\}$$

It can be simplified as follows:

$$\left\{ \begin{array}{l} \overline{F}_{bx}^{-y} = \dfrac{G}{2r_b} u^f + \rho c_s \dot{u}^f \\[3mm] \overline{F}_{by}^{-y} = \dfrac{E}{2r_b} v^f + \rho c_p \dot{v}^f + \dfrac{\lambda}{c_p} \left[\dot{w}_0 \left(t - \dfrac{z}{c_p} \right) - \dot{w}_0 \left(t - \dfrac{2H - z}{c_p} \right) \right] \\[3mm] \overline{F}_{bz}^{-y} = \dfrac{G}{2r_b} w^f + \rho c_s \dot{w}^f + \dfrac{\mu}{c_s} \left[\dot{v}_0 \left(t - \dfrac{z}{c_s} \right) - \dot{v}_0 \left(t - \dfrac{2H - z}{c_s} \right) \right] \end{array} \right. \tag{8.38}$$

b. Similarly in the side boundary $y = y_b$; $l = 0$; $m = 1$; $n = 0$; using $+y$ to express, there will be the following:

$$\left\{ \begin{array}{l} \overline{F}_{bx}^{+y} = \dfrac{G}{2r_b} u^f + \rho c_s \dot{u}^f \\[3mm] \overline{F}_{by}^{+y} = \dfrac{E}{2r_b} v^f + \rho c_p \dot{v}^f - \dfrac{\lambda}{c_p} \left[\dot{w}_0 \left(t - \dfrac{z}{c_p} \right) - \dot{w}_0 \left(t - \dfrac{2H - z}{c_p} \right) \right] \\[3mm] \overline{F}_{bz}^{+y} = \dfrac{G}{2r_b} w^f + \rho c_s \dot{w}^f - \dfrac{\mu}{c_s} \left[\dot{v}_0 \left(t - \dfrac{z}{c_s} \right) - \dot{v}_0 \left(t - \dfrac{2H - z}{c_s} \right) \right] \end{array} \right. \tag{8.39}$$

8.5.1.3 *Viscoelastic boundary virtual work forms*

The FEPG system stipulates the peculiar finite element language used to express the partial differential equation formula. For this reason, this system requires the users to use this kind of language to write out the PDE document. In terms of the forms of virtual displacement principle (the finite element method needs this kind of forms) the partial differential equation expression formula of the finite element problems by the users should be written on the document in the extended name of PDE, and the PDE document can be used by the system systematically to generate the unit subprogram required by the finite element problems.

In order to conveniently apply the FEPG system tools, the following will give out the mentioned virtual work forms including the viscoelastic boundary dynamics equation virtual work form. The following taking the bottom side $z = 0$, and input shear wave u_o^f as an example, an introduction is carried out:

It can be known from Eq. (8.33) that

$$[k_b^{-z}] = \begin{bmatrix} G/2r_b & 0 & 0 \\ 0 & G/2r_b & 0 \\ 0 & 0 & E/2r_b \end{bmatrix}; \quad [c_b^{-z}] = \begin{bmatrix} \rho c_s & 0 & 0 \\ 0 & \rho c_s & 0 \\ 0 & 0 & \rho c_p \end{bmatrix};$$

In order to express it in simplicity, the tensile form is adopted; we can obtain:

$$\iiint_V [D_{ijkl}\varepsilon_{kl}\delta\varepsilon_{ij} + \rho\ddot{u}_i\delta u_i + \alpha\dot{u}_i\delta u_i]dV + \iint_{\Gamma^f}[\frac{G}{2r_b}u\,\delta u + c_s\dot{u}\,\delta u]dS$$
$$= \iiint_V f_i\,\delta u_i\,dV + \iint_{\Gamma^f}\bar{F}_{bx}^{-z}\,\delta u\,dS$$

where r_b is scattering sources to the artificial boundary distance.

If in the bottom side $z = 0$; $l = 0$; $m = 0$; $n = -1$ and at the same time, inputting free waves u_0^f, v_0^f, and w_0^f, each boundary surface force will be given out by Eq. (8.34), and Eqs (8.36)–(8.39). In the viscoelastic boundary, the general equation of virtual work equation of dynamic equation can be written as follows:

$$\iiint_V [D_{ijkl}\varepsilon_{kl}\delta\varepsilon_{ij} + \rho\ddot{u}_i\delta u_i + \alpha\dot{u}_i\delta u_i]dV + \iint_{S^{vz}}[k_{ij}^m u_j\,\delta u_i + c_{ij}^m \dot{u}_j\,\delta u_i]dS^m$$
$$= \iiint_V f_i\,\delta u_i\,dV + \iint_{S^{vz}}\bar{F}_i^m\,\delta u_i\,dS^m \tag{8.40}$$

where f_i is body loading; $\bar{X}_i^m$ is free displacement; u_i^f yield surface force:

$$m = \pm x, \pm y, -z; i, j, k, l = 1, 2, 3$$

$$[k^{\pm x}] = \begin{bmatrix} E/2r_b & 0 & 0 \\ 0 & G/2r_b & 0 \\ 0 & 0 & G/2r_b \end{bmatrix}; \quad [k^{\pm y}] = \begin{bmatrix} G/2r_b & 0 & 0 \\ 0 & E/2r_b & 0 \\ 0 & 0 & G/2r_b \end{bmatrix}; \quad [k^{-z}] = \begin{bmatrix} G/2r_b & 0 & 0 \\ 0 & G/2r_b & 0 \\ 0 & 0 & E/2r_b \end{bmatrix};$$

$$[c^{\pm x}] = \begin{bmatrix} \rho c_p & 0 & 0 \\ 0 & \rho c_s & 0 \\ 0 & 0 & \rho c_s \end{bmatrix}; \quad [c^{\pm y}] = \begin{bmatrix} \rho c_s & 0 & 0 \\ 0 & \rho c_p & 0 \\ 0 & 0 & \rho c_s \end{bmatrix}; \quad [c^{-z}] = \begin{bmatrix} \rho c_s & 0 & 0 \\ 0 & \rho c_s & 0 \\ 0 & 0 & \rho c_p \end{bmatrix} \quad (8.41)$$

8.5.2 REALIZATION OF ARTIFICIAL VISCOELASTIC BOUNDARY IN FEPG

8.5.2.1 Descriptive document of partial differential equation (PDE-type document)

In the FEPG system, the scenario document of weak solution types used to describe the principle of virtual work includes PDE, VDE, FDE, CDE, and FBC of five types of forms. PDE is the short form for partial differential equation. This type of document is the scenario document for partial differential equation expression formula; PDE document is the descriptive document for virtual work equation component form with PDE as the extension name. VDE document is the virtual work equation tensor form descriptive document with VDE as the extension name. FDE document takes FDE to extend its name to describe the operator form virtual work equation. CDE document takes CDE to extend its name to describe the complex number form virtual work equation. FBC is the short form for force boundary condition. This type of document takes FBC to extend its name to describe the boundary integral item of virtual work equation, and it is stipulated that the documents be paired with PDE or VDE documents of the same name in use. PDE and VDE documents can be directly used by the FEPG system to generate FORTRAN source program while FDE and CDE document can first operate FDE* and CDE* documents to convert them into PDE documents. The FEPG system can generate the unit rigidity matrix, unit mass matrix, unit damping matrix, unit loading vector, and so on. The unit subprograms from PDE type of documents, and in the case of general conditions, PDE or VDE document forms are more frequently used. The following is only to introduce the writing method of PDE, VDE, and FBC documents.

The PDE document consists of three sections: the definition section, function section, and expression formula section. The expression formula section consists of four paragraph contents: the unit mass matrix, unit damping matrix, unit rigidity matrix, and unit loading vector, and they take DEFI, EFUC, STIF, MASS, DAMP, LOAD as the information paragraph keywords, respectively.

The FEPG system stipulates that each segment of information or message (hereafter message segment is called for its short form) with its keyword as first term and null line as the ending. All the message segments are filled, the ending word END needs to be filled to indicate the ending. The system also allows the users to insert their own compiled FORTRAN source program, that is, with "$C6," "$C0," and "$CV" word symbol leading the manner to insert into the message segment specified. In addition, to the word "end" after, that is, the key word FORTAS the first term. Specified letter "\" (skew) indicates continuation line number. Character after continuation line number "\" will be skipped over by this system and the next line will be the continuation line of this line so that skew "\" can be used as interpretation line. Therefore, the users can add the first letter symbol of the line "\".

1. PDE document

   The following takes the virtual work Eq. (8.40) for the solution to the dynamics equation as an example, to account for how to fill the PDE document of a practical problem as well as the process of generating the unit subprogram. Based on Eq. (8.40), compile writing PDE document disp. PDE is as follows:

```
defi                              \segment identification; statement variable; define finite element
                                     parameters. etc
disp u v w                        \unknown function name
coor x y z                         \define coordinate system
func exx eyy ezz eyz exz exy  \define strain function
mate pe pv fu fv fw rou alpha 2.4e7;0.2;0.0;0.0;0.0;1000.0;0.0;    \material parameter
shap c8                           \function segment: c is 6-face cubic unit, c8 express 8 node 6 face
                                     cubic unit
gaus 2                            \in each coordinate direction having 2 integral nodes, if filling unit
                                     type symbol c denoting top node integral
mass c rou                        \mass density is ec unit mass density
damp c rou*alpha                   \damping coefficient
\null one line identification end
$c6 fact=pe/(1.0+pv)/(1.0-2.0*pv)
                                  \null one line, the system grammar requirement
exx=+[u/x]
\null one line, the system grammar requirement, the following is the same.
eyy=+[v/y]

ezz=+[w/z]

eyz=+[w/y]+[v/z]

exz=+[w/x]+[u/z]

exy=+[u/y]+[v/x]

stif                              \regidity segment identification
dist=+[exx;exx]*(1.0-pv)*fact+[exx;eyy]*pv*fact
+[exx;ezz]*pv*fact+[eyy;exx]*pv*fact+[eyy;eyy]*(1.0-pv)*fact
+[eyy;ezz]*pv*fact+[ezz;exx]*pv*fact+[ezz;eyy]*pv*fact
+[ezz;ezz]*(1.0-pv)*fact+[eyz;eyz]*fact*(0.5-pv)
+[exz;exz]*fact*(0.5-pv)+[exy;exy]*fact*(0.5-pv)    \rigidity matrix expression, with component
\null one line identification dist end;;                       \express Eq (8-40) left 1st side
line load=+[u]*fu+[v]*fv+[w]*fw    \unit body force expression
\null one line identification load end
End                               \PDE end
```

2. VDE document

The contents of VDE document and PDE document are basically same, but rigidity expression formula in VDE document is simple. Equation (8.40) VDE document disp. VDE is:

```
defi
disp u v w
coor x y z
func exx eyy ezz eyz exz exy
mate pe pv fu fv fw rou alpha 2.4e7;0.2;0.0;0.0;0.0;1000.0;0.0; \material parameters
shap %1 %2                    \number of unit type symbol and unit nodes will be by GIO or MDI
                                document
gaus %3                       \unit inner integral mode will be given by GIO or MDI document
mass %1 rou                   \unit mass density
damp %1 rou*alpha             \damping coefficient
vect u u v w                  \using key word vect to define Vetor u, shoes components are u,v,w,
                                each of them is separated by blank
vect f fu fv fw               \using key word vect to define Vector f, whose components are fu,fv,fw
vect ev exx eyy ezz           \using key word vect to define Vector ev, whose components are
                                exx,eyy,ezz
vect eg eyz exz exy           \using key word vect to define Vetcor eg, whose components are
                                eyz,exz,exy
matrix dv 3 3                 \using key word matr to define matric dv (3, 3)
(1.0-pv) pv pv
pv (1.0-pv) pv
pv pv (1.0-pv)
\null one line identification defi end
$c6 fact=pe/(1.0+pv)/(1.0-2.0*pv)
\null one line, system grammar requirement
func                         \To define function
exx=+[u/x]
\null one line system grammar requirement, the following is the same
eyy=+[v/y]

ezz=+[w/z]

eyz=+[w/y]+[v/z]

exz=+[w/x]+[u/z]

exy=+[u/y]+[v/x]

stif
dist=+[ev_i;ev_j]*dv_i_j*fact+[eg_i;eg_i]*fact*(0.5-pv)    \rigidity matrix expression with tensor
                                                            mode
\null one line identification distend;                      \to express eq   (8-40) the left 1ˢᵗ item
load=+[u_i]*f_i                                             \loading expression, to express eq(8-40) the right 1ˢᵗ item
\null one line identification load end
end                                                        \VDE end
```

Attention paid in VDE document, shap % 1 % 2, gaus % 3, and mass % 1 adopt the substitution symbol. This only matching with GIO document in application is of significance and using GIO command to generate the FORTRAN source program can be effective. If VDE command is singly used to generate the unit subprogram, the concrete establishment form in the PDE document must be adopted. In the FEPG system implementing VDE command, that is, VDE disp. generates disp. PDE; Implementing PDE; Implementing PDE command, that is, PDE disp. generates aec8g2.ges document. Finally, implementing GES command, that is, GES aec8g2 generates the unit subprogram aec8g2.for.

In fact, the compiled PPE document and GCN and GIO (MDI) documents are the same with the general linear elastic dynamic solid-state problems. It is only the handling method for viscoelastic finite element boundary conditions in this chapter that needs writing out FBC document to have realization in accordance with the corresponding grammar.

3. FBC document

PDE or VDE document is used to describe the volume integral part in virtual work equation while FBC document is used to describe the boundary integral item in virtual work equation. The main contents of the FBC document are to fill the differential equation expression formula of boundary conditions (the second and the third groups of boundary conditions). By using this document, the finite element program generator can generate the calculation boundary condition unit rigidity matrix, mass matrix, damping matrix, loading vector subprograms, and so on. The filling method of PBC document is basically the same as that of PDE document, but the FBC document is the boundary unit differential equation expression formula so that it is one-dimension lower than the corresponding body unit. In addition, this system stipulates that the prefix name of the PBC document should be the same as the prefix name of PDE (VDE) so as to make both be used in realignment. The following writes out Eq. (8.40) FBC document to illustrate how to write or compile the FBC document.

FBC document disp. FBC is written or compiled in terms of Eq. (8.40) integral equation as follows:

```
defi                          \key word
disp u v w                    \To define unknownquantity
coor x y                      \boundary conditions belong to 2-D space
shap %1 %2                    \unit type symbols and unit node numbers are given out by MDI
                                document
gaus %3                       \GIO or MDI document gives unit inner integral mode
mass %1 0.0                   \unit mass density is 0
damp %1 rou*cs rou*cs rou*cp \3 directional damping coefficients
mate pe pv rb rou  2.4E+7; 0.2;50.0;1000.0;  \material parameters
$c6 cs=sqrt(pe/2.0/(1.0+pv)/rou)   \To define shear wave speed
$c6 cp=sqrt((1.0-pv)*pe/(1.0+pv)/(1.0-2.0*pv)/rou)   \To define compression wave speed
$c6 fx = ...                  \ face loading is given out by Eqs. (8-34) and Eq. (8-36), (8-39)
$c6 fy= ...
$c6 fz= ...
\null one line identification end
stif                          \key word, the following is the surface unit rigidity matrix
dist=+[u;u]*pg/2.0/rb+[v;v]*pg/2.0/rb+[w;w]*pe/2.0/rb \unit rigidity expression formula
                                \null one line the system grammar requirement
Load =+[u]*fx+[v]*fy+[w]*fz    \unit surface loading expression formula
                                \null one line the system grammar requirement
End                           \FBC end
```

8.5.2.2 Algorithms document (NFE document)

NFE is the short form for nonlinear finite element, but it can not only aim at the algorithm document for nonlinear finite element problems, but also refer to all the finite element algorithms documents. NFE algorithm document application makes GCN system generate unit computation E element program and the postprocessing computation U element program. The FEPG system gives out NFE algorithms bank for the general definite form mathematic and physical equations such as elliptic equation, parabolic equation, wave motion equation, and so on.

There are stipulations for NFE document nomenclature in the NFE algorithm bank: ell – ellipse; par – parabola; wave – wave motion (the adoption is only to seek displacement and speed algorithms of one kind); newmark – wave motion (Newmark algorithm is used to seek solution to displacement, velocity, and acceleration) and giving a symbol n before algorithm document can express the nonlinearity. Accordingly, the system NFE algorithm bank can provide algorithm documents as follows:

1. *ELL.NFE*: Algorithm program for seeking solution to linear elliptic type equation.
2. *PAR.NFE*: Algorithm program for seeking solution to linear parabolic type equation. Crank–Nicolson format is often used for time discreteness.
3. *PARR.NFE*: Algorithm program for seeking solution to linear parabolic type equation. Backward difference scheme is adopted for time discreteness.
4. *WAVE.NFE*: Algorithm program for seeking solution to linear wave motion equation. Wave speed format is adopted for time discreteness.
5. *NEWMARK.NFE*: Algorithm program for seeking solution to linear wave motion equation. Newmark format is adopted for time discreteness.
6. *NELL.NFE*: Algorithm program for seeking solution to nonlinear elliptic type equation.
7. *NPAR NFE*: Algorithm program for seeking solution to nonlinear parabolic type equation. Crank–Nicolson format is adopted for time discreteness.
8. *HPARB.NFE*: Algorithm program for seeking solution to nonparabolic type equation. The backward difference scheme is adopted for the time discreteness.
9. *NWAVE.NFE*: Algorithm program for nonlinear wave motion equation. Wave speed method format is adopted time discreteness.
10. *NNW.NFE*: Algorithm program for seeking solution to nonlinear wave motion equation. Newmark format is adopted for the time discreteness.
11. *STR. NFE*: Algorithm program for seeking solution to stress field of known displacement field. The least squares method is adopted.

Viscoelastic boundary condition integral in Eq. (8.45) is, in essence, a linear wave motion equation derived through disp. FBC introducing the right end items from which the additional loading item is generated, while the Newmark method is adopted to obtain the solution to the equation. Therefore, NEWMARK.NFE in FEPG algorithm bank is adopted to obtain the problem algorithm document, and Newmark format is adopted for the time discreteness.

defi	\To define segment identification
stif S	\S for rigidity matrix name
mass M	\M for mass matrix name
load F	\F loading column Vector name
type p	\p indicating parabolic type problem
mdty 1	\1 indicating concentrated mass matrix
step 0	\0 indicating that there is no memory of the previous time step rigidity matrix

```
\null one line indicating definition segment end
equation                                    \giving out the linear algebraic equation group matrix
                                expression formula and the expression formula of the right end
   vect u1,v1,w1                            \To define Vector u  1,v1,w1, whose freedom is in
                                coincidence with the freedom of the original problems
read(s,unod) u1,v1,w1                       \Read previous step broad displacement
matrix = [s]+[m]*a0+[c]*a1                  \To seek solution to linear equation coefficient matrix

forc=[f]+[m*u1]*a0+[m*v1]*a2+[m]*[w1]*a3+[c*u1]*a1+[c]*[v1]*a4
     +[c]*[w1]*a5                           \To seek solution to linear equation right end item
\null one line, indicating equation segment end
solution u     \Via this segment into specifications the system obtains algebra equation group
                                solution after to generate post processing program
vect u,v,w,,u1,v1,w1                        \To define u,v,w,,u1,v1,w1, freedom in concert with the
                                original problem freedom
read(s,unod) u1,v1,w1                       \Read out previous iteration step displacement
[w]=([u]-[u1])*a0-[v1]*a2-[w1]*a3
[v] =[v1]+[w]*a7+[w1]*a6
Write (s,unod) u,v,w                        \Writing the  current broad sense displacement in
                                document
\null one line, indicating solution segment end
fortran
@subfort                \Adding  Newmark method  inserted  value  coefficient into  the  program
                                beginning part
                                Delta =0.5
          aa=0.25*(0.5+ delta)**2
          a0=1.0/(dt*dt*aa)
          a1= delta /(aa*dt)
          a2=1.0/(aa*dt)
          a3=1.0/(2*aa)-1.0
          a4= delta /aa-1.0
          a5=dt/2.0*( delta /aa-2.0)
                                a6=dt*(1.0- delta )
                                a7=dt* delta
          \null one line
End
```

If the problems met with are rather complex, and there will be no corresponding algorithm document in the FEPG.NFE bank, it will be necessary to design an algorithm by oneself, and again to write the corresponding NFE document in terms of grammar rules. However, in the case of general conditions, one can find the similar algorithm in the bank; only an appropriate modification is made, can one obtain the corresponding needed algorithm document.

8.5.2.3 *Seeking solution stream and physical coupling document*
1. Seeking solution stream GCIV document

   GCN is the generate command stream and NFE document as well as the short form for generate command stream and NFE.GCN document gives out the coupling mode of each physical field, single field problem algorithm, and seeking solution stream. GCN document takes GCN document as extended name, whose filling contents consist of two segment information: the first segment information is algorithm segment mainly to give out the algorithm of each physical field after the solution coupling, and the second is the command stream segment to give out the multifield coupling computation stream.

The VSDYN.GCN document is as follows:

```
def;          \to definite segment identification
a NEWMARK &   \a field solution displacement, corresponding to a field algorithm document is
                  NEWMARK.NFE; & indicates the generate document without nomenclature
                  or rename. The output document is unod or else unoda below through
                  command stream to organize the computation flow.
STARTilu a        \null one line, indicating definition segment end
                  \corresponding to preprocessing conjugate gradient method without existing
                   single rigidity mode displacement field initialization

if exist stop del stop

:1

bft               \side value computation
solvilu a             \pre-processing conjugate gradient method solutes a field displacement
post a                \output result in terms of time steps
if not exist stop goto I
```

2. Physical field coupling GIO document

GIO is the generate IO actual corresponding to multiphysical field module input. GIO document must take GIO document as the extended name, whose filling contents consist of two segment information: the first segment information gives out PDE (or VDE) document names of each physical field after solution coupling; the second segment information gives out the body unit type message and coordinative system in the solution zones.

The GIO document is named as VSDYN.GIO being the same name as GCN.

```
disp              \a field corresponding to the unit calculation document is disp. PDE
                  \null one line
# elemtype c8g2   \unit type, 8 nodes ,hexahedron, 2segment Gauss' integral
3dxyz             \3-Dimensional right angle coordinative system
```

Since GIO document can only generate one unit type, the FEPG system needs to be improved, with MDI document to replace GIO document function. MDI is GIO extension function program, that is, multidisciplinary input, meaning is the multiphysical field module input. VSDYN.MDI document has the following forms:

```
3dxyz y               \3-dimensional  right angle coordinative system; y indicates
                          having material parameters

#a 0 3 u v w          \physical field name is a; 0 initial value; 3 unknown quantities

vde disp c8g2   1c8g2     \from disp.PDEgenerating aec8g2.for and ae1c8g2.for

fbc disp q4g2             \from disp .FBC generating aeq4g2.for

#
```

Such an operation of VSDYN MDI can directly from disp. PDE generate aec8g2 for and aelc8g2. for two subprograms, from disp FBC generate aeq4g2 for, whereby omitting modification work on the program generated by the GIO document.

This gives out all the FEPG scenario documents with viscoelastic boundary linear dynamic problems. The realization by using viscoelastic boundary in FEPG has only introduced the application of the FEPG system in developing the finite element program and the main process and steps of computations. If more details come to be known, it is necessary to look into the users' manual.

Now, based on the compilation of disp. PDE (VDE), disp. FBC, VSDYN.GCN, and VSDYN.GIO(MDI) command can regenerate a set of complete similar linear dynamic problem computation program as indicated in Fig. 8.1b. If material parameters are complicated, * PRE parameter document regenerates the program needed to be revised. And then, operating preproc essor to enter into FEPG.GID can do the preprocessing work on the finite element computation. Final work is to carry out the postprocessing of computation results.

At present, the FEPG system element program is basically designed with an aim at the implicit program. Also, other solvers provided in the FEPG bank can be selected in terms of concrete features or performances of the solvable problems. However, if the explicit algorithm is adopted, it is necessary to redesign the program structure.

8.5.2.4 Computation example test of viscoelastic boundary conditions

The following is the computation example by the earlier compilation of the FEPG scenario document to generate the computation program, taking side length of 50 cubes with bottom surface of viscoelastic boundary and top surface freedom. Its material shear modulus is 20 MPa, Poisson's ratio is 0.2, mass density is 2000 kg/m^3, and damping ratio is set as zero. The shear wave speed of this material through calculation is 100 m/s. The shear wave speed pulse is $\dot{u}_t = \frac{1}{2}[1 - \cos(2\pi ft)]$ of which $f = 4.0$, $0 \le t \le 0.25$; $\dot{u}_t = 0$, $t \ge 0.25$. Based on the wave traveling theory, this wave response process is that the wave incidence is from the bottom surface, traveling upward and still reaching the top surface afterward, and the reflection occurs on the freedom surface, whose speed doubles the incidence speed as much, and the reflect wave travels downward, thus rendering each node or point on the body to double the displacement by orders. After the reflect wave reaches the bottom surface, viscoelastic boundary absorbs the reflect wave so that the displacement of each node on the body is gradually becoming zero. This speed wave can be converted into the surface force input at the viscoelastic boundary. The 5 m cubic unit with angular can be adopted, totaling 1331 nodes and 1000 units, and the boundary nodes surrounding the cubic unit are vertically applied with constraints. The material-damping coefficient takes zero. The cubic bottom, middle, and top part speed time processing is shown in Fig. 8.5.

In this computation example, the corresponding input speed time processing is corresponding to the input displacement time processing at the same time, whose expression formula is $u_t = \frac{1}{2}[t - \frac{1}{2\pi f}\sin(2\pi ft)]$ $(0 \le t \le 0.25)$; $u_t = 0.125(t > 0.25)$. The displacement varying conditions are as follows: displacement wave incidence is from bottom surface, thus traveling upward gradually to reach the displacement peak value of 0.25 m and after 0.5 s to the top surface. Since the top end is the freedom surface, the wave occurs to have the reflection so that the displace value doubles the input values, and the reflect wave travels downward, and each node on the cube occurs to have the double displacement in the orders. After one second, the reflect wave reaches the cubic bottom. It is just at this time that the viscoelastic boundary can produce the energy absorption functions. For this reason, the

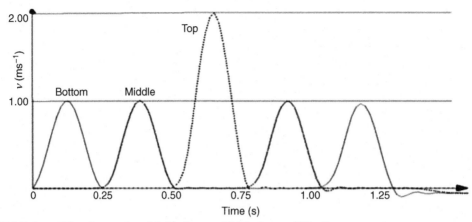

FIGURE 8.5 Speed Time Processing of Cubic Bottom, Middle Part, and Top Part

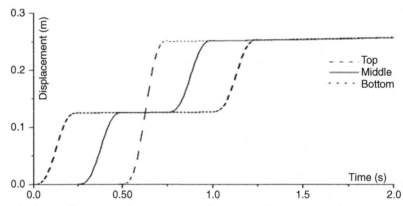

FIGURE 8.6 Displacement Time Processing Curves of Cubic Bottom, Middle Part, and Top Part

wave no longer produces any reflection so that each node displacement on the cube gradually tends to become the input stable value of 0.25 m. Figure 8.6 is the displacement time processing curves of the cubic bottom middle part and top part so that it can be seen that there is a complete agreement between the theoretical solution and actual computation.

8.5.3 REALIZATION OF ARTIFICIAL HOMOLOGY OR TRANSMISSION BOUNDARY IN FEPG

When artificial homology boundary is adopted (Zhengpeng, 2002), it consists of the dam body, foundation, and artificial boundary of three parts in terms of the finite element method discrete model. Apart from artificial boundary outside nodes serving as the inner nodes, the explicit finite element computation should be adopted for dam body and inner nodes. While the artificial homology boundary zones with three to five layers of meshes need to be set up in the outside of near regional foundation. The boundary node displacement modification should be made in terms of the computation method introduced in 60402 so as to realize the scattering wave transmission to the far region.

In terms of the finite element explicit computation format in structure seismic response analysis given in Section 8.3.1 and the motion contact nonlinear Lagrange's method of multipliers described in Section 8.3.3 and Section 8.3.4, FEMsky.gpg and FEMsky.epg element programs should be developed. The FEPG element program document consists of two parts, of which *.epg document is the program main body, while *.gpg is the memory allocation document, whose contents are located in the beginning part of generating the original program.

The explicit solution FEPG element program has integrated some of the original functional program modules. bft elements are fused into the main program and E program so that bft element cannot be adopted independently in forms. Owing to adopting an explicit computation format, the program needs not require a solver, whose solve element computation stream can be incorporated into the E element program. The explicit solution method E element program includes the original E element, U element, and bft element functions. These element programs are put into the FEPG system. Again based on FEPG finite unit element language, FEPG scenario document for structure seismic response analysis can be compiled and written. The method is similar to the computation example set in Section 8.4. The following is only to give out NFE document and PDE (FBC) document for structure seismic response analysis. (In the document, the variables and formats are not marked because they are the same as the notes in the previous examples.)

8.5.3.1 PDE document

```
dfi
disp u v w
coor x y z
func exx eyy ezz eyz exz exy
shap %1 %2
gaus %3
                                  \null one line
mate pe pv rou alpha 1.0e10;0.3;0;0;
mass %1 rou*vol
damp %1 rou*vol
$c6 pe=pe*1.3d0                    \ dynamic calculation, elastic modulus is 1.3 times
                                    as much as static elastic modulus.

$c6 fact = pe/(1.+pv)/(1.-pv*2.d0)
$c6 shear = (0.5d0-pv)
\null one line
func
$c6 vol = 1.0
$c6 factvol = fact*vol
$c6 if(imate.eq.25)then            \dam body material No.25,   g= -9.8,  the rest of
                                    materials without gravity,  g=0.0m/s²
$c6 eg=-9.8
$c6 else
$c6 eg=0.0
$c6 endif
$c6 ft = pe/(1.-2.*pv)*alpha*ut/1.3d0    \computation temperature loading
exx=+[u/x]
                                  \ null one line
```

```
eyy=+[v/y]
\ null one line

ezz=+[w/z]
\ null one line
eyz=+[v/z]+[w/y]
\ null one line

exz=+[u/z]+[w/x]
\ null one line

exy=+[u/y]+[v/x]
\ null one line

stif                                    \rigidity matrix
dist=+[exx;exx]*(1.-pv)*factvol+[exx;eyy]*pv*factvol
+[exx;ezz]*pv*factvol+[eyy;exx]*pv*factvol+[eyy;eyy]*(1.-pv)*factvol
+[eyy;ezz]*pv*factvol+[ezz;exx]*pv*factvol+[ezz;eyy]*pv*factvol
+[ezz;ezz]*(1.-pv)*factvol+[eyz;eyz]*shear*factvol
+[exz;exz]*shear*factvol+[exy;exy]*shear*factvol
                                            \ null one line

load=+[u]*0.0+[v]*0.0+[w]*rou*eg          \gravity added to z negative direction
 +[exx]*ft+[eyy]*ft+[ezz]*ft              \Temperature loading
                                           \ null one line
end
```

8.5.3.2 FBC document

```
dfi
disp u,v,w
coor x,y
coef gx,gy,gz                           \node whole coordinative
shap %1 %2
gaus %3
mass %1 0.
damp %1 0.
mate fw 0;
\null one line
                                        \null one line
stif
$c6 pw=0.0
$c6 if(gz.le.1240.0)then                 \calculation of water pressure and sediment loading
$c6  pw=pw-1.0e3*9.8*(1240.0-gz)
$c6  if(gz.le.1097.0)then
$c6   pw=pw-620.0*9.8*(1097.0-gz)
$c6  endif
$c6 endif
dist=+[u;u]*0.0
                                        \null one line
load=+[u]*0.0+[v]*0.0+[w]*pw            \face loading added to face normal direction
```

8.5.3.3 NFE document

```
defi
stif s
mass m
damp c
load f
type w
mdty 1
step 0
                              \null one line
equation
var ut
vect uu,vv,uu1,uu2,masd,seep,us,usls,usls1
read(s,unods)us,usls,usls1        \read access dam body static calculation results used in
                                   dynamic modification
read(s,unod) uu,vv,uu1,uu2        \read access initial values
read(s,emas0)masd                 \read access additional mass
read(s,temp0)ut                   \read access temperature
read(s,seep0)seep                 \read access center force
matrix=[s]
l,mass=[m]
forc=[f]
solution u                        \null one line
vect u,u1,v,u2
[v]=[u]/dt-[u1]dt                 \solution to previous step speed, adopting backward
                                   differentiation,  $\{v\}_{t+\Delta t} = (\{u\}_{t+\Delta t} - \{u\}_t )/\Delta t$
[u2]=[u1]                         \store t time displacement into u2 number group
[u1]=[u]                          \store previous step displacement into u1 number group
                                  \null one line
fortran                           \Insert FORTRAN source program
@inreal                           \variable initial valuation
    a0=0.628                      \Rayleigh damping coefficient
    b0=0.0015                     \Rayleigh damping coefficient
    nblm=984                      \node pair control parameters
    xlb=3.6e6                     \node pair initial strength
    iwave=1                       \marking dynamic calculation
@begin                            \number initial valuation
    do i=1,knode                  \pair node circulation
    do j=1,kdgof                  \pair node freedom circulation
eu(i,j)=eus(i,j)                  \with static displacement as dynamic displacement
                                   initial value

    ev(i,j)=evv(i,j)
    eu1(i,j)=eus(i,j)
    eu2(i,j)= eus(i,j)
    enddo
    enddo
    itm=3                         \state marking,3,for seismic response analysis
                                  \null one line
end
```

Viscoelestic boundary conditions can be regulated via e * for the program, and realized in terms of FVC document generating subprogram, without individual introduction of viscoelastic boundary conditions, while the introduction of homology or transmission boundary condition needs modifying E program after generating the FORTRAN source program. In E program, it is necessary to read in reflecting homology boundary zone three to five layers mesh node data and material parameters in terms of specified format, and at the same time, it is necessary to regulate the homology boundary special subprogram (bdwave for) to modify the displacement on the artificial homology boundary.

8.6 PARALLEL COMPUTATION PROGRAM DEVELOPMENT BASED ON PEFPG SYSTEM HIGH ARCH DAM SEISMIC RESPONSE

In different parallel computers and on parallel realization platform, the realization mode of the parallel program is different. Open MP and MPI are two means of parallel programming. In spite of aiming at the specified problem, both can realize the parallel algorithm, and their adaptable hardware environments are different so that the applied methods are also different. Open MP adopts shared storage mode, and MPI adopts distributed storage mode. Open MP adopts the implicit data allocation mode, and MPI employs the explicit data allocation mode. Open MP adopting shared storage mode means only adapting to SMP. DSM machines but unadapting the machine group system; and although MPI is adaptable to machines of various types, the programming module is very complex. In addition, debugging the MPI program needs considering two main problems of large communication delay and loading unbalance, whose program reliability is not so high that one process is out of the question, and all programs will occur to have errors. Open MP has the poor expansion, but higher requirement for machines. If we want to improve the speed of parallel computation, we must pay a high cost for the hardware. It is just because of the consideration concerning economy and system expansibility that the MPI parallel mechanism research and development can be selected to develop the structure dynamic parallel computation program.

8.6.1 MESSAGE PASSING PROGRAMMING INTERFACE

MPI is a message passing programming standard established by the world's industrial community, scientific research institutions, and governmental departments in union, whose main purpose is to provide a highly effective, expansible, and unified programming environment based on message passing parallel program design. It has become the most universal parallel programming mode as well as the distributed mode parallel system main programming environment. MPT is a kind of message passing programming model and the representative and standard of this type of programming. Although MPI is vast, its final objective is only to serve the communication among the processes MPI standard defines a group of function interfaces, used in message passing among the processes. The concrete realization of these functions will be implemented by the computer manufacturers or scientific research departments.

MPI standardization began in the message passing standard Forum on distribution storage environment held in Williamsburg, Virginia, USA, on April 29, in 1992. The initial scheme suggested by Dongarra, Hempel, Hey, and Walker was popped out in November 1992 and the revised edition was completed in February 1993. This is MPII.O. In order to promote MPI development, one nongovernmental organization called as the MPI forum was born. This forum has played an important part in MPI development. In June 1995, MPI's new edition MPII.1 was popped out, and a further revision completion and expansion were made to the original MPI. However, when the MPI standard was popped out in order to be able to make it realized and accepted as fast as possible, many important but difficult to implement functions were not defined. For instance, I/O, after MPI has been widely accepted, it is urgent that the requirement for raising MPI function become higher and higher. In July 1997, on the basis of extension of the original MPI, MPI extension part MPI-2 was popped out, while the original MPI editions are called as MPI-1. MPI-2 has many extended contents, but they are mainly in three aspects, that is, I/O, far storage, and access and dynamic progress management.

MPICH is, at present, the most extensively used MPI system free of charge. It supports most of Linux/Unix, Windows 9x, NT, 2000, and XP systems. MPICH is an important MPI realization, and it can be obtained from the network free of charge. What is most important is that MPICH is one edition developed synchronically with MPI-1 specifications. When MPI pops out a new edition, there will be a corresponding MPICH realized edition. Using MPICH can either establish MPI program debugging environment on a single microcomputer or work station, applying multiprocessing to simulate operation of the MPI parallel program, or set up a practical parallel computation environment on the SMP system or machine group environment. At present, it is the principal parallel environment for most of the machine groups' system.

One MPI parallel program consists of a group of operating on the same or different computers or computing node process or line process. These processes or line processes can operate on different processors. In MPI program, one independent participating communication individual can be called as a process. One MPI process is usually corresponding to one common process or line process, but in shared storage/message passing mixture model program, one MPI process is likely to represent a group of UNIX line process.

In one MPI program, one ordered set consisting of part or the whole process is called one process group. Each process in the process group is endued with a serial number to identity the process in that group, being called as the process number. The keying number of process number starts from zero.

Communication, synchronization, and so on among the processes in MPI program are carried out via communication devices. MPI communication devices have two kinds of inner-field communication device and in-between field communication device; the former belongs to the communication between the processes of the same process group, and the latter is used in the communication between the processes belonging to two different process groups.

One communication device constitutes its inclusion of process group and the related attribute group. The communication can provide the basic environment for the communication between the processes. Accordingly, all the communication in the MPI program must be completed in the specified communication device. When the MPI program starts, it can automatically create two communication devices. One is called as MPI-COMM-WORID; it includes all the processes in the program. Another is called as MPI-COMM-SELF; it consists of every process independently and only includes its own communication device.

In the MPI program, one MPI process has one communication device (or process group) and the process in this communication device (or process group) has the process number serving as the only identified mark. The process number is relative to the communication device or the process group. The same process in the different communication device (or process group) may have different process numbers. The process number is endowed when the communication device (or process group) is created. The MPI system provides a special process number MPI-PROC-NULL. It represents empty process (there is no process), with MPI-PROC-NULL communication, being rather one empty operation, there is no effect upon the program operation, and its introduction can be convenient for some program compilations.

In the MPI program, communications between the processes are completed via the message receiving and sending or the synchronous operation. One message is one data exchange between the processes.

In the MPI program, one message consists of a communication device, source address, objective address message label, and data. Each process of one MPI program carries out communication through calling MPI function to coordinate one computation task.

8.6.2 PFEPG SYSTEM, ITS STRUCTURE AND WORK MODE

The PFEPG is an additional subprogram of network edition IFEPG by the finite element generator system. The users have installed the IFEPG system after being delegated; the users can use PEFPG functions. The users can generate all parallel finite element computation programs through the network on any user's computer, and these programs can be passed onto any parallel computer to carry out the compilation operation to obtain computation results. In the case of the general condition, the parallel program and serial program can use the same element document (PDE document and algorithm document) to generate. So long as the formula documents are right, two types of programs are right. Therefore, so long as the users debug these element documents on a single computer serial program, the correct finite element parallel program can be generated. The concrete operation is very simple, because PFEPG is based on the MPI parallel finite element autogenerator system and the generated program also takes MPI as the message passing environment so that it is necessary to operate the MPI command to produce the whole finite element parallel computation program.

8.6.2.1 PFEPG data structure

The serial program generated by FEPG can read and write the data document through document parameter mode in the operating process so as to realize the data transmission among each element program. In order to avoid the complexity brought about by the parallel I/O to the program design, the system stipulates that all the disk file I/O operation be completed by its main process from the process of parallel program without participating disk file I/O.

In the case of parallel program without file I/O, the PEFG system (Guoping, 1990) used the GPG system to realize the internal storage dynamic allocation. In the PFEPG (parallel finite element program generator) system, PPG system is used to realize the correspondence of data document to the internal storage; and the data needed to be preserved should be written into the preopen internal memory segment to realize the corresponding relation between the data document and the specified internal memory segment, being stored with data groups. Single data can be taken as the common variable can be passed among each element program. In the PPG system, some special words used in the original documents of the element program can achieve the documents to the internal memory correspondence. Generally speaking, the magnitudes or sizes of data group to be stored should be preknown or calculated in advance. Accordingly, it is completely necessary to open memory field in terms of the sizes of documents dynamically and to make the arrangement of document storage positions so as to be convenient for management in terms of sequences of opening the documents. That is to say, it is necessary to give out the initial position of the document internal memory in such a way that the segment internal storage can be read and written to realize the data read-in and memory.

8.6.2.2 PFEPG system working model

The PFEPG system serves as an important extension function of the IFEPG system (FEPG system network edition), whose working model is roughly similar to the IFEPG system (Fig. 8.7).

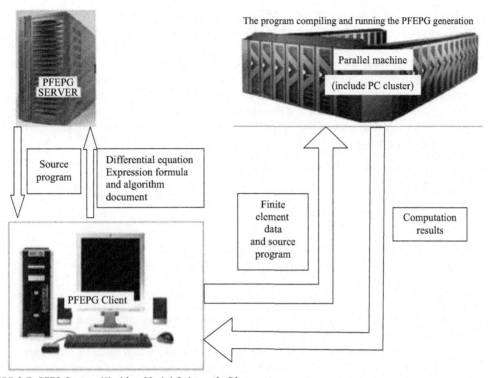

FIGURE 8.7 PFEG System Working Model Schematic Diagram

First of all, this system can be installed on any server on the Internet. The users can install one system user terminal program in the local computer, use this program and the users can land on the servers installed on the system through the Internet remotely and use the program boundary surface to fill and write differential equation express formula and algorithm document. They are mainly GCN, GIO, PDE, and so on, documents and then these text files are uploaded to the servers through the customer's terminal boundary surface, and then the users can use IFEPG system commands so to obtain the serial computation on the local computers so as to guarantee the correctness of various kinds of the finite element expression formulas and generated program filled by the users. Then, the PFEPG system can generate the parallel finite element computation source program, being transmitted to parallel computations and finally, the computation results are sent back to the users' terminal to carry out the results display. Since the parallel program debug is very difficult, this working mode of PFEPG system has decreased the difficulty of the program debug at the maximum degree.

The finite element expression formula uploaded by the users and the computation source program generated by the servers are the text file forms, their sizes are within look and their transmission speeds on the Internet are very fast. In general, it can take several tens of seconds to generate the whole source programs required by the users and sent to the user's local computers.

8.6.2.3 Generation and operation of PFEPG system parallel program

Based on the similarity to developing the FEPG finite element serial program, first of all, it is necessary to compile and write VDE document and algorithm (NFE, GCN, and GIO) scenario documents in terms of weak solution integral of physical equations of physical problems, and then, these documents are used to generate a parallel program and at the same time to generate the single computer serial finite element program. The system developers suggest that in the case of general conditions, the serial program be debugged and tested to confirm that this method is right without errors, and that the generation, improvement, and debugging work of the parallel program be implemented again. In general, there are the following steps for applying the PFEPG system.

1. Filling partition document.

    There are some constants needed in the preprocessing partition program in partition and data document, whose contents are in general as follows:

10	1	blocking number	Partition Algorithm
5	2	1	x-direction profile score, y-direction profile score z-direction profile score

    In general, only blocking number needs to be modified, and the modification can reach the number of what is needed (the minimum is 2) of which, the first integer 10 in the first line denotes the total blocking number; the second integer 1 indicates the adaptation partition algorithm (1 indicates the adaptation of graph theory algorithm; 2 indicates the use of coordinate partition algorithm). The first integer 5 in the second line expresses x-direction profile score or number; the second integer 2 indicates y-direction profile score. The first number in the first line is the product of three numbers in the second line. In doing the large-scale computation, it is necessary to modify partition data document to increase the blocking numbers, but this relation must be kept. If this document is not filled, the system will produce the default files. If the graph theory partition algorithm is adopted, the partition will not accept the second line data restriction.

2. The compilation of filling the files or documents (VDE, GCN, GIO, etc.) and sending them to the servers through the boundary surface, operation MPI commands are as follows:
    Impi gen file prefix names.
    That is, to produce the parallel finite element source program, and to return or transfer back to the customers' computers.

3. The following files are transferred to parallel computers.
    The whole parallel FORTRAN source program (*.f file), fegen.b/makefile, runmpi/ mpd. hosts, time 0. nbefile, partition. dat.*.io.
    After this work is completed, with an aim at solving the concrete problems, bftm, bftm and *. f files, and so on, should be revised so as to be adaptable to the solution to the problem parameter transmission and computation control needs.

4. Running command csh-x fegen b. on the parallel computers or adopting a macro command makes, the parallel implementing program is generated through the compiler.

5. csh preproc. mpi operation preprocessing can generate the finite element model input and data control file.

6. Revise hosts documents

In hosts (mpd. hosts) documents, the computer names in participating computation must be assigned. For instance:

console	machine name
co101	machine name
co102	machine name

This is the parallel machines at each node machine name.

7. The following are the operation commands or instructions for the parallel machines:

mpdboot-n 10	10 is the number of parallel machine nodes
mpiexec-n 11 stdy	11 is the blocking number+1 , stdy is the parallel implementing
	programming nameconducing parallel computation

8. csh postproce, mpi gio file names

Generated result file names are: gio file name, flavia msh, and gio file names, flavia.res, and then the two files are downloaded onto the single machine, running gid, entering the postprocessing, and then, point-and-click opening to find out two files' catalogs, opening any of the two files and seeing the postprocessing model.

8.6.2.4 *PFEPG-generated parallel program structure*

The parallel program and the serial program are different in nature in the process of operation. The serial program has only one code so that in the course of operation, only one process can be started to run this program code; the parallel program has also only one code, but in the course of operation and in terms of the users' demands, the system can start the multiprocess to run the multicopies of this code. The program code implements the different program functions based on the process number. In general, the process called as the zero code is the main process, and the rest of the processes are the slave process (or subprogram); the main process and the slave process implement different program functions.

Based on regional decomposition method features, Fig. 8.8 gives out the flowchart of the general linear dynamic problem solution program in the PFEPG system. It can be seen from the flowchart that apart from solv element program, in PFEPG generated program, there are no communications among slave processes, so that message exchanges can only take place between the main process and the slave process.

Preprocessing partition module consists of the partition, getpart, spart, and segetpart element programs. This module function is based on the preprocessing data provided by the users rationally to classify the whole solution region into several subregions in accordance with the users requirements, of which partition element is run by the main process, being responsible for the node partitions in the whole region, giving out each subregion including overall node number, local node numbering, and overall node numbering corresponding relations and sending some preprocessing messages including node numbers, time information, and so on, in each subregion and each region including node coordinate

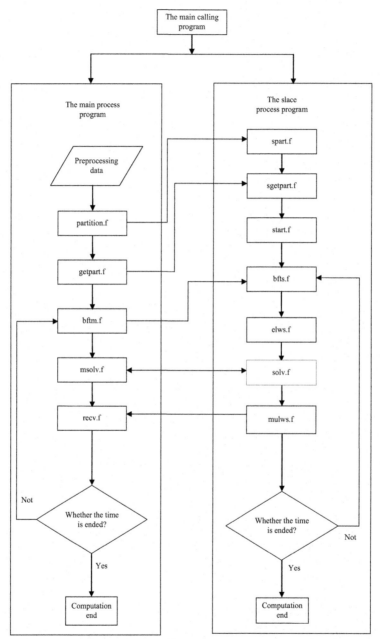

FIGURE 8.8 A General Linear Dynamic Problem Solution Program Stream Framework in PFHPG System

value blockings to each subregion corresponding to slave process, while the slave process to operate spart element receives the partition element coordinate value coming from the main process; getpart element is also run by the main process. Its main function is based on partition element to provide each subregion including node messages from which each local unit message initial boundary value message, node ID message and each freedom of each subregion node corresponding whole equation number, and so on, are obtained and sent to the corresponding slave process. The slave process can run the corresponding agetpart element to carry out the acceptance.

After the preprocessing partitioned module, the corresponding preprocessing data for each subregion have been obtained and each computation module for each subregion begins to be implemented. The first implementation is the start element program. This element gives out each freedom of each node and the algebraic equation group variables to be formed in the future (i.e., equation number) in corresponding relation (which node and which freedom will be corresponding to which variable of the equation group) so as to obtain the storage format of the whole rigidity matrix. Accordingly, this element program can only be implemented in slave process.

The bft element consists of the two programs bftm and bfts, being implemented by the main process and the slave process. bftm is implemented by the main process, mainly giving out the boundary value at every moment solution, renewing the time, judging the end or not, and sending the judgment results and new time and solution boundary values of each subregion to the slave process. bfts is implemented by the slave process, receiving new time and solution boundary values sent by the main process.

The solv element is the solver element, consisting of the two programs solv and msolve, of which the solv program is implemented by the slave process. The algebraic group coefficient matrix formed after through E element in the subregion and the right end items should be sorted out so as to satisfy the data format required by the corresponding parallel solver, and then the corresponding parallel solver is called to obtain the solution to each subregion. It is known from the regional decomposition method that the solution computation of each subregional equation group is dependent on the public boundary solution value of the neighboring subregion, while in order to make the program structure clear, the intercommunication of the solv element program among each process does not need the users to participate. The msolv element is the main process solver that manages the coordination work of communication.

The E element program is adopted in unit computation, including unit rigidity matrix, unit mass matrix, and unit loading computation, and in converting their node freedom expression into the variable expression by algebraic equation group, and at the same time, processing boundary constraint conditions has formed the coefficient matrix and right end item of the algebraic equation group.

The U element is used to convert the variable solution of the equation group derived by the solver into the displacement values of each node freedom and to conduct other postprocessing calculations, and finally to send each node displacements to the main process to convert them into each node of each freedom displacement in the whole region and this element program is implemented by the slave process. The main process has the receiving and sorting elements corresponding to that element, that is, the recv element program. This element receives the solutions coming from each slave process subregion, and sorts them out to obtain the solution to the whole region in terms of the corresponding relation between the local nodes and the whole nodes and carries out the disk memory processing.

The calling of these element programs is completed by the slave process program and the main process program. The two main programs are automatically generated in accordance with the gen document filled by the users and is called by a general calling program called the mainmpi in terms of the process number.

8.6.3 EXPLICIT FORMAT PARALLEL PROGRAM DEVELOPMENT BASED ON FEPG SYSTEM DYNAMIC EQUATION

The existing element program in the FEPG system aims at the implicit solution. The problems dealt with in Sections 8.4 and 8.5 have adopted the implicit algorithm, while the algorithm dealt with in Section 8.3.1 is the explicit algorithm format. The explicit algorithm need not seek a solution to the simultaneous equation group with a small quantity of computation. If the structure seismic response analysis parallel computation program is developed based on the FEPG software system, it is necessary to compile the FEPG element program.

In accordance with the finite element explicit computation format of the structure seismic response analysis given in Section 8.3.1 and motion contact nonlinear Lagrange's method of multipliers described in Sections 8.3.3 and 8.3.4, the following element programs have been developed:

1. Partition_ lm. gpg, partition_ lm. epg: conducting physical zone decomposition.
2. Partition. gpg, partlm. epg: processing node messages, carrying out contact node blocking of information data and sending them to each slave process node.
3. rdatf. gpg, radtf. epg: processing seepage pressure data, additional mass data, temperature data, and seismic wave data, and blocking them and sending them to each slave process node.
4. mulm_ exp. gpg, mulm_ exp.epg: parallel U element program.
5. etlm_exp. mpg: parallel E element program, accounting for explicit algorithm and processing Lagrange's method of multipliers of motion contact problem.
6. startexp. ppg: data initialized element program.

Obviously, the element program of the parallel program is different from the serial element program developed in Section 8.5.3. In addition, the parallel program is rather complex in introducing artificial homology boundary conditions. For this reason, mulwsc, for mdf document has been compiled. The present mdf document in this document can be divided into three paragraphs: the first paragraph is the computation equation numbers to be initialized; the second paragraph calls bdwave.f program. In terms of the artificial homology boundary realization method described in Section 6.4, the displacement modification on the artificial boundary must be completed. The third paragraph sends the displacement-modified displacement back to the mulwsc.f computation program.

Combining the mentioned developed element programs with other basic element programs and commands in PFEPG can generate the considering contact nonlinear high arch dam seismic response parallel finite element program as shown in Fig. 8.9. The structure seismic dynamic response problem parallel computation program includes dynamic contact nonlinear problems and the artificial boundary condition processing or handling, whose contact nonlinear part is included in the course of seeking a solution to the equation group. Therefore, there has been not much difference in the general linear dynamic problem parallel solution stream (Fig. 8.8), in the PFEPG system. It is only the newly developed parallel program that adopts the explicit format similar to the serial explicit program without needing a linear equation group solver, but it is necessary to integrate some original parallel function modules. The following introduces the newly developed every parallel program module main functions through the analysis and comparison of the two sets of programs:

1. *bft element*: Based on the performances of explicit solution algorithm adopted by the program, it is necessary to integrate the bftm and bfts element parts into the main program and E program so that the independent bft element is not adopted in forms, which cannot affect the program

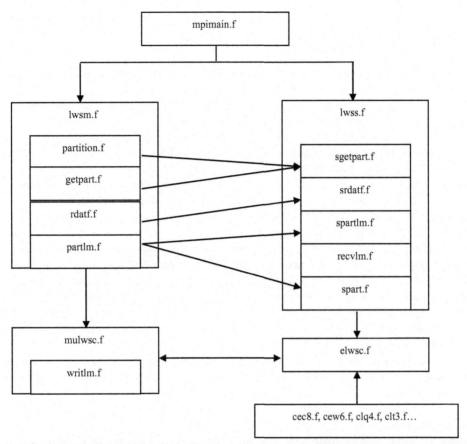

FIGURE 8.9 High Arch Dam Contact Nonlinear Seismic Response Parallel Finite Element Program Structure

development thinking and computation results, but is only to save the redundant operation and to raise the computation speed.

2. *solv element*: Owing to the adaptation of dynamic explicit computation format, there is no need for a solver in the program, whose computation stream can be converged into the main program and E program.

3. *E element*: Figure 8.9 indicates that the structure seismic dynamic response parallel program has adopted the implicit solution mode. Only in accordance with FEPG finite element grammar rules to write the corresponding text document, and based on the software platform element bank, can the required functional parallel program be generated, whereas in terms of dynamic explicit calculation format and dynamic contact problem processing method and in considering the artificial boundary treatment, it is necessary to develop the corresponding PFEPG element program. The work is also the core work to be introduced in this chapter.

4. *U element*: In correspondence with E element, the contact message passing and save processing is increased, and after the computation is completed, the revised results of contact message can be obtained.

Based on the structure dynamic response parallel computation solution thinking way, every program name and the functions in the developed high arch dam contact nonlinear seismic response parallel finite element program can be accounted for as follows:

1. mpimaint: The main control program operates in the main nodes and specified in the implementation course of the whole parallel program.
2. lwsm.f: The main program running on the main node, controlling scheduling main node program stream to coordinate communication relation between the main and slave nodes.
3. lwss.f: The main program running on the slave nodes, corresponding to lwsm.f program on the main nodes, accepting the data passed from the main nodes and specifying the sequence of implementing program from nodes.
4. partition.f, getpart.f: Iwsm.f is called on, being implemented on the main node. In terms of the users demands, numbers of parallel process are classified to determine the corresponding relations of blocking data and slave nodes, and then, the initial message (including the unit message, node constraint messages, and initial displacement data) are sent back to slave nodes.
5. sgetpart.f: lwss.f is called on, accepting the initial message passed from the main nodes.
6. rdatf.f: Implemented on the main node and called on by lwsm.f corresponding to the slave process running node message. The blocking of seepage pressure data, additional mass data, temperature data, and seismic wave data is carried out and sending the data to corresponding slave nodes.
7. srdatf.f: Implemented on the slave node, and called on by lwss.f. in corresponding to the main node running rdatf.f, accepting seepage pressure data, additional mass data, temperature data, and seismic wave data.
8. partlm.f: Operates on the main node, based on the classification of process quantity and the processed node message from each slave node, the contact messages are blocked and sent to every slave node.
9. spartlm.f, recvlm.f: Operates on the slave node, accepting the contact message data passed from the main node partlm.f.
10. mulwsc.f: Operates on the main node, corresponding to elwsc.f running on the slave node, conducting the operation of writing disk operation and judging whether the time ends or not. If ending, writlm.f is implemented to carry out statistic operation state, thus forming report document; if not ending, the time message is renewed, thus making the slave node elwsc.f continue to implement.
11. elwsc.f: Operates on the slave node. The main computation program can conduct the finite element computation based on the passed data and calls on the unit subprogram cec8.f, cew6.f, clq4.f, clt3.f and so on, to form the linear equation group. The explicit mode is adopted to carry out the solution. The computation results are sent back to the main node to accept and process them by mulwse. Accordingly, such contents as structure self-weight, seepage flow pressure, additional mass, temperature, seismic waves, dam surface water, and sediment pressure, loading, and concrete computation of contact problems are all included in this program.
12. writlm.f: Operates on the main node, and to be used after the ending of the program operation to record the contact message variations.

In developing the concrete high dam engineering contact nonlinear seismic response parallel computation program, the introduction modes of homology boundary conditions and viscoelastic boundary conditions into the parallel program are different.

Viscoelastic boundary processing is included in the elwsc.f program. Data blocking and passing can be realized. In partition.f. get part.f. so that viscoelastic boundary condition processing is also the parallel mode. In fact the viscoeleastic boundary processing method applies the surface loading to the boundary. Rigidity and damping can also be generated through the unit subprogram, and integrated into the E program at last. The same as the common force boundary processing method can be realized in parallel computation on every computation node.

Homology boundary method needs differential computation to filtrate scattering waves within the boundary three to five lager meshes, and the parallel comparison on every computation node is complicated. Therefore, when the homology boundary method is adopted in this chapter, the displacement modification on the artificial homology boundary is included into the mulwsc.f program. Owing to mulwsc.f operating on the main node, the boundary condition processing is still the serial mode, whose speed cannot be compared with the parallel mode, thus causing the waiting time to be prolonged of the slave node to the main node data, in such a way that the problem of boundary condition occupying a large percentage will affect the parallel computation speed.

8.6.4 HIGH ARCH DAM STRUCTURE DYNAMIC PARALLEL COMPUTATION STREAM

8.6.4.1 Loading process simulation

In accordance with the construction process of high arch dam engineering, the computation process can also be divided into three successive loading processes:

1. Operational situation I, the a field solution: It is necessary to calculate mountain mass gravity and not to calculate dam body gravity, water and sediment loading, and temperature influences. In calculation, the dam body is taken as the dead unit in its treatment, in such a way that it is only to calculate mountain mass gravity and not to calculate dam body effect. At the same time, in calculation when after the contact surface appears to have sliding motion, node pair coherence force valuation is zero. After the calculation is ended, the contact force revised node on the contact surface is obtained via calculation of node force ($N = N + \lambda$, if $N < 0$, letting $N = 0$) and the node corresponding to coherence force is reevaluated with its original value.

2. Operational situation II, the b field solution: It is necessary to calculate the dam body gravity and not to calculate the mountain mass gravity in considering water and sediment loading and temperature impact, and in addition, seepage pressure within the foundation is also taken into account, water and sediment loading is added to the face unit, and the foundation seepage pressure is added to through node force. In calculation, when after the contact surface appears to have a sliding motion, the node pair coherence force valuation is zero, and at the same time, it is necessary to consider in this operational situation that the dam foundation interface is likely to be tensile open. When the contact force on the dam foundation interface is larger than the strength of the dam foundation interface, as to the dam foundation interface node pairs without being tensile open, the initial pressure force of the dam foundation interface should subtract the dam foundation interface strength ($(N = N - Q$, letting Q be the dam foundation interface strength). Because of the initial node pairs information $N = Q$ on the dam foundation interface, the initial pressure force valuation of node pairs on the dam foundation interface can be zero directly, and at the same time, the node pairs equal area is changed into the negative value serving as the already tensile open identification. As for the already tensile open dam foundation interface node pairs, the node pairs area has negative value so that the initial pressure force of node pairs will not be

treated. When the calculation is ended, the node pairs initial pressure force will be treated as the same condition in operation at situation I. Particularly, as far as the dam interface is concerned, and owing to considering the situation of being tensile open, preparation must be made for the next step dynamic calculation so that the tensile-open node pairs on the dam foundation interface is $N = \lambda$ (when $\lambda > 0$), and the nontensile open node pairs initial pressure force is $N = 1.3N + \lambda$.

3. Operational situation III, the c field solution: In calculating the dam body to the dynamic response to the seismic wave inputs, it is necessary to consider the gravity between the dam body and mountain mass as well as water sediment loading and temperature, to consider water effect upon the dam body dynamic response, and to take it as the dam body surface additional mass into full account. In calculation, every step boundary can be calculated through the homology boundary calculation program. In the course of calculation, the nontensile open node pairs on the dam foundation interface need treatment to be done with the same as that in operational situation II. When node pairs are tensile open, the initial pressure force of nontensile open node pairs on the interface can be subtracted by 1.3 times of the strength of dam foundation interface ($N = N - 1.3Q$). Node pairs equal area is changed into negative value serving as the tensile-open identification. It is considered that in dynamic calculation, the elastic modulus of materials is 1.3 times in static calculation. For this reason, the elastic modulus of material parameters in calculation can be multiplied by 1.3 times. Static calculation results can serve as the initial value to be introduced into the equation so that the dynamic equation can be revised into $M\ddot{U} + C\dot{U} + 1.3K_s U + B\lambda = F_d' + F_s' - 0.3K_s U_s$, where U_s is static loading (including temperature loading) to yield the displacement column matrix, K_s is the static rigidity matrix. Coherence force treatment is the same as that in operational situation II.

The previous two operational situations are mainly used in calculating mountain mass self-weight, dam body self-weight, and the effects or impacts upon the whole model contact surface of contact situations, whereby making the preparations for the calculation of the third operational situations.

8.6.4.2 The principal project documents

The parallel computation program structure seismic response analysis NFE documents and PDE (FBC) documents describe the contents the same as those described in the documents set in Section 8.5.3. The following gives out the project GCN and GIO documents:

1. lws.GCN document:

Defi	\key words
a hill &	\ a field seeking solution to hill mass gravity, corresponding algorithm document hill.nfe
b dam &	\b field seeking solution to dam body gravity, corresponding algorithms document dam.nfe
c earth &	\ c field seeking solution to seismic response, corresponding algorithm document earth.nfe
	\null one line system grammar requirement

partlm a	\processing node pair message
rdatf a	\processing additional message
STARTexp a	\data initialization
SOLVlm_exp a	\solution a field
SOLVlm_exp b	\solution b field
SOLVlm_exp c	\solution c field

2. lws.GIO document:

hill	\ a field corresponding unit computation document：hill.pde & hill.fbc
dam	\b field corresponding unit computation document：dam.pde & dam.fbc
earth #obj bdwave	\c field corresponding unit computation document：earth.pde & earh.fbc

These documents adopt the homology boundary, of which "#obj bdwave" expresses c field unit subprogram chain connection homology boundary differential program bdwave.for.

Parallel computation should be conducted in terms of the mentioned Xiaowan arch dam computation stream sequence of three operational situations, and they are connected with each other through node pairs of message documents. Geometric models, equations, and the algorithm of the three operational situations are basically in agreement with each other, but their loadings and boundary conditions are somewhat different. For this reason, through a set of element program, document (PDE document) and algorithm document (NFE document) can be described with the different differential equations in order to generate the parallel computation source programs for three operational situations. In generating the parallel program, the parallel program can be inserted with the homology boundary-processing program. While the homology boundary processing is carried out on the main node of the parallel machine, the internal region solution parallel computation is carried out on the slave nodes.

In dynamic computation, after the additional mass to dam body surface nodes can be accumulated to general mass matrix, the mass matrix is no longer the diagonal matrix so that the computation solution cannot be directly carried out through the explicit algorithm. With an aim at such a situation, and as for the node without additional mass, the explicit algorithm can still be adopted to calculate node pair multipliers. As for the node with additional mass, the multipliers are obtained through the nondiagonal additional mass matrix pointwise solution in Section 8.3.4 (Tang Juzhen, 2006). The pointwise solution will be conducted, and this method can only meet with the needs of calculation with additional mass and also reach the purpose of parallel computation.

ENGINEERING REAL EXAMPLE ANALYSIS OF PARALLEL COMPUTATION

9.1 PARALLEL COMPUTATION ANALYSIS OF SEISMIC MOTION RESPONSES TO XIAOWAN ARCH DAM

The Xiaowan hydropower station, located in the boundary area between Nanjian and Fengqing counties in Dianxi, is the second cascade in the midstream and downstream river section cascade planning of the Lancang River. The installed capacity is 4.2 million kW with an annual power generation of 18.853 billion kWh after the hydropower station is completed.

The Xiaowan hydropower station was formally launched in January 20, 2002. After the completion of this hydropower station, a reservoir with a storage capacity of 14.914 billion m^3 was formed with the priority given to power generation holding the comprehensive benefits of flood control, irrigation, sediment control, navigation, and so on, being a "dragon head reservoir" in the midstream and downstream river valley of the Lancang River. The total dam height is 294.5 m with the capacity to regulate water resources for years, being a higher hydropower station among all the cascade power stations in the midstream and downstream of the river. The large dam consists of a concrete double-curved arch dam, rear dam water cushion pond and secondary dam, one flood discharge tunnel on the left bank, and the underground water-diverting hydropower station on the right bank.

9.1.1 XIAOWAN ARCH DAM COMPUTATION MODEL

Dam body geometric sizes, dam body, and dam foundation material performances are based on the technical data provided by Kunming Investigation Survey and Design Academy. Dam bottom altitude is 953 m above sea level; dam crest altitude is 1245 m above sea level; crest arch arc length is 935 m long; dam crest thickness is 12 m thick; and dam bottom thickness is 73 m thick. The dam body concrete elastic modulus is $E = 21$ GPa; density is $\rho = 2400$ kg/m^3; and Poisson's ratio is $v = 0.189$. Foundation rock mass is divided into 24 kinds of materials. The material parameters value is shown in Table 9.1, of which the twenty-third kind of material is the soft-weak sandwich material among the hill masses. The modulus of dynamic elasticity takes 1.3 times of the modulus of static elasticity; and damping ratio is 0.05.

The water level of the dam upstream normal impounding level is 1240 m, and low water level is 1181 m above sea level. Sedimentation altitude is 11,097 m above sea level. Sediment submerged weight is 0.4 t/m^3; the minimum distance between the transverse cracks, and 17 transverse cracks are set up. Temperature loading in computation is obtained from the analysis of temperature data provided by Kuming Investigation Survey and Design Academy.

In accordance with the requirements of the "Seismic Design Code of Hydraulic Structure" (DL 5073-200) in our country, the region with basic earthquake intensity of VI above, and the large-sized

Table 9.1 Foundation Rock Mass Material Parameters of Xiaowan Arch Dam Computation Model

Material No.	Elastic Modulus	Poisson's Ratio	Density (kg/m³)
1	25.00	0.220	2630.0
2	20.0	0.250	2630.0
3	3.50	0.300	1900.0
4	1.40	0.310	1900.0
5	3.90	0.300	2000.0
6	9.00	0.265	2630.0
7	2.00	0.300	1900.0
8	1.02	0.310	1900.0
9	3.00	0.350	1900.0
10	5.50	0.300	1900.0
11	8.00	0.290	1900.0
12	17.00	0.260	2630.0
13	5.50	0.300	2630.0
14	23.00	0.230	2630.0
15	25.00	0.220	2630.0
16	20.00	0.250	2630.0
17	22.00	0.240	2630.0
18	26.00	0.210	2630.0
19	21.00	0.250	2630.0
20	19.00	0.250	2630.0
21	18.00	0.250	2630.0
22	21.00	0.180	2400.0
23	21.00×10^{-9}	0.180	24.0
24	10.00	0.250	2400.0

reservoir with storage capacity reaching 10 billion m³ should be based on the special probability theory of the achievements of earthquake hazard evaluation to determine the design seismic acceleration. As far as backwater structure is concerned, it can take the datum level transcendental probability of 0.02 seismic acceleration representative values within 100 years. Accordingly, as for the Xiaowan arch dam in terms of design and prevention of intensity scale of IX, bedrock horizontal design seismic acceleration is $0.308g$, and vertical design seismic acceleration takes the horizontal two-thirds, being $0.205g$. The seismic wave adopted in computation is the artificial seismic wave generated by designing response spectrum as the objective spectrum.

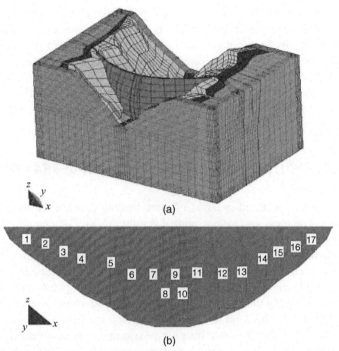

FIGURE 9.1 Xiaowan Arch Dam System 3-D Discrete Mesh

(a) Finite element mesh and (b) transverse crack distribution situation.

As shown in Fig. 9.1, the Xiaowan arch dam system is a three-dimensional discrete mesh, having 28.285 nodes, 1075 contact node pairs, 23.022 hexahedron units, 746 triangular prism units in total, and the minimum side length of the unit is only 1.92 m long.

In view of the transmission course, the computation model has the effect upon the elastic wave magnification so that the input wave can be reduced by half (Houqun, 2006) as the real input data. The displacement time process of three directions of the adopted artificial seismic wave is shown in Fig. 9.2.

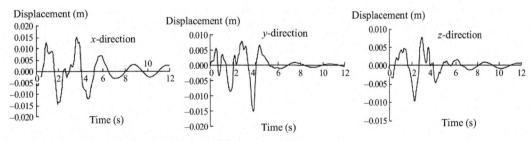

FIGURE 9.2 Xiaowan Arch Dam Input Seismic Displacement Waves

Parallel computation is conducted on the Association & Deep Dump 1800 computer group of the Engineering Aseismic Center of the Chinese Academy of Water Resources. Apart from the main control nodes, this system has five computation nodes. The even node has two main frequencies with 3.0 GHz CPU, being shared 2 GB internal memory, and the Kilomaga Ethernet switching system communication is adopted. The software environment is the Red Hat Advanced Server V4.0 operating system, Intel MPI Library 2.0.1.012 parallel programming environment, Intel Fortran 9.0.031/C Complier 9.0.030 complier, together with PFEPG parallel finite element program to generate the system platform automatically.

9.1.2 PARALLEL COMPUTATION UNDER THE CONDITION OF TRANSMITTING BOUNDARY

The computation course is conducted in terms of three loading courses, that is, conducting in three operational conditions. The previous two operational conditions can calculate the hill mass dam body and water pressure, displacement field and stress field under the action of static loading, and static loading applied to the dam body foundation system with the step function forms. The quasidynamic method is adopted; dynamic computation course is employed to simulate static loading course until the displacement value is stable, and then the third operation condition computation can be carried out, that is, seismic waves can be input by the bedrock, and the wave motion response analysis computation for the system can be carried out. This kind of simulation method is called the static-dynamic combination computation method. The previous two operational conditions in near-field foundation boundary adopts the usual fixed boundary, that is, boundary displacement is zero, while the seismic waves are input by the bedrock in terms of artificial boundary processing.

1. *Operational condition I*: It is necessary to calculate hill mass gravity but not to calculate dam body gravity, water and sediment loading, and temperature influence, calculating in 10 s and time increment steps are 0.0001 s, totaling 100,000 steps.
2. *Operation condition II*: It is necessary to calculate dam body gravity but not to calculate the hill mass gravity considering water and sediment loading, and temperature impact or influence and in addition, taking seepage pressure with the foundation into full account, calculating 6 s and time increment steps are 0.0002 s, totaling 30,000 steps.
3. *Operational condition III*: It is necessary to calculate dam body upon seismic wave input dynamic response. It is only to consider the transmitting boundary but not to take the dam body and hill body gravity, water, sediment loading, and temperature effect into full account, and at the same, considering water effect upon dam body dynamic response. Therefore, it is here that water serving as the additional mass to the dam body surface can be considered, calculating 10 s and the time increment steps are 0.0001 s, totaling 100,000 steps.

The artificial seismic displacement waves are input into the foundation boundary nodes (see Fig. 9.2), and one calculation node, two calculation nodes, till five calculation nodes are adopted to carry out dynamic analysis of the Xiaowan arch dam, whose calculated efficiency is shown in Table 9.2.

It can be seen from Table 9.2 that with an increase in calculation nodes, the calculation time and efficiency decrease, but with five nodes, 10 CPU parallel calculation efficiency still remains over 62%.

Table 9.2 Parallel Computation Efficiency

Calculate Node No.	Calculate Time (s)	Calculate Efficiency (%)	Calculate Node No.	Calculate Time (s)	Calculate Efficiency (%)
1 (2CPU)	17,621	100	4 (8CPU)	6,337	72
2 (4CPU)	10,277	86	5 (10CPU)	5,645	62
3 (6CPU)	7,560	78			

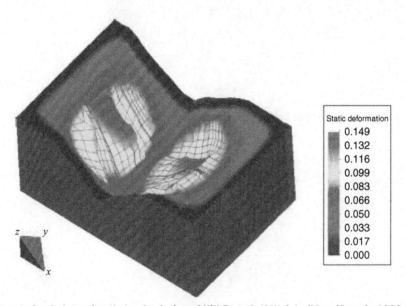

FIGURE 9.3 Foundation Deformation Under the Action of Hill Body Self-Weight (Magnificatoin 1000×)

Figure 9.3 is the foundation variation cloud diagram produced by the hill gravity. The calculated results are used to determine the initial pressure force on the foundation boundary face nodes. Figure 9.4 is the dam body and foundation variation cloud diagram under the action of static loading (dam body gravity, water and sediment loading, temperature, and seepage pressure within the foundation). Figure 9.5 is the dam body maximum and minimum main stress cloud diagram under the action of static loading. The maximum main tensile stress at the local zone of the arch crown beam dam heel is about 7.46 MPa. The larger main pressure stress is about 11.56 MPa. It can be seen from Fig. 9.5 that this zone variation is larger. Figure 9.6 is the dam body whole time process main stress enveloping cloud diagram under the action of combination of static–dynamic loading. The main pressure stress value is within 0.64~17.55 MPa. The maximum main tensile stress is 9.76 MPa, occurring at the arch crown beam dam heel. Figure 9.7 gives out the displacement response curves of the arch crown beam (the upstream surface) top nodes. It can be seen from Fig. 9.7a that seismic response to oscillation on the static balance, so that seismic loading disappears return to the static balance position. Figure 9.7b is the pure seismic loading response curves. Obviously, the accordant river response is very strong.

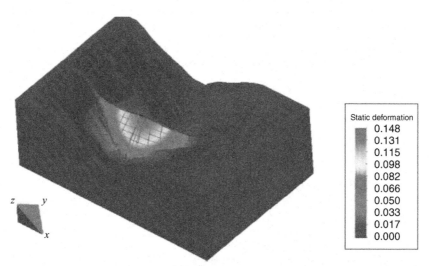

FIGURE 9.4 Dam Body and Foundation Deformation Under Action of Static Loading (Magnifying 1000 Times)

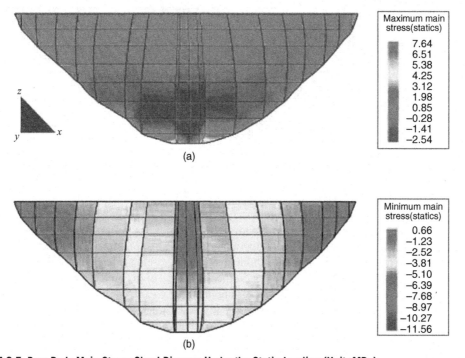

FIGURE 9.5 Dam Body Main Stress Cloud Diagram Under the Static Loading (Unit: MPa)

(a) Dam body maximum main stress distribution and (b) dam body minimum main stress distribution.

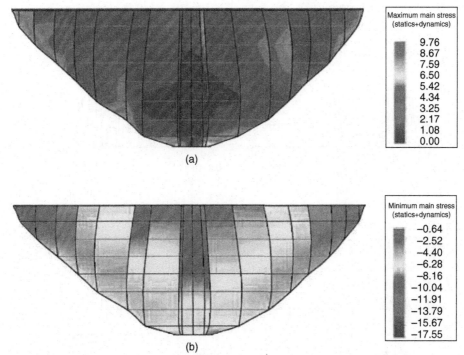

FIGURE 9.6 Dam Body Whole Time Process Main Stress Enveloping Cloud Diagram (Unit: MPa)

(a) Dam body maximum main stress distribution and (b) dam body minimum main stress distribution.

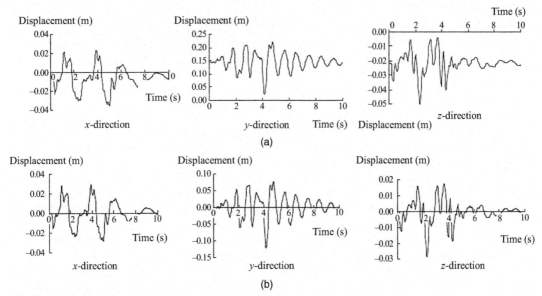

FIGURE 9.7 Displacement Response Curses of Arch Crown Beam (Upstream Surface)

(a) Displacement response under joint action of static-dynamics and (b) displacement response under seismic loading action.

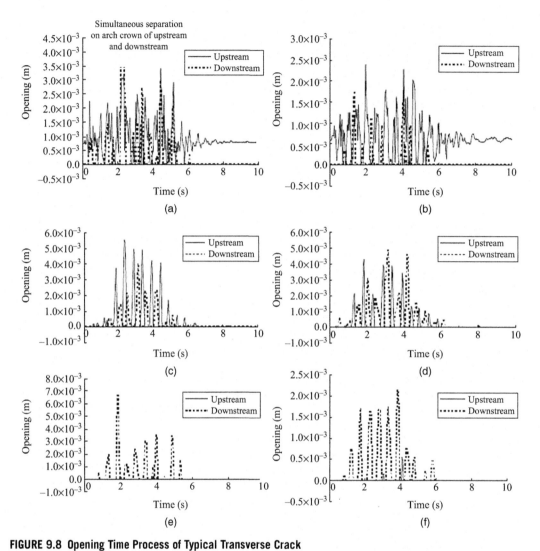

FIGURE 9.8 Opening Time Process of Typical Transverse Crack

(a) No. 1 node pair, (b) No. 17 node pair, (c) No. 4 node pair, (d) No. 14 node pair, (e) No. 7 node pair, and (f) No. 11 node pair.

Seismic displacement response maximum value reaches 11.69 cm so that the input maximum value is magnified by as much 7.8 times.

Figure 9.8 gives out the opening time process of typical transverse cracks in the arch crown. In most of cases, one crack opening and closing on the upstream and downstream faces appear to be alternative, whereby illustrating that one crack cannot appear to have the case of the whole contact surface separation at the same moment in general, but when there is about 2.5 s the arch crown upstream and downstream faces separate from each other, with the maximum peak value of 4 mm.

9.1.3 PARALLEL COMPUTATION UNDER ARTIFICIAL VISCOELASTIC BOUNDARY CONDITIONS

9.1.3.1 Parallel computation efficiency analysis

It is here that seismic dynamic response of one kind of operational condition is calculated in the case of normal water level. The scheme is the same as the parallel computation under the transmitting boundary condition in Section 9.1.2. The scheme is divided into three steps to conduct the calculation, but what is different is that in the first operational condition, the quasidynamic method is adopted to calculate the hill mass gravity function in modifying contact node pair messages within the hill mass, and the fixed boundary is usually adopted for the near-field foundation boundary, while in the other two kinds of operational conditions, the near-field foundation boundary is treated in terms of the artificial viscoelastic boundary. The computation analysis is as follows:

1. *Computation of structure response under the action of the foundation bedrock mass gravity*: At this time, only gravity is applied to the foundation rock mass, and the rest of other loadings are the zero, with an aim at modifying the contact node pair messages within the hill mass. As to the foundation rock mass having no sliding move engineering, there is no need to have this step calculation. This step calculation in nature belongs to the utilization of dynamic calculation source to simulate statics in comparison with serial calculation results. It is found that in computing only 6.0 s time, two kinds of calculation modes have realized the complete stability, and the calculation results are also in agreement with each other. In serial computation, the computer configuration used is the double core master frequency of 3.4 GHz and internal memory of 2 GB. Time step length in calculation takes 0.00005 s, with the time consumption of 7 h 7 min. The Association & Deep Dump 1800 computer system is adopted in parallel calculation, and time step takes 0.00005 s. In calculation, the problem is divided into 12 processes and the calculation completion takes 20 min 48 s, without considering the differences in computer hardware as viewed from the time spent on calculation.

2. *Computation of structure static displacement under the action of dam body gravity*: This computation step includes two courses: (1) Static computation under the action of dam body gravity. At this time, there is only gravity function on the dam body; the other loadings are zero. Based on the combination of static and dynamic computation method principle, this computation is relatively an independent step. This step uses the contact message Step 1 result. The computation purpose is to obtain structure shape displacement under the action of dam body gravity; the viscoelastic boundary condition is adopted to compute the middle boundary, in order to compute the preparatory data of the final static and dynamic stress combination. (2) Computation of static displacement of foundation rock mass and dam body without being subject to gravity action. This step computation belongs to serving as a key link between past and future. The Step 1 revised contact node pair messages are used, and the function loading includes temperature loading, water and sediment static pressure, seepage flow pressure, and the viscoelastic boundary condition are adopted to complete the middle boundary. This step needs preparing stabilized each node static displacement as well as contact message for dynamic calculation. The time process of this step computation still is 6 s; the time step length also adopts 0.00005 s. Computing 6.0 s reaches complete stability, computing 120,000 steps in total. Single machine serial computation time is 15 h 25 min, while the parallel computation time is 59 min 50 s so that the parallel computation is about 1/15 of the serial computation time.

Table 9.3 The Conditions and Time Consumption Between Parallel Computation and Serial Computation of Xiaowan Arch Dam Seismic Dynamic Analysis

Computation Mode		Serial Computation	Parallel Computation
Basic configuration		3.4 GHz master frequency, 2GB inner memory	Single nodes two single-core CPU 3.0 GHz master frequency, 2 GB internal memory, total 5 + 1 node
Step (1)	Computation conditions	Fixed boundary, set up inner damping. After structure stabilized to obtain static solution computing 6 s time, time step length is 0.00005 s	Fixed boundary, set up inner damping. After structure stabilized to obtain static solution computing 6 s time, time step length is 0.00005 s
	Time spending	7 h 7 min	20 min 48 s
Step (2)	Computation conditions	Viscoelastic boundary computing 6 s time, time step length is 0.00005 s	Viscoelastic boundary condition computing 6 s time, time step length is 0.00005 s
	Time spending	15 h 25 min	59 min 50 s
Step (3)	Computation conditions	Transmitting boundary computing 10 s time, time step length is 0.00005 s	Viscoelastic boundary condition computing 10 s time, time step length is 0.00005 s
	Time spending	62 h 56 min	2 h 50 min

3. Computation of structure displacement response under the action of seismic loading: this step is the core part of structure dynamic static computation. At this time, the structure bearing the static loading includes all the loading of the earlier Step 2 computation, the static displacement obtained Step 2 computation as well as revised node pair messages serving as the initial values. Hydrodynamic pressure and sediment dynamic pressure need to be converted into additional mass to dam surface nodes based on the method suggested by Westergaard in terms of requirements by hydraulic structure aseismic design codes. This step adopts viscoelastic boundary condition, from which seismic waves are input. This computation course is compared with the adaptation of transmitting boundary condition (Jin et al., 2000) single machine serial program with artificial seismic waves as shown in Fig. 9.2 input. Computing 10 s seismic response, time step length is 0.00005 s. Computing 200,000 steps, the single machine serial consuming time is 62 h 56 min, but the computation on Association & Deep Dump 1800 parallel computers spending time is only 2 h 50 min, so that the parallel computation time is only 1/22 of the serial computation time. Table 9.3 sums up the conditions and time computation between the parallel computation and the serial computation of the Xiaowan arch dam seismic dynamic analysis, whereby the efficiency of the two kinds of computation methods can clearly be compared.

9.1.3.2 Analysis of parallel computation results of seismic dynamic response of Xiaowan arch dam

It can be seen from Table 9.3 that the parallel computation has raised the computation speed of seismic dynamic response greatly but the guarantee of accuracy of computation results is an important prerequisite of the practical application of parallel computation technology to the structure seismic dynamic

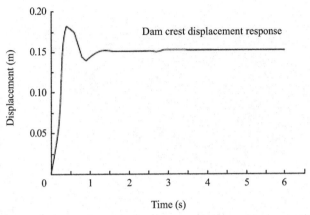

FIGURE 9.9 Dam Crest Nodes Stabilizing Course Under the Action of Dam Body Gravity

response analysis. If the correctness of computation results cannot be guaranteed, the improvement of computation speed is of no use. Though the parallel computation program compiled in this book is tested through the analytical computation examples, there will be more mature practical application programs through engineering checks for further references because of the complexity of actual engineering works and their results must be checked and identified.

Figure 9.9 is the dam crest nodes stabilizing course under the action of dam body gravity. The static course is simulated by the dynamic method in order to stabilize the structure as quickly as possible so that the dam damping when taken in computation has been increased by 10 times as much as the original. As viewed from the dam body response course, each node in the dam body is stabilized after 3 s. As a result, the results computed to 6 s, are enough to be the static displacement results of each node under the action of dam body gravity.

In comparison of Fig. 9.6 with Fig. 9.10, the distribution situations of the maximum tensile stress and the compressive stress in the dam body can be seen. In the case of two kinds of boundary conditions, the tensile stress distribution zones and the magnitude of the maximum tensile stress as well as the maximum tensile stress distribution position are in uniformity; the compressive stress distribution zones are basically the same, but the maximum compressive stress values have had some differences, and the viscoelastic maximum is 12.67 MPa, the transmitting boundary maximum compressive stress is 17.55 MPa. In addition, in the case of artificial viscoelastic boundary, deformation obtained via the parallel computation and the displacement time process curves under the action of seismic waves as well as the calculated results in the case of transmitting boundary conditions are well in good agreement with each other.

9.2 PARALLEL COMPUTATION ANALYSIS OF SEISMIC MOTION DYNAMIC RESPONSE OF XILUODU ARCH DAM

The Xiluodu hydropower station located in Xiluodu gorge in Leipo County of Sichuan Province, neighboring Yongshan County of Yunnan Province, is the largest of the huge type of four hydropower stations in the lower reaches of the Jinsha River. The installed capacity is rather the same as that of

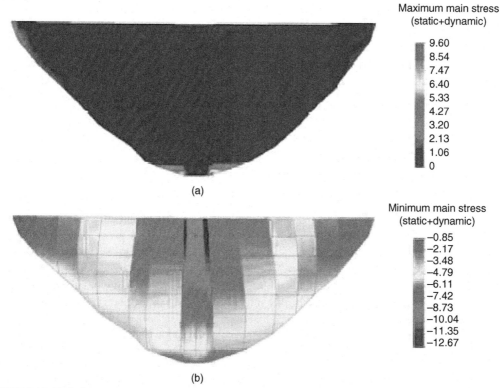

FIGURE 9.10 Main Stress Envelope Cloud Diagram for Static–Dynamic Loading Whole Time Process Under the Viscoelastic Boundary Conditions (Unit: MPa)

(a) Whole time process maximum main stress envelope cloud diagram and (b) whole time process minimum main stress envelope cloud diagram.

Yitaip hydropower station on the Balana River. Total installed capacity reaches 126 million kW and it is the second largest hydropower station next to the Three Gorge Project hydropower station in our country. The Xiluodu hydropower station junction engineering works consist of river dam, flood discharge engineering works, water diversion works, power generation system, and so on. The river dam is a concrete double-curved arch dam; the dam crest altitude is 610 m above sea level and the maximum dam height is 278 m high. Dam crest center line arc length is 648.0 m long; the underground power houses are laid out on the left and right banks, in each of which nine single units of hydrogenerator sets with the installed capacity of 0.7 million kW can be installed with an annual power generation of 571 million~640 million kWh. The Xiluodu reservoir normal storage water level is 600 m; the regulation reservoir capacity is 6.46 billion m^3. The reservoir of the hydropower station is 208 km in length. The normal water strong level is 600 m; total reservoir water storage capacity is 12.67 billion m^3; the dead water level is 540 m, corresponding to reservoir storage capacity of 11.57 billion m^3; regulation storage capacity is 6.46 billion m^3; flood control reservoir capacity is 4.65 billion m^3, with a strong flood control capacity. This huge hydropower project was formally launched in December 2005, and in November 2007, the river enclosure was completed. It was planned that the first set of power generating units be installed in 2013, and that all engineering projects will be completed in 2015.

9.2.1 BASIC DATA FOR COMPUTATION

Dam body geometric sizes or magnitudes and dam body as well as foundation material performances are based on the technology data provided by Chengdu Investigation Survey and Design Academy. The Xiluodu dam-site river valley is basically to appear as U-shaped. Dam crest arch crown is 14 m in thickness and arch crown bottom is 69 m in thickness. Dam body concrete elastic modulus is $E = 26$ GPa; density is $\rho = 2600$ kg/m^3; Poisson's ratio is $v = 0.167$. Material numbered is I starting from No. 2 is the foundation rock mass materials. The foundation rock mass is divided into 17 kinds of materials. The values taken for the materials are shown in Table 9.4. The dynamic computation is the same as that of the Xiaowan arch dam. The dynamic elastic modulus takes 1.3 times of the static elastic modulus. Dumping ratio takes 0.05. In calculation, temperature loading is obtained from the temperature data provided by Chengdu Investigation Survey and Design Academy. There are 30 transverse cracks set up in the whole dam body.

Based on the requirement of "Seismic Design Code of Hydraulic Structure" (DL 5073-2000) in China, and in terms of design and prevention intensity of IX for the Xiluodu arch dam, bedrock

Table 9.4 Foundation Rock Mass Material Parameters for Xiluodu Arch Dam Computation Model

Material No.	Elastic Modulus (GPa)	Poisson's Ratio v	Density ρ (kg/m^3)
2	13.28	0.220	2600.0
3	12.96	0.220	2600.0
4	12.51	0.220	2600.0
5	13.54	0.220	2600.0
6	16.27	0.220	2600.0
7	17.85	0.220	2600.0
8	15.70	0.220	2600.0
9	14.37	0.220	2600.0
10	13.07	0.220	2600.0
11	13.88	0.220	2600.0
12	15.45	0.220	2600.0
13	17.60	0.220	2600.0
14	18.06	0.220	2600.0
15	15.23	0.220	2600.0
16	13.38	0.220	2600.0
17	19.50	0.220	2600.0
18	19.50	0.220	2600.0

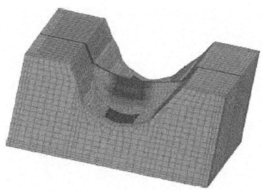

FIGURE 9.11 Three-Dimensional Finite Element Discrete Mesh for Xiluodu Arch Dam Seismic Response Analysis

horizontal design seismic acceleration is $0.321g$, vertical design seismic acceleration taking two-third of horizontal acceleration is $0.214g$. In computation, the seismic wave adopted in the study is the artificial seismic wave generated with the response spectrum as the objective spectrum.

Figure 9.11 is a three-dimensional finite element discrete mesh for the Xiluodu arch dam to carry out seismic dynamic response analysis, having 146,769 nodes, 9,298 contact node pairs being all the nodes on the dam body transverse cracks, 117,002 hexahedron units, and 1,826 triangular prism units. Owing to this shape with an opening detailed build, the dam body is divided very finely so that most of the calculating nodes concentrate on the dam body, but the unit sizes on the dam body are very small. The minimum side length of the unit is only 0.48 m.

In computation, seismic displacement waves are input from the artificial viscoelastic boundary, being similar to the analysis of the Xiaowan arch dam. The input wave reduced to half can serve as the real input data. The displacement time process of artificial seismic wave three directions is shown in Fig. 9.12.

9.2.2 PARALLEL COMPUTATION OF XILUODU ARCH SEISMIC DYNAMIC RESPONSE

Being different from the Xiaowan arch dam dynamic computation, there are no sliding move zones on the Xiluodu arch dam foundation bedrock mass so that there is no need to revise the contact message of contact nodes on the dam foundation bedrock mass. Also, there is no need to compute structure response under the action of hill mass gravity as the Xiaowan arch dam does. The whole dynamic computation is divided into two steps to carry out as follows:

1. *Computation of structure static displacement under the action of dam body gravity (the same condition of computation steps with those of the Xiaowan arch dam)*: The step includes two courses: (1) Static computation under the action of dam body gravity. At this time, the dam body only has had gravity function and the other loadings are zero. Based on the principle of combination of static and dynamic computation method, it is here that computation is a relatively independent step, and the computation purpose is to obtain the displacement of structure build or shape under the action of dam body gravity. In computation, the viscoelastic boundary condition is adopted for the boundary in order to compute the preparatory data for the final combination of static and dynamic stress. (2) Computation of foundation bedrock mass and dam body without

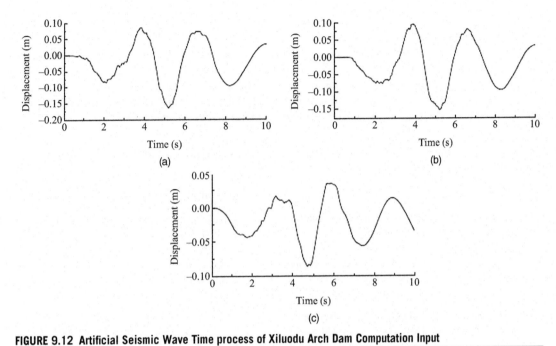

FIGURE 9.12 Artificial Seismic Wave Time process of Xiluodu Arch Dam Computation Input

(a) *x*-Direction displacement course, (b) *y*-direction displacement course, and (c) *z*-direction displacement course.

being subject to static displacement by the gravity action. This step function loading includes temperature loading, seepage flow pressure, and the boundary adoption of viscoelastic boundary condition in calculation. This step needs to prepare the static displacement and contact messages of every stabilized node for dynamic computation. The step computation time is 6 s, and time step length adopts 0.00005 s; Steps 1 and 2 computation steps and similar conditions have no big difference in time consumption. The computing time by single computer serial program under the viscoelastic boundary condition is about 111 h 40 min, while the time spent on parallel computation is 4 h 45 min.

2. *Computation of structure displacement response under the action of seismic loading*: At this time, the structure bearing static loading includes all the static loadings in the mentioned course 2 computation. Also, the static displacement obtained in course 2 and revised node pair messages should be input. Hydrodynamic pressure and sediment dynamic pressure must be converted into the additional mass to dam surface nodes. Viscoelastic boundary conditions are adopted in computation, and seismic waves are input from the boundary. The computed results should be compared with those of single computer serial program by adopting transmitting boundary conditions (Jin et al., 2000). The artificial seismic waves in Fig. 9.12 should be input, computing 10 s seismic response; time step length is 0.00005 s. The time spent, computed on Association & Deep Dump 1800 parallel computers, is 10 h 50 min, and the transmitting boundary single serial computer computes 9.5 s structure response. Consuming time is 505 h 12 min so that parallel computation saves much time.

Table 9.5 summarizes the conditions and time consumption of the Xiluodu arch dam serial computation and parallel computation, whereby comparing the efficiency of two kinds of calculations.

Figure 9.13 is the displacement time process curves of dam upstream surface arch crown nodes under the action of seismic waves. It can be seen from the Fig. 9.13 that despite seismic waves traveling and reaching the dam crest after, and owing to the action of ongoing opening and closing of two sides of contact cracks, the displacement response to the dam crest and the input seismic wave motion cannot keep the complete agreement with each other, but the general trend remains much better.

Table 9.5 Conditions and Time Consumption of Xiluodu Arch Dam Serial Computation and Parallel Computation

Computation Mode		Serial Computation	Parallel Computation
Basic configuration		3.4 GHz master frequency, 2 GB internal memory	Single nodes two single-core CPU 3.0 GHz master frequency, 2 GB internal memory total 5+1 node
Step (1)	Computation conditions	Viscoelastic boundary computing 6 s time. Time step length is 0.00005 s	Viscoelastic boundary conditions 6 s time. Time step length is 0.00005 s
	Time spending	111 h 40 min	4 h 54 min
Step (2)	Computation conditions	Transmitting boundary computing 9.55 s time. Time step length is 0.00005 s	Viscoelastic boundary condition computing 10.5 s time. Time step length is 0.00005 s
	Time spending	505 h 12 min	10 h 48 min

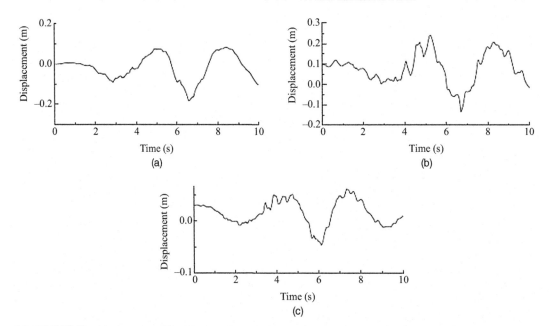

FIGURE 9.13 The Displacement Time Process Curves of Dam Upstream Surface Arch Crown Nodes Under the Action of Seismic Waves

(a) x-Direction displacement response, (b) y-direction displacement response, and (c) z-direction displacement response.

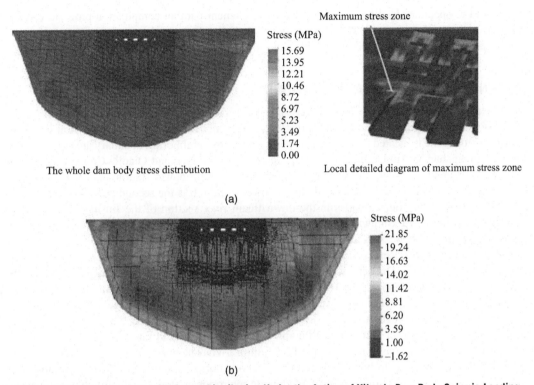

(a)

(b)

FIGURE 9.14 The Maximum Tensile Stress Distribution Under the Action of Xiluodu Dam Body Seismic Loading

(a) Artificial viscoelastic boundary and (b) artificial transmitting boundary

Figure 9.14 is the upstream surface tensile stress distribution cloud diagram obtained through the computation of two kinds of boundary conditions under the action of seismic loading of the Xiluodu dam body. It can be seen from the Fig. 9.14 that there are some differences in the tensile stress slope and magnitude in the case of boundary conditions of two kinds. The stress obtained by adopting the transmitting boundary computation program is not only large numerically, reaching 21.69 MPa, but also has a broad distribution zone of high stress values. The maximum tensile stress obtained adopting the viscoelastic boundary computation program is very small, only being 15.69 MPa, whose high tensile stress zone is very small, only occurring near some transverse crack contact node on the arch crown upstream surface, whose results are closer to the engineering experiences.

9.3 PARALLEL COMPUTATION ANALYSIS OF BAIHETAN LEFT BANK SIDE SLOPE SLIDING BLOCK STABILITY

In hydraulic and hydroelectric engineering, the research objectives, apart from big dams, include side slope stability, abutment stability, underground caverns, arch dam abutment stability problems, and so on. The finite element parallel computation program developed by the authors is based on the continuum or continuous deformation model. Computation course can satisfy the stress balance and displacement coordinative conditions. In considering dynamic contact, hydrodynamic pressure, seepage flow pressure, and temperature loading problems, this finite element parallel computation program can be used in the multistructure dynamic response analysis. The following uses this program to carry out the dynamic response analysis of the left bank side slope sliding blocks of Baihetan engineering.

9.3.1 ENGINEERING GENERAL CONDITIONS

The Baihetan hydropower station was surveyed and designed by China East Survey and Design Academy of the National Hydroelectric Ministry in 1992, and launched into formal construction in 2008. The Baihetan dam site is located in the Jinsha River gorge of the neighboring boundary area of Ningnan County of Liangshan Yi Autonomous Prefecture in Sichuan Province and Qiaojia County of Yunnan Province. Its upstream is connected with Wudongde cascade hydropower station and its downstream tail water is linked with the Xiluodu cascade hydropower station. It is the second cascade hydropower station developed of the four cascades in the downstream river section of the Jinsha River (Yalong river mouth – Ninbin), 75 km away from Ningnan County town. This hydropower project gives the priority generation and holds sediment check, flood control and prevention, navigation, and irrigation of comprehensive benefits. Reservoir normal storage level is 820 m, corresponding to reservoir storage capacity of 17.9 billion m³. The dead water level is 760 m under which reservoir storage capacity is 7.9 billion m³. Total reservoir capacity is 18.8 billion m³. During the flood season, water level is 790 m, prereserved flood control storage capacity is 5.6 billion m³. The regulating storage capacity is 10 billion m³, with the seasonal regulation capacity, whereby increasing the ensured 2.2 million km of such four cascades of hydropower stations as Xiluodu, Xiangjiaba, TGP, and Guozhouba downstream in the dry season, with an increase in electricity of 5.5 billion kWh. The upstream return water of 180 km connects with Wudongde hydropower station. The reservoir normal water storage level overlaps with the tail water level (805.5 m) of 14.5 m of Wudongde hydropower station, which is the water head overlapping the largest reservoir in this river section.

The engineering complex or junction consists of a diversion dam, flood discharge facilities, and water-diversion system for power generation. The diversion dam is a double-carved dam with a height of 277 m, dam crest altitude is 827 m above sea level; the crest width is 13 m wide. The maximum dam bottom width is 72 m wide. The 16 sets of 0.75 million kW mixed flow turbine generating sets are installed into the underground powerhouse. Total installed capacity is 12 million kW, and annual power generation is 51.5 billion kWh. After the upstream tiger jump gorge dragon head reservoir is completed, the capacity can be extended to 15 million kW, and the annual power generation will reach 56.87 billion kWh.

9.3.2 COMPUTATION OF BASIC DATA

Two banks of the Baihetan dam site are seriously weathered with strong unloading function and the left bank weathered with the unloading depth being widely larger than that in the right bank. The left bank strong unloading distributes on the left bank at an altitude of over 690 m above sea level. The level depth is, in general, 0~44 m, with the deepest depth reaching 93 m. The level depth of the weak unloading belt is 26~102 m, and the lower limit buried depth in the weak weathered lower section is 43~132 m; the right bank strong unloading level depth is 0~3 m with some local depth reaching 49 m. The weak unloading distributes at an altitude of over 660 m above sea level, and the level buried depth is 5.5~62 m. The low limit level buried depth of the weak weathered lower section is 43.5~74 m. The

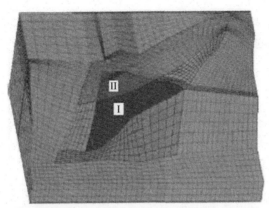

FIGURE 9.15 Mesh Division of Sliding Block Computation Zone of Baihetan Left Bank Side Slope

low limit vertical depth of the river bed weak weathered lower section is 20~28 m. The left bank side slope rock mass in the mid-dam site includes the internal layer and interlayer disturbed belt with less inclination near the vertical fault and unloading fracture. Internal layer and interlayer disturbed belt being less inclined upstream deviated to the right bank is the latent bottom sliding face. Under the cutting-off fault and unloading fracture, several possible sliding bodies will be formed.

In order to ensure the safety of the left side slope of Baihetan hydropower station in construction and operation, it is essential that an evaluation should be made of the left bank side slope dynamic stability under the seismic action, and the corresponding consolidation treatment measures should be adopted.

Since the Baihetan mid-dam-site left bank rock strata are inclined upward and riverbed, the faults, interstrata, and internal layer disturbed belt have developed to the full extent so that side slope is easy to yield the loose unloading phenomenon. Accordingly, the left bank side slope is the controlling factor for dam line selection. The schemes with different dam lines may be subject to the different degrees of each main side slope influence. The following analyzes the stability of the most dangerous sliding block among all the sliding blocks. The selection of sliding blocks and mesh divisions of model zones are shown in Fig. 9.15. I and II are two master modules. In accordance with the landform map of Baihetan left bank side slope and bedrock material natures, the finite element mesh profile technology is adopted to generate the finite element mesh for the Baihetan left side slope. Total number of the finite element mesh hexahedron units are 24,906; total number of nodes is 28,787; total number of freedoms is 86,361; and the contact node pair number is 316, being used to simulate faults. In computation, materials are divided into five kinds, and the parameters of each material are shown in Table 9.6 (Figs. 9.16–9.21).

At present, the usually used methods for analyzing the side-slope stability include the limit balance method, the numerical analysis method (Bofang, 1998), and the probability method (Haifei et al., 2008), etc. of which the limit balance method belongs to the classical method. The limit balance method is to use the static theory to analyze the side-slope stability without taking the side-slope deformation and failure process into account, on the basis of the bulk of assumptions, the solution is obtained in terms of force or moment balance. The strength reduction method is the better method to use the numerical computation technology to seek for the slope-body

Table 9.6 Material Parameters of Baihetan Left Bank Side-Slope Computation Model

Material No.	Elastic Modulus E (GPa)	Poisson's Ratio v	Density ρ (kg/m³)
1	4.00	0.310	2551.0
2	2.00	0.330	2551.0
3	8.00	0.270	2653.0
4	11.50	0.250	2806.0
5	14.50	0.230	2857.0

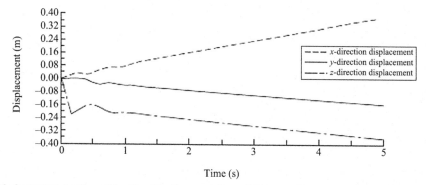

FIGURE 9.16 Control Node Three-Direction Displacement Curved Diagram (Strength Reduction Factor is 1.2)

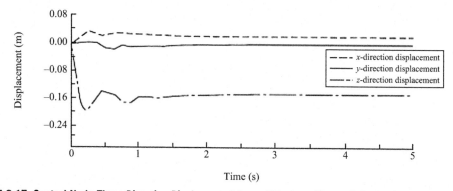

FIGURE 9.17 Control Node Three-Direction Displacement Curved Diagram (Strength Reduction Factor is 1.1)

safety coefficient, which has found a wide application at present in many aspects. The so-called strength reduction method is gradually to lower its shear strength based on the real shear strength of slope-body (whereby the real shear strength is divided by a coefficient, i.e., the strength reduction coefficient so as to achieve the purpose of strength reduction) until the slope-body reaches the critical balance state. It is just at this time that the reduction coefficient can be considered as

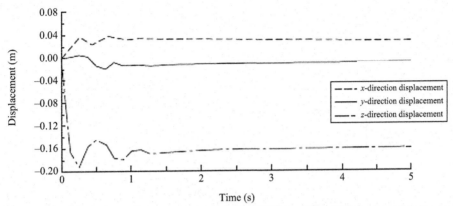

FIGURE 9.18 Control Node Three-Direction Displacement Curved Diagram (Strength Reduction Factor is 1.15)

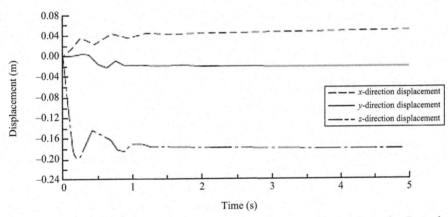

FIGURE 9.19 Control Node Three-Direction Displacement Curved Diagram (Strength Reduction Factor is 1.17)

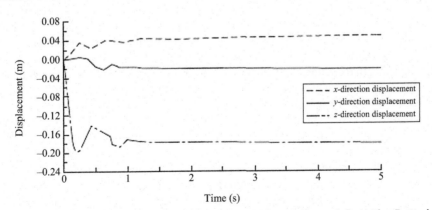

FIGURE 9.20 Control Node Three-Direction Displacement Curved Diagram (Strength Reduction Factor is 1.19)

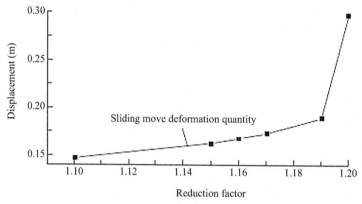

FIGURE 9.21 Relation Curve Between Strength Reduction Factor and Sliding Move Quantity

the side-slope safety coefficient. Accordingly, the application of the strength reduction technology to seek for the method for the side-slope safety coefficient is the strength reduction method (Hua et al., 2009).

9.3.3 STATIC STABILITY COMPUTATION

As far as the side-slope safety under the action of self-weight is concerned, the strength reduction method is adopted to determine its static safety coefficient: the parallel computation (associated deep gallop 1800 computer system, with 6 computation nodes, 12 CPU, CPU main frequency 3 GHz. 2 GB internal memory) 5 s response time process reaches stability. Time step is 0.000025 s, computing 200,000 steps needs 62 min. In the case of same computing conditions, single computer (Association Kaitian B6650, 3.4 GHz main frequency, 2 GB internal memory) needs 12 h 24 min. The computing process of determining static safety coefficient is as follows: the static computation of strength reduction coefficient is 5 in total, being 1.2, 1.1, 1.15, 1.17, and 1.19, respectively.

1. When taking strength reduction coefficient as 1.2, the displacement curve of controlling point is shown in Fig. 9.16. In a total computing time, the displacement on sliding block in three directions continues to enlarge. This means that in the case of this strength reduction coefficient, the sliding block under its self-weight action, it will slide off.
2. When taking strength reduction coefficient as 1.1, the displacement curve of controlling point is shown in Fig. 9.17. With an increase in computing time, the displacement on sliding block in the three directions will not increase. The displacement is able to realize stable convergence, whereby indicating that in the case of strength reduction coefficient, the sliding block is safe under its own self-weight action, and it will no longer slide off. Accordingly, the static safety coefficient of side-slope is within the range of 1.1–1.2.
3. When the strength reduction coefficient increases by 1.15, the displacement of controlling point is shown in Fig. 9.18. In the case of this strength reduction, the side-slope still remains stable so that the sliding block will no longer slide off under its own weight action. Accordingly the static safety coefficient of side-slope will be within the range of 1.15–1.2.

4. When the strength reduction coefficient continues to increase by 1.17, the displacement curve of controlling point is shown in Fig. 9.19. In the case of this strength reduction coefficient, the displacement is able to become stable and the sliding block will no longer slide off. The static safety coefficient of side-slope determined by this is within the range of 1.17–1.2.
5. When the strength reduction coefficient increases by 1.19, the displacement curve of controlled point is shown in Fig. 9.20. In the case of this strength reduction coefficient, the displacement is able to become stable and the sliding block will not slide off.

It can be known from the above computations that the side-slope static coefficient should be within the range of 1.17–1.20. Figure 9.21 is the side-slope sliding deformation quantities in the case of different strength reduction coefficients (in order to clearly discover the laws of side-slope sliding deformation quantities in the case of different reduction coefficients, the computation with 1.16 of reduction coefficient increases). It can be seen that when the reduction coefficient is 1.19, the sliding deformation quantities can appear to have a mutation so that the determination of side-slope static safety coefficient is 1.19.

9.3.4 DYNAMIC STABILITY COMPUTATION

To investigate the safety of the left bank side slope sliding blocks under the seismic action, it is necessary to compute the sliding block response under the action of seismic loading so as to determine the safety coefficient. In corresponding to static safety recheck, the strength reduction method is adopted to determine the safety coefficient in dynamic computation. Owing to the seismic loading action, it is difficult to judge whether displacement response to each node on the structure is in the stable state. For this reason, we can by no means determine whether the sliding blocks slide falling down in terms of node absolute displacement values on sliding blocks, but we can judge the sliding move state of sliding blocks in terms of relative displacement growing conditions of sliding blocks. It is necessary to compute 30 s response time process; time step length is 0.000025 s. Parallel computation (adopting the same computer group system as the static computing system) of 1.2 million steps spends 13 h 13 min. In the same case of computation conditions, single computer computation (adopting the same computer as the static computing system) consumes 156 h 24 min. The strength reduction factors are computed with five, and they are 1.0, 1.12, 1.06, 1.08, and 1.07, respectively.

1. Taking strength reduction factor is 1.0 (side slope response in the case of no reduction in strength); and the relative displacement curve of contact node is seen in Fig. 9.22. It can be seen that the wave motion of contact node relative displacement within 30 s displacement input time process always remains very small, whereby illustrating that in the case of no reduction in strength, side slope is safe, and the sliding blocks will not slide down.
2. Increasing strength reduction factor is 1.12; and the relative displacement curve of contact node is seen in Fig. 9.23. It can be seen that the relative displacement of the contact node within 30 s displacement input time process continue to grow, whereby illustrating that in the case of strength reduction factors, the sliding blocks on the side slope will slide down. Accordingly, the dynamic safety coefficient should be in the range of 1.0~1.12.
3. Lowering strength reduction factor is 1.06, and the relative displacement curve of the contact node is seen in Fig. 9.24. It can be seen that the relative displacement of the contact node within 30 s displacement input time section waves in one small zone and shows no trend to appear to

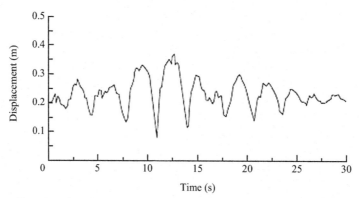

FIGURE 9.22 Contact Node Relative Displacement Curve (Strength Reduction Factor is 1.0)

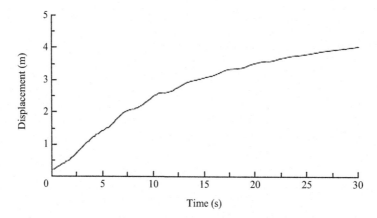

FIGURE 9.23 Contact Node Relative Displacement Curve (Strength Reduction Factor is 1.12)

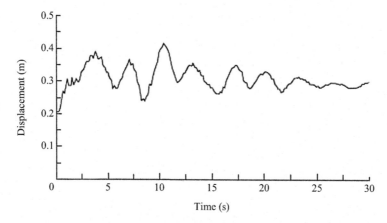

FIGURE 9.24 Contact Node Relative Displacement Curve (Strength Reduction Factor is 1.06)

stop growing, whereby illustrating that in the case of strength reduction factor, side slope is safe and sliding blocks will not slide down. Accordingly, side slope dynamic safety factors should be within the range of 1.06~1.12.

4. Reincreasing strength reduction factor to 1.08; the contact node relative displacement curve is seen in Fig. 9.25. It can be seen that the contact node relative displacement within 30 s wave motion input time section continues to grow, whereby illustrating that in the case of this strength reduction factor, side slope is unsafe, so that the sliding block will slide down. Side slope dynamic safety coefficient should be within the range of 1.06~1.08.

5. Changing strength reduction factor takes 1.07, and contact node relative displacement curve is seen in Fig. 9.26. It can be seen that the node relative displacement within 30 s wave motion input time section does not appear to have the trend to stop growing, but it can be seen from the curve that the displacement varying morphology between two nodes has not been as the case of strength reduction factor of 1.06 conditions. It is just at this time that side slope is close to critical state.

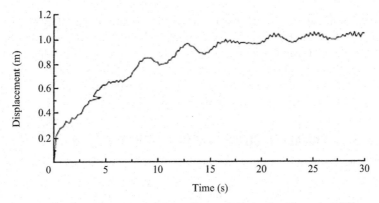

FIGURE 9.25 Contact Node Relative Displacement Curve (Strength Reduction Factor is 1.08)

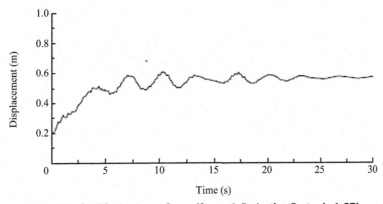

FIGURE 9.26 Contact Node Relative Displacement Curve (Strength Reduction Factor is 1.07)

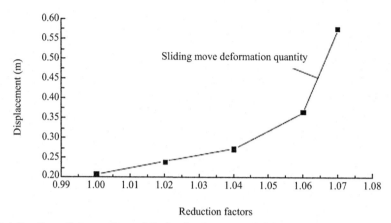

FIGURE 9.27 Relation Curve Between Strength Reduction Factor and Sliding Move Quantity

It can be known from the mentioned computations that the side slope safety coefficient should be within the range of 1.06~1.08. Figure 9.27 is the sliding move deformation quantity of the side slope under the different reduction factors (in order to better discover the laws of sliding move deformation quantity under different reduction factors, reincreasing reduction factors is 1.02 in computation). It can be seen that when the reduction factor is 1.07, the sliding move deformation quantity appears to have the mutation so that the side slope safety factor determined should be 1.06.

9.3.5 ANALYSIS OF PARALLEL COMPUTATION EFFICIENCY

The strength reduction method is employed to analyze the side slope stability, and determining safety coefficient or factor of side slope sliding blocks needs to carry out the vast computation work. Determination of one safety coefficient needs to compute at least 5~6 strength reduction coefficients. As to the complicated side slope safety recheck with a large number of freedoms, a single-computer serial program must be adopted to carry out computation so that the time cost is hard or difficult to bear. As far as the stability analysis of the left bank side slope sliding blocks of Baihetan is concerned, if only five reduction factors are computed in static and dynamic (this is the realistic condition in real computation, but as to the practical engineering works, the number of computed reduction factors will surpass this number), the parallel computation method is adopted, it will save a large amount of time. Table 9.7 shows the time consumed in the static and dynamic computation by adopting the parallel computation method and the serial computation method. If all the static and dynamic safety factors or coefficients are computed by adopting the parallel computation scheme, it needs only one week, while the traditional serial scheme that is used to determine the dynamic safety factors need over one month. Therefore, the high efficiency of the parallel computation can be shown to the fullest extent. It is necessary to point out that with an increasing complexity and a growing improvement of the engineering problem involved, the computation scales will also increase correspondingly; the time cost on many problems by the serial computation cannot be borne. At the same time, the parallel computation characterized by the expansion will gradually display in such a way that in the case of increasing not much hardware input, the parallel computation can improve the computation speed, increase the scale

Table 9.7 Time Comparison Between the Parallel and Serial Stability Analysis of Baihetan Left Bank Side Slope Sliding Blocks

Computation Mode	Parallel Computation		Serial Computation	
System configuration	Association & Deep Dump 1800 computer group system six computation nodes, 12 CPU, CPU master frequency 3.0 GHz single node 2 GB internal memory		Association kaitian B6650 3.4 GHz master frequency 2 GB internal memory	
Numbers of freedoms to the problem-solving (pieces)	86,361			
Single reduction factor to consume time	Static computation	Dynamic computation	Static computation	Dynamic computation
	62 min	13 h 28 min	12 h 24 min	156 h 24 min
Five reduction factors to consume time	5 h 10 min	67 h 20 min	62 h	782 h

of solutions to the problems, which will become the best choice for the large-scale and complicated structural static and dynamic analysis.

This analysis indicates that the developed dynamic parallel computation program can be used to solve the problem of the side slope static and dynamic stability. As to the engineering real examples needed to carry out comparative analysis of large schemes, the advantages of parallel computation saving enormous time can be embodied to the fullest extent. Apart from improving computation speed, adoption of parallel computation can greatly increase the solution scale so that the problems that cannot be solved by the single-computer serial program can be solved.

In addition, the debug development and engineering application of the program of the developed parallel computation are being carried out on the Association & Deep Dump 1800 computer group system purchased in 2005 by China Hydraulic and Hydropower Academy. The computer system has only included 6 nodes and 12 computation units, that is to say, in order to achieve the best computation speed. Most of the problems can only implement 12 parallel courses synchronously. With the progress made in hardware technology, the rapid improvement has been made in the parallel computer performances in recent years. The Association high performance parallel computer group system has been upgraded to Deep Dump 7000 serial. In order to investigate the latent performances of the developed hydraulic structure dynamic analysis parallel computation program, this program is run or operated on the current advanced Association & Deep Dump 7000 parallel computer system. The adopted hardware condition is the single-node two-CPU system and every CPU distributed with four cores. CPU time clock frequency is 2.67 GHz, single node shares to use 32 GB internal memory. The infinite and message passing pattern is adopted among the nodes. The Xiluodu arch dam dynamic response is selected as the test engineering. Time step length and computed response time are the same as those in Section 9.2, being 0.00005 s and 10 s. In order to investigate the decreased conditions of computation efficiency when the computation nodes increase, different numbers of computation nodes are used to operate this program. The computation efficiency and spent time are shown in Table 9.8.

Table 9.8 Association & Deep Dump 7000 Parallel Computation Efficiency and Spent Time

No. of Nodes to Join in Computation (Piece)	Single Operating Consumed Time (s)	Total Computer Step Numbers (Step)	Parallel Computer Efficiency	Problem Spent Total Time (h)
1	0.0627	200,000	1.000	3.49
2	0.0303	200,000	1.034	1.68
3	0.0195	200,000	1.066	1.09
4	0.0162	200,000	0.965	0.90

It is especially necessary to point out that the single-node computation of this problem needs the same time. (It is not the serial program with the parallel isomorphic PC computer to compute in wasting time. The main cause lies in the fact that node computation unit for the parallel computer group is designed separately. It is difficult to find the same configuration of the common computer to examine this program, while as to the different configuration of a single computer, the corresponding program in spending time can be used to define the parallel efficiency but the corresponding bases cannot be found.) for benchmark to define the parallel computation efficiency. It can be seen from Table 9.8 that the time spent on Association & Deep Dump 7000 single-node in parallel computation of seismic response of the Xiluodu arch dam is already extremely lower than the time required by Association & Deep Dump 1800 on six nodes. When four nodes are used, the computation needing 10 h 48 min to be completed originally on the Association & Deep Dump 1800 computer group now only needs 54 min to end the computation. It can be seen from Table 9.8 that the parallel efficiency of the developed parallel program is still high and that with an increase in node numbers, the average efficiency drop on each node is not significant. Accordingly, using this program to increase computation nodes to improve the computation speed can hold great potential. So long as there are enough hardware resources, the computation time can be greatly shortened. This has pointed out the technological way for the following-up whole structure to carry out material damage nonlinear analysis and create the basic conditions for the future development.

DYNAMIC TESTING, NUMERICAL SIMULATION AND MECHANISMS FOR DAM CONCRETE

10 RESEARCH PROGRESS ON DYNAMIC MECHANICAL BEHAVIOR OF HIGH ARCH DAM CONCRETE 239

11 DYNAMIC FLEXURAL EXPERIMENTAL RESEARCH ON DAM CONCRETE 257

12 THE EXPERIMENTAL RESEARCH ON THE DYNAMIC AND STATIC MECHANICAL CHARACTERISTICS OF DAM
   CONCRETE AND THE CONSTITUTIVE MATERIALS 297

13 EXPERIMENTAL STUDY OF DYNAMIC AND STATIC DAMAGE FAILURE OF CONCRETE DAM BASED
   ON ACOUSTIC EMISSION TECHNOLOGY 321

14 TESTING RESEARCH ON LARGE DAM CONCRETE DYNAMIC-STATIC DAMAGE AND FAILURE BASED
   ON CT TECHNOLOGY 395

15 RESEARCH ON NUMERICAL ANALYSIS OF FULL GRADATION LARGE DAM CONCRETE DYNAMIC BEHAVIORS . . 491

Under the strong earthquake, the dynamic response and failure process of concrete dams are extremely complicated. The failure of high arch dam is reflected as the serious cracking and crack extension of dam concrete, which results in the loss of water retaining capacity. Therefore, whether the seismic design of dam structures is safe and reasonable depends not only on the dynamic input of earthquake and the accuracy of seismic analysis methods, but also on the dynamic mechanical properties of dam concrete in designed structures. The determination of dynamic mechanic parameters for dam concrete is a necessary part of seismic safety and evaluation. However, little progress has been made in this research area, and a large number of fundamental issues need to be investigated and researched. This has become the bottleneck in the seismic safety evaluation of dam structures.

The difference between large size dam concrete and normal concrete in building area is the utilization of multigrade aggregates with the maximum aggregate size of 80 mm or even 150 mm. At present, the characteristics of dam concrete and its mechanical properties are determined by the wet-sieved concrete with the aggregates larger than 40 mm being sieved. The composition, mix proportions, and specimen size of wet-sieved concrete specimens are different from the original dam concrete, which influences the true characteristics of dynamic and static mechanical properties. The research on mechanical properties of full-graded dam concrete is scarce, especially on the dynamic characteristics and resistances.

There exists a large surplus for the compressive strength of high arch dam concrete, and this is different from the reinforced concrete in building engineering. The cracking damage of dam structures, especially under the strong earthquake, is controlled by the tensile strength of concrete. The tensile strength of concrete is far below its compressive strength. Therefore, the dynamic tensile strength of dam concrete is the key factor influencing the seismic safety of dams, which is one of the major contents for the dynamic experimental researches.

The stress state of dam is complicated and in the conditions of multiaxial stresses. The tensile condition is shown as shearing and flexural tension, which is similar to the flexural conditions of split-tensile and rupture test. The Federal Emergency Management Agency (FEMA) and the Design Specification for Concrete Arch Dams in China (DL/T 5346-2006) specifies that the tensile strength of concrete should adopt the value of splitting tensile or flexural tensile strength. Raphael (1984) has carried out extensive research on tensile tests of dam cores. Based on the achievements by Raphael (1984), Bureau of Reclamation (2006) employs the splitting tensile of cylinder specimen as the design value of tensile strength for concrete dams. As the stress-strain curves of concrete soften near the peak stress, the apparent strength calculated by the critical strain and elastic modulus is higher than the measured peak strength value. During the linear-elastic analysis of dam structures, the elastic modulus of concrete is deemed as constant value. Therefore, the apparent strength in the splitting tensile tests should be adopted when the stress is checking. The experimental results by Raphael showed that the flexural strength determined by rupture test is closed to the apparent strength in the splitting tensile strength. The Bureau of Reclamation employs the peak strength of dam concrete rupture tests as the design value of apparent tensile strength for dam structures. Therefore, the peak strength of dam concrete in rupture test is significantly important for the seismic response analysis of concrete dam.

The dam usually undertakes the action of earthquake during the normal operation, while the existing experiments for dynamic mechanical properties of concrete are carried out without the consideration of initial static loads. The dynamic characteristics and resistances of concrete under certain initial loads are the key problems for the engineers. Therefore, for the concrete used in high arch dams, the significant task to assure the seismic safety of high arch dams is the experiments for dynamic rupture tests of full-graded concrete specimens based on the variable amplitude cyclic loading, whose frequency is closed to fundamental frequency of dam structures.

In addition, for quantification of the global failure rules of concrete dams under extreme earthquakes, it is necessary to investigate its damage process and failure mechanisms. But the experimental results for the full tensile stress-strain curves of dam concrete, which is the fundamental data in this research, are scarce.

For investigating the mechanisms of damage extension of dam concrete which results in the catastrophic and global failure of dam structures and analyzing the macro experimental results, it is necessary to consider concrete as the composite structures composed of aggregate, mortar, and the interface between them. And the nonlinear dynamic numerical analysis based on three-dimensional mesomechanics should be carried out to investigate the way for numerical concrete in the area of high dams.

Based on the three-dimensional mechanics of concrete, the corresponding experimental research on the internal damage process and failure mechanisms of concrete is still necessary. The X ray and CT techniques are chosen as the most perspective methods. The real time CT scanning can be conducted on the damage failure process of concrete specimens under dynamic loading, and the three-dimensional display figures for the cracking and extension of concrete specimens can be output. The three-dimensional dynamic figures can be revealed based on these displayed figures. Furthermore, the dynamic acoustic emission technique can be used to understand the internal damage state of concrete specimens.

The content demonstrated in this part focuses on the need by the practical engineering. The main innovative achievements are based on the above technology thoughts and approaches in combination of the guiding thinking of "production, learning, and research." The detailed contents are described in the Chapters 10–15, respectively.

The researching progress in the dynamic mechanical behavior of concrete is introduced in Chapter 10, which includes dynamic strength of concrete, deformation, influencing factors, macro experiments, X ray, and CT techniques. The mechanisms for strain rate effect on concrete are also discussed. Combining the working state of the high arch dam under earthquakes and the main characteristics of dam concrete, the dynamic tensile strength, strain rate effect, and initial static loading effects on dynamic tensile strength and elastic modulus are mainly demonstrated. The engineering background in the following chapters is demonstrated, and the research points and technology approaches are clarified.

Chapter 11 introduces the flexural testing results for full-graded concrete specimens, including the pouring and forming of specimens, installation, dynamic loading methods, data acquisition, and processing. The effect of initial static loading and dynamic testing method on dynamic flexural mechanical characteristics of dam concrete is investigated, and the dynamic flexural behavior of full-graded specimens is compared with that of wet-sieved specimens. The fundamental rules and mechanisms for dynamic flexural strength of dam concrete are preliminarily discussed.

Chapter 12 introduces the achievements and results of the research on the mechanical characteristics of dam concrete and its constituents. The experimental techniques for determining the full uniaxial tensile stress–strain curves of concrete under dynamic or static loading are proposed, and the strain effects on dam concrete, cement mortar, and the interface between them are investigated. These achievements supply fundamental data on the constitutive relationship of concrete during dam's earthquake damaging.

Chapter 13 demonstrates the experimental results of dynamic mechanical testing of dam concrete based on acoustic emission techniques. The correlation between the parameters and the positioning systems of the sources of acoustic emission is investigated. For uniaxial tensile behavior of concrete and its constitutive materials under the different loading rates and testing methods, the relationship between the mechanical behavior of materials and the activity of acoustic emission, including the analyzing and distinguishing on the characteristic parameters of acoustic emission, wave spectrum, and failure mechanisms, is investigated. The Kaiser and Felicity effects on the softening stage during the unloading and reloading of cyclic loads are also researched.

Chapter 14 is the experimental research on dynamic failure of dam concrete based on CT techniques. The differences between the scanning principles of medical and industrial CT machines, and the dynamic loading system and testing procedure matched with medical CT machines, are introduced. The CT experiments on the static and dynamic tensile or compressive processes are analyzed. The theories on the failure partition of concrete are proposed. For the CT images of concrete, the classification of the supporting vector machine, the fractal theory, and the corresponding methods are introduced. The calculation and analysis on the fractal dimensions of the CT images of concrete under uniaxial compressive loading are carried out. Based on the CT images, the equations for damage evolution process and the constitutive relationship of concrete are proposed. Accordingly, the numerical simulation on the failure process of concrete is conducted.

Chapter 15 describes the numerical simulation on the dynamic mechanical behavior of dam concrete. The main contents contain the mesomechanical numerical methods of dam concrete, the space occupying and rejecting methods for forming the three-dimensional random aggregate model, the generation of the random convex polyhedral aggregate model and the corresponding mesomethods for subdivision of the finite element mesh. The strengthening of strain rate effect and damage evolution of concrete is introduced, the influence of initial static loading is considered, and the integral and differential forms of finite element equation of concrete materials under static and dynamic loading are established. The physical meaning of strengthening parameters for concrete is discussed. The effects of loading rate, initial static loads, inhomogeneity of materials (the random characteristics of mechanical parameters and proportioning grading), and morphology on the dynamic flexural strength of concrete are simulated systematically. Based on double scale methods, the equivalent simplified methods for the fine aggregate of full-graded concrete are investigated, and the calculated methods have been applied to the prediction of mesoparameters of full-graded concrete. The other important contents are the parallel computing methods based on PFEPG parallel computing platform. The script literature of FEPG (the GIO commanding flow file, GCN commanding flow file, NFE algorithm files, and PDE/VDE files) for mesonumerical simulation of concrete is established. The parallel computing program for three-dimensional mesosimulation models is generated, and the corresponding calculating analysis for the three-dimensional mesomodels under the action of initial static loads is conducted.

RESEARCH PROGRESS ON DYNAMIC MECHANICAL BEHAVIOR OF HIGH ARCH DAM CONCRETE

Although there exist obvious differences in the size, volume, aggregate gradation, mix proportions, construction method, and working conditions of concrete used in high arch dam and common concrete, the same characteristics also exist. Therefore, research progress on the dynamic mechanical behavior of concrete is outlined, including the working condition of high arch dams under an earthquake and the main characteristics of dynamic mechanical behavior of dam concrete. Based on the mentioned characteristics, the dynamic testing methods for dam concrete, numerical simulation, and corresponding mechanisms are outlined. These contents will be described in detail in the following sections.

10.1 DYNAMIC MECHANICAL BEHAVIOR OF NORMAL CONCRETE

10.1.1 DYNAMIC STRENGTH, DEFORMATION, AND ITS INFLUENCING FACTORS OF CONCRETE

10.1.1.1 Dynamic strength characteristics of concrete

1. Dynamic uniaxial and flexural strength of concrete

   Presently, there are lots of research results on the dynamic uniaxial strength of concrete. Bischoff and Javier summarized the dynamic compressive and tensile behavior of concrete under different strain rates.

   Bischoff et al. summarized the compressive behavior of concrete under different strain rates, and the relationship between dynamic compressive strength of concrete and strain rate is given in Fig. 10.1. It can be seen from the figure that the dynamic compressive strength of concrete increases with an increase in strain rates, and the $5 \times 10^{0} \mathrm{s}^{-1}$ is a critical value. When the strain rate is higher than this critical value, the increase rate for strength becomes obviously quicker.

   Compared with the compressive tests, the requirements for the testing equipment of tensile testing of concrete are more serious and the success probability is much lower. The testing data for uniaxial tensile behavior of concrete under earthquake strain rate are scarce. Through exerting an impact loading on the front of the specimen ends and obtaining the tensile wave reflected on the other end, Mellinger (1966) carried out the tensile testing of concrete with the strain rate of 0.57×10^{-6}, 0.2×10^{2}, and $0.23 \times 10^{2} \mathrm{~s}^{-1}$ respectively. The same impact testing was also fulfilled by Ross et. al (1995, 1996) with Hopkinson bar and Antoun (1991) with Taylor plate, whose strain rate is as high as $10^{2} \mathrm{~s}^{-1}$.

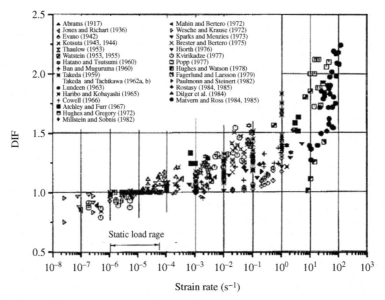

FIGURE 10.1 The Effect of Strain Rate on the Increase of Dynamic Compressive Strength

Javier Malvar et al. (1998) summarized the strain rate effects on the tensile behavior of concrete, and the relationship between dynamic tensile strength and strain rate is given in Fig. 10.2. It can be seen from this figure that the dynamic tensile strength of concrete increases with an increase in strain rate, and the strain rate 1 s^{-1} is a critical value. When the strain rate is over this critical value, the strength rapidly increases with an increase in strain rate.

From the existing data for normal concrete, it can be seen that there exist large variances among the experimental results obtained by different researchers. However, the following conclusions could be clarified: the strength of concrete increases with an increase in strain rate; the strain rate sensitivity of uniaxial tensile strength is higher than that of the uniaxial compressive strength; the factors of strength level, humidity, initial static loads, and confined pressure all influence the strain rate sensitivity; for strain rates ranging from 10^{-5} s^{-1} to 10^0 s^{-1}, the uniaxial compressive strength of concrete increases from 4% to 20%, and the uniaxial tensile strength of concrete increases 15%. When the strain rate ranges from 10^0–10^2 s^{-1}, there is a critical state in the increasing of the uniaxial compressive and tensile behavior of concrete. Usually, the uniaxial compressive strength of concrete could increase the highest to four times of the static value, and the uniaxial tensile strength could increase up from 3 to 12 times.

Based on the experimental research, the European Code (CEB-FIP Model Code 1990: Design Code) proposes the calculated formula for dynamic strength of normal concrete.

Shang (1994) has carried out the dynamic uniaxial compressive and tensile mechanical testing on concrete specimens with variable sections, and the corresponding strain rate ranges from $1 \times 10^{-5} \text{ s}^{-1}$ to $2 \times 10^{-2} \text{ s}^{-1}$. The experimental results showed that the compressive and tensile strength increase by 12–20% and 25–50%, respectively. After analyzing the experimental results under explosion loads (strain rates range from 10^1 s^{-1} to 10^3 s^{-1}), Malvar et al. (1998) thought that the compressive strength could increase by 100%, and tensile strength could increase by over 600%.

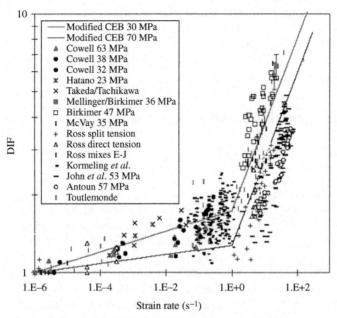

FIGURE 10.2 The Effect of Strain Rate on the Increase of Dynamic Tensile Strength

2. Dynamic multiaxial strength of concrete

   The concrete structures usually are in the state of complicated stress, and it needs to consider the multiaxial strength of concrete materials. Due to the limitation of the experimental instrument, the result from dynamic multiaxial tests is scarce, especially for the tensile behavior.

   By applying constant confining pressure (0–30 MPa) on concrete specimens, Takeda et al. (1974) conducted dynamic triaxial tests with the strain rate in the main axis $(0.2–2) \times 10^{-5}$, $(0.2–2) \times 10^{-2}$ and $(0.2–2) \times 10^{0} \text{ s}^{-1}$. The experimental results revealed that both confining pressure and axial loading rate affected compressive strength and maximum axial strain. By using split Hopkinson pressure bar (SHPB) instruments, Grote et al. (2001) carried out impact tests for concrete under strain rates $(0.25–1.7) \times 10^{3} \text{ s}^{-1}$. Their experimental results showed that there existed a sharp increasing zone near the strain rate $0.4 \times 10^{3} \text{ s}^{-1}$, and that the dynamic compressive strength could increase up to four times of static value at the strain rate of $1.7 \times 10^{3} \text{ s}^{-1}$. With the help of a plate impact instrument, they found that dynamic compressive strength of concrete could increase up to 50 times under the static water pressure 1–1.5 GPa and axial strain rate 10^{4} s^{-1}. And they discussed that the 42% of increase was due to the strain rate effect, and 58% was the influence of pressure. By modifying the classical SHPB experimental system with the lateral compressed clamp, Jacob (2002) carried out dynamic triaxial tests on concrete specimens under strain rate of $0.6 \times 10^{2} \text{ s}^{-1}$ and $1.2 \times 10^{2} \text{ s}^{-1}$, and the compressive strength increases four to five times the static value compared with the nonpressure condition.

   Xiao conducted the dynamic and static splitting tensile tests of concrete under lateral pressure. The experimental results showed that the lateral pressure and loading rate had significant influence on splitting tensile of concrete. Under the constant lateral pressure, the average strength was in proportion to the loading rate. And under the constant loading rate, the average strength decreased with the increasing of lateral pressure. Based on the results from the dynamic compressive characteristics of concrete under uniaxial and triaxial stress state, Yan (2006) proposed that with an increase in lateral pressure, the influence of strain rate on the dynamic strength of concrete became insignificant.

10.1.1.2 Dynamic deformation properties of concrete

At present, the experimental results focus on the influence of strain rate on the strength characteristics of concrete materials, while the research achievement of deformation characteristics is scarce and the conclusions are not consistent with the results by various researchers.

1. Elastic modulus of concrete

   Based on the existing experimental results, it is usually deemed that the elastic modulus of concrete increases with an increase in strain rate. The secant modulus under certain strain or stress value increases accordingly, while the increase percentage obtained by various research is different. For the dynamic mechanical experiments by Dhir (1972), the secant modulus under 50% strength increased 22% when the strain rates ranged from $5.0 \times 10^{-5}\,\mathrm{s}^{-1}$ to $2.5 \times 10^{-4}\,\mathrm{s}^{-1}$. The results by Shang (1994) showed that the elastic modulus was up 10–15% when strain rate was 10^{-3}–$10^{-2}\,\mathrm{s}^{-1}$, whose value was lower than the increase in strength. Yon et al. (1992) conducted dynamic compressive and tensile tests on concrete. The dynamic compressive and tensile modulus was the same as the static ones when the strain rate was below $3 \times 10^{-3}\,\mathrm{s}^{-1}$. When the strain rate was higher than $3 \times 10^{-3}\,\mathrm{s}^{-1}$, the elastic modulus increased as the strain rate increased. When the strain rate was $0.24\,\mathrm{s}^{-1}$, the tensile and compressive elastic modulus increased 60% and 40%, respectively. Bischoff et al. (1995) carried out the experiments on concrete with strength levels of 30 and 50 MPa. Their results showed that the secant modulus increased 60% and 13% as the strain rate increased from $1.0 \times 10^{-5}\,\mathrm{s}^{-1}$ to $(0.5–1.0) \times 10^{-1}\mathrm{s}^{-1}$, respectively. Atchley et al. (1967) and Sparks et al. (1973) thought that the stiffness of concrete was in proportion to strain rates, while there was no definite relationship between stiffness increase and concrete strength.

   There exists an inconstancy in the effect of strain rate on the initial tangent modulus. Watstein (1953), Dong et al. (1997), and Xiao et al. (2002) thought that the initial elastic modulus increased slightly as the strain rate increased, the tensile and compressive elastic modulus was up 15–30%, whose increase was lower than that of strength. However, Takeda et al. (1972) and Bischoff et al. (1995) deemed that there was no obvious correlation between the initial elastic modulus and strain rate.

   The European Code (CEB-FIP Model 1990: Deign Code) also suggests the calculating formula for dynamic elastic modulus of concrete.

2. Peak strain of concrete

   The peak strain of concrete is defined as the strain at the peak stress of concrete, which is an important parameter to reflect the deformation characteristic of concrete. Since Watstein (1953) reported first on the results that the dynamic peak strain of concrete under quick loading was higher than that under static loading, many researchers have carried out investigations into this aspect. However, there exists obvious inconsistency among the results of various investigators.

Watstein (1953), Takeda (1971), Kvirikadze (1977) and Yan (2006) all thought that the peak strain of concrete increased as the strain rate increased. On the other hand found that the peak strain remained unchanged with an increase in strain rates. However, Dhir (1972), reported that the peak strain decreased with an increase in strain rate, and the peak strain reduced by 30% at most.

After analyzing the relevant experimental results, obtained a fundamental tendency that the peak compressive strain increased slightly as the strain rate increased and the increasing amplitude ranged from 0% to 40%. The peak strain increased by 10–30% until the strain rate increased to 10 s^{-1}. The European Code (CEB-FIP Model Code 1990: Design Code) also proposes the calculation formula for dynamic peak strain of concrete.

There are many factors leading to the variance of peak strain of concrete with a change in strain rate, such as the difference in the loading methods. For example, by using a hydraulic servo machine, Watstein (1953) reported that the peak strain decreased with an increase in strain rate, while the peak strain increased as the strain rate increased by using a drop hammer as the loading instrument.

3. Poisson's ratio of concrete

For the relationship between Poisson's ratio and strain rate, there is inconsistency among the few research results. Takeda (1962, 1972, 1974), all thought that the compressive and tensile Poisson's ratio decreased as an increase in strain rates, and the maximum reduced amplitude could be up to 40%. However, there are also different conclusions obtained by other experiments. Paulmann thought that the Poisson's ratio remained unchanged until the strain rate increased to 0.2 s^{-1}, and the same conclusions were obtained by Yan (2006), while reported that the Poisson's ratio was proportional to the strain rate.

There are many factors leading to the variances obtained by experimental results of deformation characteristics, such as the static strength of materials, the aggregate type, experimental requirements, strain rates, and so on. Comparing with the dynamic strength behavior of concrete, the dynamic deformation characteristic is not too sensitive to strain rates and the effects are not obvious, which may be obscured by the difference in experimental conditions. Furthermore, Poisson's ratio of concrete is closely related to the stress state and the discreteness of the data is large. Therefore, it is hard to determine the relationship between Poisson's ratio and strain rates. It is usually deemed that Poisson's ratio is independent of the strain rate, and this conclusion is adopted by the European Code CEB (CEB-FIP Model 1990: Design Code).

10.1.1.3 The influencing factor of dynamic strength of concrete

Based on experimental results, many researchers and engineers have reached the consensuses that the dynamic mechanical properties of concrete increase as the strain rates increase. However, there exists large discreteness in the dynamic increase factor of concrete strength from various experimental results. Therefore, the influencing factors on the dynamic strength of concrete have been investigated extensively. Presently, the following aspects have been investigated, including the static strength of concrete, coarse aggregate, water/cement ratio, curing, humidity, age, temperature, size effect, testing instruments and dynamic loading methods, and so on.

1. The static strength of concrete

It has been widely held that the static strength of concrete has an effect on increasing rate of dynamic strength, which is that the rate sensitivity of low static strength concrete is higher than that of high static strength concrete. This achievement has been adopted into the relevant formula in the European Code (CEB-FIP Model Code 1990: Design Code). In the relevant experimental investigations, Watstein (1953) conducted the uniaxial compressive tests on cylindrical concrete with two static strength values (17.4 and 45.1 MPa), and the corresponding strain rate ranged from 10^{-6} s^{-1} to 10^{1} s^{-1}. The experimental results show that the strength of concrete increases by 84% and 85%, respectively, and that the static strength has no effect on dynamic strength. However, the results obtained from Evans (1968) show that the concrete with low static strength has had much more significant strain rate sensitivity during a larger range of strain rates.

2. Coarse aggregate

The coarse aggregate is the main constituent for concrete, and the materials, shape, grade, and content have an effect on the strain rate sensitivity of the dynamic mechanical behavior of concrete. Until now the relevant experiments are very seldom. Sparks et al. (1973) have carried out dynamic compressive testing on concrete with three different aggregates. Hughes et al. conducted drop hammer tests on concrete with two different aggregates, and their results showed no effects. By using the SHPB technique, carried out the experimental researches on the relationship between dynamic increase factor and aggregate size. The results showed that the dynamic increase factor of normal concrete with different aggregate changed a little under a certain strain rate. For strong impact load (large strain rate), it was found from the tests that the compressive strength of specimens with large aggregate size would not increase when external load increased.

3. Water–cement ratio

The water–cement ratio is the main factor influencing the static mechanical behavior of concrete. But the research about the water–cement ratio effect on the dynamic mechanical characteristic of concrete is seldom. The results by Rossi et al. show that the dynamic growing factor of tensile strength of concrete increases as the water–cement ratio increases, and that the absolute strength increment ($f_{t_{dyn}} - f_{t_{stat}}$) is not relevant to the water–cement ratio.

4. Curing and humidity conditions

The curing condition can be divided into such three kinds of conditions as water curing, standard, and laboratory curing. The effect of curing conditions on the rate sensitivity of the dynamic mechanical behavior is the focus of many researchers. The relevant conclusions obtained by various investigators (Watstein, 1953; Dhir et al., 1972; Ross et al., 1996) are very consistent. The effect of loading rate on concrete strength is different under various curing regimes, and an increase for concrete cured in moist environments is relatively large. The comparison of specimen strength under various curing conditions is shown as follows: water curing > curing in chamber > laboratory curing. The curing regimes are closely related to the humidity of the specimen, while the specimen humidity under testing has a significant effect on strength. The relevant research results are much. It is usually considered that dynamic increasing value of wet concrete is higher than that of dry specimens. But there are some different experimental results. The experiments by Zielinski et al. (1981) showed that the humidity condition of concrete had no effect on the dynamic tensile strength of concrete. Harris et al. (2000) thought that the dynamic compressive strength of wet specimens decreased, while the dynamic splitting tensile strength increased.

5. Temperature

   The temperature can change the existing state and amount of free water, and even the internal structure of concrete. However, the corresponding research is very seldom. Yan (2006) carried out the dynamic tensile testing of concrete under low temperature ($-30°C$), and the results showed that the rate sensitivity of dynamic strength decreased under low temperature.

6. Age

   Largely owing to the progress made in the hydration reaction for the main cementitious materials in concrete, static and dynamic strength and elastic modulus increase with a growth in age, but the rules are affected by such factors as cement types, quality, additional agents, external conditions, admixture, and so on. As far as the dynamic mechanical property of concrete is concerned, the main point is the effect of age on the strain rate sensitivity of concrete. It is hard to determine the effect of concrete age on the rate sensitivity from the limited experimental results. Dhir et al. found that the rate sensitivity increased as the age increased. Cowell reported that the concrete strength was in proportion to age, but the rate sensitivity was weakened. After comparison on the relevant researches, Soroushian et al. (1986) thought that there was no connection between strain rate sensitivity of concrete and age under the same humidity conditions.

7. Size effect

   The nominal strength of materials is related to the characteristic length of the specimen. The research results about the size effect on the dynamic strength of concrete are limited. Only Shang and Elfahal have conducted the relevant investigations. The testing results from Shang (1994) revealed that there was no size effect in the dynamic flexural strength of concrete, while the impact experimental results by Elfahal showed that the size effect of the dynamic strength of concrete was obvious. It is clear that the two research conclusions are inconsistent.

8. Experimental methods

   The dynamic testing instruments are only suitable for some certain range of loading rate, such as hydraulic servo machine and drop hammer for the range of intermediate strain rates. The experimental results are different due to the variance of loading equipment. In addition, the measuring instruments also influence the experimental results.

9. Dynamic loading system

   The loading system refers to the applied form, such as impact wave, triangle wave, pulse wave, variable amplitude wave, variable frequency wave, and so on. It can be seen from the existing results that the dynamic loading system has a significant effect on the dynamic strength of concrete (Yan, 2006).

10.1.2 EXPERIMENTAL TECHNIQUES FOR DYNAMIC MECHANICAL CHARACTERISTICS OF CONCRETE

10.1.2.1 Dynamic loading instruments for concrete

Presently, the commonly used instruments for testing dynamic mechanical behavior of concrete include electrohydraulic servo system, drop hammer loading system, SHPB, light gas gun, plate impact loading system, and so on.

The instruments of SHPB, light gas gun, and plate impact are mainly used for the high strain rate (10^1–10^4 s^{-1}) mechanical tests for concrete (Gary et al., 1998; Rome, 2002; Shi et al., 2000;

He et al., 1992; Grote et al., 2001). For the concrete under the earthquake action, the electrohydraulic servo system and drop hammer system are mainly adopted.

1. The electrohydraulic servo loading system

   The electrohydraulic servo loading system that appeared in the 1970s can carry out uniaxial and triaxial experiments for brittle materials, and the total curve for axial loads, axial deformation, transverse deformation, and volumetric deformation can be measured. The electro-hydraulic servo loading system is suitable for the tests with the strain rate below 10^{-2} s^{-1}. The dynamic characteristic testing researches have been extensively conducted by using the electro-hydraulic servo loading system at home and aboard (Dong et al., 1997). The machine consists of an electro-hydraulic servo-loading instrument, controlling system, and hydraulic pressure source. The representative instruments are MTS in USA, INSTRON in UK, and so on.

2. Drop hammer impacting tests

   The drop hammer impact machine is a loading device with simplicity, reliability, and good repeatability. The drop hammer loading system consists of a vertical track, the freely falling material, and the measuring system for height. The impact load on the specimens is from the impact of a piston rod by free falling drop hammer. The changing of load wave pulse can be obtained by various falling distance, drop weight, and backing plate material (Tian et al., 1998; Lu et al., 1995). The suitable strain rate for drop hammer tests ranges from 10^{-5} s^{-1} to 10^1 s^{-1}.

10.1.2.2 Dynamic measuring technique for concrete

1. Conventional measuring technology

   The conventional measures refer to the deformation measuring owned by the loading system, loads measurement, resistance strain gauge, and displacement meter. In dynamic experiments, the measured data should be timely recorded by high-speed acquisition instruments.

2. Acoustic emission monitoring

   Under the action of external loads, the partial storage energy induced by the formation, extension, and connection of concrete cracks will release and the acoustic emission (AE) happens. Through the measurement and analysis of the acoustic emission signal, the change and failure condition of the internal structure of concrete under the external forces can be investigated. This technology has been used extensively, but the research achievements about the acoustic emission monitoring dynamic loads are scarce.

3. Noncontact measuring system

   The noncontact measuring system is based on the optical principle, and the best advantage for this method is that the whole measuring progress is not influenced by the measuring system and instruments. The failure of the specimen does not damage the measuring equipment. The noncontact measuring system mainly includes X-ray, CT technique, high-speed camera, and nondestructive microwave monitoring techniques.

   a. *Laser extensometer*: The strain can be measured directly by processing the laser reflected from the specimen surface. This measuring method has high resolution and free drift. Among the existing techniques, the Doppler laser extensometer is mainly applied in the high-speed tensile tests with deformation velocity up to 50 ms^{-1}, which is suitable for the strain measurement for the direct tensile system under intermediate strain rates. By using the noncontact methods, the specimen is not influenced by the measuring equipment, and the fracture of the specimen does not damage the equipment.

b. *High speed camera system*: This system can suit investigating the dynamic direct tensile tests of brittle materials. The elastic deformation of the specimen during the whole loading procedure, crack appearance, extension, and the fracture process can all be investigated. The corresponding mechanisms can be analyzed directly.

c. *Nondestructive microwave monitoring technique*: The nondestructive microwave monitoring technique can monitor the invisible defect and damage with high precision. This method is usually suggested to monitor the detachment between FRP and concrete. If the scanning velocity of this method can be further developed, it will be adopted in dynamic tests of concrete with nondestructive high-speed image monitoring (Feng, 2006). The microwave detector with the type GAP-CAT-1200 can be imaged timely on the computer screen for saving and comparison.

10.1.2.3 Application of X-ray CT technique

For the investigation on the internal damage of materials, X-ray CT technique is mainly applied in the static compressive testing of rock specimens, which includes the image recognition, width measurement, evolution rules, the reconstruction of three-dimensional images, and so on. Research on the static loading equipment especially for CT experiments has achieved a series of progresses (Kawakata et al., 1999; Ueta et al., 2000; Ding et al., 2000). But for the research about the internal structural damage of inhomogeneous concrete materials, the application of X-ray CT technique is still seldom, and, especially, little has been reported on the CT experimental results on dynamic tensile testing of concrete hitherto.

10.1.3 MESOMECHANICAL NUMERICAL ANALYSIS OF CONCRETE

For investigation of the mechanism for internal damaging failure and explanation of the results for macro experiments, the concrete has been viewed as a composite material consisted of different-sized aggregate, mortar, and the interface between the two, and the spatial position of aggregate and the strength of composition are randomly distributed. Based on the mesomechanical numerical analysis, the cracking extension, and the failure morphology can be indicated directly. The concept of the so-called numerical concrete research is induced by studying damage mechanisms deeply. However, three-dimensional mesomechanical numerical analysis of concrete on multigraded concrete in practical engineering is challengeable. There are no relevant investigation results except for the results reported in this book.

10.1.3.1 Mesomechanical mathematical model for concrete

For the lattice model, the continuous medium is viewed as the lattice system connected by the discrete bar elements or beam elements, and every element represents a little part of materials. Schlangen (1991, 1997) and Van Mier et al. (1999) adopted the lattice model to investigate the coalescence in cracking surface of concrete. Van Mier (1997) used this model to simulate the uniaxial tensile, joint tensile-shear, and uniaxial compressive experiments. The relevant researches indicate that the simulation by this model can be effective in the fracture process induced by tensile failure, while the results from simulating the compressive loads are not so ideal. This model is hard to reflect the real deformation of elements under cyclic loading, and is limited to the plane problem. It cannot suit the investigation of dynamic mechanical behavior of concrete under an earthquake.

For random aggregate and mechanics characteristics model, concrete is viewed as a three-phase inhomogeneous composite material consisting of aggregate, hardened cement paste, and the interface between them. With the transformation from the Fuller gradation curve of three-dimensional aggregate to Walraven gradation equations of two-dimensional aggregate, the particle numbers of aggregate can be determined (Walraven et al., 1991). The spatial distribution model for aggregate inside the specimen can be formed on the basis of the Monte Carlo method. The finite element mesh is projected to the aggregate structure, or the finite element meshes are directly generated for the coarse aggregate and cement mortar matrix. The medium characteristics of elements can be judged by the aggregate's location in the mesh, and the corresponding materials characteristics are given. Using the nonlinear finite element methods by considering the degradation of elastic modulus and strength of concrete, the simulated cracking extension process can directly reflect the whole damage and fracture process of specimens. The key point to the random aggregate model is to choose the constitutive relationship and damage evolution model of the elements. The current research is limited basically to the plane problem. Liu et al. (1996) used the random aggregate model to simulate the fracture of concrete materials. Wang (1997) calculated the constitutive behavior for uniaxial tension and compression, and the strength under biaxial loading, and the failure progress for splitting tension. The strength criterion was introduced and the progress of cracking extension under various stresses was simulated. Li et al. (2001) conducted the mesodamage fracture of roller compacted concrete. The static mechanical characteristics of roller compacted concrete and size effect on concrete were simulated accordingly. In the above research, the models for aggregate are all assumed to be spherical. For simulating the morphology of aggregate more realistically, Wang (2000) conducted the numerical simulation on strain softening and localization progress of concrete by using a convex polygon model and orthotropic damage constitutive relationship. Gao et al. (2003) carried further investigations on the packing algorithm of two-dimensional polygon random aggregate, the judgment criterion of aggregate invasion, and the forming methods for convex polygon aggregate model were determined in terms of the area scale, and the packing algorithm for two-dimensional aggregate was established based on the results. Liu et al. (2003) proposed the random packing algorithm for three-dimensional aggregate based on the volume scale.

In the mentioned random aggregate model, the random properties for the spatial distribution of the mechanical behavior of each phase are not considered. For considering this random property, proposed the random mechanical model, and the material characteristics of each continuum materials were determined by some chosen Weill distribution. Tang et al. described the damaging evolution of meso-elements in terms of elastic damage constitutive relationship, and the critical conditions for tensile and shear damage were determined by the maximum tensile stress (or tensile stress criterion) and Mohr–Coulomb criterion. The uniaxial tensile, biaxial tensile-compressive, flexural and tensile (I) type fracture, and shear fracture behavior were numerically simulated systematically. However, the random property for each gradation aggregate inside the specimens was not accounted for in their researches.

Until now, the numerical simulation of concrete based on meso level has been limited to the plane static problem, and only for one or two gradations and small-sized concrete specimens. Most of the literatures are focused on the analysis of the failure process.

10.1.3.2 The material mechanical characteristics for each medium in the meso-mechanical models of concrete

In the mesomechanical numerical simulation of concrete, the concrete is viewed as a composite material composed of mortar, aggregate, and the interface between them. Thus, the key question is to determine the mechanical characteristics of the three phases. However, the experimental data about the material properties of the three phases are very few. In our country, Wu et al. (1997) first carried out the bond capacity between concrete and cement mortar. In their experiments, four bond types were designed, the splitting strength and fracture energy of the bond face between marble aggregate and cement matrix were measured, and the influence of strengthening bond between mortar and aggregate was also discussed. Liu et al. (1996) presented the tensile strength, elastic modulus of coarse aggregate, cement matrix, and the interface. Schlangen (1991, 1997), Van Mier (1997), and Horsch et al. (2001) reported the data about the mechanical characteristics of constitutive materials.

10.1.4 DISCUSSION ON THE MECHANISMS FOR STRAIN RATE EFFECT OF CONCRETE

The dynamic mechanical behavior of concrete is mainly reflected as its strain rate effect. There are lots of literatures about the discussion on the mechanisms for strain rate effect of concrete, while the authoritative explanation has not been obtained. From the aspect of practical application of concrete materials, the following points should be addressed.

10.1.4.1 Preliminary consensus of strain rate effects revealed by existing experimental results

Although there is obvious variance among the experimental results of strain rate effects, the following consensus has been obtained.

As an inhomogeneous material for concrete, the strain rate effect of concrete is more obvious than the homogeneous materials, such as metals. The strain rate effects for tensile behavior of concrete are more obvious than that for compressive behavior, and the strain rate effects for concrete strength are more significant than that for elastic modulus. Under the same strain rates, the increasing value for dynamic strength of wet concrete is higher than that of dry concrete. The strain rate effect increases sharply when the strain rate is up to the range of 10^0–10^2 s^{-1}.

The discussion on strain rate effects of concrete is mainly based on the explanation on the earlier experimental results with basic consensus. For the dynamic mechanical characteristics of concrete structures under an earthquake, the maximum strain rate under an earthquake ranges from 10^{-3} s^{-1} to 10^{-2} s^{-1}. The explanation on the mechanisms for strain rate effects can be summarized as follows:

1. Energy dissipation mechanisms

   Based on the view of fracture mechanics, and compared to the energy needed for crack development, the energy needed for the forming process of crack is far higher than that during crack development. For example, the number of crack increases as the loading rate increases, thus more energy is dissipated (Eibl et al., 1989; Rossi, 1991). As the dynamic loading time is short, the weakest part cannot be tracked to release cracking energy, which also induces the increase of the number of cracks.

2. The influence of free water viscosity in concrete

By experimental monitoring and theoretical assumption, Rossi et al. (1990, 1991, 1994) pointed out that the mechanical characteristics of concrete under high strain rates can be explained by Stefan effect and the mutual action during the cracking process of materials. Stefan effect shows that, when a thin liquid membrane exists between two parallel plates with h distance and the plates separate with a certain velocity, the resistance will be induced. And the resistance is in proportion to the velocity. The Stefan effects mainly explain why the strain rate effect in the strength of wet concrete is more obvious than that of dry concrete, and the experimental results also indicate that the dry concrete shows strain rate effect as well.

3. Thermal activation and macroviscous mechanisms

Based on the experimental results from Zhurkov (1979), Qi and Qian (2003) thought that the relationship between the strain rate and strength of materials was controlled by the mechanisms of thermal activation during the small strain rate range. With a further increase in strain rate, the macroviscous damping of the materials appears and becomes dominant. Under the region of high strain rate, the influence of material inertia becomes significant. The dependence of strength on strain rate is the result due to the parallel existence and mutual competition of the thermal activation and macroviscosity mechanisms. These two mechanisms dominate in different strain rate ranges.

10.1.4.2 The problems in strain rate effects mechanisms for concrete

It is widely accepted that the sharp increase for strain rate effects of concrete is due to the inertia effects. However, explanation is lacking on how the inertia influences strain rate effects, and the verification for the sharply changing critical stage is also not convincing. Because the mechanical characteristics should be measured by the corresponding specimens and testing method, some opinions suggested that the rate sensitivity of concrete under high strain rate loading is not the real property of concrete, but is due to the testing methods (Brace et al., 1971). From the concept of engineering structures, the response to structure under static loading is reflected by the stiffness item related to the displacement. However, the structural response under dynamic loading includes the damping item related with velocity and the inertia item related with acceleration. The higher the loading rate is, the more significant the inertia item in structural resistance will be. For the structural response as the specimen, the proportion of the inertia item changes with the ratio of structural natural frequency to the loading frequency. When loaded under low velocity, the structural natural frequency is much higher than the loading frequency. The action of inertia is small, and the measured dynamic resistance reflects the strain rate effect of material characteristics. When loaded under high velocity, the loading frequency is near the structural natural frequency. As the damping value of concrete specimens during dynamic tests is usually small, the inertia item in measured dynamic resistance of specimen structure will suddenly increase. Obviously, the measured dynamic resistance is exactly influenced by the specimen size and testing methods. Therefore, the application of experimental results should consider these factors.

The mechanisms for strain rate effects are very complicated. The theoretical explanation on the experimental results still needs to be further investigated, especially for the mechanisms on the sudden change of strain rate effect in critical strain rates. Fortunately, it has a small impact on the dynamic characteristics of concrete under low-velocity earthquake.

10.2 **DYNAMIC MECHANICAL CHARACTERISTICS OF DAM CONCRETE**

The content described in Section 10.1 is the research progress on the seismic dynamic mechanical characteristics of normal concrete. The dynamic experimental research on dam concrete materials is relatively seldom at home and aboard. In the later stage in the 1950s, for seismic design of concrete dams, carried out the systematical investigation into the dynamic compressive and tensile strength of concrete. Then, the Bureau of Reclamation in the United States conducted a series of investigations on the dynamic mechanical characteristic of dam concrete.

10.2.1 **THE MAIN FEATURES OF DAM CONCRETE**

As we know, the dam concrete usually uses three-graded or four-graded aggregates; the content of coarse aggregate makes up the 60–70% of concrete mass, and the cementitious materials only make up 8–10%. Due to the limitation of the experimental condition, the mechanical research on dam concrete usually adopts the wet sieving methods. The so-called wet sieving method is where the aggregate larger than 40 mm of the dam concrete is sieved and that the small shaped specimens are measured by various testing methods. In this way, the coarse aggregate content will be reduced to 1/2 to 1/3, and the cementitious content will increase by one to two times. The materials composition and mix proportions of the raw concrete are changed a lot, especially the ratio between the cement mortar content and aggregate content. In addition, the microcracking induced by the hardening process of concrete differs due to the difference of specimen and aggregate size. Due to the impact of mix proportion variation and size effect, the experimental results from the wet sieving specimen will result in the difference from the results from full-graded concrete specimens. The full-graded concrete specimen can reflect the true properties of dam concrete, and bring significant influence on dam design and the science and safety in construction.

The United States has begun the investigations into dam full-graded concrete since the 1940s. Till the middle of the 1970s, the designed codes for concrete arch dam and gravity dam regulated by the US Bureau of Reclamation suggested that the full-graded aggregates should be adopted in the behavior testing for the strength or elastic modulus of dam concrete, and the smallest size for the specimen should be larger than three times of the maximum aggregate particle size. The former Soviet Union also carried out the experimental investigation into dam full-graded concrete in the 1980s. In the middle of the 1960s, the Chinese Institute of Water Resources and Hydropower Research carried out the preliminary tests on the full-graded concrete specimens, and the effect of specimens on compressive strength, flexural strength, uniaxial tensile strength, maximum tensile strain, elastic modulus, and Poisson's ratio was also investigated. Thereafter, Tsinghua University, Dalian University of Technology, Hohai University, and Chengdu Engineering Corporation Limited also conducted the relevant researches on the full-graded aggregate concrete. The Test Code for Hydraulic Concrete (DL/T 5150-2001) was promulgated in 2002 with the static experimental standard for full-graded concrete, but without dynamic testing regulation. Based on the experimental results of full-graded concrete in the Ertan and Dongjiang engineering projects, the working group for the Unified Standard for Reliability Design of Hydraulic Engineering Structures (GB50199-94, 1994) proposed that the strength ratio between the full-graded concrete specimen and wet-sieved small specimen ranged from 0.49 to 0.72, and the strength of large full-graded concrete specimens represented the resistance of materials in dam structures basically, although even for the large full-graded concrete specimens, there also exist differences from the practical massive concrete due to the size effect. The most important fact is that the mix proportion of full-graded aggregate concrete specimens is the same as that of dam concrete used in practical dam structures. As for the secondary effect from the size difference between full-graded concrete and practical dam concrete, the effect cannot be evaluated without the prototype tests, and can only be considered in the margin of safety.

Therefore, in the present dam design, the standard value for the resistance of dam concrete materials can be determined by the strength of full-graded concrete specimen. However, the experimental investigation into full-graded dam concrete is scarce, and the relevant experimental shaping and conditions need to be further researched.

10.2.2 DYNAMIC TENSILE STRENGTH OF DAM CONCRETE

10.2.2.1 Determination of dynamic tensile strength of concrete in dam design

To prevent the cracking damage on a dam under an earthquake, the key point is dependent on the tensile strength of concrete, and the tensile strength of concrete is the main characteristic of dam concrete. The experimental methods for measuring the tensile strength of concrete usually include direct tension, splitting tension, and flexure tests, while the results obtained from these testing methods have large discreteness. Therefore, the first thing is to determine which experimental results are taken as the standard value of tensile resistance of dam concrete. The strength of concrete materials is related to the stress condition and actions. And the stressed conditions of the dam are complex and under the multiaxial stresses. The tensile conditions usually are shear flexural tension and flexural tension. The US Bureau of Reclamation (2006) thought that the tensile strength based on splitting tension tests was more reasonable than the direct tensile strength.

Since the 1960s, the United States has carried out relatively many experimental researches on the dynamic and static characteristics of dam concrete. Raphael (1984) and Harris et al. (2000) carried out the static and dynamic mechanical experiments on dam concrete cored from five and nine dams, respectively. Based on the respective results, Harris et al. (2000) suggested that the ratio of splitting tensile strength to direct tensile strength is approximately 1.75, while Raphael gave the ratio of 2.0. Raphael suggested that the following formula can be used to calculate the static splitting tensile strength from its compressive strength without experimental data:

$$f_{st} = 0.326 f_c^{2/3} (\text{MPa}) \tag{10.1}$$

The ratio of the apparent splitting tensile strength f_{st} to the experimental strength can take the value of 1.33, and the apparent splitting tensile strength f_{st} can be obtained from flexural tests. This value can also be used as the tensile strength of dam structures when checking the cracking of dam structures in the analysis of dam stresses. However, when the stress analysis is conducted on dam structures as the linear elastic body, the constant chord modulus, not the secant modulus, is adopted. These suggestions from Raphael were adopted in the dam design code issued by the US Bureau of Reclamation (2006).

Therefore, the flexural strength obtained by flexural tests is suitable as the standard value of tensile strength of dam structures.

By statistical analysis on the data at home and abroad, the working group for the Unified Standard for Reliability Design of Hydraulic Engineering Structures (GB50199-94, 1994) suggested that the average value of splitting tensile strength of dam concrete can be obtained by the following equation:

$$f_{st} = 0.299 f_c^{0.752} (\text{MPa}) \tag{10.2}$$

According to the experimental results from Raphael, the flexural strength can be obtained by splitting tensile strength multiplied by 1.33. Equations (10.1) and (10.2) show that the relationship between tensile strength and compressive strength of dam concrete is not linear. In the range of C20 to C45 grade concrete, the ratio of flexural strength to uniaxial compressive strength of dam concrete ranges from 0.12 to 0.17 according to Eq. (10.1). The higher the concrete grade is, the smaller the ratio of tensile to compressive strength becomes. Considering the cylinder compressive strength used in the United States is 0.8 times of the cubic strength in our country, the corresponding ratio of flexural strength to cubic compressive strength should be multiplied by 0.86, and the value ranges from 0.11 to 0.15. According to Eq. (10.2), the ratio of splitting tensile strength to cubic compressive strength ranges from 0.12 to 0.14, and the ratio with conversed flexural strength is even larger.

The specifications for seismic design of hydraulic structures regulates that the ratio of flexural strength to compressive strength takes the value of 0.10.

However, it should be pointed out that the ratio values from (10.2) are also obtained from small-sized specimens. In the experimental results adopted by the US Bureau of Reclamation, the specimens are cored directly from the existing dam structures. Although the mix proportion is the same as that of practical dam structures, the small cylindrical specimens with 6 in. diameter can be used. The particle size of concrete aggregate is larger than 3 in., which does not satisfy the requirements that the specimen size should be larger than three times the maximum aggregate particle size. Obviously, the ratio of tensile to compressive strength for full-graded large specimens would be smaller than that of wet-sieved specimens, which is due to the higher mortar content in wet-sieved specimens than in full-graded concrete specimens. The recent experimental results from Xiluodu arch dam in China have shown that the ratio of splitting tensile strength to compressive strength of full-graded concrete was lower by 18.6% than that of wet-sieved specimens, and the flexural strength of large full-graded concrete specimens can range from 57% to 62% of the value for small wet-sieved specimens (Zhu et al., 2010). The experimental research on this aspect is still very seldom and should be investigated further.

10.2.2.2 *The strain rate effects for dynamic strength of dam concrete*

Raphael achieved the maximum strength within the loading time of 0.05 s (equal to the vibration frequency of 5 Hz), and found that the dynamic compressive strength of dam concrete raised 60% compared to the static strength, while the experimental results have certain discreteness. He suggested that the dynamic tensile and compressive strength of concrete increased 56% and 31%, respectively. Harris et al. (2000) carried out the static and dynamic testing on dam concrete under the strain rate of 1×10^{-3} s^{-1}, and gave the results that the dynamic tensile and compressive strength of dam concrete respectively increased 14.3% and 7% compared to the static strength. Their experimental results were more scattered and without definite regularity, and this may be due to the influence of composition, aggregate, and dry or wet environment on the dynamic mechanical behavior of dam concrete. The experimental results from Harris et al. (2000) showed that, compared to the strength of the air-dried specimens, the static and dynamic compressive strength of wet specimens decreased, while the static and dynamic splitting tensile strength increased. In the dam design, the US Bureau of Reclamation regulated the ratio of dynamic to static tensile and compressive strength as 1.5 and 1.2, respectively.

The investigations into the dynamic mechanical behavior of dam concrete in China are relatively late. In the "Eighth Five-Year" plan period for economic construction, Ji (1995) studied the effect of loading rate on the dynamic strength of dam concrete, and the experimental results showed that the dynamic uniaxial tensile and compressive strength of concrete respectively increased by 27% and 35% with the 0.25 s failure times.

In the current specifications for seismic design of hydraulic structures (DL 5073-2000), the experimental data and specifications from the United States and Japan are referred to, and the ratio of dynamic to tensile and compressive strength is taken as the value of 1.3. Obviously, it is necessary to carry out dynamic flexural tests on full-graded specimens.

Furthermore, due to the limitation of experimental instruments, the dynamic multiaxial experimental data are very scarce, especially for the tensile tests. It is more difficult for conducting the tests on full-graded concrete as there are more serious requirements. Currently, the dam design codes from all countries except Japan do not consider the influence of multiaxial strength. Even in the nonlinear dynamic and static stress analysis, the constitutive relationships of materials under complex stress conditions are mainly based on the results from uniaxial tests.

10.2.2.3 Dynamic tensile strength of dam concrete with the initial static loads

For the mentioned experimental results, the ratio of dynamic to static compressive and tensile strength of dam concrete are all obtained from the pure dynamic or static loading. As the dams are operated under the action of an earthquake, the engineers should focus the dynamic mechanical characteristics of full-graded dam concrete under a certain static loading. Currently, the experimental data on the dynamic compressive and tensile behavior of full-graded dam concrete are few, especially for the flexural tests. The data on the influence of initial loads on the ratio of dynamic to static strength are even much less.

During the "Ninth Five-Year" plan period for economic construction, under the key project about the seismic safety of high arch dams, the Chinese Institute of Water Resources and Hydropower Research and the Kunming Hydropower Investigation Design & Research Institute, combined with the practical Xiaowan high arch dam engineering project, carried out the dynamic compressive and flexural testing on full-graded specimens made from the raw materials in the engineering sites. The test results showed that different initial loads do not have any negative effects on the dynamic behavior of full-graded concrete specimens. The investigated results supply the basis for choosing the material parameters in the seismic design of the Xiaowan arch dam. However, due to the restriction of experimental conditions, the loading type for impact loads was different from the cyclic loading under an earthquake, and the sample number was limited. Thus, the obtained experimental results had a larger discreteness, and it was hard to achieve quantitative rules. More investigated results combined with more practical engineering projects are needed to be verified.

10.2.3 DYNAMIC ELASTIC MODULUS OF DAM CONCRETE

The designed elastic modulus of dam concrete is related to the designed strength. In the design of a dam, the designed static elastic modulus for dam concrete does not take the value from the static testing in the laboratory. The static loads in the laboratory usually induce the specimens into failure within several minutes, and the US Bureau of Reclamation suggested this value as the experimental static modulus. As the elastic modulus is low in the initial small stress, the determination of static elastic modulus in the linear stage does not include the initial stage. The practical dam concrete sustains static loads, and the elastic modulus would decrease and then be stable due to the influence of creep effects. When the dam was conducted by stress analysis under static loads, the US Bureau of Reclamation adopted the stable elastic modulus after long-term loading, and this value was 33% lower than that of the static chord modulus measured in the laboratory. In the dynamic tests under an earthquake, the specimens usually fail within several seconds. The increase in dynamic strength is the so-called strain rate effect. The dynamic elastic modulus under dynamic loads does not show the phenomenon of low static elastic modulus in the small stress condition. Based on the experimental results of Harris (1998), the ratio of dynamic elastic modulus to the laboratory measured static chord modulus of dam concrete ranged from 0.7 to 1.20, and the average value was 1.0. Thus, the US Bureau of Reclamation suggested that, if without the experimental results, the dynamic modulus can be taken as equal to the laboratory static chord modulus, which is 1.5 times the designed static elastic modulus. It can be seen that, in the seismic design, the increase of dynamic elastic modulus is not induced by strain effects, but is the result from the decrease of elastic modulus due to long period static loads.

In the current specification for the seismic design of hydraulic structures, the dynamic strength and elastic modulus are regulated to increase by 30%, but the mechanisms for these two are obviously different.

10.3 THE KEY PROBLEM AND TECHNICAL WAY IN THE DYNAMIC MECHANICAL CHARACTERISTIC STUDY

Based on the above research progress on the dynamic mechanical characteristics of concrete, the following key problems and technical methods on this area can be summarized as follows:

1. Although there is significant progress on the dynamic mechanical characteristics of normal concrete, the research on the property of large-sized, multigraded aggregate dam concrete is still seldom. There are some fundamental problems to be investigated and further studied.
2. Under the action of an earthquake, the key point for the dynamic mechanical characteristic of dam concrete is the tensile behavior, especially for the dynamic flexural tensile strength and elastic modulus. The ratio of dynamic flexural strength to elastic modulus is the key parameter in the seismic design of high arch dams. The investigation on the dynamic flexural tests of full-graded concrete, especially for the influence of initial static loads on dynamic flexural strength, is still in the blank.
3. For the investigation on the internal damage mechanisms and explanation of the results of macro tests, more attention has been paid to the mesomechanical numerical analysis of concrete. The researches on the numerical concrete have a significant theoretical meaning and extensive application prospect. But the current research limits to the two-dimensional static problems. For satisfying the requirements of the three-dimensional dynamic meso-numerical simulation on the multigraded dam concrete, there are a lot of challenges to be faced. There is no open report on the relevant achievements. In the further study on meso-mechanics of dam concrete, the experimental research on the mechanical characteristics of the composite materials is the fundamental key technical problem that needs to be solved.
4. X-ray CT technique is the direct, feasible, and effective method to reveal the mechanism of internal damage, explain the macro experimental results, and verify meso-mechanical analysis. The current researches are mainly about the static compressive tests on rock materials based on medical CT instrument. For the dynamic tensile tests on inhomogeneous concrete materials, the

application of CT technique needs to conquer a lot of difficulties, such as loading instruments, testing methods, the data acquisition, and image analysis. From the aspect of development, the investigation on specialized industrial CT instruments suitable for the full-graded dam concrete and matched with dynamic and static material experimental instruments is a direction worthy of great attention.

5. The strain rate effect is the key point to the dynamic mechanical characteristics of concrete, and is a basic property for inhomogeneous concrete material. The rules are statistically summarized by experimental results in the current study. Although there are lots of results on the mechanisms, the consensus is still not formed. Among them, the strain rate effect obtained by different testing methods is hard to be clarified as the natural characteristic, or the influence of dynamic response from specimen structures. This problem is an important premise for engineering application and theoretical research.

The mentioned contents are the analysis on the research trends and progress on the dynamic mechanical characteristics of normal and dam concrete, and provide the engineering background, key problem, research keynote, and main technical ways for the research in the following chapters.

DYNAMIC FLEXURAL EXPERIMENTAL RESEARCH ON DAM CONCRETE

The experiments on dynamic mechanical characteristics of dam concrete are the basis for studying the dynamic resistance of dam structures. As stated previously, the standard value for dynamic tensile strength of dam structures adopts the experimental results from dynamic flexural tests; thus the four-point flexural testing is the fundamental content in this chapter. The main goals in this test are focused on the key points in the seismic design of high arch dams and are listed as follows:

1. The effect of initial loads and dynamic loading mode on the dynamic flexural mechanical characteristics of full-graded dam concrete.
2. The influence of size effect of full-graded and wet-sieved specimens on the dynamic flexural mechanical behavior.
3. The influence of strain rate effect on dynamic flexural strength of dam concrete.

The main way in tests is combined with the practical engineering project of Xiaowan and Dagang-shan high arch dams, and the specimens are made with the concrete of the same raw materials, mix proportions, and ages as that used in these projects.

11.1 TESTING METHODS FOR DYNAMIC FLEXURAL CHARACTERISTICS OF DAM CONCRETE

The regulation on the dynamic flexural tests of dam concrete is absent in China; therefore, the specimen making and instruments can only be referred to the static flexural tests in the Test Code for Hydraulic Concrete (DL/T 5150-2001). For achieving the stated goals, the determination of dynamic testing methods is the first premise. In the testing on the dynamic mechanical characteristics of dam concrete, the usual method is the impact loading. The impact loading frequency is taken as the value of one-fourth of the fundamental frequency when the materials achieve the maximum strength. The fundamental frequency for a full reservoir dam ranges from 0.1 s to 0.5 s, and the concrete grade in high stress zone is between C30 and C45 with the approximate strain rate of $(1-10) \times 10^{-3}\,\mathrm{s}^{-1}$. In the chosen range, the load wave of impact loads is far different from the practical earthquake wave, which cannot reflect the low-cycle fatigue effects of materials induced by an earthquake. Therefore, the dynamic loading mode of variable triangle cyclic wave is adopted to simulate the earthquake cyclic reaction. It is necessary to use the electro hydraulic servo loading system, develop the experimental instruments suitable for dynamic flexural tests of large full-graded specimens, and choose feasible measured instruments, operation control, data acquisition, processing, and so on. The according experimental methods and procedures are needed to be determined.

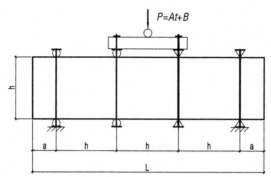

FIGURE 11.1 The Simply Supported Three Points Beam Loading Method

11.1.1 DYNAMIC FLEXURAL INSTRUMENTS, FIXTURES, AND THE REQUIREMENTS

Referring to the Test Code for Hydraulic Concrete (DL/T 5150-2001), the dynamic flexural tests by a simply supported beam as shown in Fig. 11.1 are adopted. The height of specimens for wet-sieved concrete is 150 and 300 mm, and 450 mm for three-graded and four-graded specimens, respectively.

Because the large material-testing machine with large tonnage for satisfying four graded high strength dam concrete is not available, the dynamic and static loading instruments consisted of the hydraulic servo actuator and large frame are designed. The maximum output load for the actuator is 1000 kN, the maximum displacement is 500 mm, and the working stage for frequencies ranges from 0.01 Hz to 10 Hz. With the random wave procedures, the controlling system with the computer can determine the loading procedure by displacement, force, or strain rate of the specimen. Figure 11.2 is the developed dynamic and static hydraulic servo loading instruments. The MTS322 testing machine is used for wet-sieved specimen.

The fundamental requirements for providing cyclic tensile and compressive loads are that the loading fixture can transfer the tensile–compressive cyclic action. The fixture consists of a support, clip, and bolt. There are two types of supports, rolling support and fixed support, which consist of a round steel block, respectively. The cyclic loads are transferred by bolt and support. The loading instruments for wet-sieved and full-graded concrete specimens are shown in Fig. 11.3. For ensuring the failure of the specimen in the pure flexural region, the loading direction of the actuator should be vertical to the specimen plane and pass the center of this plane, and the axial line for the bolt should be in line with the location of three dividing points. The distance between the two end supports should be equal to that between the axial line of the actuator, and the force line is perpendicular to their axial line. The error for the installation should be smaller than the value of 1 mm, and each bolt is under the condition of pretension. Figure 11.4 shows the failure conditions for the pure flexural region of the specimens.

11.1.2 DYNAMIC LOADING MODE, DATA ACQUISITION, AND PROCESSING

11.1.2.1 Dynamic loading mode

According to the requirements in the Test Code for Hydraulic Concrete (DL/T 5150-2001), the loading rate for the static tests was 250 N/s. The dynamic loading uses two loading types, which are impact and variable amplitude.

FIGURE 11.2 Hydraulic Servo Loading System

 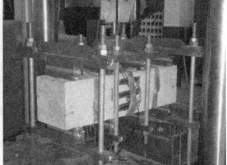

FIGURE 11.3 The Loading Equipment for Concrete Specimens

(a)   (b)

FIGURE 11.4 The Failure Condition for the Pure Flexural Region in Specimen

(a) Three-graded concrete specimen and (b) wet-sieved specimen

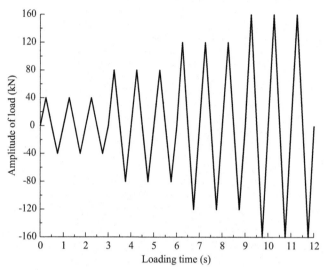

FIGURE 11.5 The Triangle Loading Wave

Applying the impact loads should preevaluate the failure loads and failure strain. The loading rate can be determined based on the time from starting of the load to the specimen failure. The variable cyclic wave loads adopt the triangle wave with the same frequency as the dam structure, and the amplitude of the wave increases step by step until the specimen fails, as shown in Fig. 11.5. These two loading modes have the corresponding programs and are automatically controlled by a computer.

11.1.2.2 Acquisition of experimental data

1. Acquisition of experimental data

   In the experiments, the load value is measured by the high-precision loading sensor. The vertical displacement on the top and bottom surface of the specimen is determined by a high-precision displacement sensor. The top, bottom, and side surfaces are bonded with strain gauges to measure the elastic modulus, Poisson's ratio, stress–strain relationship, damage evolution law, and so on.

   Three groups of horizontal and vertical strain gauges are bonded on the top and bottom surfaces of specimens, and three groups of vertical strain gauges are bonded on the lateral surface. To satisfying the effective measuring distance larger than three times the particle size of coarse aggregate, the vertical strain gauges are bonded continuously by 100 mm length gauges, and the length of the horizontal strain gauges is 50 mm. Taking the three-graded concrete specimen as example, the bond locations for strain gauges are shown in Fig. 11.6.

2. Processing of experimental data

   The experimental data are automatically collected by a computer. The common experimental results, such as flexural strength of concrete, maximum flexural strain, elastic modulus, Poisson's ratio, and so on, can be analyzed by the methods reported in the Test Code for Hydraulic Concrete (DL/T 5150-2001), while the following questions should be paid attention to the processing of dynamic experimental data.

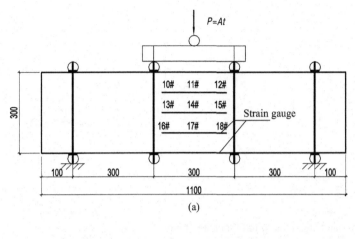

(a)

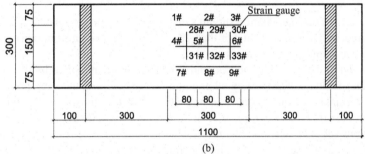

(b)

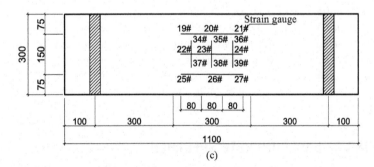

(c)

FIGURE 11.6 The Bonded Locations for Strain Gauges (unit: mm)

(a) specimen façade (b) top surface (c) bottom surface

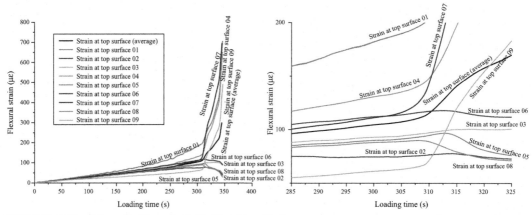

FIGURE 11.7 The Schematic Figure for Initial Cracking Strain

The frequency for data acquisition should satisfy that in the pre-evaluated failure time enough data are acquired, and it is usually higher than 200 Hz. The same data from one experiment can be acquired with two sampling frequencies.

As the amount of data is large, the load and deformation are measured by the loading system, while the strain should be measured by other equipment. Therefore, the data should be dispersed into the same format based on the time and frequency for data acquisition, which is processed by a professional software (such as Origin). Then the key points are in alignment, and the different collected systems should be installed with synchronous methods to ensure the simultaneous acquisition of load and measured data. The matching problems among different systems should also be paid attention to.

3. Determination of initial cracking strain and stress

   For the determination of the initial cracking strain and stress, the top and bottom surfaces should be arranged in a staggered manner with the strain gauges with the same direction. The judgment rules for the cracking appearance is when the strain value for one certain gauge is increased suddenly, and the value for other gauges is quickly decreased, then the average value for these three strains can be taken as the initial cracking strain. The corresponding stress is the initial cracking stress, as shown in Fig. 11.7.

11.1.3 CASTING OF DAM CONCRETE SPECIMEN

The equipment for casting concrete include the mixer (250 L, 350 L, the setting type), vibration rod (diameter of 5 cm, length of 50 cm), dressing sieve, screen (4 cm of pore diameter), the instrument for measuring air content of concrete, vibration table, steel formwork, and so on.

Before the casting of the specimens, the material testing according to the relevant codes (DL/T 5150-2001, DL/T 5151-2001) is conducted, including the grain gradation of sand and the water absorption of aggregate. The casting procedure is referred to the requirements from the test code for hydraulic concrete to ensure the construction quality.

The part of the constructed specimens is shown in Fig. 11.8. The demolded specimens are covered by a plastic film or sacks, and water cured timely. Then the specimens are naturally cured in the laboratory.

(a)   (b)

FIGURE 11.8 Full Graded Concrete and Dry or Wet-Sieved Concrete Specimens

(a) Prismatic specimens and (b) cubic specimens

11.2 TESTING ON DYNAMIC MECHANICAL CHARACTERISTICS OF DAM CONCRETE

11.2.1 EXPERIMENTAL MATERIALS

The effect of initial static loads and dynamic loading mode is the focus questions in the seismic design for high arch dams, and the research results are very seldom. These experiment results are obtained combined with the 295-m high Xiaowan arch dam and the 210-m high Dagangshan arch dam projects. For the Xiaowan arch dam, the three-graded concrete of $R_{180}400$ dam concrete was tested. For Dagangshan arch dam, the four-graded concrete of $R_{180}360$ dam concrete was tested.

The mixing proportions for Xiaowan arch dam is shown in Table 11.1. The composition for aggregate is mixed manmade aggregate with the mixed ratio 7.5:2.5 of the biotite granite gneiss to the hornblende plagioclase. The gradation for aggregate is big stone (80–40 mm):middle stone (40–20 mm):small stone (20–5 mm) = 4:3:3. The 42.5-type Portland cement made by Hongta Western Yunnan Company was used. The Class I fly ash from the power plant in Xuanwei was adopted.

Table 11.1 The Designed Mix Proportions for Xiaowan Arch Dams										
Strength Grade	Grada-tion	w/c	Water (kg/m³)	Cement (kg/m³)	Flyash (kg/m³)	Water Reducing Agent ZB-1A (%)	Air en-training Agent AEA (‰)	Slump (cm)	Air Content (%)	Sand Ratio (%)
$R_{180}400$	3	0.42	104	173.3	74.3	0.70	1.4	6.4	4.49	30

Table 11.2 The Designed Mix Proportions for Dagangshan Arch Dams

Strength Grade	Water-binder Ratio	Theoretical Bulk Density (kg/m³)	Quantity of Materials for Unit Cube Concrete (kg)						
			Water	Cement	Flyash	Sand	Aggre-gate	JM-II (C)	JM-2000
$R_{180}360$	0.45	2418	86	134	57	534	1607	1.43	0.0105

The mixing proportions for Dagangshan arch dam is shown in Table 11.2. The aggregate used is the Cloud Two granite. The aggregates were sieved with especially large stone (80–150 mm), big stone (40–80 mm), middle stone (20–40 mm), and small stone (5–20 mm), the ratio being 35:25:20:20. The type 42.5 moderate heat Portland cement from Emeishan Company was used. The fly ash was the Class I fly ash in Guangan, and the admixture is JM-II (C) retarder and JM-2000 air-entraining agent.

11.2.2 LOADING SCHEME

The experimental research was carried out according to the different initial static loads and dynamic loading mode. According to the requirement of engineering design, the experimental objects were the specimen at the age of 180 d, aiming to study the influence of age on the compressive strength of full-graded concrete. For the concrete used in the Dagangshan arch dam, the compressive tests with the age of 28, 90, 180, and 360 d were also carried out. The experiments were using the three loading waves discussed next.

11.2.2.1 Static load

According to the code for testing hydraulic concrete, the average loading rate for static test is 250 N/s. The monotonic loading wave was programmed. The cubic compressive tests were carried out on the full-graded and wet-sieved concrete specimens.

11.2.2.2 Impact wave loading

1. Xiaowan arch dam engineering project

   The fundamental frequency for the full reservoir Xiaowan arch dam is 1 Hz, and the controlled failure time for one entire load is within 0.25 s. Through the evaluation of the maximum strength of concrete, the loading rate is 600 kN/s, and the maximum load value is 180 kN. The 0, 40, and 80% of the maximum loads were used in the impact loads, and the corresponding initial loads were 0, 40, and 80 kN, respectively. The static load with 250 N/s was first applied and then the impact loads were added.

   The peak value for wet-sieved concrete specimens was preset as 80 kN. The initial static loads were the 0, 40, 80, and 90% of peak loads, respectively. The corresponding value was 0, 18.6, 37.2, and 41.8 kN. During the loading process, the loading rate for initial static loads was 250 N/s.

   The wet-sieved specimens were conducted with dynamic flexural characteristic experiments with strain rates of 1×10^{-6}, 1×10^{-5}, 0.9×10^{-4}, and 0.9×10^{-3} s^{-1}, respectively.

2. Dagangshan arch dam engineering project

   The fundamental frequency for the full reservoir dam is 1.67 Hz, and the failure time for one entire load is about 0.15 s. The analyzed loading rate is 1800 kN/s. The preset value for peak load was 270 kN, and the initial static load was not applied with the impact loads.

The loading rate for wet-sieved concrete specimens was adopted as 450 kN/s, and the peak value for loads was controlled as 80 kN. The initial static loads were 0, 40, and 80% of the maximum loads, and the corresponding static loads were 0, 19.4, and 38.8 kN, respectively. The static load with 250 N/s was first applied and then the impact loads were added.

The flexural strength testing on wet-sieved specimens was carried out with the strain rates of 1×10^{-6}, 1×10^{-5}, 1×10^{-4}, and 1×10^{-3} s^{-1}, respectively.

The splitting strength testing on wet-sieved cubic specimens was carried out with the strain rates of 1.3×10^{-7}, 1.3×10^{-6}, 1.3×10^{-5}, and 1.2×10^{-4} s^{-1}, respectively.

11.2.2.3 The cyclic loads with variable triangle wave

1. Xiaowan arch dam

   The frequency for the cyclic loads with variable triangle wave was taken as 1 Hz, and the value for peak loads was assumed to be 160 kN. The initial loads with 0, 40, and 80% were considered.
2. Dagangshan arch dam

   The frequency for the cyclic loads with variable triangle wave was taken as 1.67 Hz, and the value of peak loads was assumed to be 350 kN. The initial loads with 0, 40, and 80% were considered.

11.2.3 EXPERIMENTAL RESULTS

11.2.3.1 Compressive strength tests

The average compressive static strength for full-graded and wet-sieved cubic specimens in Xiaowan arch dam was 41.87 and 47.81 MPa, respectively.

The average compressive static strength for full-graded and wet-sieved cubic specimens in Dagangshan arch dam was 40.37 and 51.79 MPa, respectively.

11.2.3.2 Splitting tensile strength tests

For wet-sieved concrete of Dagangshan arch dam, the splitting tensile strength for cubic specimens under different loading rate is shown in Table 11.3.

11.2.3.3 Flexural strength tests

1. Three-graded and wet-sieved concrete for Xiaowan arch dam

   The dynamic and static flexural strength of three-graded and wet-sieved concrete for Xiaowan arch dam are listed in Tables 11.4 and 11.5, respectively.

   The influence of loading modes on the dynamic increase factor (DIF) of three-graded and wet-sieved concrete is shown in Figs 11.9 and 11.10, respectively.
2. Four-graded and wet-sieved concrete for Dagangshan arch dam

   The dynamic and static flexural strengths of four-graded and wet-sieved concrete for Dagangshan arch dam are listed in Tables 11.6 and 11.7, respectively.

   The influence of various initial static loads and dynamic loading modes on the dynamic increase factor of flexural strength of four-graded and wet-sieved concrete for Dagangshan arch dam is shown in Figs 11.11 and 11.12, respectively.

11.2.3.4 The flexural deformation characteristics

The flexural deformation characteristics include the stress–strain curves, Poisson's ratio, initial cracking strain and stress, maximum tensile strain, the deformation distribution across the normal section,

Table 11.3 The Splitting Tensile Strength for Wet-Sieved Cubic Concrete Specimens

Loading Rate (kN/s)	Number of Specimen	Loading Time (s)	Ultimate Load (kN)	Average (kN)	Strength (MPa)	Ratio of Dynamic and Static Splitting Tensile Strength	Stress Rate (MPa/s)	Corresponding strain rate (/s)
0.25	w101107	449.68	122.7	125.8	3.56	1	7×10^{-3}	1.3×10^{-7}
	w101201	535.03	133.7					
	w101304	472.37	121.0					
2.5	w101407	51.03	143.0	135.6	3.83	1.08	7×10^{-2}	1.3×10^{-6}
	w101103	56.66	141.5					
	w100812	48.96	122.2					
25	w100807	5.72	142.4	153.7	4.35	1.22	7×10^{-1}	1.3×10^{-5}
	w101203	6.72	167.1					
	w101804	6.10	151.6					
250	w101802	0.42	145.7	170.8	4.84	1.36	7×10^{-0}	1.2×10^{-4}
	w101206	0.64	180.1					
	w101703	0.51	186.6					

Table 11.4 Dynamic and Static Flexural Strength of Three-Graded Concrete for Xiaowan Arch Dam

Loading Mode	Initial Load Ratio (%)	Flexural Strength (MPa)	Dynamic Increase Factor (DIF)
Static load	100	3.58	1
Impact	0	4.20	1.17
	40	4.60	1.29
	80	4.92	1.37
Triangle wave	0	4.19	1.17
	40	4.48	1.25
	80	4.70	1.31

elastic modulus, the cyclic loading effects, and so on. From the requirements of the code for testing hydraulic concrete, the elastic modulus adopts the tangent modulus at the origin point. The elastic modulus adopted in this work was the linearly fitted slope from 0% to 50% of the peak stress in the stress–strain curves.

Table 11.5 Dynamic and Static Flexural Strength of Wet-Sieved Concrete for Xiaowan Arch Dam

Loading Mode	Initial Load Ratio (%)	Flexural Strength (MPa)	Dynamic Increase Factor (DIF)
Static load	100	6.19	1.00
Impact	0	8.54	1.39
	40	8.77	1.42
	80	9.24	1.49
	90	8.99	1.45
Triangle wave	0	7.80	1.26
	40	7.93	1.28
	80	7.37	1.19
	90	7.67	1.24

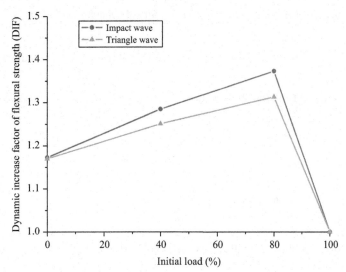

FIGURE 11.9 The Influence of Loading Modes on Dynamic Increase Factor (DIF) of Three Graded Concrete for Xiaowan Arch Dam

1. Xiaowan arch dam concrete
   a. Three-graded concrete
      - The flexural stress–strain curves of the typical specimen described based on the experimental measured data are shown in Figs 11.13–11.16.
      - The elastic modulus obtained from the stress–strain curves and the average strain rate calculated from the measured strain–time curve before failure occurs are shown in Table 11.8. With the existence of dynamic loads and initial static loads, the dynamic and static elastic modulus and Poisson's ratio refer to the corresponding value under

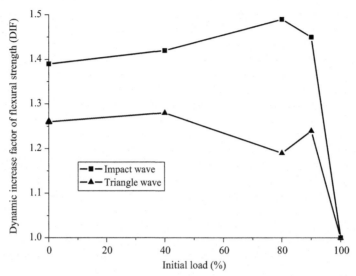

FIGURE 11.10 The Influence of Loading Mode on DIF of Wet-Sieved Concrete for Xiaowan Arch Dam

Table 11.6 Dynamic and Static Flexural Strength of Four-Graded Concrete for Dagangshan Arch Dam

Loading Mode	Initial Load Ratio (%)	Flexural Strength (MPa)	Dynamic Increase Factor (DIF)
Static load	100	2.97	1
Impact	0	4.18	1.41
Triangle wave	0	3.49	1.18
	40	3.97	1.34
	80	3.99	1.34

Table 11.7 Dynamic and Static Flexural Strength of Wet-Sieved Concrete for Dagangshan Arch Dam

Loading Mode	Initial Load Ratio (%)	Flexural Strength (MPa)	Dynamic Increase Factor (DIF)
Static load	100	6.46	1.00
Impact	0	8.42	1.30
	40	7.70	1.19
	80	8.21	1.27
Triangle wave	0	7.17	1.11
	40	7.83	1.21
	80	8.32	1.29

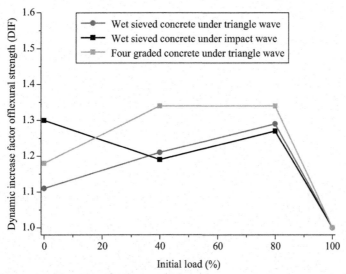

FIGURE 11.11 The Influence of Various Loading Modes on Dynamic Increase Factor of Flexural Strength of Four-Graded Concrete for Dagangshan Arch Dam

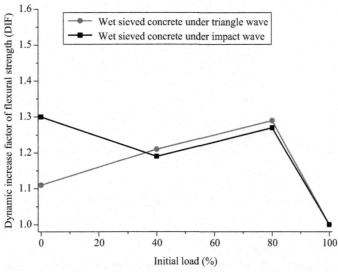

FIGURE 11.12 The Influence of Various Loading Modes on Dynamic Increase Factor of Flexural Strength of Wet-Sieved Concrete for Dagangshan Arch Dam

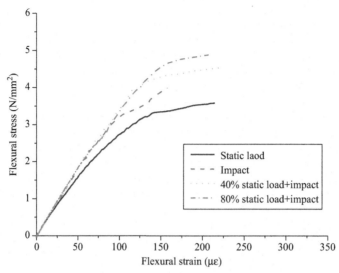

FIGURE 11.13 The Flexural Stress–Strain Curves for Combined Initial Static Loads and Impact Loads

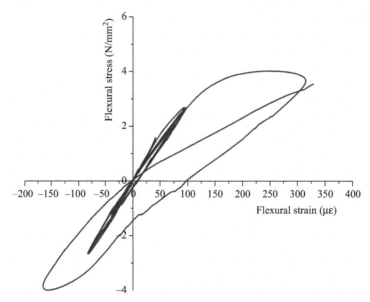

FIGURE 11.14 The Stress–Strain Curve of Specimen No. 11033 (0% Static Loads + Triangular Loading Wave)

dynamic and static conditions. The dynamic value under the triangle wave is based on the corresponding initial tensile stage.
– Under the triangle cyclic loading, the changing of elastic modulus of concrete with loading number is shown in Table 11.9 and Fig. 11.17. It can be found that the strain rate is increased with the increase of the loading number.

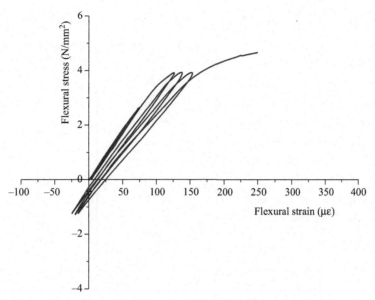

FIGURE 11.15 The Stress–Strain Curve of Specimen No. 11035 (40% Static Loads + Triangular Loading Wave)

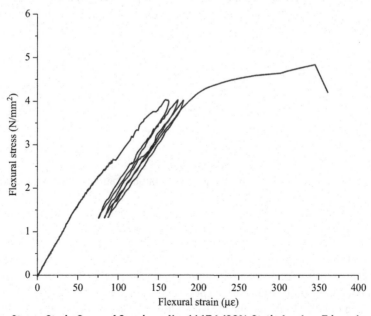

FIGURE 11.16 The Stress–Strain Curve of Specimen No. 11174 (80% Static Loads + Triangular Loading Wave)

Table 11.8 The Dynamic and Static Flexural Elastic Modulus and Poisson's Ratio for Concrete

Conditions	Loading Mode	Elastic Modulus of Flexural Tensile (GPa)	Poisson Ratio of Flexural Tensile	Strain Rate (s^{-1})
Static	Static load	35.03	0.17	1×10^{-6}
	Impact wave + 40% static load	37.23	0.20	
	Impact wave + 80% static load	39.10	0.14	
	Triangle wave + 40% static load	35.75	0.11	
	Triangle wave + 80% static load	33.50	0.26	
	Average	36.12	0.18	
Dynamic	Impact wave +0% static load	37.64	0.15	200×10^{-6}
	Triangle wave +0% static load	37.10	0.20	
	Average	37.37	0.18	
Dynamic increase ratio (%)		3.5	0	

Table 11.9 The Changing of Dynamic Tensile Elastic Modulus of Three-Graded Concrete Under Triangle Wave

Number of Specimens	Dynamic Tensile Elastic Modulus of Every Loading Cycle (GPa)									
	1	2	3	4	5	6	7	8	9	10
11024	36.62	36.77	36.23	36.32	31.52	31.02	30.14			
11033	34.41	33.18	33.12	33.03	28.42	27.22	27.31	13.38		
11051	38.02	39.06	37.50	37.39	33.35	33.03	32.86	27.05	26.58	26.09
Strain rate ($\times 10^{-6}$ s^{-1})	200	200	200	400	400	400	700	700	700	900

- The flexural initial cracking stress and strain, and the ratio of dynamic to static results for concrete are shown in Table 11.10.
- The maximum flexural tensile strain under dynamic and static loading is determined based on the regulation of the test code for hydraulic concrete. The flexural stress–strain curves are plotted first, the line parallel to the horizontal axis is drawn, and the linear part of the measured stress–strain curve before the abrupt change occurs is extended to the horizontal line just drawn. The corresponding strain of the intersection point of the two lines is the maximum flexural strain value. The stress–strain curves for the typical specimens are shown in Fig. 11.18, and the measured flexural maximum flexural strain is listed in Table 11.11.

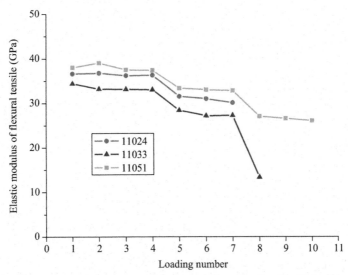

FIGURE 11.17 The Changing of Elastic Modulus of Concrete with Loading Number Under Triangle Cyclic Wave

Table 11.10 The Flexural Initial Cracking Stress and Strain, and the Ratio of Dynamic to Static Results for Concrete

Loading Mode	Ratio of Initial Static Load (%)	Flexural Strain of Crack Initiation ($\times 10^{-6}$)	Ratio of Dynamic and Static Crack Initiation Strain	Flexural Stress of Crack Initiation (N/mm²)	Ratio of Dynamic and Static Crack Initiation Stress
Static load	100	108	/	2.83	1.00
Impact wave	0	109	1.01	3.27	1.16
	40	108	1.00	3.45	1.22
	80	114	1.06	3.65	1.29
Triangle wave	0	110	1.02	3.54	1.25
	40	103	0.95	3.40	1.20
	80	/	/	/	/

    – The deformation characteristics of normal section in the failure region. The measured strain in the top, bottom, and lateral surfaces of the specimens under dynamic and static loading is plotted. The results for static loading are shown in Fig. 11.19, and the results for dynamic loading are shown in Fig. 11.20.

b. Wet-sieved concrete
    – The dynamic and static stress–strain curves under different loading conditions are shown in Figs 11.21–11.25, respectively.

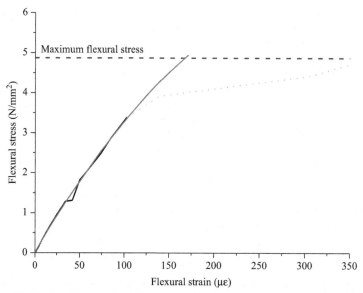

FIGURE 11.18 The Fitted Flexural Stress–Strain Curves for Specimen No. 11076

Table 11.11 The Maximum Flexural Strain of Three-Graded Concrete Under Different Loading Modes

Loading Mode	Ratio of Initial Static Load (%)	Maximum Flexural Strain ($\times 10^{-6}$)	Ratio of Dynamic and Static Maximum Flexural Strain
Static load	100	187	1.00
Impact wave	0	153	0.82
	40	159	0.85
	80	164	0.88
Triangle wave	0	173	0.93
	40	184	0.98
	80	213	1.14

- The dynamic and static elastic modulus and Poisson's ratio. The measured elastic modulus and Poisson's ratio from the stress–strain curves in tensile surface are listed in Table 11.12.
- The maximum flexural strain under dynamic and static loading. The measured flexural strain under dynamic and static loading of the wet sieved concrete is shown in Table 11.13.
- The deformation characteristics of the normal section in the failure region. The measured strain in the top, bottom, and lateral surfaces under typical dynamic and static loadings are shown in Figs 11.26 and 11.27, respectively.

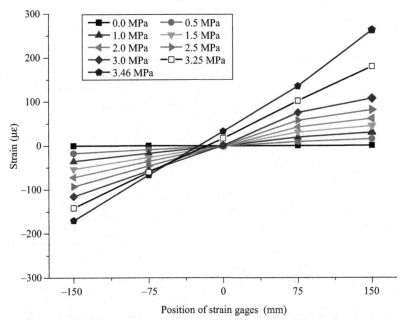

FIGURE 11.19 **The Static Deformation of the Normal Section in Flexural Region of Concrete (Specimen No. 11073)**

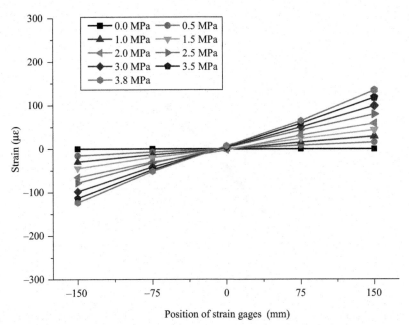

FIGURE 11.20 **The Dynamic Deformation of the Normal Section in Flexural Region of Concrete (Specimen No. 11036)**

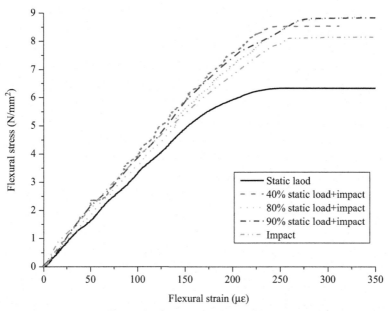

FIGURE 11.21 The Measured Flexural Stress–Strain Curves (Initial Static Loads + Impact Loads)

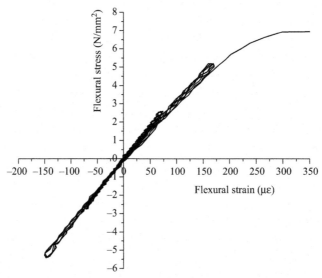

FIGURE 11.22 The Flexural Stress–Strain Curves of Specimen No. 11021 (the Variable Amplitude Triangle Wave)

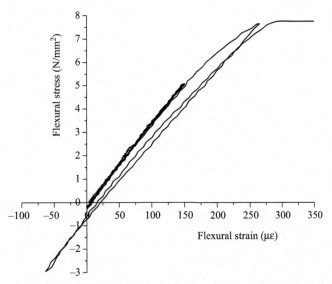

FIGURE 11.23 The Flexural Stress–Strain Curve of Specimen No. 11096 (40% Initial Static Loads + Variable Amplitude Triangle Wave)

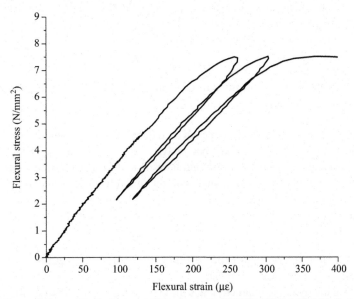

FIGURE 11.24 The Flexural Stress–Strain Curves of Specimen No. 11055 (80% Initial Static Loads + Variable Amplitude Triangle Wave)

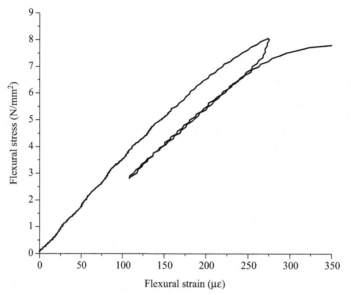

FIGURE 11.25 The Flexural Stress–Strain Curves of Specimen No. 11133 (90% Initial Static Loads + Variable Amplitude Triangle Wave)

2. Dagangshan arch dam concrete
   a. Four-graded concrete
      - The stress–strain curves of four-graded concrete for Dagangshan arch dam under static and dynamic loadings are shown in Figs 11.28 and 11.29, respectively.
      - The dynamic and static elastic modulus and Poisson's ratio. The elastic modulus and Poisson's ratio obtained from the initial linear elastic condition of the stress–strain curves in the flexural surface are listed in Table 11.14.
      - The maximum dynamic and static flexural strain value. The measured maximum flexural strain value is listed in Table 11.15.
   b. Wet-sieved concrete
      - The stress–strain curves of Dagangshan arch dam under static and dynamic impact loading are shown in Figs 11.30–11.33, respectively.
      - Dynamic and static elastic modulus and Poisson's ratio. The elastic modulus and Poisson's ratio of wet-sieved concrete obtained from the initial linear elastic condition on the tensile surface are shown in Table 11.16.
      - The maximum flexural strain value. The measured maximum flexural strain value is listed in Table 11.17.

11.2.3.5 The experiments on strain rate effects on flexural characteristics

1. Wet-sieved concrete for Xiaowan arch dam
   In the research on the dynamic characteristics of Xiaowan arch dam, the flexural testing on the wet-sieved concrete was conducted under the strain rates of 1×10^{-6}, 1×10^{-5}, 1×10^{-4}, and 1×10^{-3} s^{-1}. The obtained results are shown as follows:

Table 11.12 The Elastic Modulus and Poisson's Ratio of Wet-Sieved Concrete Under Various Initial Static Loads and Loading Conditions

Condition	Loading Mode	Strain Rate (s⁻¹)	Elastic Modulus of Flexural Tensile (GPa)	Average of Elastic Modulus (GPa)	Ratio of Dynamic and Static Elastic Modulus	Poisson Ratio	Average of Poisson's Ratio	Ratio of Dynamic and Static Poisson Ratio
Static	Static load	1.0×10^{-6}	36.35	36.72	1.00	0.22	0.20	1.00
	40% static load + impact wave		37.52			0.20		
	80% static load + impact wave		37.43			0.21		
	90% static load + impact wave		37.65			0.19		
	40% static load + triangle wave		35.21			0.19		
	80% static load + triangle wave		37.00			0.20		
	90% static load + triangle wave		35.91			0.19		
Dynamic	triangle wave	270×10^{-6}	38.85	38.85	1.06	0.19	0.19	0.95
	impact wave	1000×10^{-6}	40.90	40.90	1.11	0.20	0.20	1.00

Table 11.13 The Measured Flexural Strain of Wet-Sieved Concrete

Loading Mode	Ratio of Initial Static Load (%)	Flexural Strain ($\times 10^{-6}$)	Ratio of Dynamic and Static Flexural Strain
Static load	100	262	1.00
Impact wave	0	250	0.95
	40	252	0.96
	80	258	0.98
	90	260	0.99
Triangle wave	0	262	1.00
	40	268	1.02
	80	273	1.04
	90	297	1.13

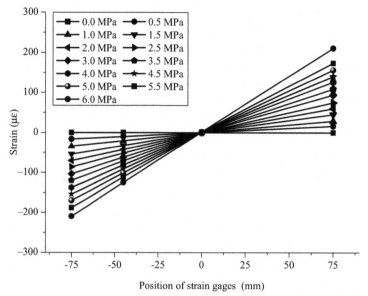

FIGURE 11.26 The Static Deformation of the Normal Section in the Pure Flexural Region of Concrete

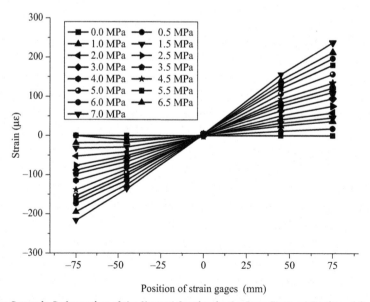

FIGURE 11.27 The Dynamic Deformation of the Normal Section in the Pure Flexural Region of Concrete

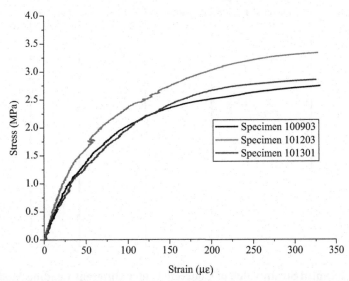

FIGURE 11.28 The Flexural Stress–Strain Curves of Four-Graded Concrete for Dagangshan Arch Dam Under Static Loading

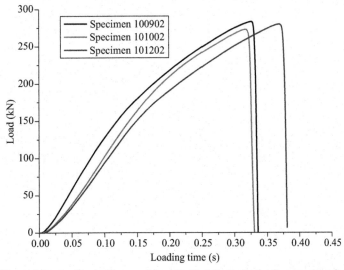

FIGURE 11.29 The Flexural Stress–Strain Curves of Four-Graded Concrete for Dagangshan Arch Dam Under Dynamic Impact Loading

Table 11.14 The Dynamic and Static Flexural Elastic Modulus and Poisson's Ratio of Concrete

Condition	Loading Mode	Elastic Modulus of Flexural Tensile (GPa)	Average (GPa)	Ratio of Dynamic and Static Elastic Modulus	Poisson Ratio	Average	Ratio of Dynamic and Static Poisson's Ratio
Static	Static load	38.90			/		
	Triangle wave + 40% static load	40.07	38.83	1.00	0.14	0.15	1.00
	Triangle wave +80% static load	37.53			0.15		
Dynamic	Impact load + 0% static load	53.73	53.73	1.38	0.20	0.20	1.33
	Triangle wave +0% static load	35.20	37.25	0.96	0.15	0.15	1.00

Table 11.15 The Flexural Strain Value of Concrete Under Different Loading Modes

Loading Mode	Ratio of Initial Static Load (%)	Flexural Strain ($\times 10^{-6}\text{s}^{-1}$)	Ratio of Dynamic and Static Flexural Strain
Static load	100	159	1.00
Impact wave	0	173	1.09
Triangle wave	0	195	1.23
	40	176	1.11
	80	199	1.25

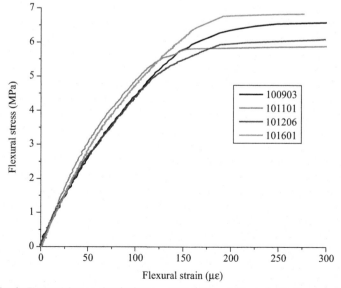

FIGURE 11.30 The Static Flexural Stress–Strain Curves of Wet-Sieved Concrete for Dagangshan Arch Dam

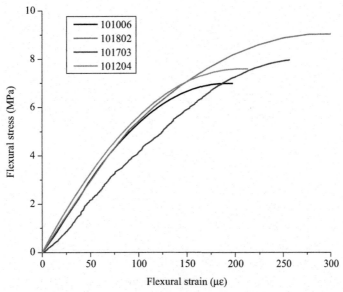

FIGURE 11.31 The Dynamic Flexural Stress–Strain Curves of Wet-Sieved Concrete for Dagangshan Arch Dam (0% Initial Static Loads + Impact Loading)

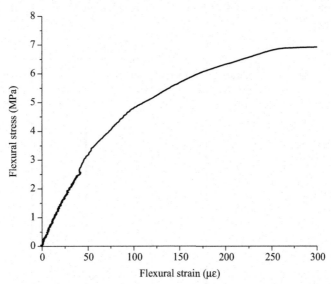

FIGURE 11.32 The Dynamic Flexural Stress–Strain Curves of Wet-Sieved Concrete for Dagangshan Arch Dam (40% Initial Static Loads + Impact Loading, Specimen No. 101707)

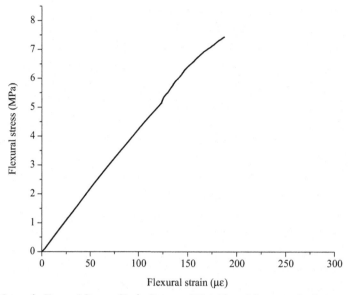

FIGURE 11.33 The Dynamic Flexural Stress–Strain Curves of Wet-Sieved Concrete for Dagangshan Arch Dam (80% Initial Static Loads + Impact Loading, Specimen No. 101705)

Table 11.16 The Elastic Modulus of Wet-Sieved Concrete Under Various Initial Static Loads and Loading Modes

Condition	Loading Mode	Elastic Modulus of Flexural Tensile (GPa)	Average of Elastic Modulus (GPa)	Ratio of Dynamic and Static Elastic Modulus	Poisson's Ratio	Average of Poisson Ratio	Ratio of Dynamic and Static Poisson's Ratio
Static	Static load	53.10	50.00	1.00	0.30	0.20	1.00
	40% static load+ impact wave	51.86			0.17		
	80% static load+ impact wave	54.77			0.16		
	40% static load+ triangle wave	46.00			0.23		
	80% static load+ triangle wave	43.40			0.18		
Dynamic	Triangle wave	44.80		0.90	0.17		0.85
	Impact wave	58.18		1.16	0.18		0.90

Table 11.17 The Flexural Strain Value of Wet-Sieved Concrete

Loading Mode	Ratio of Initial Static Load (%)	Flexural Strain ($\times 10^{-6} s^{-1}$)	Ratio of Dynamic and Static Flexural Strain
Static load	100	159	1.00
Impact wave	0	191	1.20
	40	187	1.18
	80	186	1.17
Triangle wave	0	188	1.18
	40	177	1.11
	80	194	1.22

Table 11.18 The Flexural Strength and Dynamic Increase Factor of Wet-Sieved Concrete Under Different Strain Rates

Measured Strain Rate (s^{-1})	Average of Flexural Strength (MPa)	Standard Deviation (MPa)	Ratio of Dynamic and Static Flexural Strength
1.0×10^{-6}	6.19	0.27	1.00
1.0×10^{-5}	7.19	0.15	1.16
0.9×10^{-4}	8.26	0.56	1.33
0.9×10^{-3}	8.54	0.47	1.38

a. *Flexural strength*: The flexural strength of concrete under different strain rates is seen in Table 11.18.
b. *Dynamic and static flexural elastic modulus*: According to the requirement of the code for hydraulic concrete, besides the tangent modulus obtained at the original point of the stress–strain curves by linear fitting, the measured peak secant modulus of each specimen is also calculated. In the flexural testing, the tangent modulus at the original point and the peak secant modulus of concrete under different strain rates are listed in Tables 11.19 and 11.20, respectively.

Table 11.19 The Tangent Modulus at the Original Point of Wet-Sieved Concrete Under Different Strain Rates

Measured Strain Rate (s^{-1})	Average of Flexural Elastic Modulus (GPa)	Standard Deviation (GPa)	Ratio of Dynamic and Static Elastic Modulus
1×10^{-6}	36.35	3.16	1
1×10^{-5}	36.93	2.42	1.02
0.9×10^{-4}	38.00	0.90	1.05
0.9×10^{-3}	40.90	1.05	1.13

Table 11.20 The Flexural Peak Chord Modulus of Wet-Sieved Concrete Under Different Strain Rates

Measured Strain Rate (s⁻¹)	Average of Flexural Peak Chord Modulus (GPa)	Standard Deviation (GPa)	Ratio of Dynamic and Static Flexural Peak Chord Modulus
1×10^{-6}	26.11	3.77	1
1×10^{-5}	27.93	2.93	1.07
0.9×10^{-4}	32.42	2.90	1.24
0.9×10^{-3}	34.63	0.18	1.33

Table 11.21 Poisson's Ratio of Wet-Sieved Concrete Under Different Strain Rates

Strain rate (s⁻¹)	1×10^{-6}	1×10^{-5}	0.9×10^{-4}	0.9×10^{-3}
Average of Poisson ratio	0.22	0.18	0.19	0.20
Standard deviation	0.012	0.010	0.026	0.023

Table 11.22 The Maximum Flexural Strain Value of Wet-Sieved Concrete Under Different Strain Rates

Strain Rate (s⁻¹)	Average of Flexural Strain ($\times 10^{-6} s^{-1}$)	Standard Deviation ($\times 10^{-6} s^{-1}$)	Ratio of Dynamic and Static Flexural Strain
1×10^{-6}	262	36.2	1
1×10^{-5}	259	23.5	0.99
0.9×10^{-4}	255	10.7	0.97
0.9×10^{-3}	250	13.6	0.95

    c. *The dynamic and static Poisson's ratio of concrete*: The Poisson's ratio of wet-sieved concrete under different strain rates is shown in Table 11.21.

    d. *The maximum flexural strain value*: The maximum flexural strain value of wet-sieved concrete under different strain rates is shown in Table 11.22.

2. Wet-sieved concrete for Dagangshan arch dam

The experimental results on the flexural strength testing of wet-sieved concrete under different strain rate conditions are seen in Table 11.23.

11.2.3.6 The experiments for age effects on flexural characteristics

The compressive testing results on the four-graded and wet-sieved concrete for Dagangshan arch dam are seen in Tables 11.24–11.31, respectively.

Table 11.23 The Dynamic and Static Flexural Strength of Wet-Sieved Concrete Under Different Strain Rates

Strain Rate (s⁻¹)	Number of Specimen	Measured Strain Rate ($\times 10^{-6}$ s⁻¹)	Ultimate Load (kN)	Flexural Strength (MPa)	Average if Flexural Strength (MPa)	Ratio of Dynamic and Static Flexural Strength	Elastic Modulus of Flexural Strength (GPa)
10^{-6} (Static load)	w101101	1.01	45.7	6.10	6.46	1.00	53.1
	w101206	1.11	46.1	6.15			
	w101601	1.02	51.4	6.86			
	w100903	0.96	50.6	6.74			
10^{-5}	w101105	8.3	54.9	7.32	7.34	1.14	53.6
	w100801	8.2	56.4	7.52			
	w101002	9.3	53.9	7.19			
10^{-4}	w100901	72	63.9	8.52	7.72	1.20	55.4
	w102202	92	58.1	7.75			
	w101106	86	51.6	6.88			
10^{-3}	w101802	1606	69.0	9.19	8.42	1.30	58.2
	w101204	1690	64.2	8.56			
	w101006	1130	58.9	7.85			
	w101703	1870	60.5	8.07			

Table 11.24 The 28 d Cubic Compressive Strength of Wet Sieved Concrete for Dagangshan Arch Dam

Number of Specimen	Ultimate Load (kN)	Compressive Strength (MPa)	Average of Strength (MPa)
W101801	895	39.78	39.56
W101701	885	39.34	
W101601	1077	47.87 (reject)	

Table 11.25 The 90 d Cubic Compressive Strength of Wet-Sieved Concrete for Dagangshan arch Dam

Number of Specimen	Ultimate Load (kN)	Compressive Strength (MPa)	Average of Strength (MPa)
W101006	1093	48.58	50.26
W101003	1173	52.14	
W101403	1126	50.05	

Table 11.26 The 90 d Cubic Compressive Strength of Full-Graded Concrete for Dagangshan Arch Dam

Number of Specimen	Ultimate Load (kN)	Compressive Strength (MPa)	Average of Strength (MPa)
F101601	7443	36.76	34.71
F101603	6282	31.02	
F101801	7362	36.36	

Table 11.27 The 180 d Cubic Compressive Strength of Wet-Sieved Concrete for Dagangshan Arch Dam

Number of Specimen	Ultimate Load (kN)	Compressive Strength (MPa)	Average of Strength (MPa)
W100805	1174	52.18	51.79
W100808	1160	51.56	
W100809	1072	47.64	
W100902	1194	53.07	
W100906	1160	51.56	
W100404	959	42.62 (reject)	
W101005	1107	49.20	
W100811	1103	49.02	
W101602	1289	57.29	
W101307	1228	54.58	

Synthesizing the findings finally, the comparison between the cubic compressive strength of full-graded and wet-sieved concrete can be seen in Table 11.31.

11.3 DISCUSSIONS ON EXPERIMENTAL RESULTS

From the dynamic and static flexural tests on the two different grades of dam concrete, and analyzing the flexural characteristics, the following conclusions can be summarized as follows.

11.3.1 THE RATIO FOR STATIC STRENGTH OF FULL GRADED CONCRETE TO WET-SIEVED CONCRETE

1. Static cubic compressive strength

   The strength grades of dam concrete are determined by the cubic compressive strength of wet-sieved concrete specimens. For the multigraded dam concrete, as the coarse aggregate larger than 40 mm and the cement mortar bonded on such aggregate are sieved in the wet-sieved concrete

Table 11.28 The 180 d Cubic Compressive Strength of Full-Graded Concrete

Number of Specimen	Ultimate Load (kN)	Compressive Strength (MPa)	Average of Strength (MPa)
F100801	8233	40.66	40.37
F100802	8098	39.99	
F100803	8779	43.35	
F100901	7220	35.65	
F100902	8165	40.32	
F100903	8152	40.26	
F101001	8023	39.62	
F101003	8671	42.82	
F101103	9049	44.69	
F101202	7051	34.82	
F101301	8894	43.92	
F101403	7922	39.12	
F101502	6714	33.16 (reject)	
F101503	10933	53.99 (reject)	
F101602	7578	37.42	
F101702	8935	44.12	
F101703	7537	37.22	
F101803	8496	41.96	

Table 11.29 The 360-d Cubic Compressive Strength of Wet-Sieved Concrete

Number of Specimen	Ultimate Load (kN)	Compressive Strength (MPa)	Average of Strength (MPa)
W101106	1152	51.20	52.34
W101208	1183	52.58	
W101303	1198	53.24	

specimens, the practical mixing proportion is changed. This effect on the strength characteristics is more significant than the specimen size effect. Therefore, in the limit state equation expressed by partial factors, as the standard value for the resistance in the dam materials, the full-graded cubic specimens need to be reflected by the real compressive strength. However, only in recent years, the strength testing on the full-graded specimens could be conducted in dam engineering. In testing on wet-sieved concrete specimens, the conversion factor 0.67 multiplying the strength of wet-sieved concrete specimens can be taken as the standard value for the resistance value of materials.

Table 11.30 The 360 d Cubic Compressive Strength of Full-Graded Concrete

Number of Specimen	Ultimate Load (kN)	Compressive Strength (MPa)	Average of Strength (MPa)
F101203	10622	52.45	53.35
F101302	11608	57.32	
F101501	7362	50.29	

Table 11.31 The 360 d Cubic Compressive Strength of Full-Graded Concrete

Concrete Type	Age (d)			
	28	90	180	360
Full-graded (MPa)	/	34.71	40.37	53.34
Wet-sieved (MPa)	39.56	50.26	51.79	52.35
Full-graded/ Wet sieved	/	0.69	0.78	1.02

Based on the experimental results, the average 180 d cubic compressive strength for three-graded and four-graded concrete specimens in Xiaowan arch dam and Dagangshan arch dam is 47.81 and 51.79 MPa, respectively. Considering the guarantee rate of 80%, the measured strengths have exceeded the strength for grade $R_{180}400$, and $R_{360}360$. The cubic compressive strength ratio for full-graded to wet-sieved concrete of Xiaowan and Dagangshan arch dam is 0.87 and 0.78, respectively. This is close to the value of the Xiluodu arch dam reported (Zhu et al., 2010). The cubic compressive strength ratio of full-graded to wet-sieved concrete specimens is larger than 0.67. This can be thought that in the determination of the standard value, the size effect between specimen and dam concrete, the transportation influence of dam concrete, and the vibrated variation between laboratory specimen and dam concrete, and so on, should be considered.

In the designed code for concrete dam, for the determination of the strength grade in dam concrete, the determination of the standard value by multiplying the factor of 0.67 by the specimen measured value is relatively safe.

2. Static flexural strength
The 180 d designed static flexural strength ratio of full-graded to wet-sieved concrete is 0.58 and 0.46, respectively.

Besides, as reported by the literature (Zhu et al., 2006), the splitting tensile strength ratio of full-graded to wet-sieved concrete during 90 to 180 days varies in a small range, and approximately equals 0.70.

3. Dynamic flexural strength
The dynamic flexural strength ratio of full-graded to wet-sieved concrete for Xiaowan and Dagangshan arch dam after 180 d is 0.49 and 0.50 under impact loading. For the triangle cyclic wave dynamic loading, the ratio is 0.54 and 0.49, respectively. Generally speaking, the dynamic flexural strength ratio of full-graded to wet-sieved concrete is approximately 0.5.

From the analysis on the experimental results, the strength ratio of full-graded to wet-sieved concrete, for the cubic compressive strength, splitting tensile strength, and flexural strength, is not the same value. Therefore, when the determination of the standard value for the resistance value of dam materials, it is needed to multiply the corresponding factor with the tensile and compressive strength of dam concrete. In the seismic design, the static compressive strength and dynamic tensile strength should be multiplied by the reduction factor of 0.67 and 0.5, respectively.

11.3.2 THE RATIO OF THE STATIC FLEXURAL TENSILE STRENGTH TO CUBIC COMPRESSIVE AND SPLITTING TENSILE STRENGTH

The experimental results showed that, in the failure zone of the pure bending in the flexural testing, the strain across the height of the normal section is linearly distributed, which satisfies the assumption for plane section in the structural mechanics. Thus, the flexural strength can be calculated by the load based on the beam theory.

The static flexural strength value of the wet-sieved concrete in Xiaowan and Dagangshan arch dam at the 180 d age is 6.19 and 6.46 MPa, respectively. The ratio of flexural strength to cubic strength for these two dams is 0.129 and 0.125, respectively.

The static flexural strength of full-graded concrete in Xiaowan and Dagangshan arch dam is 3.58 and 2.97 MPa, respectively. The ratio of flexural strength to cubic compressive strength is 0.086 and 0.074, respectively.

Obviously, the ratio of flexural strength to cubic compressive strength of full-graded concrete specimens is lower than that of wet-sieved specimens.

Based on the experimental results from the 6 in. diameter dam concrete cores, the US Bureau of Reclamation established the relationship formula between splitting tensile strength f_{st} and cubic compressive strength f_c as $f_{st} = 1.7 f_c^{2/3}$ (Bureau of Reclamation US, 2006). The apparent splitting tensile strength is taken as the tensile strength, the value is 1.33 times the measured results, and is approximately equal to the flexural strength f_{mr}. Based on the designed 180 d cubic compressive strength of wet-sieved concrete for Xiaowan and Dagangshan arch dams, the flexural strength calculated from this equation is 5.79 and 6.10 MPa, respectively. These calculated values are close to the experimental value for wet-sieved concrete. Based on the designed 180 d cubic compressive strength of full-graded concrete for Xiaowan and Dagangshan arch dams, the calculated flexural strength is 3.98 and 6.10 MPa, respectively, and is higher than the experimental values of 3.58 and 2.97 MPa, respectively, for the full-graded concrete specimens.

The splitting tensile tests for the wet-sieved concrete specimens for Dagangshan arch dam are also carried out, and the static splitting tensile strength is 3.56 MPa. Based on the formula proposed by the US Bureau of Reclamation, the calculated cylinder splitting tensile strength is 4.59 MPa, which is much higher than the experimental value. And the experimental value of flexural strength to splitting tensile strength is 1.81 and is also larger than the value calculated by the US Bureau of Reclamation, which is 1.33. This may be due to the difference between the cubic and cylinder specimens, and the difference between the wet-sieved specimens and dam cores. Some further investigations should be carried out to get more experimental data and do more deep analysis.

From the preceding demonstration, it can be thought that, for important high concrete dams, the determination of tensile strength should be conducted with the flexural tests on full-graded specimens. In the current seismic design code for dams, for the project without carrying out the flexural tests on full-graded concrete specimens, the standard value for wet-sieved concrete should be taken as 10% of the cubic compressive strength.

11.3.3 STRAIN RATE EFFECTS ON FLEXURAL STRENGTH AND THE EFFECTS OF DYNAMIC LOADING MODE

Under the variable triangle cyclic dynamic loading similar to earthquake motion, the designed 180 d dynamic flexural strength of full-graded concrete for Xiaowan and Dagangshan arch dams is 4.19 and 3.49 MPa, respectively, and the flexural strength of wet-sieved concrete is 7.80 and 7.17 MPa, respectively. The strain rate effects are the ratio of dynamic to static flexural strength. The value is 1.17 and 1.18 for full-graded concrete, and 1.26 and 1.11 for wet-sieved concrete. Generally, the dynamic flexural strength is less than 20% larger than static flexural strength.

In the impact dynamic loading, the designed 180 d dynamic flexural strength for Xiaowan and Dagangshan arch dams are 4.20 and 4.18 MPa, respectively, and the dynamic flexural strength for wet-sieved concrete are 8.54 and 8.42 MPa, respectively. The strain rate effects are the ratio of dynamic flexural strength to static flexural strength. For full-graded concrete, the value is 1.17 and 1.41, and for wet-sieved concrete, the value is 1.38 and 1.40. Generally, the dynamic flexural strength is 30% larger than static flexural strength.

Because of the low cyclic amplitude fatigue effect under triangle cyclic loading, the dynamic flexural strength of concrete is lower than the value obtained under impact loading mode. For the 180 d full-graded concrete of Xiaowan and Dagangshan arch dams, the dynamic flexural strength ratio under triangle cyclic dynamic loading to the impact loading is 1.0 and 0.83, respectively. For the wet-sieved concrete, the corresponding ratio is 0.91 and 0.85. Generally, the dynamic flexural strength under triangle cyclic dynamic loading is smaller 10–15% than that of impact loading.

The dynamic compressive and tensile strength of dam concrete used by the US Bureau of Reclamation is 20% and 50% higher than that of static value. The growing amplitude for the tensile strength is obviously a little too large. This mainly refers to the splitting tensile testing of dam cores under impact loading. The above testing results from the full-graded concrete showed that the dynamic flexural strength was higher 20% than the static strength. The current seismic design code regulated that the dynamic compressive and tensile strength is 30% higher than that of static value, this mainly refers to the experimental results from the wet-sieved specimens under impact loading. Considering the cyclic effect of earthquake motion is more close to the experimental results of full-graded concrete specimens, the increasing amplitude of 20% for dynamic tensile and compressive strength is more suitable.

11.3.4 THE EFFECT OF INITIAL LOADS ON DYNAMIC FLEXURAL STRENGTH

Figure 11.34 summarized the effect of initial loads on the flexural strength of full-graded and wet-sieved concrete for the Xiaowan and Dagangshan arch dams. There is some difference in the influence of initial static loads and dynamic loading modes on full-graded and wet-sieved concrete for two dams. For lack of experimental data, it is difficult to get the definite rules for these different factors. But from the general tendency, when the initial static loads do not exceed 80% of the maximum loads, the initial loads would induce the increase of flexural strength. Only the strength of wet-sieved concrete for Dagangshan arch dam under impact loading, and the wet-sieved concrete for Xiaowan arch dam under triangle dynamic loading are decreased.

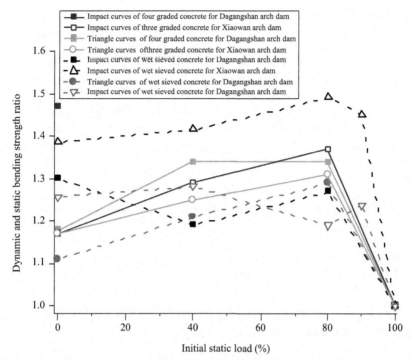

FIGURE 11.34 Impact of Initial Static Load on Dynamic and Static Bending Strength Ratio

Under dynamic loading, the mechanisms for damage and strain rate effects are very complicated, and there is no consistent explanation. The preliminary analysis thought that, for concrete during loading, the damage induced by the development of microcracks results in the decrease of strength and increase of deformation. Under dynamic loading, the increase of deformation would strengthen the strain rate effects and make the dynamic strength increase. Under a certain initial static loading, when the strain rate effects exceed the damage evolution effects, the dynamic strength would increase. However, when the initial static loads exceed the 80% of maximum loads, the damage develops quickly, and the influence exceeds the strain rate effect under earthquake loading, thus the dynamic strength decreases quickly. The explanation to this experimental result has been verified in the three-dimensional mesomechanical analyses for specimens considering damage evolution and strain rate effects (seen in Chapter 15), but it still needs to be further investigated.

From the viewpoint of engineering, as the dams have relatively large safety margins, with the static load much less than the 80% of maximum load, thus for the common engineering project without carrying out the flexural tests on full-graded flexural tests, the seismic design without considering the initial loads effect is relatively safe.

11.3.5 THE INFLUENCE OF AGE ON DAM CONCRETE STRENGTH

The strength of a dam increases with age, but the test results on the change laws for the ratio of strength for full-graded to wet-sieved concrete is relatively scarce. The earlier few testing results on the Dagangshan arch dam show that the cubic compressive strength of full-graded and wet-sieved concrete become close as the age increases, and the ratio for the two was 0.68, 0.78, and 1.0 at the age of 90 d, 180 d, and 360 d, respectively (see Table 11.31). However, as reported by Zhu et al. (2006), the experimental results reported for the Xiluodu arch dam showed that the compressive ratio of full-graded to wet-sieved concrete ranges from 0.85 to 0.88; this is close to the results of 0.68 and 0.78 for Dagangshan arch dam. But the experimental data for 360 d are missing for Xiluodu arch dam concrete.

The investigation on the influence of long age on the mechanical characteristics of dam concrete has a significant meaning for the seismic design for high concrete dam, because a strong earthquake usually happens after a long time of the engineering project running. But the investigations on this aspect are not enough. As the influence of raw materials and mix proportions, the influence of long age on the mechanical characteristics of concrete for each dam would have a certain difference. It is necessary to accumulate more experimental data and investigation for considering the long age effects on the dam concrete strength. Therefore, the experimental researches on the practical mechanical characteristics of dam concrete in the existing dams should be strengthened, and this is also important for the durability evaluation of old dams.

It is a pity that the experimental data of long age on the dynamic to static flexural strength ratio of full-graded and wet-sieved concrete are absent.

11.3.6 THE ANALYSIS OF THE CHARACTERISTICS OF FLEXURAL DEFORMATION AND THE VALUE OF DYNAMIC ELASTIC MODULUS

From the flexural stress–strain curves from the flexural tests on dam concrete, there is a linear relationship in a large stress range, and the nonlinear characteristics only appear near the peak stress, which indicates the delay of internal cracking developments. The yield limit for the dynamic flexural strength of dam concrete is larger than 2/3 of maximum tensile strength. Besides, some residual strain is shown under the cyclic action of the variable triangle wave.

The flexural tests on Xiaowan and Dagangshan arch dam showed that, whether the full-graded or wet-sieved concrete is used, with or without the initial static loads and under different loading modes, the change for the dynamic–static ratio for flexural elastic modulus and Poisson's ratio is not obvious. The change of the maximum flexural tensile value is also not large. Generally, the maximum flexural tensile value of wet-sieved concrete is a little higher than that of full-graded concrete. The maximum flexural strain of dam concrete for Xiaowan arch dam is higher than that for Dagangshan arch dam; this may be related to the used aggregates and the mix proportions of concrete.

It is needed to point out that, in the current dam design, the elastic modulus is always assumed to be constant when the stress does not achieve the peak strength of dam concrete. Considering the creep effects of concrete under long-term static action, the static elastic modulus is lower than the measured elastic modulus in the laboratory. The US Bureau of Reclamation takes the designed elastic modulus as two-thirds of the laboratory value. Corresponding to the impact dynamic loading under the earthquake action, the loading time is only a fraction of a second and the action time for the real strong earthquake is also only 10 s. Thus, the corresponding dynamic elastic modulus would be higher than the static value. It can be thought that, according to the experiments of the USA Bureau of Reclamation, the dynamic and static elastic modulus can be assumed to be the same from the statistical meaning.

The current seismic design code for hydraulic structures regulates that the dynamic elastic modulus of concrete is 30% higher than the static value. The static elastic modulus here refers to the value considering the long-term creep effects. Therefore, the increase of dynamic elastic modulus is not the result from the strain rate effects on dynamic strength, but is the difference of long-term creep effects and the instantaneous loading in the laboratory. Although these two are all increased by 30%, the concepts are different. For the common engineering, if without carrying out the dynamic flexural tests, the regulations for the dynamic elastic modulus of dam concrete in the current seismic codes for hydraulic structures are basically feasible.

THE EXPERIMENTAL RESEARCH ON THE DYNAMIC AND STATIC MECHANICAL CHARACTERISTICS OF DAM CONCRETE AND THE CONSTITUTIVE MATERIALS

Currently, the outstanding problem in the seismic design of high concrete dams is to investigate the damage and failure process under a strong earthquake. From the viewpoint of engineering, based on the characteristics of high concrete dams, the nonlinear finite element analysis on the damage evolution rules for dam concrete is the reasonable method. Among them, the most important thing is to establish a suitable model and determine the reasonable parameters, especially the constitutive relationship for dam concrete materials, including the full stress–strain curves with the softening stage. As the tensile strength of concrete is far lower than the compressive strength, the damage is mainly resulted from the initiation and development of initial microcracks. The cracking mainly depends on the tensile strength and deformation of materials, thus the experimental data on the dynamic and static direct tensile stress–strain curves of concrete are the most important fundamental data and premise. But the experimental testing to determine the direct tensile stress–strain curves of concrete materials is very difficult, especially for the descending part of stress–strain curves after the peak point (soften stage). There are many problems to be solved in the experimental techniques; therefore, the experimental results are very little, especially for the experimental results on the dynamic direct tensile stress–strain curves of dam concrete materials. This has become the important hindrance for deep investigation.

Besides, for the investigation on the damage and failure process and mechanisms of the internal structure of dam concrete materials and the explanation for the macro experimental results, the dam concrete should be viewed as a three-phase composite material with multigraded aggregate, cement mortar, and the interface between the mortar and aggregate, and the meso-mechanical analysis should be carried out. The analysis with this kind of meso-mechanics should be based on the dynamic mechanical properties of aggregate, cement mortar, and the interface between these two materials, while the relevant experimental results are very little, and there are no standard testing methods.

Combined with the mentioned factors, the dynamic mechanical characteristic testing on the composite materials of concrete and the experimental techniques for the direct tensile stress–strain curves of concrete have been summarized in this chapter. As the experimental techniques are very complicated, they should be improved further. The obtained preliminary experimental results are needed to be further verified and investigated deeply in the future.

12.1 THE EXPERIMENTAL TECHNIQUES FOR THE DYNAMIC AND STATIC DIRECT TENSILE STRESS–STRAIN CURVES OF CONCRETE

12.1.1 THE STABLE FRACTURE CONDITIONS FOR QUASIBRITTLE MATERIALS

To measure the descending part of curves, it is needed to assure that the stiffness of the machine is large enough. If the stiffness of the machine is smaller than that of the specimens, the recovered deformation would be larger than the tensile deformation for specimens. Thus, the released elastic energy by the recovery deformation of machines would be far beyond the needed energy inducing the stable fracture of the specimens, and the specimen fails suddenly at the peak loads. This results in the failure for obtaining the descending part of stress–strain curves.

For achieving the stable fracture, the derivative of the total energy assumption with respect to any deformation δ should be positive. When the specimen is loaded under the force F, the energy induced by the deformation is W_s, the storage elastic energy for the machines during the synchronous loading is W_M, thus the total deformation energy for these systems (specimens and machine) is

$$W = W_s + W_M = \int_0^\delta F(\delta) \mathrm{d}\delta + \frac{F^2}{2K_M} \tag{12.1}$$

For obtaining the stable fracture, the derivative of the total energy assumption with respect to any deformation δ should be positive, which is

$$\frac{\mathrm{d}W}{\mathrm{d}\delta} = \frac{\mathrm{d}(W_s + W_M)}{\mathrm{d}\delta} = F + \frac{F}{K_M} \frac{\mathrm{d}F}{\mathrm{d}\delta} \tag{12.2}$$

For obtaining stable fracture, $K_M > -(\mathrm{d}F / \mathrm{d}\delta)$, that means the stiffness of the machine should be larger than the slope in any point of the load deformation curves for the specimens. For the ascending stage of the stress–strain curves of the specimens, this formula is always satisfied. The key lies in that the stiffness of the machine should be larger than the slope in any point of the descending part of full stress–strain curves.

For the quasibrittle materials like concrete, as the sudden dropping of the descending part of the stress–strain curves due to the instant damage, the slope of the load deformation in the descending part would be large. Therefore, it is needed to increase the stiffness of the machine to obtain the full stress–strain curves of concrete. Currently, the direct tensile tests usually adopt the hydraulic servo machine, and the experimental techniques for dynamic direct tensile stress–strain curves of concrete is complicated, which need to be solved in a further study.

12.1.2 THE EXPERIMENTAL TECHNIQUES FOR THE DYNAMIC UNIAXIAL TENSILE STRESS–STRAIN CURVES OF CONCRETE MATERIALS

12.1.2.1 The experimental research on the stiffness of the loading system

The MTS322 hydraulic servo machine is adopted for the loading equipment. The stiffness of the loading system includes the two parts of machine and fixture. The fixture is consisted of the spherical hinge and steel cap bonded in the end of specimens. The designed experimental program is shown in Table 12.1, and a part of the experimental instrument is illustrated in Fig. 12.1.

Table 12.1 Test Scheme About Stiffness Influence of Loading System

Scheme	1#	2#	3#	4#	5#	6#	7#
The form of spherical hinge	None	Large spherical hinge+steel cap	Large spherical hinge+without steel cap	Small spherical hinge+steel cap	Small spherical hinge+without steel cap	Without spherical hinge+steel cap	Without spherical hinge+without steel cap

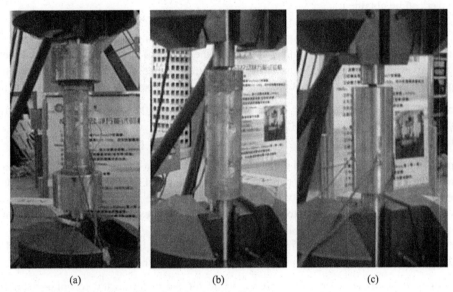

(a)   (b)   (c)

FIGURE 12.1 Axial Tensile Device of Standard Component Test

(a) Large spherical hinge+steel cap (b) Without spherical hinge+steel cap (c) Without spherical hinge+without steel cap.

The steel standard specimens (materials of No. 45 steel) with the same diameter as the concrete specimen is used, and a structural adhesive is used to connect the steel cap and specimen. The diameter for the steel cap is 68 mm, and the thickness is 35 mm; the size for big spherical hinge is 2.5 times of that for the small one.

The experimental procedures are shown as follows:

1. The steel bar with 40 mm diameter is carried out with direct tension, and the measured deformation of the actuator is the deformation for the entire loading system, then the deformation–load relationship for the experimental machine systems can be obtained.
2. The direct tension is conducted after the two spherical hinges are contacted without gap. The tightness for the spherical hinge is adjusted before every tension. Taking the measured deformation of the actuator as the deformation of the spherical hinge and the displacement of the machine, the deformation–load relationship can be obtained.

3. The standard specimens (without bonding steel cap) are connected with the machine by the 40-mm-diameter bolt. Taking the measured displacement of the actuator as the deformation of the standard specimens and the displacement of the machine, the deformation–load relationship for the standard specimens can be obtained.
4. The standard specimen with bonding steel caps in the two ends (including the glue layer) is connected with the machine by using a 40-mm-diameter bolt. Taking the measured displacement of the actuator as the deformation of the specimens, the displacement of the machine and the deformation of the glue layer, then the deformation–load relationship between the steel cap and glue layer can be obtained.
5. The standard specimens with bonding steel caps in two ends (including the glue layer) are connected with the entire loading system by the spherical hinge. Taking the measured displacement of the actuator as the displacement of the spherical hinge, the displacement between the steel cap and glue layer, the deformation of the standard specimens and the displacement of the machine, then the total deformation including all parts can be obtained.

The relationships between each part of the loading system and the loads can be obtained from the tests, and the results are shown in Fig. 12.2 (Table 12.2).

The test results showed that

1. The proportion for the steel cap and glue layer takes the smallest proportion, except the small deformation for the glue layer. This part has insignificant impact on the stiffness of the system.
2. There are no linear relationships between the deformation of the spherical hinge and loads, and it takes the largest part of the system total deformation. The stiffness of the large spherical hinge is increased 50% compared to the small spherical hinge.
3. The calculated value for system stiffness has a major difference from the measured value. This may be due to the difference in the installation process of the spherical hinge.

The testing results have important reference value for the experimental techniques for the full stress–strain curves of concrete. Through the choosing of the suitable experimental clamp methods,

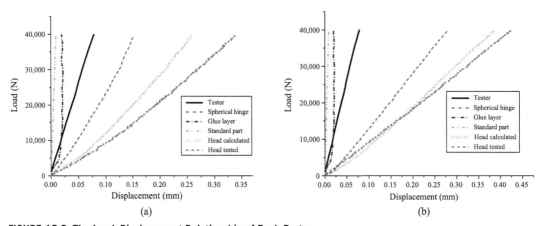

FIGURE 12.2 The Load–Displacement Relationship of Each Parts

(a) Large spherical image (b) Small spherical hinge.

Table 12.2 The Analysis Testing on the Stiffness of the Loading System

The Form of Spherical Hinge	Steel Standard Parts	Testing Machine	Spherical Hinge (Including Anchor Rod)	Steel Cap and Glue Layer	The Whole Loading System (Including Test Machine, Spherical Hinge, Glue Layer)	
					Calculation	Test
Small spherical hinge	4450	507.5	171.3	∞	128.1	95.8
Large spherical hinge	4450	507.5	262.3	∞	172.9	127.9

regulating the procedure, and increasing the stiffness of the system, the requirements for full stress–strain curves can be satisfied.

12.1.2.2 Making of the specimen

1. *The choosing of the specimen shape*: In our country, the shapes of concrete and cement mortars used in the direct tensile tests are usually prismatic or dumbbell shaped, while the specimens used in Europe and the United States take the cylinder shape. Each shape has its own advantages and disadvantages. The cylinder and prisms are easy for making specimens, while the dumbbell shape is not. The materials for prisms and dumbbell specimens have large discreteness, while the cylinder specimens cored from one matrix has little discreteness, and is easy for analysis.

   For comparison of the various section shape effects, the elastic three-dimensional finite element calculation is carried out with the ANSYS software. The 230-mm height cylinder (100-mm diameter) and prism (side length 100 mm) under the bonded tension are analyzed for the stress distribution. When the concentrated load is applied on the steel cap, and the average stress is 1.5 MPa, the axial stresses across the specimen end and in the axial direction are shown

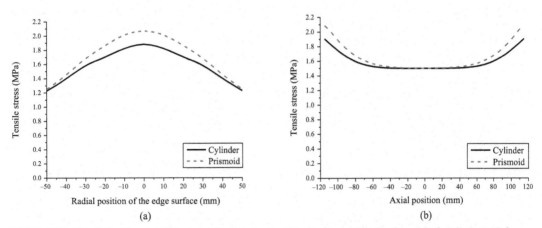

(a)   (b)

FIGURE 12.3 The Finite Element Calculated Results for the Tensile Stress Distribution of Cylindrical and Prism Specimen

(a) Radial distribution of the tensile stress in the specimen end (b) Distribution of the tensile strength in the axial line.

Table 12.3 The Influence of Clamping Way of Specimen on the Test Success Rate

Material	Shape	Bonded (Thickness of Steel Cap = 15 mm)		Bonded (Thickness of Steel Cap ≥1/3 D)		Preburied	
		Total	Success rate	Total	Success rate	Total	Success rate
Concrete	Cylinder	6	40%	6	84%	/	/
	Plane dumbbell shape	3	0%	/	/	10	60%
Mortar	Cylinder	6	50%	6	100%	/	/
	Plane dumbbell shape	3	0%	/	/	6	67%
Granite	Cylinder	3	0%	3	100%	/	/

in Fig. 12.3. The analyzed results showed that, for the cylinder specimens, the uneven degree induced by the end restraint is smaller than that for the prism specimens, and the length of the uniform stress region is relatively long. The test results showed that the stress condition of the cylindrical specimens is relatively fine, and this is helpful for choosing a suitable specimen clamping method and increasing the success rate.

2. *Specimen clamping modes*: From the analysis, it can be known that the specimen clamping mode affects the success rate of the tests. Compared with the bonded mode, the effective fracture zone of the preburied mode is much smaller. Therefore, the tests with the bonded mode have higher success rate. For reducing the failure between the specimen and the steel cap, the uneven degree of the stress distribution on the glue surface should be lowered. For this purpose, the influence of the thickness of the steel cap on the end restraint is investigated by experiment and finite element methods.

For comparison of the success rates for different specimen shapes and clamping modes, each specimen in Table 12.3 was fabricated and tested in direct tension in a normal mechanical machine, and the specimen end was connected to the machine by the spherical hinge. The material used for the steel cap is No. 45 steel.

The testing results showed that for the specimens cored from the large samples, the discreteness for strength is smaller than the other construction methods. The success of the experiments is taken when the distance between the end and the failure section of the specimen is larger than one-third of the diameter or the length of the side. In Table 12.3, the success rate for the cylinder bonded with the steel cap of thickness larger than one-third of specimen diameter is high, even up to 100%.

The elastic three-dimensional finite element analysis on the influence of steel cap thickness on the end restraint is carried out. The diameter for the steel cap takes 100 mm, and the thickness is 20, 30, and 40 mm, respectively. The calculated results are shown in Fig. 12.4. The test results showed that the concentrated degree for the stress in the specimen end decreased as the thickness of the steel cap increased, and at the same time, the length of the uniform stress region also increased.

When the strength of the material is relatively high, the thickness for the end of the steel cap should be increased properly, and the finite element analysis can be carried out if it is needed.

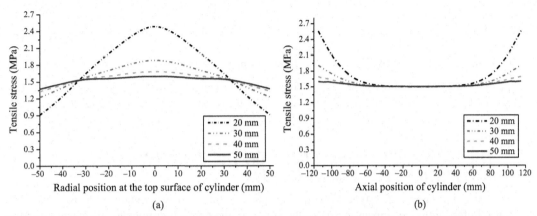

FIGURE 12.4 The Calculated Results for the Influence of the Thickness of the Steel Cap on the End Restraint of Specimens

(a) The radial distribution of tensile stress on the edge surface (b) The axial distribution of tensile stress.

3. *Specimen processing*: In the processing of the specimens, a strict control should be conducted by a core-drilling and cutting machine, and the flatness of the surface of the cylinder and the perpendicularity of two end surfaces assured. For concrete and mortar specimens, the 2-cm-thick laitance should be cut, and only the roughening of the surface is not enough.

   The steel cap is bonded to the end of the specimens, and its diameter should not be smaller than the diameter of the specimen, and 2 cm larger than the diameter. Otherwise, the centering cannot be easily achieved.

   The bond materials are chosen as the epoxy resin structural glue, and the bond strength should be larger than two times the static tensile strength of the materials to satisfy the requirement by the dynamic strength. In the tests the strength for the structural adhesive is approximately 20 MPa. In the bonding process the glue layer should be made uniformly, and the thickness ranges from 0.3 mm to 0.8 mm. To ensure the glue layer without slip before consolidation, the specimen should be fixed with the steel cap by the glue. The bond of the steel cap cannot be carried out simultaneously; that is, the structural glue at one end should be bonded after the consolidation of the other end.

4. *Loading modes*: The two ends of the specimens are connected with the spherical hinges by the center screw in the steel cap. Adjusting the clearance between the specimen and the screw, the centering is conducted. Meanwhile, the spherical hinge should not be tightened to hinder its free rotation. The unbalanced force is adjusted automatically. The 20% of maximum loads are preapplied, and the eccentricity is calculated by reading of the strain gauge, which should not be larger than 5%; otherwise, the adjustment should be made again, and the examination of each instrument is conducted. Then the load is dropped to zero, and the spherical hinge is tightened and reloading starts.

   For carrying out the entire process of the tensile experiments for the concrete materials, the strain-controlling mode should be adopted.

5. *Data acquisition*: For measurement of the strain, the resistance strain gauge and extensometer are used, which are arranged with equal spacing around the circumference in the middle of the

specimen. The length of the strain gauge should be larger than three times the diameter of the aggregate, and smaller than three-fourth height of the specimens. Through the self-designed hoop, the extensometer is fixed on the two sides of the specimens. When the specimen fractures occur even outside the strain gauge, the data for the softening stage also can be obtained.

The measurement of the deformation is recorded timely by the dynamic strain meter. Based on the numbers of experiments, the value of 20 kHz can satisfy the requirement of dynamic loading.

12.2 THE EXPERIMENTAL RESEARCH ON THE DYNAMIC TENSILE CHARACTERISTICS FOR CEMENT MORTAR

The cement used was the 32.5-type Portland cement, and the sand was middle sand (particle size < 5 mm). Referring to the cement matrix in the concrete, the mix proportions for the cement mortar is W:C:S = 1:2:4. The large samples were constructed first and the cored specimens were made from them.

The loading system adopts the MTS322 electrohydraulic servo machine, and the loadings were carried out with the deformation-controlled modes. The measuring of deformation was recorded timely with the dynamic strain meter.

The monotonic loading with four loading rates was designed, and the ranges for the strain rate are from 10^{-6} s^{-1} to 10^{-3} s^{-1}.

The test results showed the following:

1. The strain rate has no influence on the tensile strength of mortar.
   During experiments, most of the specimens were fractured during the strain rate ranges, and the success rate was above 80%. The photos for the failure specimens are shown in Fig. 12.5. The strain rate used is the average strain rate in the ascending part of the stress–strain curves, which is obtained by the strain at the peak stress divided by the time for achieving peak value. The average strain rate and the maximum tensile strength for each specimen are shown in Table 12.4.

   The experimental results showed that the maximum tensile strength of cement mortar increased with the increasing of the strain rate. The standard deviation for the experimental results is small. The cement mortar is the main composite material for concrete. Referring to the formula (12.3) in the European Concrete Code (CEB), and fitting the experimental data, the

FIGURE 12.5 The Failure Photos for the Cement Mortar

Table 12.4 The Tensile Strength of Cement Mortar Under Different Strain Rates

Working Conditions	Total Time (s)	Descending Stage Time (s)	Ascent Stage Average Strain Rate (s^{-1})	Average Strain Rate (s^{-1})	Tensile Stress (MPa)	Average Stress (MPa)	Standard Deviation (MPa)	Dynamic Strength Increase Factor (Ratio to 10^{-6}s^{-1})
1	11.99	0.005	3.7×10^{-6}	3.9×10^{-6}	2.19	2.31	0.10	1
	11.82	0.003	4.0×10^{-6}		2.44			
	11.54	0.004	4.1×10^{-6}		2.29			
2	1.598	0.0024	3.6×10^{-5}	4.7×10^{-5}	3.50	3.40	0.21	1.47
	0.900	0.0014	7.2×10^{-5}		3.06			
	1.548	0.0012	4.0×10^{-5}		3.61			
	1.557	0.0012	4.0×10^{-5}		3.46			
3	0.167	0.0015	4.1×10^{-4}	3.9×10^{-4}	3.49	3.65	0.29	1.58
	0.165	0.0011	4.4×10^{-4}		4.07			
	0.161	0.0014	3.0×10^{-4}		3.00 (Reject)			
	0.167	0.0014	3.2×10^{-4}		3.40			
4	0.039	0.0011	1.8×10^{-3}	2.6×10^{-3}	4.56	4.49	0.07	1.94
	0.031	0.0013	1.3×10^{-3}		3.2 (Reject)			
	0.033	0.0012	1.8×10^{-3}		4.52			
	0.020	0.0011	4.1×10^{-3}		4.4			

empirical formula for dynamic tensile strength of cement mortar under low strain rate is obtained as Eq. (12.4), with the correlation factor 0.96. The equations are as follows:

$$\frac{f_{td}}{f_{ts}} = \left(\frac{\dot{\varepsilon}}{\dot{\varepsilon}_s}\right)^{1.016\delta} \tag{12.3}$$

$$\delta = \frac{1}{10 + \dfrac{6f_c'}{f_{co}'}}$$

where, f_{td} is the dynamic tensile strength of concrete; f_{ts} is the static tensile strength of dry concrete; $\dot{\varepsilon}$ is the dynamic strain rate, ≤ 30 s^{-1}; $\dot{\varepsilon}_s$ is the static strain rate, 3×10^{-6} s^{-1}; f_c' is the cylindrical compressive strength of concrete; f_{co}' is the strength parameter, 10 MPa.

$$\frac{f_{mtd}}{f_{mts}} = \left(\frac{\dot{\varepsilon}}{\dot{\varepsilon}_s}\right)^{2.213\delta} \tag{12.4}$$

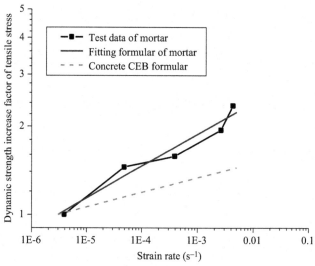

FIGURE 12.6 Dynamic Strength Increase Factor of Tensile Stress

$$\delta = \frac{1}{10 + \dfrac{6 f_{mcu}}{f_{mcc}}}$$

where f_{mtd} is the dynamic tensile strength of cement mortar; f_{mts} is the static tensile strength of cement mortar; $\dot{\varepsilon}$ is dynamic strain rates, $\leq 5 \times 10^{-3} \, s^{-1}$; $\dot{\varepsilon}_s$ is the static strain rate, $3 \times 10^{-6} \, s^{-1}$; f_{mcu} is the cubic compressive strength of cement mortar; f_{mcc} is the reference strength, 10 MPa.

The measured compressive strength (side length 70.7 mm) is 17.4 MPa. As the strain rate for the experimental data is $\dot{\varepsilon} \leq 5 \times 10^{-3} \, s^{-1}$, the Eq. (12.4) can be suitable for this strain rate range.

The experimental results and the dynamic increase factor for cement mortar and concrete calculated by Eq. (12.4) are illustrated in Fig. 12.6. These two are basically consistent, and the strain rate effects for cement mortar are more obvious than that of concrete.

2. The influence of strain rate on the deformation characteristics of cement mortar.
   a. *Elastic modulus*: Based on the experimental results, the brittleness for mortar is larger than that of concrete. The ascending part of stress–strain curves before 80% of the peak stress is linear. After that a bit of nonlinearity exists, and then the sudden fracture happens without any obvious presage. According to the code for testing of hydraulic concrete (DL/T 5150-2001), the secant modulus at the 50% of the strength can be taken as the representative value for elastic modulus. The elastic modulus for each specimen is shown in Table 12.5.

   The test results showed that, when the strain rate ranges from $3.9 \times 10^{-6} \, s^{-1}$ to $2.6 \times 10^{-3} \, s^{-1}$, the elastic modulus of cement mortar is increased as the strain rate increased, which is shown in Fig. 12.7.

Table 12.5 The Tensile Elastic Modulus of Cement Mortar Under Different Strain Rates

Working Conditions	Average Strain Rate (s⁻¹)	Tensile Elastic Modulus (GPa)	Mean Square Error (GPa)	Dynamic Increase Factor of Elastic Modulus
1	3.9×10^{-6}	53.0	0.9	1.00
2	4.7×10^{-5}	55.3	3.1	1.04
3	3.9×10^{-4}	57.0	0.7	1.08
4	2.6×10^{-3}	60.0	3.6	1.13

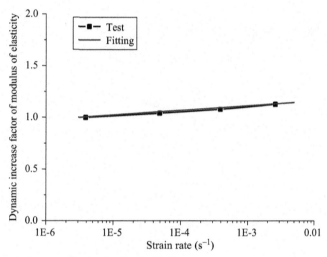

FIGURE 12.7 The Dynamic Increase Factor for Elastic Modulus

The data analysis for the elastic modulus of cement mortar is carried out, and the following equations are obtained:

$$\frac{E_{td}}{E_{ts}} = 1 + 0.0455 \lg \frac{\dot{\varepsilon}}{\dot{\varepsilon}_s} \tag{12.5}$$

where, E_{td} is the dynamic tensile elastic modulus of cement mortar; E_{ts} is the static elastic modulus of cement mortar; $\dot{\varepsilon}$ is the dynamic strain rates, $\leq 5 \times 10^{-3}$ s⁻¹; $\dot{\varepsilon}_s$ is the static strain rate, 3×10^{-6} s⁻¹.

b. *Peak tensile strain*: The experimental results for the strain at the peak stress (abbre. peak tensile strain) are shown in Table 12.6.

The peak tensile strain increases as the strain rate increases. When the strain rate increases from 10^{-6} s⁻¹ to 10^{-5} s⁻¹, the peak tensile strain has a significant increase, then there is no obvious increasing of peak tensile strain with the increase of the strain rate, as shown in Fig. 12.8. The range for the peak tensile strain is from 40 με to 80 με.

Table 12.6 The Peak Tensile Strain and Ultimate Strain for Cement Mortar Under Different Strain Rates

Working Conditions	Average Strain Rate (s⁻¹)	Average of Peak Tensile Strain (με)	Standard Deviation (με)	Dynamic Increase Factor of Peak Tensile Strain
1	3.9×10^{-6}	47.7	0.41	1
2	4.7×10^{-5}	62.6	1.38	1.31
3	3.9×10^{-4}	70.3	8.0	1.47
4	2.6×10^{-3}	72.4	8.36	1.52

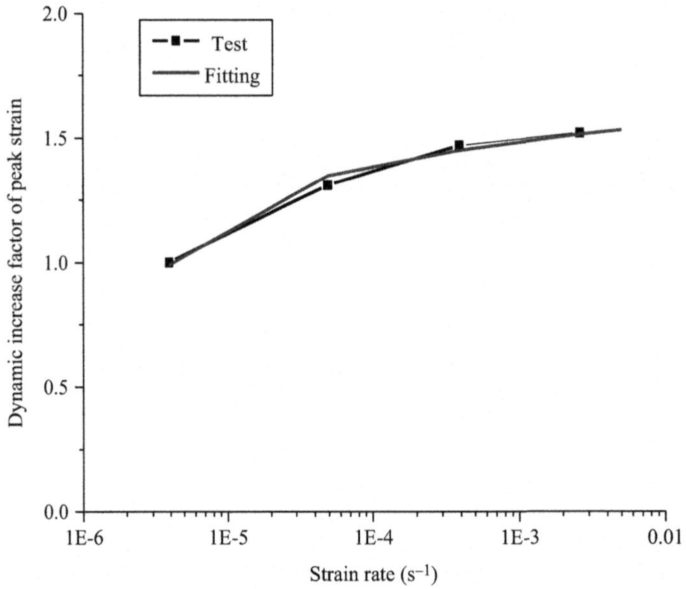

FIGURE 12.8 Dynamic Increase Factor for Peak Tensile Strain

The data analysis was carried out on the peak tensile strain of cement mortar, and the equations can be obtained as follows:

$$\frac{\varepsilon_{tpd}}{\varepsilon_{tps}} = 1.314 \left(\lg \frac{\dot{\varepsilon}}{\dot{\varepsilon}_s} \right)^{0.1309} \tag{12.6}$$

where ε_{tpd} is the dynamic peak tensile strain for cement mortar; ε_{tps} is the static peak tensile strain for cement mortar; $\dot{\varepsilon}$ is the dynamic strain rate, $\leq 5 \times 10^{-3}\,\text{s}^{-1}$; $\dot{\varepsilon}_s$ is the static strain rate, $3 \times 10^{-6}\,\text{s}^{-1}$.

  c. *Dynamic and static stress–strain curves*: The data for the stress and strain curves under different strain rates are obtained experimentally. The corresponding stress–strain curves are

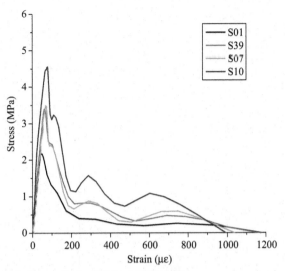

FIGURE 12.9 The Stress–Strain Curves Under Different Strain Rates

shown in Fig. 12.9. It can be seen from Table 12.4 that the time for the descending part under different strain rates is short and the strain rate is higher than the loading strain rates, which shows that the descending part of the stress–strain curves is not stable.

The research results showed that the tensile strength, elastic modulus, and peak tensile strain increased as the strain rate increased, while the increase for elastic modulus is not obvious.

12.3 THE EXPERIMENTAL RESEARCH ON DYNAMIC DIRECT TENSILE CHARACTERISTICS OF AGGREGATES

The loading instruments adopt the MTS322 electrical hydroservo machine, and the strain rate range could be up from 10^{-6} s^{-1} to 10^{-3} s^{-1}.

The granite materials were cored, cut, and buffed to the specimens with the diameter of 68 mm and the height of 160 mm. The machining accuracy could refer to paragraph two to five of Article 4.4.2 of the code of rock testing in hydroelectric engineering. The two ends of the specimens are connected with the 30-mm-thick steel cap (steel materials type 45) by the epoxy resin structural glue.

Two 5 mm × 100 mm strain gauges were bonded to the surface of the specimens to measure the longitudinal strain. Also, the 5 mm × 30 mm transversal strain gauges were used to measure the transversal strain.

The test results showed the following:

1. The influence of strain rate on tensile strength.
   The maximum tensile strength for the aggregate is shown in Table 12.7. The part of failure specimens is shown in Fig. 12.10, and the change of the dynamic strength increase factor with average strain rate is shown in Fig. 12.11. The results showed that the tensile strength of

Table 12.7 The Ultimate Tensile Strength of Granite Under Different Strain Rates

Working Conditions	Number of Specimen	Total Time (s)	Average Strain Rate at Ascent Stage (s^{-1})	Average of Strain Rate (s^{-1})	Tensile Stress (MPa)	Average Strength (MPa)	Increase Factor of Strength
1	A01	59.20	3.7×10^{-6}	4.4×10^{-6}	9.11	9.22	1
	A02	36.36	5.1×10^{-6}		9.33		
	A03	33.06	3.2×10^{-6}		6.60 (Reject)		
	A06	25.35	5.0×10^{-6}		7.94 (Reject)		
2	A04	3.302	5.5×10^{-5}	5.9×10^{-5}	9.38	9.40	1.02
	A05	3.294	6.3×10^{-5}		9.42		
	A13	2.431			5.59 (Reject)		
3	A07	0.280	4.1×10^{-4}	4.6×10^{-4}	7.61 (Reject)	10.54	1.14
	A08	0.340	6.2×10^{-4}		8.38 (Reject)		
	A09	0.417	4.7×10^{-4}		11.16		
	A35	0.346	4.4×10^{-4}		9.92		
4	A10	0.051	4.0×10^{-3}	3.9×10^{-3}	12.09	12.12	1.31
	A11	0.058	3.7×10^{-3}		12.20		
	A12	0.050	4.0×10^{-3}		12.07		
	A36	0.037	7.1×10^{-3}		6.64 (Reject)		

FIGURE 12.10 The Failure Photos for the Granite Specimens

the aggregate has a large discreteness, and in general the strength increased as the strain rate increased.

2. The influence of strain rate on deformation characteristics.

   a. *Tensile elastic modulus*: Based on the experimental results, the tensile elastic modulus of granite below 40% strength is linearly increased with the strain rate, and then the obvious nonlinearity shows. This is different from the deformation characteristics of mortar. The tensile elastic modulus for each specimen is shown in Table 12.8. The test results showed that the elastic modulus of aggregate is not sensitive to the change of strain rates.

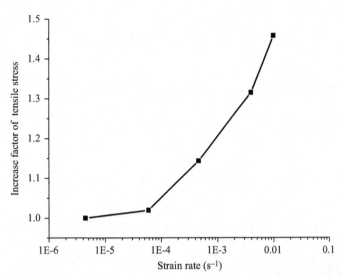

FIGURE 12.11 The Dynamic Increase Factor for Tensile Strength Versus Average Strain Rate

Table 12.8 The Tensile Elastic Modulus of Granite Under Different Strain Rates

Working Conditions	Average Strain Rate (s^{-1})	Average of Tensile Elastic Modulus (GPa)	Standard Deviation (GPa)
1	4.4×10^{-6}	80.0	5.35
2	5.9×10^{-5}	75.0	5.66
3	4.6×10^{-4}	81.3	2.52
4	3.9×10^{-3}	75.7	4.73

Table 12.9 The Peak Tensile Strain of Granite Under Different Strain Rates

Working Conditions	Average Strain Rate (s^{-1})	Average ($\mu\varepsilon$)	Standard Deviation ($\mu\varepsilon$)
1	4.4×10^{-6}	159	52
2	5.9×10^{-5}	194	17
3	4.6×10^{-4}	176	56
4	3.9×10^{-3}	218	26

b. *Peak tensile strain*: The experimental results for the tensile strain at peak stress (peak tensile strain) are shown in Table 12.9. The peak tensile strain increases as the strain rate increases, but the increase rule is not significant.

c. *Dynamic and static stress–strain curves*: The data on the stress and strain under the full loading process are obtained, and the calculated stress–strain curves are processed as shown in

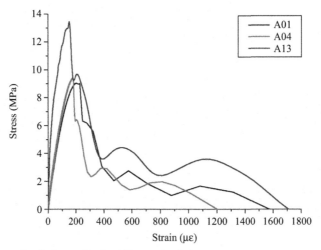

FIGURE 12.12 The Stress–Strain Curves Under Different Strain Rates

Fig. 12.12. Similar to the mortar, the time for the descending part for each loading condition is short, and the strain rate is higher than the loading strain rates. This phenomenon illustrates that the descending part of the stress–strain curve is not stable.

The experimental results show that the tensile strength and peak tensile strain of aggregate increase with the strain rates, and the rule for the elastic modulus is not obvious.

12.4 THE DIRECT TENSILE TESTS ON THE DYNAMIC MECHANICAL BEHAVIOR OF THE INTERFACE BETWEEN THE MORTAR AND AGGREGATE

The experiments adopt the cylindrical constructed specimens with aggregate at one end and mortar at the other end (if the two ends are mortar simultaneously, as the air entrains, the bond between mortar and aggregate would not be dense), the section of the aggregate is in the natural condition (reflecting the true interface characteristics), and the two ends are bonded by a steel plate with holes. The schematic figure for the specimen is shown in Fig. 12.13. The part of fracture specimen is shown in Fig. 12.14.

The experimental results are listed in Table 12.10, and the dynamic increase factor for the interface with the change of strain rate is shown in Fig. 12.15.

12.5 THE EXPERIMENTAL RESEARCH ON THE DYNAMIC TENSILE CHARACTERISTICS OF CONCRETE FOR DAGANGSHAN ARCH DAM

The mesomechanical analysis is the important method for revealing the dynamic mechanical behavior of mass concrete. However, the numerical analysis cannot proceed without the experimental data on the mechanical characteristics of concrete materials, and the relevant data are very seldom. Therefore,

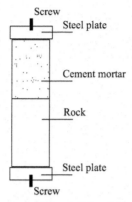

FIGURE 12.13 The Uniaxial Tensile Specimens with Bonded Steel Plates

FIGURE 12.14 A Part of Fractured Specimens

combining the raw materials, mix proportions, and the interface of the dam concrete for Dagangshan dam, the strain rate effects on dynamic strength are investigated by the direct tensile methods. The experimental instruments and methods are the same as illustrated previously. The loading strain rates have four conditions with strain rates ranging from 10^{-6} to 10^{-3} s^{-1}. The dynamic increase factor of the tensile strength of the mortar, aggregate, and the interface are listed in Table 12.11 and Figs 12.16 and 12.17.

Comparing the experimental results, the following conclusions can be obtained:

1. The tensile strength of the interface, mortar, and aggregate increased as the strain rate increased.
2. Under the same loading rate, the tensile strength for the aggregate was the largest, for the mortar the second, and for the interface the smallest.
3. Under the same loading rate, the DIF of the aggregate for the dynamic tensile strength was the largest, the mortar the second, and the interface the smallest.

Table 12.10 The Tensile Strength of Interface Under Different Strain Rates

Strain Rate (s^{-1})	Number	Ultimate Load (N)	Tensile Strength (MPa)	Average Strength (MPa)	Standard Deviation (MPa)	DIF	Status of Failure
10^{-6}	I32	8,942	2.46	2.20	0.30	1.0	Interface failure
	I33	8,670	2.39				Interface failure
	I06	8,683	2.39				Mortar failure
	I07	6,766	1.86				Interface failure
	I22	6,794	1.87				Mortar failure
10^{-5}	I03	8,344	2.30	2.30	/	1.05	Interface failure
	I04	5,469	1.51 (Reject)				Interface failure
10^{-4}	I05	9,309	2.56	2.73	0.23	1.24	Interface failure
	I09	10,482	2.89				Interface failure
10^{-3}	I13	13,048	3.59	3.52	0.37	1.60	Mortar failure
	I14	13,597	3.75				Interface failure
	I11	12,230	3.37				Mortar failure
	I12	11,242	3.10				Interface failure
	I15	11,029	3.04				Interface failure
	I16	14,534	4.00				Interface failure
	I17	13,883	3.82				Interface failure

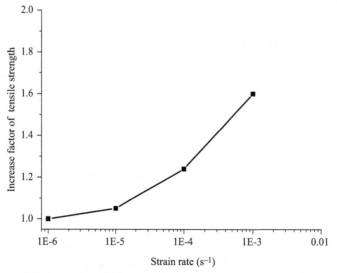

FIGURE 12.15 The Dynamic Increase Factor of the Tensile Strength of Interface Versus the Average Strain Rate

Table 12.11 Comparison of the Strain Rates on Cement Mortar, Aggregate, and the Interface

Loading Conditions	Interface (MPa)	DIF	Mortar (MPa)	DIF	Aggregate (MPa)	DIF
$10^{-6}\,\mathrm{s}^{-1}$	2.36	1.00	4.34	1	5.32	1
$10^{-5}\,\mathrm{s}^{-1}$	2.38	1.01	5.78	1.33	7.85	1.48
$10^{-4}\,\mathrm{s}^{-1}$	2.51	1.06	5.72	1.32	8.88	1.67
$10^{-3}\,\mathrm{s}^{-1}$	3.49	1.48	7.14	1.65	10.37	1.95

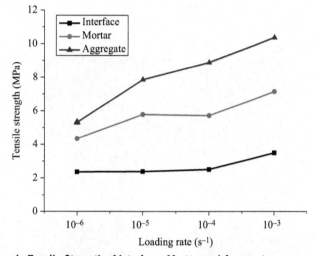

FIGURE 12.16 The Dynamic Tensile Strength of Interface, Mortar, and Aggregate

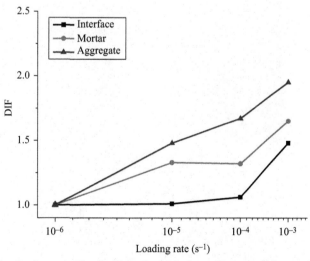

FIGURE 12.17 The DIF for the Dynamic Tensile Strength of Interface, Mortar, and Aggregate

12.6 EXPERIMENTAL RESEARCH ON THE DYNAMIC TENSILE STRESS–STRAIN CURVES OF CONCRETE

The constituent materials for concrete are all brittle materials, and there are no experimental results for the full stress–strain curves under direct tension. The experimental results obtained by some researchers provided the dynamic descending part, while it still is not the stable fracture. Therefore, a preliminary research and investigation are carried out in this part. The dynamic full stress–strain curves for concrete under direct tension of concrete are conducted successfully, but the success rate for the tests is less than 25%, which needs to be further improved.

The experimental instruments and methods are the same as previously stated. The concrete specimens adopt the plane dumbbell-type, and the experiments are shown in Fig. 12.18. The test results are listed in Table 12.12.

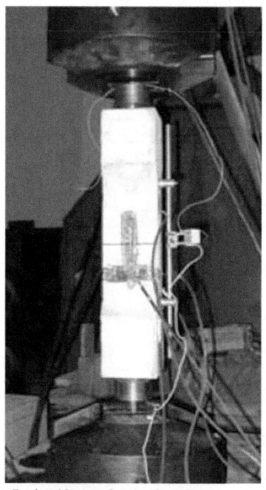

FIGURE 12.18 Dynamic Direct Tension of Concrete Complete Curve Test

Table 12.12 The Experimental Results for the Full Stress–Strain Curves of Concrete Under Dynamic Loading

Loading Conditions	Number	Tensile Strength (MPa)	Average (MPa)	Variance (MPa)	DIF	Remarks
$10^{-7} s^{-1}$	CQQX01	2.16	2.15	0.13	1.02	
	CQQX02	2.12				
	CQQX03	2.27				
	CQQX04	1.95				
	CQQX05	2.24				
	CQQX06	1.28				Complete curve
$10^{-6} s^{-1}$	CQQX07	2.05	2.11	0.17	1.00	
	CQQX08	1.98				
	CQQX09	2.31				Complete curve
$10^{-5} s^{-1}$	CQQX10	2.31	2.12	0.24	1.00	
	CQQX11	1.85				Complete curve
	CQQX18	2.21				
$10^{-4} s^{-1}$	CQQX12	2.54	2.41	0.30	1.14	
	CQQX13	0.80				
	CQQX14	2.39				
	CQQX15	1.92				
	CQQX16	2.72				
	CQQX17	2.47				Complete curve

   The full stress–strain curves for direct tensile behavior of concrete are obtained under the strain rates ranging from $10^{-6} s^{-1}$ to $10^{-3} s^{-1}$, as shown in Fig. 12.19. The load-time process line for each specimen is illustrated in Fig. 12.20.

   As seen from Fig. 12.19, after the load achieves the maximum value, it is stably decreasing and the time lasts long. Even for the conditions of $10^{-4} s^{-1}$, the time for the softening stage is 0.01 s, which is under the stable fracture condition. Among them, the condition for $10^{-5} s^{-1}$ is the most representative. During the ascending and descending stages, the specimens are loaded continuously with the rate of $10^{-5} s^{-1}$. By generalizing the four conditions, the dynamic stress–strain curves are shown in Fig. 12.21.

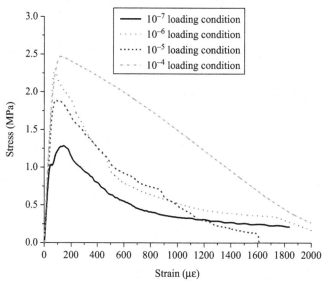

FIGURE 12.19 The Full Stress–Strain Curves for the Direct Tensile Behavior of Concrete Under Dynamic Loading

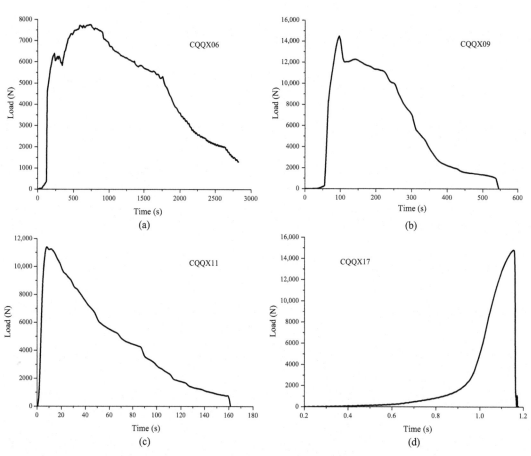

FIGURE 12.20 The Load-Time Process Curve for Specimens Under Different Strain Rates

(a) 10^{-7} s^{-1} (b) 10^{-6} s^{-1} (c) 10^{-5} s^{-1} (d) 10^{-4} s^{-1}.

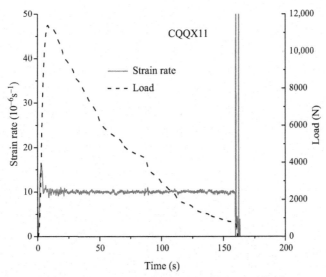

FIGURE 12.21 The Comparison of Measured Strain Rate, Load, and Time (10^{-5} s^{-1})

FIGURE 12.22 Crack Without Penetrating the Fracture Section of the Specimen

As shown in Fig. 12.22, the experiments with strain control are significantly different from the condition by force or displacement control. During the process with the strain control, as the specimen cracks, the crack extends continuously, and the crack will not penetrate the whole section of the specimens.

EXPERIMENTAL STUDY OF DYNAMIC AND STATIC DAMAGE FAILURE OF CONCRETE DAM BASED ON ACOUSTIC EMISSION TECHNOLOGY

In the process of concrete material damage, acoustic emission will be generated by releasing the energy of the microcracks initiation and expansion of the internal structure. Acoustic emission (AE) technique is a nondestructive, real-time, and effective means to detect the internal damage status of concrete and study the mechanism of its destruction. It is particularly important for deepening the study of the dynamic mechanical properties of the concrete dam. But so far up to now, there has not been a research result obtained by using the acoustic emission technique to study the dynamic damage of the concrete dam and the failure mechanism. Therefore, this chapter describes the recent study of acoustic emission technology applications in the test of the properties of concrete dams in the high arch dam project, including the strain rate effects of the whole-graded and wet-sieved concrete in the bending tensile failure process of the dynamic and static bending test on the AE parameters, the recognition and regularity of AE parameters of each concrete component in the failure process of the uniaxial tension tests, especially for the measuring of AE parameters of the softening part of the stress–strain curves of concrete under uniaxial tension test, and the discovery of "strain Kaiser effect" in the process of unloading and reloading. In addition, the preliminary results are obtained, the understanding of the process of dam concrete dynamic damage and destruction mechanisms, and the application of AE technology in the seismic safety of high arch dams are analyzed and discussed.

In the experimental studies, the SAMOS™ Series 16-channel acoustic emission instrument produced by the American Physical Acoustics Company (PAC) is used to collect and store acoustic emission signals. The dead time of the system is 300 μs, and the maximum sampling frequency and the length of wave are 3 MHz and 4096 μs, respectively. Acquisition and control software is called *AE*win™. The selection of gain of the preamplifier (model PAC-2/4/6, a bandwidth of 10 kHz to 2.0 MHz, gain range 20, 40, 60 dB), and a set of the channel threshold voltage are related with the ceiling of the preamplifier. Channel threshold and a high-pass filter frequency need to be set according to the requirements of the ambient noise level and signal-to-noise ratio signal analysis.

Two types of sensors are used in the experiments, in which the resonant high-sensitivity sensor (model R6α, the resonant frequency is 90 kHz) is used for the determination of AE parameters and three-dimensional positioning of the acoustic emission source, and the broadband sensor (model PAC-WD, bandwidth 100 kHz ~ 1.0 MHz) is used for spectral analysis of the acoustic emission waveform. When the sampling frequency is 1~3 MHz, 4096 sampling points can record wave, whose length is 1365~4096 μs. According to the sampling theorem, the sampling frequency of the signal should be not less than two times the highest frequency of the signal that is taken, so that 1~3 MHz sampling frequency is sufficient to collect and analyze the maximum frequency of the acoustic emission signal that is 500 kHz to 1.5 MHz, in order to meet the requirements of dynamic acoustic emission testing of concrete.

In the tests, Vaseline is used as a coupling agent, and with a rubber band or a piece of metal bent, specially designed, the sensor is fixed on the surface of the concrete specimen.

The experimental study using acoustic emission techniques is conducted in dynamic and static bending tests and uniaxial tensile tests with the concrete specimens of the Xiaowan arch dam and Dagangshan arch dam projects.

13.1 RESEARCH ON MONITORING OF ACOUSTIC EMISSION TECHNOLOGY OF CONCRETE MATERIALS

13.1.1 EXPERIMENTAL STUDY OF THE CORRELATION AMONG THE FEATURE PARAMETERS OF ACOUSTIC EMISSION

The recorded signals of acoustic emission testing are very complex, with numerous parameters. The basic feature parameters are seen in Fig. 13.1. The recorded acoustic emission wave is made of the impact wave that is generated by the fracture occurring within the material under the action of external force. In order to reduce the noise interference and improve the signal-to-noise ratio, it is required to set the threshold level of the waveform amplitude. Signal analysis only considers the hits with amplitude that exceeds the preset threshold. Each hit in its duration contains a number of ringing, and the times of ringing with amplitude exceeding a set threshold are called count; its maximum amplitude is called the amplitude. The time from wave exceeding the initial threshold to the time of the maximum amplitude is called the rise time. The acoustic emission energy of function characterization of sampling point voltage is called the signal strength.

According to these definitions, the feature parameters of acoustic emission can be divided into the following five categories:

1. The hit number and the count correlated with pulse times in waveform.
2. The amplitude related to the signal effective value of the voltage and average signal level.

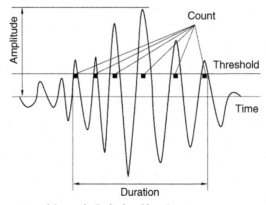

FIGURE 13.1 The Basic Parameters of Acoustic Emission Signals

3. The rise time and duration of the signal.

4. The signal strength, energy strength, and so on, associated with the actual energy of signal.

5. The average frequency associated with the signal frequency (the number of rings within the duration), initial frequency (number of rings in rise time), reverberation frequency (duration minus the number of rings in rise time), and so on.

The basic parameters of acoustic emission signals are shown in Fig. 13.1.

Each kind of parameter can only describe the features of an aspect of the signal waveform. In order to analyze the damage process and failure mechanism of materials from these parameters, it is necessary to study the correlation between these feature parameters, so that it is possible to focus on the main parameters for analysis, and if needed, it is allowed to replace some parameters by others with high correlation in the analysis process. The correlation of all kinds of the feature parameters of AE wave is not the same; therefore, it is essential to study the correlation among the feature parameters of AE wave when concrete is in the seismic dynamic action.

The correlation between two random variables can be quantitatively represented by the correlation coefficient. $(X1, X2,.., Xn)$ and $(Y1, Y2,... Yn)$ are from a sample of the total X, Y, respectively, and the correlation coefficient r_{xy} of the sample of X, Y is defined as follows:

$$r_{xy} = \frac{\sum_{i=1}^{n}(X_i - \overline{X})(Y_i - \overline{Y})}{\sqrt{\sum_{i=1}^{n}(X_i - \overline{X})^2 \sum_{i=1}^{n}(Y_i - \overline{Y})^2}} \tag{13.1}$$

The correlation test of the feature parameters of AE wave of concrete is conducted by combining with the uniaxial tension test, that is, for three specimens (WP1, WP2, WP3), three sets of data of the parameters of acoustic emission are collected in the process of the tensile failure and the parameter correlation is analyzed. Experiments use six PAC-R6α sensors. Table 13.1 shows the parameters of

Table 13.1 The Parameters of Acoustic Emission and Their Correlation Coefficient in the Three Sets of Concrete Specimens Under Uniaxial Tension Failure Process with Significant Correlation

Correlation Evaluation	WP1	WP2	WP3	Average
Amplitude–signal strength	0.74	0.36	0.61	0.57
Amplitude–rise time	0.75	0.43	0.76	0.65
Amplitude–duration	0.85	0.51	0.72	0.69
Amplitude–count	0.84	0.48	0.74	0.69
Rise time–signal strength	0.87	0.94	0.93	0.91
Rise time–duration	0.93	0.98	0.98	0.96
Rise time–count	0.91	0.98	0.98	0.96
Duration–signal strength	0.96	0.96	0.98	0.97

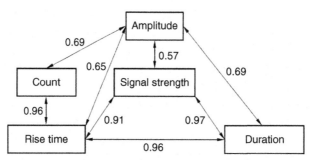

FIGURE 13.2 Schematic Diagram of Correlation Coefficient for AE Parameters in the Failure Process of Concrete

acoustic emission and their correlation coefficient in the three sets of concrete specimens under uniaxial tension failure process with significant correlation.

The sample number n of acoustic emission hit that is collected for each specimen is larger than 102, when the significance level $\alpha = 0.001$, $r(n-2)\alpha$ is 0.3211. That is to say, if the correlation coefficient is $r_{xy} \geq 0.3211$, it is considered as significant correlation; otherwise, no significant correlation can be considered.

As seen from Table 13.1, there are good correlations among the amplitude of the acoustic emission and signal intensity, rise time, duration, and number of rings. The average correlation coefficient in the three groups of concrete under uniaxial tension test is 0.57~0.69, but the correlation coefficient of the specimen WP2 is low, whereby indicating that correlation among these parameters is easily affected by individual specimen and test random bias. In addition, it also can be seen that there is good correlation among the rise time and signal intensity, duration, and number of rings, and between the duration and signal strength, and the correlation coefficients reach more than 0.90, so that each specimen is very stable. Figure 13.2 intuitively shows the relationship between AE parameters whose correlation coefficient is high (the value near the arrow is the value of the corresponding correlation coefficient).

13.1.2 IMPROVEMENT OF AE SOURCE LOCATION TECHNOLOGY

13.1.2.1 Determination of wave velocity

Acoustic emission source location is helpful to determine the position of damage and to understand the mechanism and trends of damage. In studying the process of destruction of concrete damage, most of the source localization studies use the first arrival time of P wave for position calculation, but it is necessary to study the effect of the choice of velocity on positioning accuracy.

The broken pencil can be used as artificial excitation source in measurements of wave velocity. When measured, two groups of sensors are arranged on the surface of the different directions of concrete specimens, and two lead breaking artificial excitation sources are arranged on the surface of each direction. The results of the wave velocity test are shown in Tables 13.2 and 13.3.

When using different velocity for acoustic emission source location, the distance between the coordinates of the actual excitation point and the AE source coordinates calculated based on the velocity is defined as x, y, z distance error. Relationship curves between distance error and velocity in Fig. 13.3 shows that, as can be seen from the graph, there is the lowest value of the curve points in each direction. The corresponding wave velocity of the lowest value is called the optimal value. The optimal value of velocity in each direction is not exactly the same. The positioning error in different directions and the

Table 13.2 The Results of the Wave Velocity Test of the First Group of Specimens

Sample Number	No. 1 and 2 Sensor Measurement		No. 3 and 4 Sensor Measurement		Average
	No. 1 Excitation Source	No. 2 Excitation Source	No. 3 Excitation Source	No. 4 Excitation Source	
WY2	4167.2	3502.88	3899.36	4382.14	3987.98

Table 13.3 The Results of the Wave Velocity Test of the Second Group of Specimens

Sample Number	No. 1 and 2 Sensor Measurement		No. 3 and 4 Sensor Measurement		Average
	No. 1 Excitation Source	No. 2 Excitation Source	No. 3 Excitation Source	No. 4 Excitation Source	
WY5	3581.58	3981.38	4223.43	4223.43	4002.46
WY6	3981.38	4018.77	3782.73	4264.84	4011.93

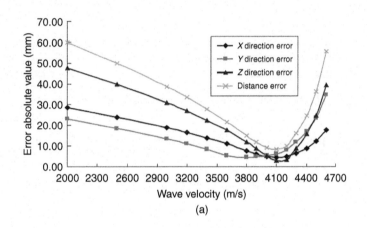

(a)

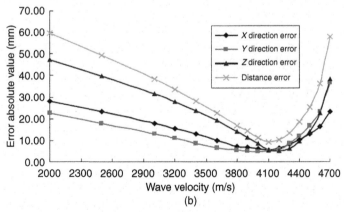

(b)

FIGURE 13.3 The Error of Source Localization Under Different Velocities

(a) WY2 specimen (b) WY5 specimen.

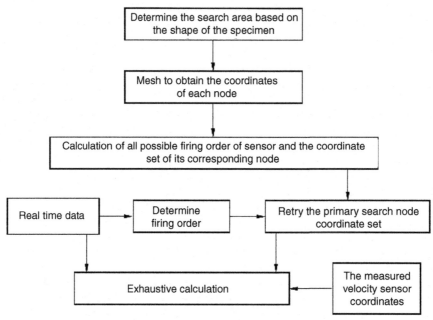

FIGURE 13.4 The Diagram of the Exhaustive Method of the Improved Calculation Process

velocities of the lowest point between acoustic emission wave velocity are called the global optimal value of AE wave velocity.

13.1.2.2 New algorithm of the three-dimensional AE source location based on the exhaustive method

The AE software often uses the least squares method to make excitation source localization and also localization by a direct search method of exhaustion (also known as grid method), but the biggest problem of the latter is that it is too time-consuming to calculate, so that it is improved in this paper. The basic idea of how to improve it is that judging the firing order of the sensor from the data, determining the region that may be related to the location of AE source, excluding the region where acoustic emission sources do not exist, and then preliminarily selecting the search node. After that, the exhaustive method is used to calculate the selected node. In this way, the efficiency of the method is improved. The calculation process of improved exhaustive method is shown in Fig. 13.4.

Using the commercial AE software, the direct search method of exhaustion and the improved method of exhaustion are used to solve the coordinate of artificial excitation source for specimen C150, and the calculation data are compared with the actual excitation position. The results are consistent. The typical results are shown in Table 13.4.

In the process of loading the specimen to its sudden rupture, the results of z direction positioning using two kinds of AE source location algorithms are shown in Fig. 13.5. It can be seen from the figure that the number of positioning events with the improved exhaustive method is 95, which is not only much higher than the number of 31 positioning events obtained by a commercial acoustic emission analysis software, but also more clearly and comprehensively reflects the positions of emergence of

Table 13.4 Artificial Excitation Source Positioning Error for Specimen C150

Evaluation Index	A Commercial AE Software		The Direct Searching Method of Exhaustion		Improved Method of Exhaustion	
	Error (mm)	Relative Error (%)	Error (mm)	Relative Error (%)	Error (mm)	Relative Error (%)
x Direction	3.07	2.0	2.76	1.8	2.80	1.9
y Direction	2.39	1.6	3.1	2.1	3.09	2.1
z Direction	6.4	3.6	3.33	2.2	3.42	2.3
Distance	8.22	5.5	6.48	4.3	6.58	4.4
The computational time (s)	–	4.90	1.12			
Mean residual (m^2)	–	6.3×10^{-6}	6.3×10^{-6}			
Data utilization rate (%)	88	100	100			

microcracks and development in concrete. Comprehensive test results show that the computational efficiency and data utilization rate of the improved method of exhaustion are high and its positioning error is small.

13.2 EXPERIMENTAL STUDY OF ACOUSTIC EMISSION CHARACTERISTICS IN THE DAMAGE PROCESS OF DAM CONCRETE UNDER DYNAMIC AND STATIC FLEXURAL FAILURE

13.2.1 SETTINGS OF ACOUSTIC EMISSION ACQUISITION SYSTEM

In the flexural tests of wet-sieving and three-graded concrete in the Xiaowan arch dam, six R6α resonant sensors and one PAC-WD broadband transducer are arranged on the surface of concrete specimens, which are fixed with couplant (Vaseline) and rubber tape. The layout scheme of acoustic emission sensors can be seen in Fig. 13.6, where No. 1–6 are the resonant sensors, and No. 7 is the broadband sensor. The thresholds of the channel of resonant and broadband sensors are set as 38 and 35 dB, respectively, according to the preliminary test and the results of the field loading equipment noise test.

In the flexural tests of wet-sieving and four-graded concrete in the Dagangshan arch dam, three R6α resonant sensors and one PAC-WD broadband transducer are arranged on the surface of concrete specimens. The acoustic emission sensors are fixed on the corresponding position on the surface of concrete, using metal-bending parts specially designed, in the test of four-graded concrete with relatively large size specimens under dynamic and static flexural tests. The layout scheme is shown in Fig. 13.7, No. 1 is the broadband sensor (PAC-WD) and No. 2–4 are the resonant sensors (PAC-R6α). The thresholds of the channel of resonant and broadband sensors are set as 38 dB, and the filter range of the band-pass filter is set within 20 kHz to 1 MHz.

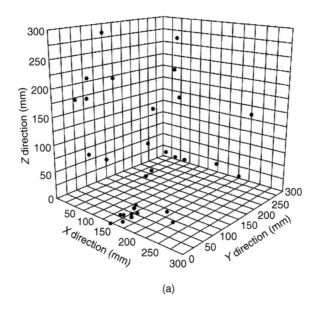

(a)

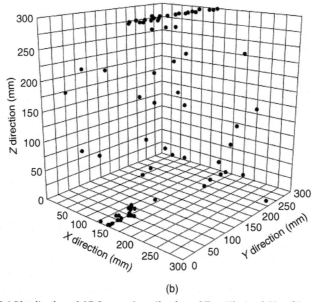

(b)

FIGURE 13.5 The Spatial Distribution of AE Source Localization of Two Kinds of Algorithms

(a) A commercial AE analysis software (event:31) (b) the improved method of exhaustion (event:95).

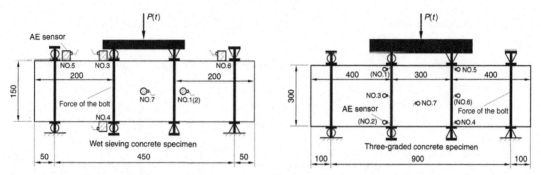

FIGURE 13.6 The Layout Plan of Acoustic Emission Sensors in the Flexural Test of Wet Sieving and Three-Graded Concrete in Xiaowan Arch Dam (Unit: mm)

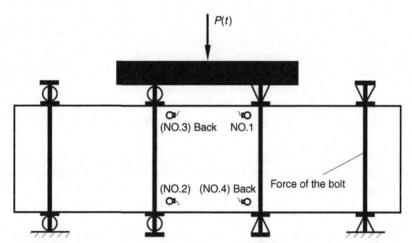

FIGURE 13.7 The Layout Plan of Acoustic Emission Sensors in the Flexural Test of Wet Sieving and Four-Graded Concrete in Dagangshan Arch Dam

The acquisition system also records the waveform data of various hits. Only the full waveform data of channel 1 for broadband sensor is collected, because the amount of full waveform data is very large. The sampling frequency of channel 1 is set as 2 MHz.

The gain of the preamplifier (PAC-2/4/6) of the acoustic emission acquisition system is set as 40 dB in these two arch dam tests.

13.2.2 STATIC BENDING TEST OF FULL-GRADED AND WET-SIEVING ARCH DAM CONCRETE SPECIMENS

13.2.2.1 Analysis of acoustic emission activity

The accumulative curve of acoustic emission hits marking the different acoustic emission activity can show the development of damage accumulation in the materials.

Table 13.5 Mechanical Property and AE Characteristics of Wet Sieving and Three-Graded Concrete in Xiaowan Arch Dam

Comparison Program		Wet-Sieved Concrete	Three-Graded Concrete
Ultimate load (kN)		49.3	133.4
Ultimate stress (MPa)		6.57	4.47
Poisson's ratio		0.21	0.24
Bending elasticity modulus (GPa)		33.21	34.85
AE hit (times)	Stage A	334 (2.92%)	10,276 (29.06%)
	Stage B	1776 (15.56%)	6,513 (19.66%)
	Stage C	9303 (81.51%)	19,188 (51.24%)

1. Acoustic emission activity of Xiaowan and Dagangshan arch dam concrete specimens
   The mechanical properties and acoustic emission characteristics in the process of flexural failure of wet-sieving and three-graded concrete in the Xiaowan and Dagangshan arch dam are shown in Tables 13.5 and 13.6. The load curve and the accumulative curve of acoustic emission hits are shown in Figs 13.8 and 13.9, respectively. The whole process can be divided into A, B, and C stages according to the shape of the accumulative curve of acoustic emission hits. In the Xiaowan

Table 13.6 Mechanical Property and AE Characteristics of Wet Sieving and Four-Graded Concrete in Dagangshan Arch Dam

Comparison Program		Wet-Sieved Concrete			Four-Grade Concrete			
		101101	101601	Average	101203	101301	100903	Average
Ultimate load (kN)		45.7	51.3	48.5	228	201	192	207
Ultimate stress (MPa)		6.1	6.68	6.48	3.38	3	2.85	3.08
Poisson's ratio		0.24	0.3	0.27	0.06	0.06	–	0.06
Bending elasticity modulus (GPa)		63.4	53.2	58.3	45.0	44.3	34.9	41.1
AE time began to appear (compared with the limit stress, %)		52.6	53.6	53.3	23.4	37.5	41.4	34.1
AE hit (times)	Stage A	122	93	108	5,588	4,052	2,712	4,117
	Stage B	107	55	81	1,978	4,967	3,456	3,467
	Stage C	3209	1745	2477	4,445	16,173	13,575	11,397
	whole process	3438	1893	2666	12,011	25,191	19,752	18,984

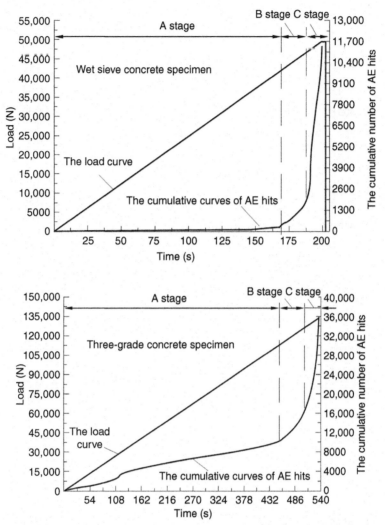

FIGURE 13.8 The Load Curve and the Accumulative Curve of Acoustic Emission Hits in the Process of Flexural Failure of Concrete in Xiaowan Arch Dam

arch dam, the A, B, and C stages correspond to 0–85%, 85–95%, and 95–100% ultimate load, respectively, and in the Dagangshan arch dam, the A, B, and C stages correspond to 0–80%, 80–90%, and 90–100% ultimate load, respectively.

2. Analysis of the acoustic emission activity of full-graded concrete and wet sieving
   For Xiaowan arch dam concrete, it can be seen in Table 13.5 that the acoustic emission hits percentage in Stage A and B of the three-grade concrete is always larger than that of the wet-sieving concrete. This indicates that the acoustic emission activity of three-grade concrete is more

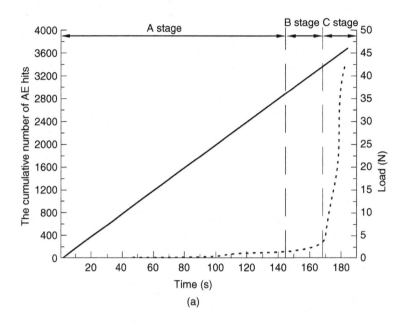

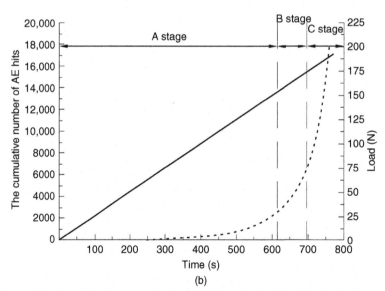

FIGURE 13.9 The Load Curve and the Accumulative Curve of Acoustic Emission Hits in the Process of Flexural Failure of Concrete in Dagangshan Arch Dam

(a) Wet sieve concrete (b) Four-grade concrete.

active than that of wet-sieved concrete. And in Stage C nearing the final destruction, the acoustic emission hits percentage of wet-sieving concrete is obviously larger than that of three-grade concrete. In addition, the slope of the cumulative curve is steep, indicating a dramatic increasing trend of the acoustic emission activities. This indicates that the acoustic emission activities related to the unstable development of microcracks of wet-sieved concrete are mainly concentrated in high stress level. It can be seen from Fig. 13.8 that the consecutive acoustic emission in three-grade concrete occurs earlier than in wet-sieved concrete.

As far as Dagangshan arch dam concrete is concerned, it can be seen in Table 13.6 that the acoustic emission hits percentage in Stages A, B, and C, and the whole process of four-grade concrete is always larger than that of the wet-sieved concrete. This indicates that the acoustic emission activity of four-grade concrete is more active than that of wet-sieved concrete. The acoustic emission hit number is nearly the same as that in Stages A and B. Consecutive acoustic emission in four-grade concrete occurs at about 34.1% ultimate load while it occurs at about 53.3% ultimate load in wet-sieved concrete.

In the flexural failure process of static bending test, the rule of acoustic emission activity of Xiaowan and Dagangshan arch dam concrete is nearly the same. Comparing with the wet-sieved concrete, there are more acoustic emission hits in the full-grade concrete, and its consecutive acoustic emission occurs earlier so that its strength is also lower. The research results by Sadowska et al. (1984) indicate that the earlier the acoustic emission single marking damage occurs, the lower the concrete strength is.

13.2.2.2 Analysis of the acoustic emission waveform spectrum

1. Acoustic emission waveform spectrum of Xiaowan and Dagangshan arch dam concrete
   Peak frequency distribution statistics of each acoustic emission waveform of wet-sieved and full-grade concrete in the Xiaowan and Dagangshan arch dam are shown in Figs 13.10 and 13.11. The statistical interval is 25 kHz. Frequency centroid distribution statistics are shown in Figs 13.12 and 13.13. The latter is measured with a broadband sensor.

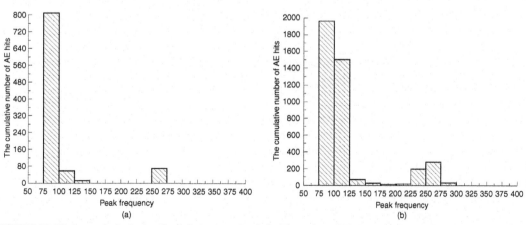

FIGURE 13.10 Acoustic Emission Peak Frequency Statistic Histogram of Xiaowan Arch Dam Concrete

(a) Wet sieve concrete (b) Four-grade concrete.

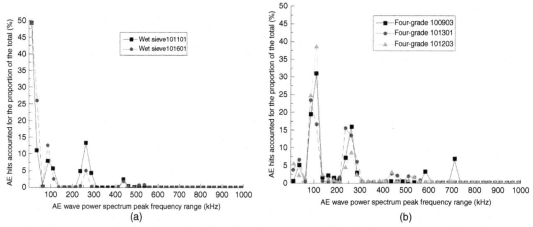

FIGURE 13.11 Peak Frequency Distribution Statistics in Dagangshan Arch Dam Wet-Sieved and Four-Grade Concrete Under Flexural Load (Broadband Sensor)

(a) Wet sieve concrete (b) Four-grade concrete.

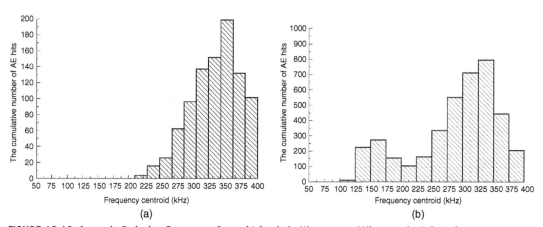

FIGURE 13.12 Acoustic Emission Frequency Centroid Statistic Histogram of Xiaowan Arch Dam Concrete

(a) Wet sieve concrete (b) Four-grade concrete.

2. Analysis of the acoustic emission waveform spectrum of wet-sieved and full-grade concrete

For three-grade concrete in the Xiaowan arch dam, there are also a number of acoustic emission hits in the frequency range of 100–125 kHz and 225–250 kHz, in addition to the general characteristics of the peak frequency distribution of the wet-sieved concrete.

For Dagangshan arch dam wet-sieved concrete, the acoustic emission hits with frequency under 75 kHz account for about 50% of total numbers in the flexural failure process, while the acoustic emission hits of four-grade concrete mainly distribute in the frequency range of 100–125 kHz and 225–250 kHz. Compared with the wet-sieved concrete, acoustic emission wave spectrum of four-grade concrete tends to distribute in the high-frequency range.

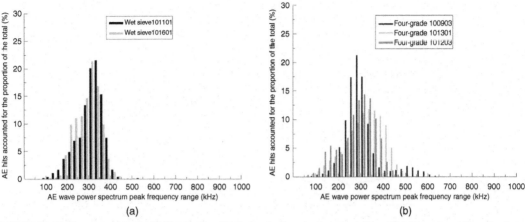

FIGURE 13.13 Frequency Centroid Distribution Statistics in Dagangshan Arch Dam Wet-Sieved and Four-Grade Concrete Under Flexural Load (Broadband Sensor)

(a) Wet sieve concrete (b) Four-grade concrete.

In the whole loading process, the acoustic emission frequency centroid of wet-sieved concrete of Xiaowan arch dam distributes in the frequency range of 200–400 kHz, while that of the three-grade concrete distributes in 100–400 kHz, which is wider.

The acoustic emission frequency centroid of wet-sieved concrete of Dagangshan arch dam distributes in the frequency range of 130–450 kHz, while for the three-grade concrete the range is 75–600 kHz, which is wider. A certain number of acoustic emission signals with frequency higher than 400 kHz are detected in the experiment, being relevant to the acoustic emission signal acquisition system with wide frequency range and use of broadband sensor. The acoustic emission signal acquisition system with wide frequency range improves the differentiation degree of acoustic emission characteristics of two kinds of concrete and is beneficial to compare the differences.

The frequency centroid of full-grade concrete in Xiaowan and Dagangshan arch dam concrete also occurs in low-frequency range. This is possibly related to its richer interface damage components than that of wet-sieved concrete. Studies have shown that the interface damage of coarse aggregate and cement paste in concrete is dominated with acoustic emission signals with low frequency (<100 kHz). Sagaidak et al. pointed out that the frequency characteristics of acoustic emission hit depended on the size and direction of the cracks, and crack growth could reduce the main frequency of acoustic emission spectrum, so that low distribution range of frequency centroid in three-grade concrete means that the size and scale of crack in the whole damage process of three-grade concrete is larger than that of wet-sieved concrete.

13.2.2.3 Failure mechanism identification based on acoustic emission parameters

Wu et al. proposed a method to separate acoustic emission signals to show the characteristics of the different fracture mechanism.

It is just here that this method is adopted to separate and identify acoustic emission characteristics and the program to process the acoustic emission data is written. The flow path of this recognition

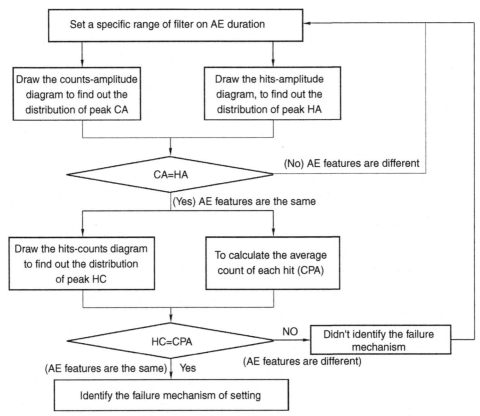

FIGURE 13.14 The Flow Chart of Mechanism Recognition Program of Concrete Damage Process Based on Acoustic Emission Characteristic Parameters

program is shown in Fig. 13.14. In the process of identification, the average count of each hit (CPA) and the peak value in hit-count figure (HC) of acoustic emission signal after filtering in certain different duration range is compared, and when the value of |HC−CPA| is not greater than 15% of their average, HC and CPA is thought to be approximately equal and is thought to be the same mechanism.

For three-grade concrete and wet-sieved concrete, 11 and 12 kinds of failure mechanisms have been obtained in the duration range under 1800 μs, as shown in Tables 11.7 and 11.8. The recognition rate can reach 7.4% and 91.9%, respectively, indicating that these mechanisms can approximately represent the main features of the whole damage process of the two kinds of concrete. After comparison, it can be found that the parameter distribution of mechanism of the two kinds of concrete is basically the same except for the mechanism 4, whereby suggesting that the failure mechanism of two kinds of concrete is basically the same.

The main difference between the two kinds of concrete is that the mechanism 4a and 4b in three-grade concrete is covered with the duration range of mechanism 4 in wet-sieved concrete. The large size aggregate in three-grade concrete may make the fracture mechanism complicated. In addition, the count of unit hit and the peak value of hit-count histogram of mechanism 4 in wet-sieved concrete is

Table 11.7 AE Features Parameter of Wet Sieving Concrete

Mechanism	The Duration Range	Amplitude Distribution Features					Hits (Times)	Account for the Proportion of the Total Hits	Counts
		Peak Value	Distribution Range	The Counts of Unit Hits	Hit-Counts for Peak Figure				
1	1~20	40	38~43	2.16	2.0	556	4.88	1,198	
2	21~41	41	38~49	3.29	3.0	1563	13.71	5,139	
3	40~85	42	38~50	4.58	5.0	1582	13.87	7,234	
4	87~187	42	38~51	6.59	6.0	1982	17.38	13,049	
5	190~239	43	38~54	10.44	10.0	647	5.67	6,743	
6	239~286	43	38~54	12.63	12.0	522	4.58	6,582	
7	286~378	45	38~59	16.77	17.0	880	7.72	14,736	
8	378~594	46	38~63	27.01	30.0	1402	12.30	37,838	
9	594~825	48	38~64	45.14	47.0	820	7.19	36,968	
10	825~1124	52	38~64	68.87	73.0	588	5.16	40,424	
11	1125~1800	58	38~70	106.48	101.0	565	4.96	60,053	

Table 11.8 AE Parameter Features of Three-Graded Concrete

Mechanism	The Duration Range	Amplitude Distribution Features					Hits (times)	Account for the Proportion of the Total Hits	Counts
		Peak Value	Distribution Range	The Counts of Unit Hits	Hit-Counts for Peak Figure				
1	1~20	40	38~48	2.27	2.0	2057	5.27	4,627	
2	20~42	40	38~49	3.44	3.0	4780	13.29	16,440	
3	40~89	42	38~56	5.69	5.0	7386	20.53	41,982	
4a	103~122	41	38~58	7.73	7.0	1536	4.27	11,861	
4b	138~192	42	38~60	11.01	10.0	3504	9.74	38,569	
5	193~249	43	38~62	14.56	14.0	2590	7.20	37,683	
6	247~274	43	38~62	18.16	19.0	1067	2.97	19,359	
7	275~374	43	38~69	21.84	19.0	2662	7.40	58,117	
8	375~594	46	38~73	31.04	27.0	3329	9.25	103,312	
9	594~825	48	38~74	47.94	49.0	1872	5.20	89,700	
10	825~1125	50	38~79	68.90	62.0	1214	3.37	83,575	
11	1125~1800	55	38~86	102.12	90.0	1057	2.94	107,842	

closer to the mechanism 4a in three-grade concrete, that is, the amplitude distribution characteristics of these two mechanisms are similar. This is likely to come from a similar mechanism. Both kinds of concrete contain coarse aggregate whose diameter is smaller than 40 mm and cement mortar; therefore, it can be concluded that mechanisms 4 and 4a could originate from interfacial damage of small-size coarse aggregate and cement paste. Also, mechanism 4b may come from interfacial damage of large-size coarse aggregate and cement mortar. But it is sure that the large-size aggregate leads to the more complicated acoustic emission mechanism.

Wu et al. found that two mechanisms in plain concrete are covered by one mechanism in mortar specimens in a certain duration range when studying the characteristics of acoustic emission in the damage procedure of mortar and plain concrete and analyzing the fracture mechanism.

13.2.2.4 Analysis of acoustic emission location

Internal microcracks in the concrete specimens are induced and extended continuously as the load is applied gradually in the process of test. When it reaches the ultimate bearing capacity of concrete, specimens suddenly break into two. The emergence and development of microcracks can be presented through the acoustic emission source localization technology in this process. Acoustic emission source location results of two kinds of concrete at different stress stage are shown in Figs 13.15 and 13.16. The acoustic emission events located in the fracture process of the two kinds of concrete occur at the stage of about 85% ultimate load. It can be seen from Figs 13.15 and 13.16 that the spatial distribution in the damage process of three-grade concrete is wider than that of the wet-sieved concrete. The development of microcracks of wet-sieved concrete mainly occurs in the largest tensile stress area, while in addition to this area, there are a certain number of distributions in other areas for three-grade concrete. This means that for the latter, microcracks are active in multiple regions, whereby indicating that there are differences between the manners of the microcrack development due to stress concentration for these two kinds of concrete. Joseph F. et al., 2001 pointed out that concrete with larger aggregate size would form a bigger microcrack localized area in the damage process, so that large-sized aggregate in three-grade concrete is the main cause of this phenomenon.

13.2.3 DYNAMIC FLEXURAL TEST OF FULL-GRADED AND WET-SIEVING ARCH DAM CONCRETE SPECIMENS

13.2.3.1 Analysis of acoustic emission activity

1. Impact dynamic load

   The mechanical properties and acoustic emission characteristics in the process of flexural failure of wet-sieved and four-graded concrete in Dagangshan arch dam are shown in Table 13.9. It can be seen from the chart that it takes a very short time for concrete specimens under the impact load from loading start to final destruction, being less than 0.5 s. Also, it can be seen by comparison with Table 13.6 that the dynamic flexural strengths of wet-sieved and four-graded concrete in Dagangshan arch dam are both obviously higher than their corresponding static values. The acoustic emission hits under dynamic load has been obviously lower than the corresponding static value, whereby indicating that there exists an evident strain rate effect.

   In addition, it can be seen from Fig. 13.17 that the accumulative curve of acoustic emission hits of wet-sieved and four-graded concrete in Dagangshan arch dam cannot present clear stage characteristics. The acoustic emission hits suddenly increase when the average stress comes

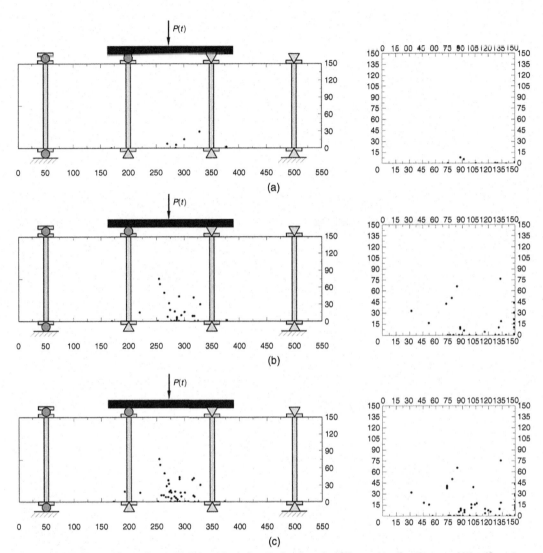

FIGURE 13.15 Three-Dimensional AE Source Location Projections in Different Loading Stages of Wet-Sieved Concrete

(a) 96% ultimate load (b) 98% ultimate load (c) 100% ultimate load.

to 61% limit load and 75% limit load. This shows that in the high rate loading, microcracks of internal concrete are difficult to develop fully in low stress level.

2. Luffing triangle wave dynamic load
   The load curve and the accumulative curve of acoustic emission hits of wet-sieved and four-graded concrete in Dagangshan arch dam under luffing triangle wave dynamic load are shown

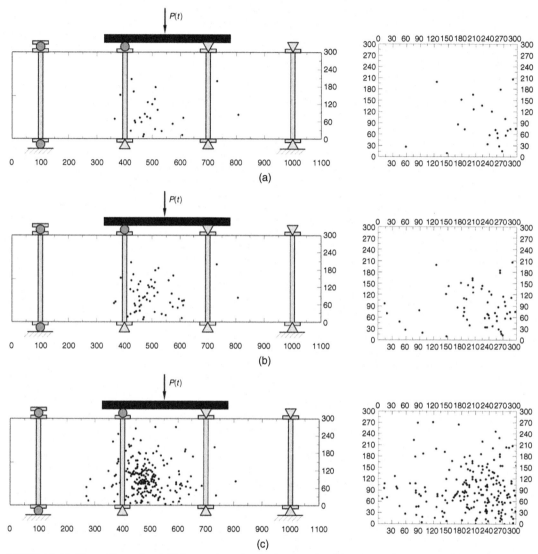

FIGURE 13.16 Three-Dimensional AE Source Location Projections in Different Loading Stages of Three-Grade Concrete

(a) 96% ultimate load (b) 98% ultimate load (c) 100% ultimate load.

in Fig. 13.18. It can be seen from Fig. 13.18 that each loading cycle includes three reciprocating loading cycles, its total duration time is about 0.6 s, and the amplitude of each loading cycle is increased every time. The acoustic emission hits and ultimate load of each loading cycle are shown in Tables 13.10 and 13.11. In Fig. 13.18, in the corresponding period of each cycle of two kinds of concrete, there is no significant new acoustic emission activity platform to indicate.

Table 13.9 The Mechanical Properties and Acoustic Emission Characteristics in the Process of Flexural Failure of Wet Sieving and Four-Graded Concrete in Dagangshan Arch Dam

Comparison Program	Wet-Sieved Concrete			Average Value	Four-Graded Concrete			Average Value
	101006	101802	101204		101002	101202	100902	
Ultimate load (kN)	58.9	51.3	64.1	64.0	273.0	279.4	283.7	278.7
Ultimate stress (MPa)	7.85	6.68	8.56	8.53	4.04	4.15	4.2	4.13
Poisson's ratio	0.11	0.3	0.18	0.18	0.17	0.17	0.26	0.20
Bending elasticity modulus (GPa)	–	66.7	65.8	66.2	51.5	54.6	69.0	58.4
Load time (s)	0.134	0.16	0.145	0.146	0.309	0.361	0.293	0.321
AE hit (times)	16	22	41	26	22	48	26	32

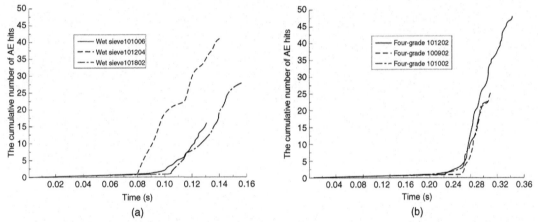

FIGURE 13.17 The Accumulative Curve of Acoustic Emission Hits of Wet Sieving and Four-Graded Concrete in Dagangshan Arch Dam Under Impacted Dynamic Load

(a) Wet sieve concrete (b) Four-grade concrete.

But as the stress increases, the platform also appears in small steps, thus indicating that there is still a certain acoustic emission activity, especially in the anterior loading cycle near destruction, within which the acoustic emission cumulative hits of each loading cycle show an increasing trend. This phenomenon can be used as a prediction of concrete near destruction and an external basis evaluation of the degree of damage of concrete.

The acoustic emission activity rules of full-graded and wet-sieved concrete are basically the same.

13.2.3.2 Analysis of the acoustic emission waveform spectrum under impact dynamic load

The flexural tensile failure process is a very short time. Compared with the static loading, under impact dynamic loading, the acoustic emission acquisition system can receive only a small amount of characteristic

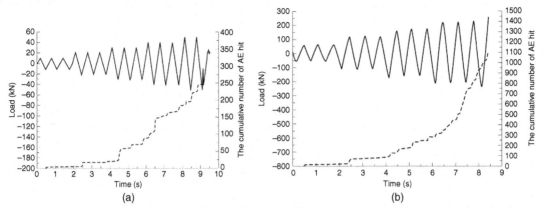

FIGURE 13.18 The Accumulative Curve of Acoustic Emission Hits of Wet-Sieved and Four-Graded Concrete in Dagangshan Arch Dam Under Luffing Triangle Wave Dynamic Load

(a) Wet sieve concrete (specimen number: wet sieve101001) (b) Four-grade concrete (specimen number: four-grade101001).

Table 13.10 The Acoustic Emission Hits and Ultimate Load of Each Loading Cycle of Wet Sieving Concrete in Dagangshan Arch Dam

Cycle Number	101001		101303		101401		Average	
	Hits Within the Cycle (time)	Hits Within the Period (time)	Hits Within the Cycle (time)	Hits Within the Period (time)	Hits Within the Cycle (time)	Hits Within the Period (time)	Hits Within the Cycle (time)	Hits Within the Period (time)
1	4	4	0	0	10	1	5	2
		0		0		8		3
		0		0		1		0
2	16	13	8	8	32	4	19	8
		0		0		23		8
		3		0		5		3
3	69	37	23	10	93	23	62	23
		12		12		36		20
		20		1		34		18
4	95	59	51	25	267	76	138	53
		15		5		92		37
		21		21		99		47
5	66 (break)	37	248	71	113 (break)	113	248 (101303)	74
		29		70				
				107				
6			78 (break)	78				
Ultimate load (kN)	50.66		50.80		51.10		50.85	

Table 13.11 The Acoustic Emission Hits and Ultimate Load of Each Loading Cycle of Four-Graded Concrete in Dagangshan Arch Dam

Cycle Number	101501		101402		101401		Average	
	Hits Within the Cycle (time)	Hits Within the Period (time)	Hits Within the Cycle (time)	Hits Within the Period (time)	Hits Within the Cycle (time)	Hits Within the Period (time)	Hits Within the Cycle (time)	Hits Within the Period (time)
1	18	13	49	23	41	0	36	12
		3		19		1		8
		2		7		39		16
2	66	51	116	23	216	10	133	28
		10		61		64		45
		5		24		138		56
3	199	80	205	100	476	75	193	85
		72		80		103		85
		47		25		282		118
4	655	83	242	43	316	87	404	71
		192		82		62		112
		38		81		146		88
5	164	225	553	145	228 (break)	146	553 (101402)	
				191		24		
				194				
6			67 (break)	137				
Ultimate load (kN)	257.22		275.90		224.08		257.22	

parameters of acoustic emission and impact wave in the process of acoustic emission characteristic parameter extraction and impact waveform acquisition, as shown in Fig. 13.9. The data are too little to carry out statistical analysis, so the acoustic emission waveform in the flexural tensile failure process of wet-sieved and four-graded concrete under impact loading conditions was fully collected.

The acoustic emission full waveform in the flexural tensile failure process of wet-sieved and four-graded concrete under the impact load is shown in Figs 13.19 and 13.20. We can see the acoustic emission full waveform in the whole bending tensile failure process of two kinds of concrete. The maximum amplitude in the diagram corresponds to the final fracture of the concrete specimens. Before that the waveform growing from small to big may be relevant to the formation and extension of microcracks of internal concrete under load. After final fracture the intermittent waveform may be relevant to the noise generated in the process of mutual friction between the concrete after fracture and loading fixture dropping off.

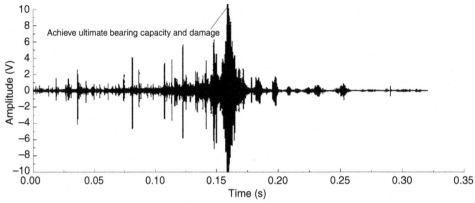

FIGURE 13.19 The Acoustic Emission Waveform in the Bending Tensile Failure Process of Wet-Sieved Concrete

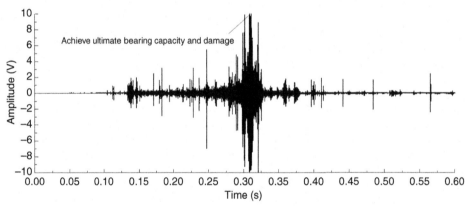

FIGURE 13.20 The Acoustic Emission Waveform in the Bending Tensile Failure Process of Four-Graded Concrete

The acoustic emission full waveform and power spectrum in the bending tensile failure process of wet-sieved and four-graded concrete under the impact load are shown in Figs 13.21 and 13.22. From the shape and distribution of the power spectrum diagram of each loading stage we can see that power spectrum diagram shapes of wet-sieved and full-graded concrete of each loading stage are different, but they are very similar under the 80–100% ultimate load. The shapes of the power spectrum diagram of each concrete are very close under the 30–60% and 60–80% ultimate load stage. For the wet-sieved concrete the shapes of the power spectrum diagram are similar under the 0–30% and 80–100% ultimate load stage.

13.2.4 IMPACT DYNAMIC BENDING TEST OF WET-SIEVED CONCRETE OF ARCH DAMS WITH DIFFERENT LOADING RATES

13.2.4.1 Analysis of acoustic emission activity

Dynamic bending test was carried out on wet-sieved concrete in Xiaowan arch dam with different loading rates, with the load time 201.10, 21.20, 2.33, and 0.29 s, respectively. The mechanical properties and acoustic emission activity characteristics in the flexural tensile failure process under different

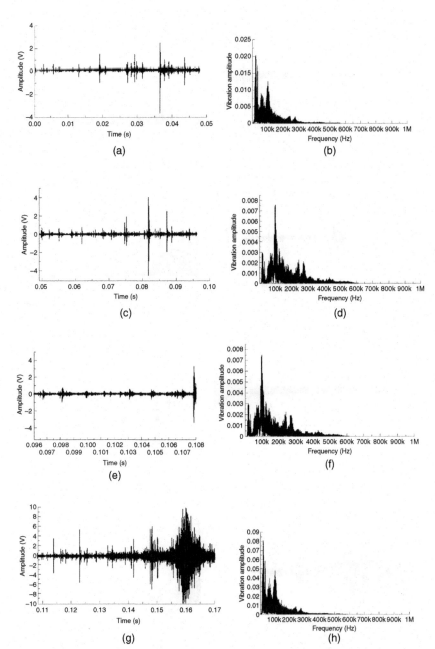

FIGURE 13.21 The Acoustic Emission Full Waveform and Power Spectrum in the Bending Tensile Failure Process of Wet-Sieved Concrete Under the Impact Load

(a) 0~0.048 s: The AE full waveform (0~30% ultimate stress). (b) 0~0.048 s: The power spectrum of AE full waveform (0~30% ultimate stress). (c) 0.048~0.096 s: The AE full waveform (30~60% ultimate stress). (d) 0.048~0.096 s: The power spectrum of AE full waveform (30~60% ultimate stress). (e) 0.096~0.108 s: The AE full waveform (60~70% ultimate stress). (f) 0.096~0.108 s: The power spectrum of AE full waveform (60~70% ultimate stress). (g) 0.108~0.16 s: The AE full waveform (after 70% ultimate stress). (h) 0.108~0.16 s: The power spectrum of AE full waveform (after 70% ultimate stress).

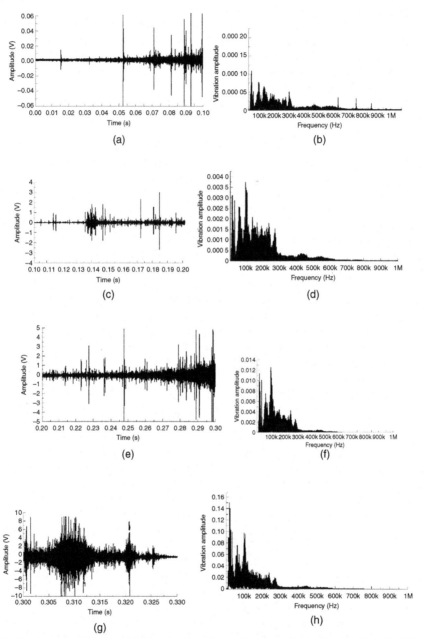

FIGURE 13.22 The Acoustic Emission Full Waveform and Power Spectrum in the Bending Tensile Failure Process of Four-Graded Concrete Under the Impact Load

(a) 0~0.1 s: The AE full waveform (0~28% ultimate stress). (b) 0~0.1 s: The power spectrum of AE full waveform (0~28% ultimate stress). (c) 0.1~0.2 s: The AE full waveform (28~55% ultimate stress). (d) 0.1~0.2 s: The power spectrum of AE full waveform (28~55% ultimate stress). (e) 0.2~0.3 s: The AE full waveform (55~83% ultimate stress). (f) 0.1~0.2 s: The power spectrum of AE full waveform (55~83% ultimate stress). (g) 0.3~0.4 s: The AE full waveform (after 83% ultimate stress). (h) 0.3~0.4 s: The power spectrum of AE full waveform (after 83% ultimate stress).

Table 13.12 The Mechanical Properties and Acoustic Emission Activity Characteristics in the Flexural Tensile Failure Process Under Different Loading Rates

Specimen Number	Loading Time (s)	Ultimate Load (kN)	The Cumulative Number of WD Hits (times)	The Cumulative Number of RD Hits (times)	Hit rate Peak (times/s)	Level Stress Occurs When AE (%)
DC-01	201.1	49.3	1577	3340	215	17.6 (8.7 kN)
DC-02	21.2	52.7	481	1245	460	48.6 (25.6 kN)
DC-03	2.33	57.3	174	433	859	73.5 (42.1 kN)
DC-04	0.29	65.6	51	69	1400	73.5 (48.2 kN)

Note: The table "WD" represents broadband sensor results; "RD" represents average value of multiple resonant sensor results.

loading rates are shown in Table 13.12. The acoustic emission hits, strength-loading rate curve, and cumulative acoustic emission hits–load curve throughout the loading process are shown in Figs 13.23 and 13.24. It can be seen in Fig. 13.23 that the resonant sensor adopts more acoustic emission hits than the broadband sensor does, primarily due to the higher sensitivity of the resonant sensor.

It can be seen from these figures and tables that the higher the loading rate is, the less the acoustic emission hits are, the higher the peak hit rate is, the later the concentration of emergence of acoustic emission hits are, and the higher the corresponding stress level is. Accordingly, with an increase of the loading rate, the whole failure time of concrete becomes shorter, and its flexural strength becomes higher. The result shows a significant strain rate effect.

Acoustic emission rate variation with time is called acoustic emission mode (Hongguang et al., 2000). The curve of acoustic emission hit rate varying with stress level under the different loading rates is shown in Fig. 13.25. It can be seen from the figure that when the loading rate increases by 10-fold of the magnitude, the peak acoustic emission hit rate increases by twofold of the magnitude. This indicates that there is a good correlation between acoustic emission hit rate and loading rate.

13.2.4.2 Analysis of acoustic emission characteristic parameters

Acoustic emission characteristics parameters of the wet-sieved concrete under the different loading rates are shown in Table 13.13. Acoustic emission amplitude distribution curve in the bending tensile failure process is shown in Fig. 13.26. It can be seen from Table 13.13 that with the increasing of loading rate, the proportion of acoustic emission signals with low amplitude (35–40 dB) increases. The average of counts, energy, and duration increases too. This is similar to the results gained by Muravin (1982) in the research of dynamic strength and acoustic emission characteristics of concrete using the electrodynamics method. The acoustic emission hits decreases under the high loading rates, but the average intensity of individual hit increases. This indicates that the higher the loading rate is, the higher the average energy releases in the development process of microcracks.

13.2.4.3 Analysis of acoustic emission source location

The region of local microcracks occurring at peak stress is known as the intrinsic process area. Acoustic emission source localization techniques can characterize the scope of this area. Figure 13.27 shows a three-dimensional projection of the acoustic emission source location. It can be seen from the scatter

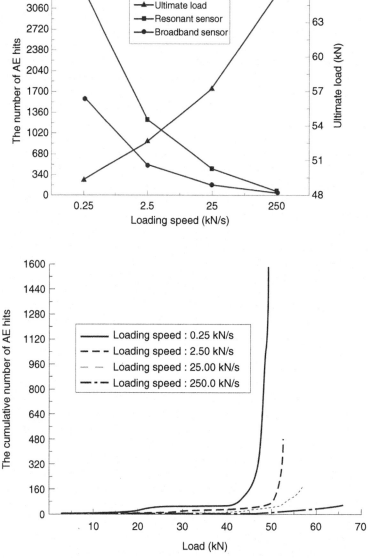

FIGURE 13.23 The Correlation Curve of AE Hits, Strength, and Loading Rate in the Flexural Tensile Failure Process of Wet-Sieved Concrete Under Different Loading Rates

range in the graph that the range of the region changes with an increase in loading rate, which means that the failure mechanism of concrete specimens is different under different loading rate. At low loading rate, there are few cracks in concrete, and its failure mechanism is controlled by a few major cracks (in the middle of the span). Only a handful of microcracks generate at other locations when the main crack is formed, most of the microcracks grow and extend surrounding the main cracks, while at high

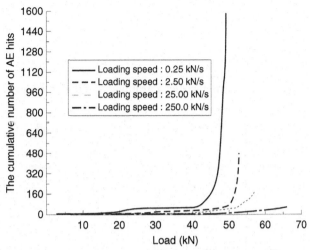

FIGURE 13.24 The Correlation Curve of the Cumulative Number of AE Hits and the Load in the Flexural Tensile Failure Process of Wet-Sieved Concrete Under Different Loading Rates

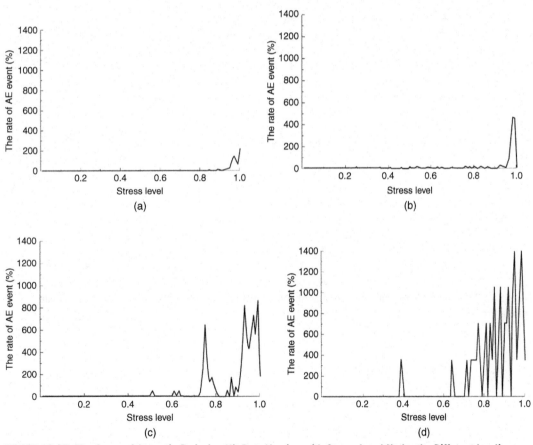

FIGURE 13.25 The Curve of Acoustic Emission Hit Rate Varying with Stress Level Under the Different Loading Rates

(a) Loading speed: 0.25 kN/s (b) Loading speed: 2.50 kN/s (c) Loading speed: 25.00 kN/s (d) Loading speed: 250.00 kN/s.

Table 13.13 Acoustic Emission Characteristics Parameters of the Wet-Sieved Concrete Under the Different Loading Rates

Specimen Number	Loading Speed (kN/s)	Hit ratio of 35~40 dB	Counts (times)		Energy Count		Duration (μs)	
			Accumulation	Average	Accumulation	Average	Accumulation	Average
DC-01	0.25	53	28,581	18	647	0.402	657,962	415
DC-02	2.50	60	16,667	35	375	0.776	406,044	844
DC-03	25.00	64	6,787	39	281	1.580	252,353	1450
DC-04	250.00	68	4,748	95	731	14.300	222,834	4457

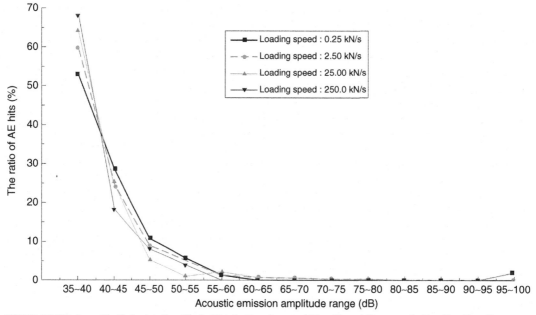

FIGURE 13.26 Acoustic Emission Amplitude Distribution Curve of Wet-Sieved Concrete in Bending Tensile Failure Process

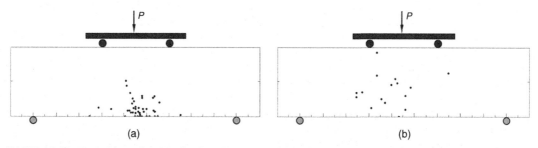

FIGURE 13.27 Three-Dimensional Projection of the Acoustic Emission Source Location

(a) Specimen DC01 (loading speed: 0.25 kN/s) (b) Specimen WDC01 (loading speed: 2.5 kN/s).

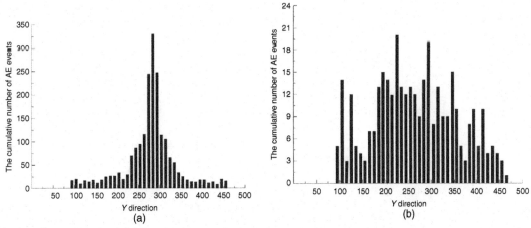

FIGURE 13.28 Acoustic Emission Source Linear Location Results

(a) Specimen DC01 (loading speed: 0.25 kN/s) (b) Specimen DC02 (loading speed: 2.5 kN/s).

loading rate microcracks distribute in a wide range. Some studies have shown that the increase in the intensity under the high strain rate is due to the resistance caused by the emergence of a large number of microcracks (Cho et al., 2003).

Acoustic emission source location results characterized by cumulative acoustic emission events under 0.25 kN/s and 2.5 kN/s loading rate are shown in Fig. 13.28. The y direction refers to the axial direction of the specimen, and the origin is at the left end. It can be seen from the figure that acoustic emission events (microcracks) concentrate in the middle of the span of the specimens, when the loading rate is 0.25 kN/s, while acoustic emission events distribute along the axis of the specimen, when the loading rate is 2.5 kN/s. It shows that the scope of the spatial distribution of acoustic emission events is more discrete under the high loading rate.

The rule that the z-coordinate of acoustic emission source location results changes with stress level increasing under two kinds of loading rates is shown in Fig. 13.29. It can be seen from Fig. 13.29 that when the loading rate is 0.25 kN/s, acoustic emission events begin to emerge from the tensile area at the bottom of the specimen and move upward to the compression zone with an increase in stress level, while microcracks appear to gradually extend downward from central part of the tensile area of concrete along with the stress increase when the loading rate is 2.5 kN/s.

13.3 AXIAL TENSION TEST OF CONCRETE AND ITS CONSTITUENTS
13.3.1 ACOUSTIC EMISSION COLLECTION SYSTEM SETTINGS

In the axial tensile test of concrete and the three constituents: aggregates, mortar, and aggregates–mortar interface, the test system sketch is shown in Fig. 13.30. Data acquisition systems of acoustic emission monitoring use the same settings.

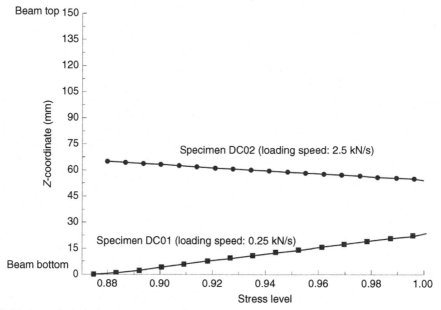

FIGURE 13.29 Acoustic Emission Events *z*-Coordinate Along with the Change of Stress Level

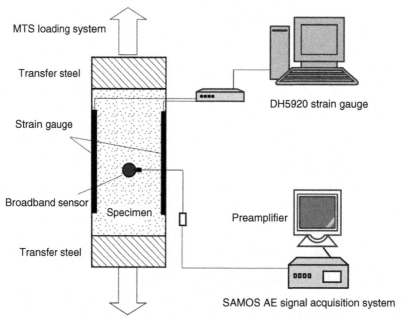

FIGURE 13.30 Experimental Arrangement for Uniaxial Tension Test and AE Measurement System

A broadband sensor (model PACWD) is fixed with rubber tape on the specimen surface in the middle of specimens, with Vaseline as a coupling agent. Preamplifier gain is set to 40 dB. According to the sampling system and sensor frequency range, the system band-pass filter is set to the range of 100~400 kHz, and the sampling rate is set to 3 MHz. For each acoustic emission hit (Hit), AE waveform of 4096 points for 1365 µs can be recorded and stored. Loading equipment noise levels are assessed by preliminary tests and field tests, and acoustic emission collection system thresholds are set to 35 dB.

On the axial tension study of the cement mortar specimens under different loading rates, six PAC-R6α sensors are arranged on the specimen surfaces (each two sensors for up, down, and side), the threshold of acquisition system and the preamplifier gain are set to 40 dB, and the sampling rate is set to 1 MHz.

13.3.2 CONCRETE

13.3.2.1 Analysis of acoustic emission activity

In the axial tension test of concrete specimen, the three-dimensional histogram of time history of acoustic emission frequency–amplitude distribution during its broken process is shown in Fig. 13.31.

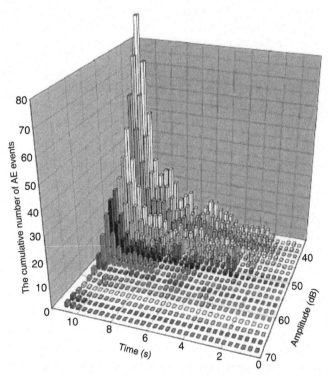

FIGURE 13.31 Time History Three-Dimensional Histogram of Acoustic Emission Frequency–Amplitude Distribution

Table 13.14 The Mechanical Properties and AE Activity Characteristics of Concrete in the Axial Tension Failure Process

Component Materials	Sample Number	Tensile Strength (MPa)	Peak Strain (με)	Elasticity Modulus (GPa)	Hits (time)	When AE Began to Continue to Appear	
						Stress (MPa)	Stress Level (%)
Concrete	CO1	2.77	88	31.5	1092	0.60	21.6
	CO2	3.28	98	33.4	559	0.50	15.2

In the axial tensile test, concrete mechanical properties and characteristics of acoustic emission activity in the failure process are shown in Table 13.14. It can be seen from the table that compared with specimen C01, specimen C02 has a lower cumulative number of acoustic emission hits, and its axial tensile strength, elastic modulus, and peak strain are correspondingly higher. Detectable AE sound appears in the failure process when stress level is about 184% on average.

In the axial tension failure process, the curve of the stress and the acoustic emission ringing accumulative number of concrete specimens versus strain is shown in Fig. 13.32. It can be seen from the diagram that acoustic emission activity gradually increases with the increasing loads, and that in the elastic phase of concrete axial deformation, acoustic emission ringing is maintained at a low level, and that the acoustic emission ringing cumulative curve has an apparent linear upward trend in the final break.

13.3.2.2 Analysis of acoustic emission rise time and amplitude

Acoustic emission rise time and amplitude are two parameters associated with the damage and fracture mechanism of the material. Rise time reflects the emergency degree of acoustic emission hits

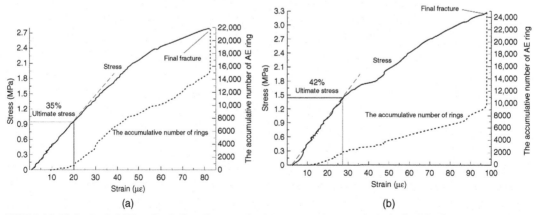

FIGURE 13.32 In the Axial Tension Failure Process the Stress and the Acoustic Emission Ringing Accumulative Frequency of Concrete Specimens Change with Strain Curve

(a) Sample number: C01 (b) Sample number: C02.

Table 13.15 AE Rise Time Statistics of Concrete in Static Uniaxial Tension Failure Process

Component Materials	Sample Number	Percentage of AE Hits Number to the Total (%)					
		≤30μɛ	31~60μɛ	61~90μɛ	91~120μɛ	121~150μɛ	≥151μɛ
Concrete	CO1	64.56	22.99	8.7	3.11	0.46	0.18
	CO2	64.22	23.43	6.98	3.4	1.79	0.18
Average value		64.39	23.21	7.84	3.26	1.13	0.18

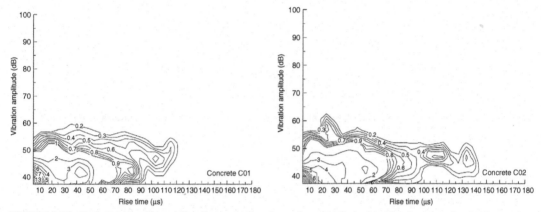

FIGURE 13.33 Cloth Contour Map of Two Concrete Specimen Acoustic Emission Hits Accounts for the Proportion of a Total Rising with the Amplitude and Rise Time

(Liu Guimin, 2006), and the shorter the rise time is, the more significant is the breaking characteristic. Table 13.15 summarizes the acoustic emission hits of concrete axial tensile specimens as a percentage of the total number of hits in different AE rise time range in the entire loading process. It can be seen from the table that the rise time of acoustic emission hits higher than 150 μs is only 0.18%, while the rise time of acoustic emission hits less than 30 μs in the entire process is 64.39%.

In the axial tension failure process, the distribution contour maps of acoustic emission hits as the proportion of the total versus the amplitude and rise time of two concrete specimens are summarized in Fig. 13.33, respectively. Contour shapes in the diagram show that acoustic emission hits with low rise time and low amplitude has a high number, while those with high rise time and high amplitude has a low number, and the whole contour map is roughly in the range of 0~140 μs rise time and 35~60 dB.

13.3.2.3 Spectrum analysis of acoustic emission waveforms

In order for it to be convenient for a quantitative analysis of the frequency characteristics of acoustic emission waveforms, in the power spectrum diagram of each acoustic emission hit waveform, the 100~400 kHz frequency band is divided into 100~150, 150~200, 200~250, 250~300, 300~350, and 350~400 kHz, six bands, respectively denoting bands 1–6. Figures 13.34 and 13.35 give the

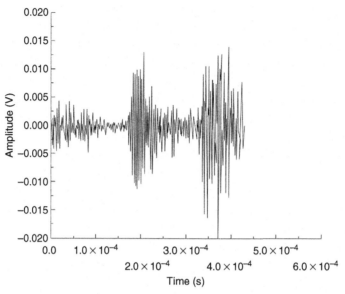

FIGURE 13.34 Typical Time-Domain Waveform of Concrete Specimens Under Axial Tensile Failure Process

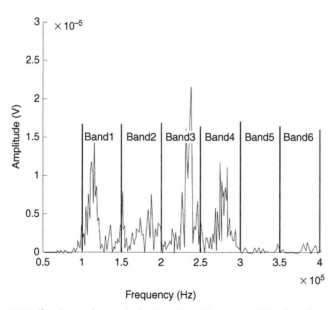

FIGURE 13.35 Its Corresponding Power Spectrum for Each Band Under Axial Tensile Failure Process

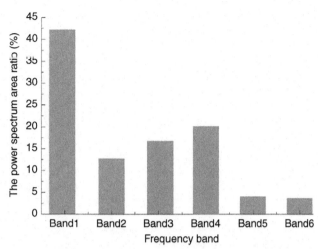

FIGURE 13.36 The Percentage of Each Area of the Power Spectrum in the Band Under the Uniaxial Tension Failure Process of Concrete Specimens

typical time-domain waveform of concrete specimens in axial tensile failure process and its corresponding power spectrum for each band.

In the failure process of concrete specimens, each power spectrum of acoustic emission hit waves is calculated, and the proportion of the area of the spectrogram in each of the six bands to the total area is counted. Figure 13.36 shows the average spectrum histogram of the proportion for each of the six bands of all the concrete specimens. It can be seen from the figure that in the failure process of axial tension, acoustic emission signal energy of concrete specimens is concentrated on bands 1~4.

Acoustic emission testing has shown that the acoustic emission signal energy of all kinds of materials tends to be concentrated in certain frequency bands. The band with the most energy concentrated is called the first frequency, and the band with the second largest energy concentrated is called the second frequency. By spectrum analysis of acoustic emission wave, and based on the bands of the first and second frequencies, the differences in material damage failure mechanism can be identified. To this end, the first frequency of band 1 spectrum is known as class A, and again in terms of the second spectrum frequency band, it can be divided into five types of A1~A5; the first frequency is outside the band 1, and according to its frequency band, it can also be divided into B~F with five total classes. Figure 13.37 gives the main acoustic emission hit power spectra of A1–A3, class B and class C.

In the axial tensile failure process of concrete specimens, for all kinds of acoustic emission hit waveforms collected, the percentage of its area to the total is shown in Tables 13.16 and 13.17, and Figs 13.38 and 13.39. It can be seen from the figure that the first frequency of concrete specimens in the axial tensile failure process are mainly distributed in bands 1, 4, and 3.

13.3.3 AGGREGATES

13.3.3.1 Analysis of acoustic emission activity

In the axial tension test of granite aggregate specimens, its mechanical properties and characteristics of acoustic emission activity are shown in Table 13.18. It can be seen from the table that compared

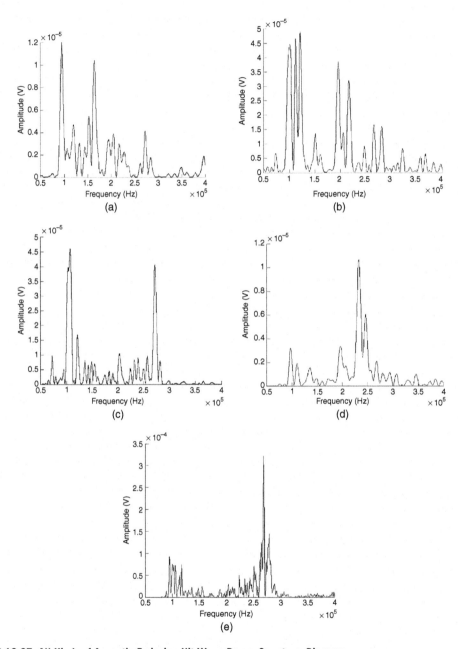

FIGURE 13.37 All Kinds of Acoustic Emission Hit Wave Power Spectrum Diagram

(a) A1 hits (the first frequency: 100~150 kHz; the second frequency: 150~200 kHz), (b) A2 hits (the first frequency: 100~150 kHz; the second frequency: 200~250 kHz), (c) A3 hits (the first frequency: 100~150 kHz; the second frequency: 125~300 kHz), (d) B hits (the first frequency: 200~250 kHz), and (e) C hits (the first frequency: 250~300 kHz).

Table 13.16 Categories of Class A of AE Hits and Various Classes of Hits Accounted for the Proportion of the Total Under Static Axial Tensile Damage of Concrete (%)

Hit Categories		A	A1	A2	A3	A4	A5
The first frequency		Band 1					
The second frequency		Random	Band 2	Band 3	Band 4	Band 5	Band 6
Concrete	C01	74.5	39	16.4	19.0	0.1	0.1
	C02	68.5	14.3	22.7	31.3	0.2	0.0
	average	71.5	26.7	19.6	25.2	0.2	0.1

Table 13.17 Categories of Class B~F of AE Hits and Various Class of Hits Accounted for the Proportion of the Total Under Static Axial Tensile Damage of Concrete (%)

Hit Categories		B	C	D	E	F
The first frequency		Band 2	Band 3	Band 4	Band 5	Band 6
The second frequency		Arbitrarily				
Concrete	CO1	2.5	6.0	15.3	0.7	1.0
	CO2	1.3	9.7	18.8	1.3	0.5
	average	1.9	7.9	17.1	1.0	0.8

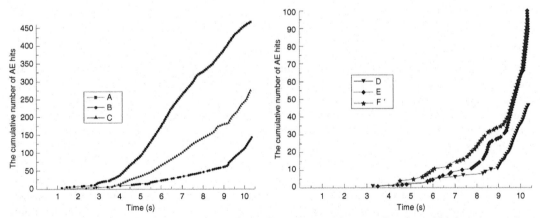

FIGURE 13.38 Acoustic Emission Hit Aggregate Time–History Curve of Concrete Specimen CO1 Spectrum Characteristics

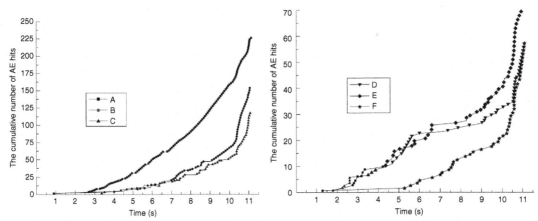

FIGURE 13.39 Acoustic Emission Hits Aggregate Time–History Curve of Concrete Specimen CO2 Spectrum Characteristics

Table 13.18 Mechanical Properties and Characteristics of Acoustic Emission Activity of Aggregates Specimens in Static Uniaxial Tension Failure Process

Component Materials	Sample Number	Tensile Strength (MPa)	Peak Strain ($\mu\varepsilon$)	Elasticity Modulus (GPa)	Hits (time)	When AE Began to Continue to Appear	
						Stress (MPa)	Stress Level (%)
Coarse aggregate (Granite)	A01	8.29	135	61.4	787	2.47	29.7
	A02	6.98			1426	1.29	18.5

with concrete specimen, the acoustic emission hits of aggregate specimens has a smaller number and their tensile strength is higher; its appearance of detectable acoustic emission signals is at the average stress level of about 24.1%, whereby indicating that its damage begins to emerge later than that of concrete. The linear relationship of concrete strength and the first acoustic emission load value established by Sadowska (1984) indicates that the first acoustic emission occurs later with a higher breaking strength.

Axial stress and acoustic emission ringing accumulative number of aggregate specimens changing with strain in axial tensile failure process are shown in Fig. 13.40. It can be seen from the diagram that acoustic emission activity gradually increases as the load increases, and that in the elastic stage of aggregate axial deformation, acoustic emission ringing number remains at low levels in the failure process, even almost no acoustic emission activity appears, and acoustic emission ringing cumulative curve shows an apparent linear upward trend in the final breakage.

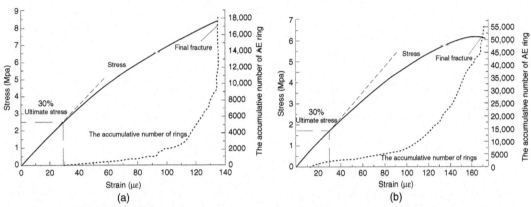

FIGURE 13.40 Aggregate Specimen Axial Stress and Acoustic Emission Ringing Accumulative Number Changes with Strain in Axial Tensile Failure Process

(a) Sample number: A01 (b) Sample number: A02.

13.3.3.2 Acoustic emission hits rise time and amplitude distribution analysis

Through the destruction process of the axial tension loading, the percentage of acoustic emission hits in rise time of aggregate specimens to the total AE hit are summarized in Table 13.19. It can be found in the damage process of aggregate specimen that the quantity distribution of the acoustic emission rise time is similar to the concrete axial tensile failure process. Acoustic emission hit number with acoustic emission rise time higher than 150 μs is very small, with only about 0.13% of the total, while acoustic emission hit number with rise time less than 30 μs is very large, with an average of 84% of the total, which is higher than the corresponding value of concrete.

The percentage of the aggregate specimen acoustic emission hit number to the total hit number changing with the magnitude and rise time in the axial tensile failure process are summarized in Fig. 13.41. The contour distribution is in the range of the rise time 0~80 μs and amplitude of 35~65 dB, and compared with concrete, mortar, and interface specimens, the contours are concentrated in a smaller range.

Table 13.19 AE Rise Time Statistics of Aggregate Specimens in Static Uniaxial Tension Failure Process (%)

Component Materials	Sample Number	AE Hit Number Accounted for the Proportion of the Total (με)					
		≤30	31~60	61~90	91~120	121~150	≥151
Aggregate	A01	87.55	11.56	0.76	0.00	0.00	0.13
	A02	80.22	14.8	3.54	0.84	0.42	0.21
Average value		83.89	13.18	2.14	0.42	0.21	0.18

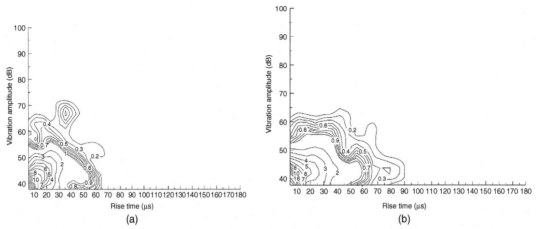

FIGURE 13.41 Distribution Summary of the Percentage of the Number of Aggregate Specimen Acoustic Emission Hit of the Total Changing with the Magnitude and Rise Time (Unit: %)

(a) Coarse aggregate A01 (b) Coarse aggregate A02.

13.3.3.3 Acoustic emission waveform spectrum analysis

Figure 13.42 shows the aggregate specimen average histogram of each band ratio. It can be seen from the figure that in the axial tensile damage process, acoustic emission signal energy produced by aggregate specimens is mainly concentrated in bands 1, 4, 3, and 2 in turn, being similar with the case of concrete specimens.

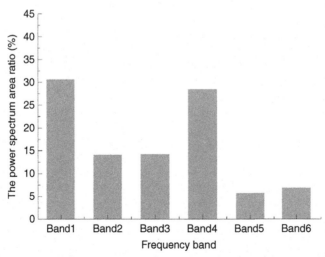

FIGURE 13.42 The Percentage of Each Area of the Power Spectrum in the Band Under the Uniaxial Tension Failure Process of Concrete Specimens

Table 13.20 Categories of Class A of AE Hits and Various Class of Hits Accounted for the Proportion of the Total Under Static Axial Tensile Damage of Aggregate Specimens (%)

Hit categories		A	A1	A2	A3	A4	A5
The first frequency		Band 1					
The second frequency		Arbitrarily	Band 2	Band 3	Band 4	Band 4	Band 5
Concrete	A01	58.1	14.4	16.3	27.1	0.1	0.3
	A02	49.6	13.9	9.9	25.4	0.1	0.3
	average	53.9	14.2	13.1	26.3	0.1	0.3

Table 13.21 Categories of Class B~F of AE Hits and Various Class of Hits Accounted for the Proportion of the Total Under Static Axial Tensile Damage of Aggregate Specimens (%)

Hit categories		B	C	D	E	F
The first frequency		Band 2	Band 3	Band 4	Band 5	Band 6
The second frequency		Arbitrarily				
Concrete	A01	1.7	7.1	30.9	0.9	1.4
	A02	1.5	5.8	40.5	0.9	1.6
	average	1.6	6.5	35.7	0.9	1.5

The percentage of the AE hit number of aggregate specimens classified by various types of acoustic emission spectrum waveform to the total hit number is summarized in Tables 13.20 and 13.21. It can be indicated from the table that acoustic emission hit number of class D accounts for the highest percentage of the total number, being 35.7%, which is significantly different from the characteristics of the concrete, mortar, and interface materials.

13.3.3.4 Aggregates acoustic emission characteristics under different loading rate of axial tension

With 0.02, 0.2, and 2 mm/s of different loading rates in the axial tensile test, the curve of aggregate specimen tensile strength and acoustic emission hits changing with the strain rate is shown in Fig. 13.43. It can be indicated from the figure that with an increase in loading rate, the acoustic emission total hits reduce and the tensile strength increases, whereby indicating that there is an obvious strain rate effect.

The curve of acoustic emission hits accumulative number changing with the axial stress in the aggregate axial tensile failure process under different loading rate is shown in Fig. 13.44. From the figure, with the increase of loading rate, the stress level at the first appearance of the acoustic emission signal increases. Under low strain rate, the stress level at the first appearance of the acoustic emission signal is low, while under high strain rate, the stress level at the first appearance of the acoustic emission signal is higher. When loading rate is 0.02 mm/s, from loading beginning to about 80% ultimate stress stage, the slope of the cumulative curve increases slowly, which shows that acoustic emission activity maintains

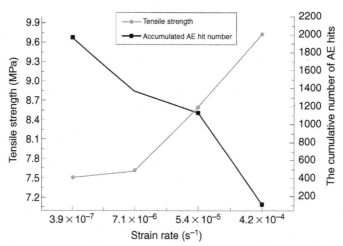

FIGURE 13.43 Curve of Aggregate Tensile Strength and Acoustic Emission Hit Changing with the Strain Rate Under Different Loading Speed

at a low level; when at over 80% limit stress stage, the acoustic emission hit number significantly increases, the cumulative curve slope begins to increase rapidly, acoustic emission activities become very active, and it shows that microcracks eventually find a weak path to release the energy from the external load (Tham et al., 2005). For rapid loading rate of 2 mm/s, acoustic emission signal begins to appear from the 32% limit stress stage, and always keeps a high slope until the near failure stage.

Figure 13.45 shows the histogram of granite acoustic emission hit rate increasing with the stress level in the axial tension failure process under different loading rates. From the figure, when the

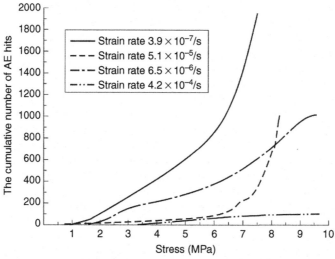

FIGURE 13.44 Curve of Acoustic Emission Hit Accumulative Number Changing with the Axial Stress in the Aggregate Axial Tensile Failure Process Under Different Loading Rate

loading rate is low, the acoustic emission rate maintains at a high level within a long load period. The higher the loading rate is, the more concentrated in high stress level the acoustic emission activities are, and before that the acoustic emission activity is rather weak and brittle fracture phenomenon is obvious. The higher the hit loading rate is, the higher the acoustic emission rate peak is, and it is from 512 times per second at 0.02 mm/s loading rate up to 766 times per second at 2.00 mm/s loading rate.

Under the conditions of different loading rates, the curves of the accumulated AE hits number and time in the range of the duration are shown in Fig. 13.46. It can be seen from the figure that AE hit of the low duration (<1000 μs) is mainly produced at the low stress level under static loading conditions, while AE

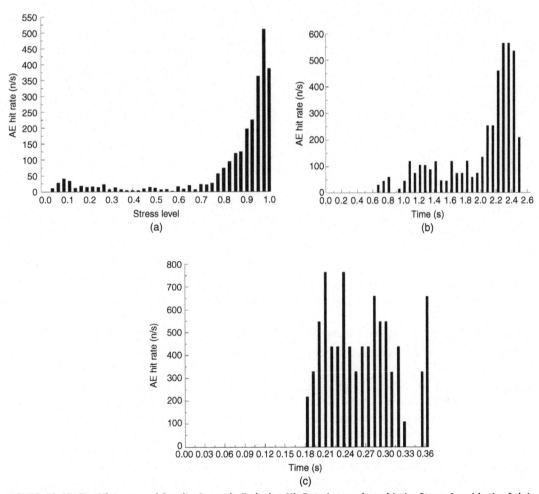

FIGURE 13.45 The Histogram of Granite Acoustic Emission Hit Rate Increasing with the Stress Level in the Axial Tension Failure Process Under Different Loading Rates

(a) Loading rate: 0.02 mm/s (b) Loading rate: 0.2 mm/s (c) Loading rate: 2.00 mm/s.

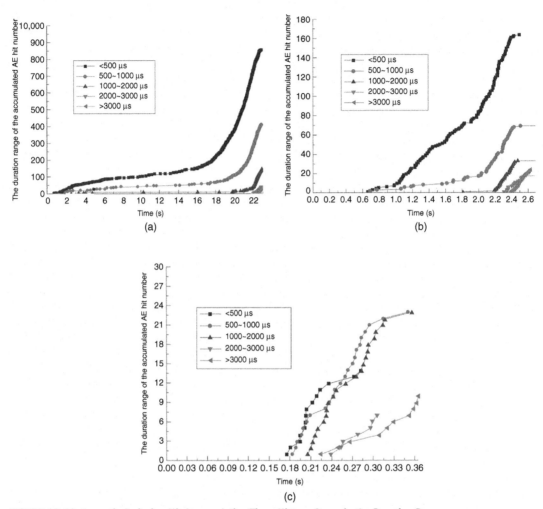

FIGURE 13.46 Acoustic Emission Hit Accumulative Time–History Curve in the Duration Range

(a) Loading rate: 0.02 mm/s (b) Loading rate: 0.2 mm/s (c) Loading rate: 2.00 mm/s.

hit of the high duration (>1000 μs) is produced at the high stress level under static loading conditions. The reasons are not that AE hits of high duration tend to produce at the same time, as shown in Fig. 13.46c.

Under different strain rates, the percentages of aggregate acoustic emission hits cumulative number to the total number in the whole process of axial tensile damage are summarized in Fig. 13.47. It can be seen from the figure that when loading rate is set to 0.02 mm/s, throughout the process of loading, within 200~800 μs, AE hits are in the largest proportion of the total, while when the loading rate increases to 2 mm/s, the acoustic emission hits within 800~1200 μs and over 3000 μs become outstanding and take the largest proportion, whereby illustrating that during the failure process under axial tension load for aggregate, the higher the loading rate is, the longer the duration of acoustic emission signal becomes, and the higher the proportion to the total signal hit number is.

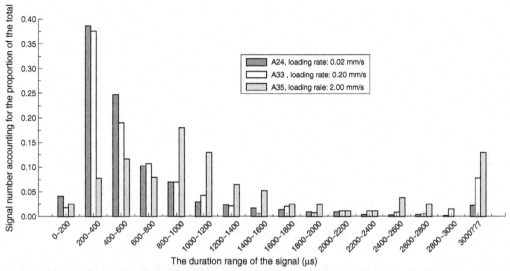

FIGURE 13.47 Duration Histogram in the Process of Aggregate Axial Tensile Damage Under Different Loading Rates the Duration of the Histogram

13.3.4 MORTAR

13.3.4.1 Mechanical properties and acoustic emission activity analysis

Mechanical properties and acoustic emission activity characteristics of mortar specimens are shown in Table 13.22. It can be seen from the table that in the case of the same sizes, the acoustic emission hit accumulative number of concrete is higher than that of mortar specimens, and mortar specimens under low stress levels have detectable acoustic emission activity.

In the axial tension test, the axial stress and acoustic emission ringing accumulative number curve along with the change of strain of mortar specimens are shown in Fig. 13.48. It is shown from the figure that in the mortar axial deformation of the elastic stage, the number of acoustic emission ringing shows a trend of gradual steady rise and the obvious linear upward trend occurs in the final fracture.

Table 13.22 Mechanical Properties and Characteristics of Acoustic Emission Activity of Mortar in Static Uniaxial Tension Failure Process

Component Materials	Sample Number	Tensile Strength (MPa)	Peak Strain (με)	Elasticity Modulus (GPa)	Hits	When AE Began to Continue to Appear	
						Stress (MPa)	Stress Level (%)
Mortar	S01	3.23	140	23.1	290	Start loading and it appeared	

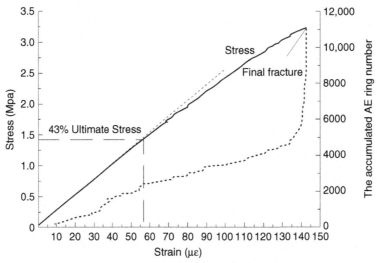

FIGURE 13.48 Shows the Axial Stress and Acoustic Emission Ringing Accumulative Number Curve Along with the Change of Strain of Mortar Specimens in Axial Tensile Damage Process (Specimen No:S01)

13.3.4.2 Acoustic emission rise time and amplitude analysis

Throughout the process of loading until damage, the percentages of acoustic emission hit numbers of mortar axial tensile specimens during each acoustic emission rise period to the total number of AE hits are summarized in Table 13.23. It is shown from the table that being similar to the case of specimens of concrete and aggregate, the acoustic emission hit number accounts for a small proportion of the total number, being only 0.34%, when the acoustic emission rise time is more than 150 μs, while the acoustic emission hit number accounts for the highest proportion of the total number, reaching about 66%, when the acoustic emission rise time is less than 30 μs; and the acoustic emission hit number is small when the acoustic emission rise time is more than 90 μs.

In the process of the axial tension damage of mortar specimens, the distribution contour plot of the percentages of acoustic emission hit number to the total varying with amplitude and rise time are summarized in Fig. 13.49. The acoustic emission hit number of low-rise time and low amplitude is higher, and the acoustic emission hit number of high rise time and high amplitude is low, which are similar to the characteristics of the damage process of concrete axial tension. The entire contour plot is in the range of 0~130 μs rise time and amplitude of 35~65 dB.

Table 13.23 AE Rise Time Statistics of Various Materials Specimens in Static Uniaxial Tension Failure Process

Component Materials	Sample Number	AE Hit Number Accounted for the Proportion of the Total n ($\mu\varepsilon$)					
		≤30	31~60	61~90	91~120	121~150	≥151
Mortar	M01	65.86	22.07	9.66	1.38	0.69	0.34

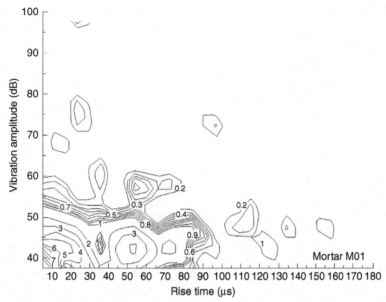

FIGURE 13.49 Distribution Contour Plot of the Percentage of the Number of Acoustic Emission Hit of the Total Along with Amplitude and Rise Time (Unit: %)

13.3.4.3 Acoustic emission waveform spectrum analysis

Figure 13.50 gives out the histogram of the average ratio of each spectrum in the test of mortar specimens. It can be seen from the figure that in the axial tensile damage process, acoustic emission signal energy produced by mortar specimens is mainly concentrated in bands 1, 4, 3, and 2 in turn.

In the mortar specimen axial tension failure process, the percentages of various types of acoustic emission hit waveform according to spectrum classification to the total are summarized in Tables 13.24 and 13.25. It can be seen from the tables that in the process of axial tensile failure, the first frequency of mortar specimens is mainly distributed in bands 1 and 4.

13.3.4.4 Mortar acoustic emission characteristics under different axial tensile loading rates

Under different loading rates, the tensile strength of mortar specimens and acoustic emission hit number are shown in Table 13.26. It can be seen from the table that with an increase in strain rate, the total number of acoustic emission hits and the time from the start of loading to reach the peak load decrease, while the tensile strength increases, whereby indicating that there is an obvious strain rate effect. This is the same as the conclusion of Khair (1995). When the strain rate is larger than 1.0 s^{-1}, for most of the axial tension mortar specimens, a small number of acoustic emission signals can only be detected in the final failure stage and acoustic emission activity has a weakening trend. In the same set of specimens from Table 13.26, there are differences in the cumulative number of acoustic emission hit before the final main fracture occurs, and the reason may be that the sensor coupling condition is different.

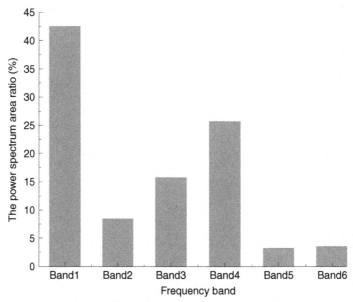

FIGURE 13.50 The Percentage of Each Area of the Power Spectrum from the Band in the Process of the Mortar Axial Tensile Damage

Table 13.24 Categories of Class A AE Hits and Various Class of Hits Accounted for the Proportion of the Total Under Static Axial Tensile Damage of Mortar Specimens (%)

Hit categories		A	A1	A2	A3	A4	A5
The first frequency		Band 1					
The second frequency		Random	Band 2	Band 3	Band 4	Band 5	Band 6
Mortar	M01	69.0	14.1	14.5	40.3	0.0	0.0

Table 13.25 Categories of Class B~F AE Hits and Various Class of Hits Accounted for the Proportion of the Total Under Static Axial Tensile Damage of Mortar Specimens (%)

Hit categories		B	C	D	E	F
The first frequency		Band 2	Band 3	Band 4	Band 5	Band 6
The second frequency		Random				
Mortar	M01	0.3	4.8	25.2	0.7	0.0

Table 13.26 Tensile Strength of Cement Mortar Specimens and Cumulative Number of AE Hits Under Different Loading Rate Tests

Loading Rate	Sample Number	The Total Load Time (s)	The Total Number of Acoustic Emission Hit (4 Channel Sum, Time)	Tensile Strength (MPa)	
0.02	S01	12.00	1075	2.18	2.46
	S02	12.35	1072	2.44	
	S03	11.87	330	2.77	
0.20	S04	1.83	2277	3.46	3.35
	S05	1.02	50	3.05	
	S06	1.65	991	3.55	
2.00	S07	0.2113	20	3.47	3.49
	S08	0.2151	53	4.02	
	S09	0.2146	6	2.98	
20.00	S10	0.0771	6	4.23	3.8
	S11	0.0737	8	3.00	
	S12	0.0752	5	4.17	
200.00	S13	0.0518	4	4.52	4.36
	S14	0.0515	8	3.69	
	S15	0.0535	4	4.86	

※ The final fracture on bonding position of the steel cap and the mortar specimen.

13.3.5 AGGREGATE–MORTAR INTERFACE

13.3.5.1 Analysis of acoustic emission activity

In the axial tension test, the mechanical properties and acoustic emission activity features of aggregate–mortar interface specimens during the failure process are shown in Table 13.27. The peak strain of the interface is the largest compared with concrete, aggregate, and mortar, which may be related to the strain gauge length of measuring the deformation of interface being shorter. In addition, in lower stress levels, aggregate–mortar interface specimens are likely to have the advent of the detected acoustic emission activity. Tensile strength and acoustic emission hit cumulative number of two aggregate–mortar axial tensile specimens are very close.

The curves of axial stress and AE ring cumulative number of aggregate–mortar interface specimen changing with strain in axial tension fracture are shown in Fig. 13.51. It can be seen from the diagram that at the elastic stage of the axial deformation of the aggregate–mortar interface, the acoustic emission ringing number in the failure process can sustain a growth trend, and that the acoustic emission ringing number cumulative curve shows an apparent linear upward trend near the final breakage.

Table 13.27 Mechanical Properties and AE Activity Features of Aggregate–Mortar Interface Specimen in Static Uniaxial Tension Failure Process

Component Materials	Sample Number	Tensile Strength(MPa)	Peak Strain (με)	Elasticity Modulus (GPa)	Hits	When AE Began to Continue to Appear	
						Stress (MPa)	Stress Level (%)
Interface	I01	2.45	437	5.6	448	Start loading and it appeared	
	I02	2.49	413	6.00	457	Start loading and it appeared	

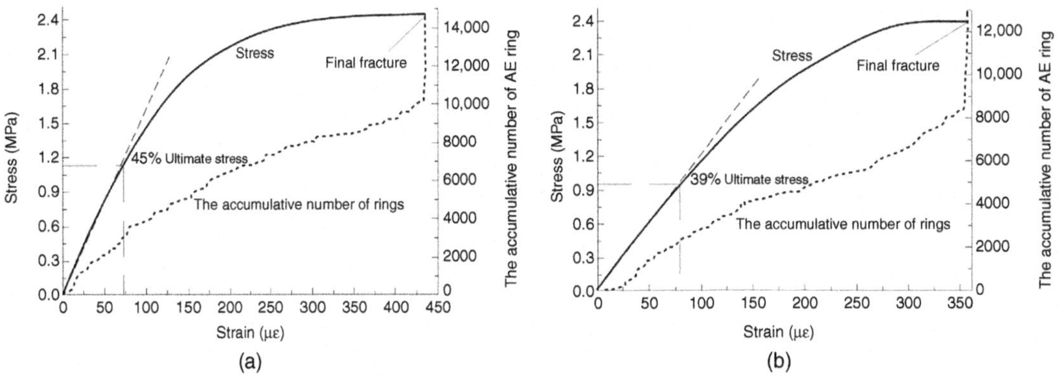

FIGURE 13.51 Axial Stress and AE Ring Cumulative Number of Aggregate–Mortar Junction Interview Pieces Changing with Strain Curve in Axial Tension Fracture

(a) Sample number: I01 (b) Sample number: I02.

13.3.5.2 Acoustic emission rise time and amplitude analysis

In the whole process of the start of loading to final damage, for cement mortar axial tensile specimen, the percentages of acoustic emission hit numbers in each acoustic emission rise period to the total AE hit number are shown in Table 13.28. It can be seen from the table that being similar to the case of specimens

Table 13.28 AE Rise Time Statistics of Aggregate–Mortar Interface Specimens in Static Uniaxial Tension Failure Process (%)

Component Materials	Sample Number	AE Hit Number Accounted for the Proportion of the Total (με)					
		≤30	31~60	61~90	91~120	121~150	≥151
Interface	I01	66.52	17.86	8.38	4.24	0.89	1.11
	T02	72.43	16.19	7.66	3.28	0.44	0.00
Average value		69.48	17.03	8.52	3.76	0.67	0.56

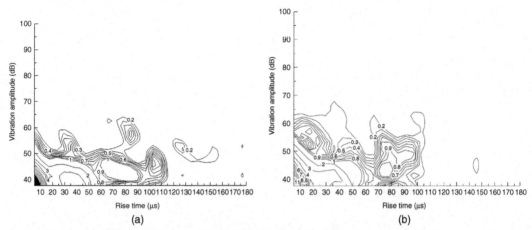

FIGURE 13.52 In the Process of the Axial Tension Damage Distribution Contour Plot of the Percentage of Aggregate–Mortar Interface Specimen Acoustic Emission Hit Number of the Total Along with Amplitude and Rise Time

(a) Interface I01 (b) Interface I02.

of concrete and aggregate, AE hit number of cement mortar specimens of the rise time longer than 150 μs has a small proportion to the total, being only 1.11%, and that AE hit number of cement mortar specimens of the shorter rise time (<30 μs) has the highest proportion to the total, reaching about 69%; the AE hit number of the rise time that is over 90 μs is a few.

In the process of the axial tension damage of aggregate–mortar interface specimens, the distribution contour plot of the percentage of acoustic emission hit number of the total with amplitude and rise time is summarized in Fig. 13.52. It can been seen from the figure that the acoustic emission hit number of low rise time and low amplitude is higher, while the acoustic emission hit number of high rise time and high amplitude is lower, and the distribution contour plot is in the range of 0~110 μs rise time and amplitude of 35~65 dB.

13.3.5.3 Spectrum analysis of acoustic emission waveforms

In the axial tension experiment of the aggregate–mortar interface specimens, the percentage of the power spectrum graph area in each band is shown in Fig. 13.53. It can be seen from the figure that in the damage process of axial tensile test, the acoustic emission signal energy generated by aggregate–mortar interface specimens and marked by the corresponding area of the power spectrum diagram is mainly concentrated in bands 1, 3, 4, and 2 in accordance with the order of sizes.

The percentages of all acoustic emission hit waveform to the total of aggregate–mortar interface specimens classified by spectrum analysis are summarized in Tables 13.29 and 13.30. It can be seen from the tables that the first frequency of aggregate–mortar interface specimens is mainly distributed in bands 1, 2, and 3, and that its characteristic is that the percentage of the number of class B acoustic emission hit to the total hit number is higher than those of aggregate and mortar.

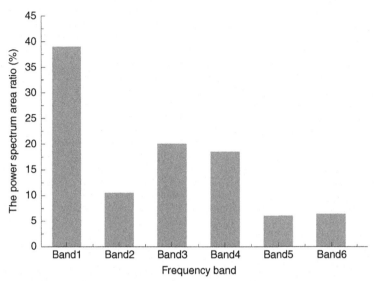

FIGURE 13.53 Power Spectrum Graph Area Percentage of Aggregate–Mortar Junction Interview Specimens in Each Band of the Axial Tension Experiment

Table 13.29 Categories of Class A AE Hits and Various Class of Hits Accounted for the Proportion of the Total Under Static Axial Tensile Damage of Aggregate–Mortar Interface (%)

Hit categories		A	A1	A2	A3	A4	A5
The first frequency		Band 1					
The second frequency		Random	Band 2	Band 3	Band 4	Band 5	Band 6
Interface	I01	64.3	15.4	29.5	19.4	0.0	0.0
	I02	63.7	23.4	29.1	10.9	0.0	0.2
	average	64	19.4	29.3	15.2	0.0	0.1

Table 13.30 Categories of Class B~F AE Hits and Various Class of Hits Accounted for the Proportion of the Total Under Static Axial Tensile Damage of Aggregate–Mortar Interface (%)

Hit categories		B	C	D	E	F
The first frequency		Band 2	Band 3	Band 4	Band 5	Band 6
The second frequency		At random (任意)				
Interface		0.9	14.7	14.1	14.1	2.0	4.9
		0.9	17.5	13.6	3.1	3.1	1.3
		0.5	16.1	13.9	2.6	2.6	3.1

13.3.6 COMPARATIVE ANALYSIS OF CONCRETE AND ITS COMPONENTS

13.3.6.1 Acoustic emission hit number cumulative curve

Acoustic emission hit accumulative number curves of concrete and its components along with the change of the axial stress are shown in Fig. 13.54. The following can be seen from the figure: (1) Under the same axial stress, the interface specimen, mortar specimen, and aggregate specimen are arranged in a descending order of the cumulative number of acoustic emission hits. The accumulative number of acoustic emission hits marks its activity and damage development degree. When the acoustic emission hits accumulative number of the interface transition zone is high, it means that a microcrack develops significantly, and the interface becomes the weakest component of concrete. (2) The acoustic emission hits for aggregate specimen appears to have the latest and the acoustic emission has very weak activity before 3.5 MPa stress level. Since the concrete strength is lower than 3.5 MPa in the test, it can be considered before reaching the peak load of concrete specimen that the damages of the mortar matrix and the aggregate–mortar interface are dominant, being consistent with the fact that there are almost no signals from the granite part in all of the acoustic emission signals collected in the whole damage process in the test and there is only a small amount of aggregate rupture in the final rupture surface of the concrete specimen. Accordingly, the stress levels from which the acoustic emission signal begins to be collected can also be taken as an auxiliary reference factor for identifying from what component the fracture is mainly originated.

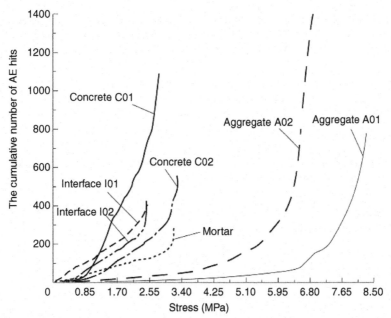

FIGURE 13.54 Acoustic Emission Hit Accumulative Number Curve of Concrete and its Various Components Along with the Change of the Axial Stress

Table 13.31 Distribution Contour Plot of the Percentage of AE Hit Numbers of the Total Along with Amplitude and Rise Time

Material	Concrete	Aggregate	Mortar	Interface
Rise ($\mu\varepsilon$)	0~140	0~80	0~130	0~110
Amplitude (dB)	35~60	35~65	35~65	35~65

13.3.6.2 Acoustic emission hit rise time

In the axial tension damage process of concrete and its three components, the distribution contour plot of the percentage of acoustic emission hit number to the total hit number versus the amplitude and rise time is summarized in Table 13.31. It can be seen from the table that the concrete and its three component specimens range in an ascending order of the rise time are as follows: aggregate, interface, mortar, and concrete. Accordingly, in identifying concrete fracture originated from which component, the rise time of acoustic emission signal hit advantage can be used as one of the references.

13.3.6.3 Types of acoustic emission wave spectrum

In the process of the axial tension damage of concrete and its three components, the major spectrum categories of AE hit are arranged according to the percentage of the total and are summarized in Table 13.32. It shows that the advantages of the acoustic emission signal band can be used as another reference of recognizing from which component the concrete fracture is originated. It can be seen from the table that the same type of frequency spectrum will be in a different band for different components. The first advantage frequency of granite aggregate is 250~300 kHz and 100~300 kHz; the first advantage frequency of mortar is 100~150 kHz and 150~300 kHz; the first advantage frequency of aggregate–mortar interface is 100~50 kHz and 150~200 kHz; the first advantage frequency of concrete is 100~150 kHz.

To sum up, the parameter analysis of the acoustic emission signal of concrete components, as well as the acoustic emission source location analysis, has an important significance on studying internal damage, failure mechanism, and strain rate effect. In particular, the acoustic emission test is with nondestructive and real-time monitoring features, which is especially important for the dynamic study on the mechanical characteristics of concrete dams. Although the current study is still at the initial stage, and it should be strengthened, improved and deepened, it has a broad application prospect.

Table 13.32 Concrete and its Component Medium Uniaxial AE Spectrum Categories (According to the Proportion of the Size of the Order)

Material Type	Concrete	Coarse Aggregate	Mortar	Aggregate–Mortar Interface
The main AE spectrum types in the failure process	A1,A3,A2,D,C	D,A3,A1,A2,C	A3,D,A2,A1	A2,A1,B,A3,C
The major categories accounted for the total proportion (%)	96.3	96	94.1	93.8

13.4 EXPERIMENTAL STUDY ON AE CHARACTERISTICS IN COMPLETE UNIAXIAL TENSILE FAILURE PROCESS OF CONCRETE CONTAINING SOFTENING STAGE

13.4.1 AE ACQUISITION SYSTEM SETTING

As earlier, the preamplifier gain is set to 40 dB. Several sensors are used in the test, as shown in Table 13.33. The acoustic emission sensor is fixed on the surface of the concrete specimen with the couplant (Vaseline) through a rubber band, and the sensor layout scheme is shown in Fig. 13.55. Through the pretest and evaluation of field loading equipment noise level, the final setting of resonant sensor threshold is 38 dB, and the broadband sensor threshold is 35 dB.

The mean P wave velocity of this batch of concrete is about 4080 m/s, tested by breaking the lead artificial excitation source. Acoustic emission waveform is filtered by band-pass filter with frequency of 20~400 kHz.

13.4.2 MONOTONIC UNIAXIAL LOAD

13.4.2.1 Analysis of the acoustic emission activity

In the uniaxial tension test, the curves of the axial stress and the cumulative number of AE hits with the strain of concrete are shown as Fig. 13.56. The complete failure process is divided into three successive stages to analyze the mechanical properties of concrete (Rossi et al., 1996; Rossi et al., 1992). The following part will take Fig. 13.56 as an example to analyze AE activity characteristics corresponding to the three typical stages.

When specimens are subjected to stress of 1.19 MPa (about 72% of ultimate stress of concrete axial tension elastic deformation stage, namely the OA stage), the initial microcracks of concrete propagate under the load, and new microcracks form, which distribute in the specimen randomly in the lowest stress area and local high stress area of specimens. Acoustic emission activity intensity is weak at this stage so that sensors mounted on various locations of the concrete specimens do not receive acoustic emission activity, which triggers the AE acquisition system.

Table 13.33 Loading System and Sensor Layout Scheme

Specimen Number	Loading System	Sensor Layout	Extensometer Guide Rod Gauge (mm)
DT01	Monotonic loading, direct tension	A broadband sensor	210
DT02	Execute load–unload process two times in the softening stage	A broadband sensor and four resonant sensors (NO. 1–5 in Fig. 13.55)	210
DT03	Execute load–unload process three times in the softening stage	A broadband sensor and five resonant sensors (NO. 1–6 in Fig. 13.55)	300
DT04	Execute load–unload process four times in the softening stage		275

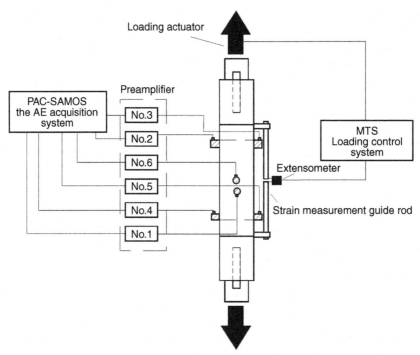

FIGURE 13.55 AE Testing Device for Monitoring Complete Concrete Uniaxial Tensile Failure Process

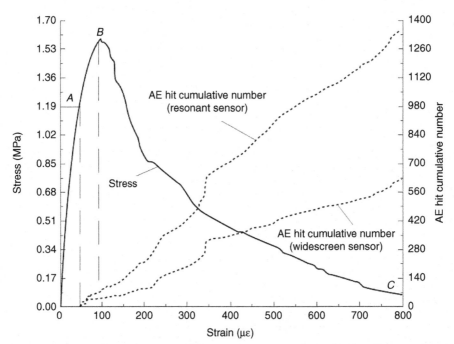

FIGURE 13.56 Curves of Tensile Stress and AE Hit Cumulative Number with Strain (Specimen Number: DT01)

At concrete uniaxial tension nonlinear deformation stage (from 1.19 MPa up to peak stress, the AB stage), the stress–strain curve shows a slight nonlinear characteristic. It is shown that further propagated microcracks in the first stage form macrocracks and localized microcracks in a specific region. The generation of macrocracks means the beginning of the third stage, at which the microcracks develop further and generate signals exceeding the threshold of the acoustic emission system and are recorded, which show that the development of acoustic emission hits accumulative curve.

At the stage that stress reaches the tensile strength of concrete, specimen fracture surface is gradually opened, but there is no failure and it still has a certain bearing capacity, while the bearing capacity decreases rapidly and the stiffness decrease shows "softening" characteristics (BC). The microcracks, width is so small that the specimen is not completely broken into two segments at the end of the test, and the surface microcracks are observed using a magnifying glass. At this stage, the macrocracks in the previous stage continue to propagate, which present the acoustic emission hit accumulative number continuing to increase and far more than that at prepeak. When strain reaches 350 µε, acoustic emission cumulative numbers increase abruptly, which is related to the internal microstructure change of concrete specimen (Landis, 1999) and possibly the localization of single critical microcracks (Li et al., 1993). But the stress–strain curve has no obvious mutation because the microcrack localization does not generate on the specimen surface mounted on the extensometer.

13.4.2.2 AE characteristic parameters and wave spectrum analysis

Under monotonic axial tensile test, the time–history curves of acoustic emission amplitude, axial stress, and duration time during the failure process of concrete are shown in Figs 13.57 and 13.58, respectively. Results indicate that AE characteristic parameters are not obviously different in different loading stages, which may be related to the sensor coupling in test.

In the complete failure process of concrete under monotonic uniaxial tensile, the peak frequency band of acoustic emission waveforms is divided into several frequency bands with 25 kHz as a

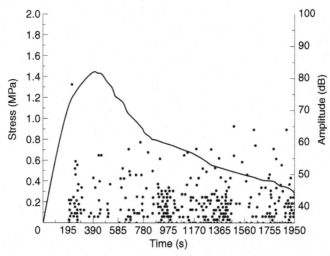

FIGURE 13.57 Time–History Curves of Tensile Stress and Acoustic Emission Amplitude in the Complete Process of Concrete Uniaxial Tensile Failure (Specimen Number: DTO2)

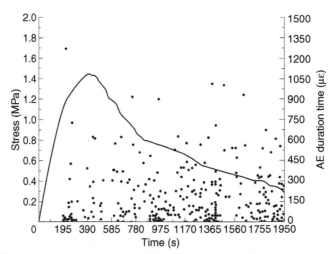

FIGURE 13.58 Time–History Curves of Tensile Stress and Acoustic Emission Duration Time in the Complete Process of Concrete Uniaxial Tensile Failure (Specimen Number: DTO2)

statistical interval used for statistics of acoustic emission hit cumulative number, and curves of each bands are shown in Fig. 13.59. It demonstrates that in frequency bands with 90~100 kHz and 100~125 kHz, acoustic emission hits occur continuously at about 400 s corresponding to the beginning of the softening stage. In frequency bands with 125~175 kHz and 200~250 kHz, acoustic emission hits occur continuously at about 1250 s, while acoustic emission hit curve gradient also

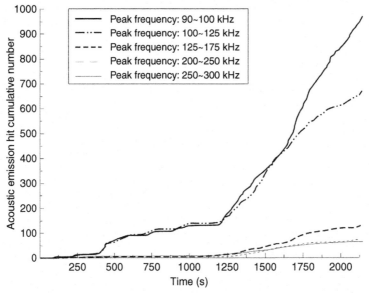

FIGURE 13.59 Time–History Curves of Acoustic Emission Peak Frequency in the Complete Failure Process of Concrete Uniaxial Tension

changes abruptly in frequency bands with 125~175 kHz and 200~250 kHz. Acoustic emission hit cumulative number curves in different frequency bands reflect the internal structure of concrete in different damage stages.

13.4.2.3 The relationship between characteristic parameters of acoustic emission and waveform spectrum

In the complete failure process of concrete under uniaxial tension load, the relationship between acoustic emission parameters and waveform frequency centroid is shown in Fig. 13.60. It can be seen that the average frequency (the average ringing number of duration), initial frequency (the average ringing number of rise time), and reflection frequency (the average ringing number of after rise time to duration) converted from the AE characteristic parameters decrease with the increasing of frequency centroid. In the range of 100~450 kHz, the correlation of frequency centroid with reflect frequency and average frequency is better than that of initial frequency.

In monotonic uniaxial tensile test, the relationship between acoustic emission amplitude and frequency centroid distribution at stress–strain curve ascending and descending stages as shown in Fig. 13.61. It can be seen that at ascending stage, the amplitude distributes in the lower level and peak frequency distributes in range between 300 kHz and 500 kHz. In the strain softening stage, acoustic emission amplitude distribution range is larger, and the acoustic emission frequency centroid decreases with an increase in acoustic emission amplitude. This indicates that in the softening stage of concrete, the microcrack propagation will produce high AE hit amplitude with low-frequency centroid. Therefore,

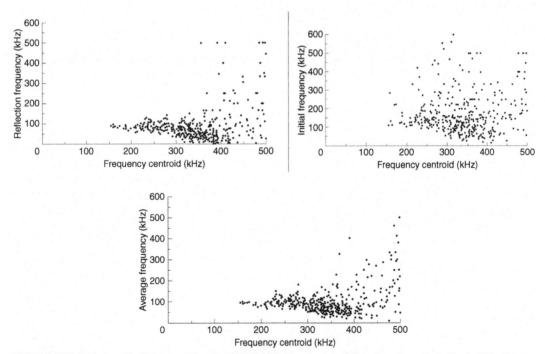

FIGURE 13.60 Relationship Between Acoustic Emission Parameters and Waveform Frequency Centroid in the Complete Failure Process of Concrete Uniaxial Tension

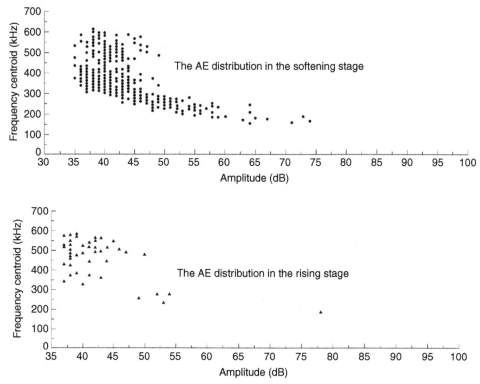

FIGURE 13.61 Relationship Between the Acoustic Emission Frequency and Acoustic Emission Magnitude in the Complete Failure Process of Concrete Uniaxial Tension

acoustic emission in the softening stage trends to low frequency. In addition, the microcracks develop continually, resulting in high-frequency components of acoustic emission signal attenuating, which leads to low-frequency tendency.

The normalized cumulative time–history curves of acoustic emission hit number, ring count, amplitude, and signal intensity collected by uniaxial tensile tests of the same concrete are summarized in Fig. 13.62. The figure shows the cumulative curve shapes of the hit number, count, and amplitude are very similar, so that they can substitute for each other in the analysis of the damage accumulation process of concrete. Compared with other parametric curves, the curve of signal intensity appears to fluctuate. Based on the signal intensity definition, signal intensity covers the acoustic emission waveform duration time and amplitude, so it can better describe the damage process.

The fracture energy of concrete can be calculated by the area under the curves of stress and average displacement. It is equal to the G_f specimen that fracture requires (Tengshanbajiu, 1997). The normalized cumulative time–history curves of fracture energy of concrete in the axial tensile failure process, along with hit number, ring count, and the signal intensity are summarized in Fig. 13.63. It can be seen that the cumulative curves of the concrete in uniaxial tensile damage process described by the several acoustic emission parameters are similar to the development trend of fracture energy, and even the correlation coefficient approximates 95%.

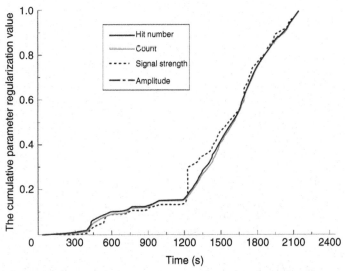

FIGURE 13.62 The Normalized Cumulative Time–History Curve of Acoustic Emission Parameters of Concrete in Uniaxial Tensile Damage Process

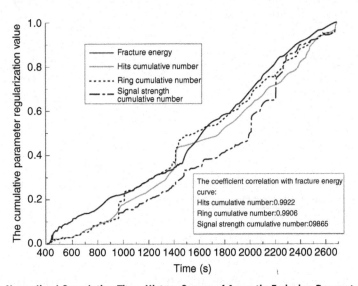

FIGURE 13.63 The Normalized Cumulative Time–History Curves of Acoustic Emission Parameters and Fracture Energy of Concrete in Uniaxial Tensile Damage Process (The Softening Stage)

13.4.3 CYCLIC LOADING IN SOFTENING STAGE

13.4.3.1 Analysis of the activity of acoustic emission

The closed-loop strain control technique in the test is used to restrain the unstable propagation of microcracks to some extent and obtain stress–strain curves including softening stage. As shown in

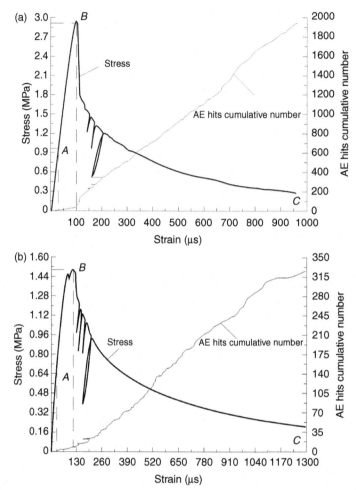

FIGURE 13.64 Curves of Tensile Stress and Acoustic Emission Hits Cumulative Number Along with Strain ((a) the specimen number: DTO3 and (b) the specimen number: DTO4)

Fig. 13.64, the cumulative number of acoustic emission hit curve suddenly rises at the strain value corresponding to suddenly dropped stress in post-peak stress–strain curve. Monitoring of acoustic emission hit accumulative curve can sensitively forecast in real time and describe the unstable development of concrete cracks.

Extremely unstable mutation appears in softening stress–strain stage of concrete under uniaxial tensile load, as shown in Fig. 13.65.

In the experiment, limited cycles of loading–unloading or variable deformation speed loading in a certain range will not affect the characteristics of the actual stress–strain curve, the envelope of which

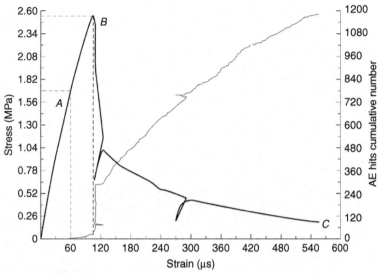

FIGURE 13.65 Curves of Tensile Stress and Acoustic Emission Hits Cumulative Number Along with Strain (the specimen number: DTO2)

is in accord with complete process curve under monotonic loading (Lu Yiyan, 1995). In the softening stage, specimens DT03 and DT04 perform three and four cycles of unloading–loading process, respectively. Here, unloading is performed by decreasing strain at high rate. As Fig. 13.64 shows, in unloading–loading cycles, the acoustic emission hit cumulative curve exhibits an approximately horizontal line, which indicates that no obvious acoustic emission activity is generated.

Figure 13.66 shows the time–history curves of acoustic emission ring number and uniaxial stress in the unloading–loading process. It can be seen that little acoustic emission activity is corresponding with the three stages, and the similar phenomenon also occurs in specimen DT04. Results illustrate that closed-loop servo system controls the axial deformation of concrete better and the development of concrete crack under cyclic loading.

13.4.3.2 Kaiser effect and felicity effect in uniaxial tensile stress–strain softening section of concrete

1. Description referred to stress

   Kaiser effect is defined as, in the fixed sensitivity, detectable AE do not appear until stress beyond that previously applied. In contrast, detectable AE appears under the stress, which is lower than the last applied stress; the phenomenon is called felicity effect. The ratio of stress at which the felicity phenomenon occurs to previously maximum applied stress is called felicity ratio (GB/T12604.4-2005). These two effects can be applied to evaluate the maximum stress concrete had suffered (Li et al., 1993).

   Kaiser effect does not always exist in the complete loading process, and it will not work at a certain stress level (Wu Shengxing, 2008), which is referred to as the felicity effect. Owing to

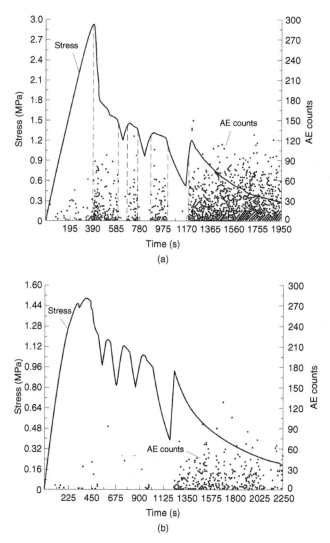

FIGURE 13.66 Time–History Curves of Tensile Stress and Acoustic Emission Counts of Concrete Under Uniaxial Tensile Load

(a) The specimen number: DT03 (b) the specimen number: DT04.

concrete uniaxial tensile stress at stress–strain softening stage under further cyclic unloading–loading may not exceed the maximum stress previously subjected (tensile strength); therefore, strictly speaking, concrete in strain softening stage does not have Kaiser effect but felicity effect.

Unloading stress of F_u and stress value of F_r corresponding to the beginning of acoustic emission activity under unloading are summarized in Table 13.34, where F_r/F_u are at about 1.0,

Table 13.34 The Characteristics of the Acoustic Emission of Concrete in the Unloading-Reloading Softening Stage

Specimen Number	nth Cycle	F_u (MPa)	F_r (MPa)	F_r/F_u	ε_u^* ($\mu\varepsilon$)	ε_r ($\mu\varepsilon$)	$\varepsilon_r/\varepsilon_u$
DT02	1	1.152	1.015	0.881	154.234	153.325	0.994
DT03	1	1.376	1.327	0.965	220.737	224.056	1.015
	2	1.266	1.192	0.941	252.109	251.810	0.999
	3	1.121	0.928	0.828	196.070	275.287	0.930
DT04	1	1.184	1.135	0.959	177.858	176.646	0.993
	2	1.138	1.068	0.938	190.417	188.982	0.992
	3	1.062	1.032	0.972	216.389	218.438	1.009
	4	0.940	0.861	0.914	254.625	244.775	0.961

The maximum strain value in history.

and F_u is not the peak load in the load history. Therefore, the acoustic emission in the softening stage during unloading–reloading cycles can record stress at which cycle begins to unload.

2. Description referred to strain (deformation)

It can be seen from Fig. 13.67, in the process of unloading–loading cycles in softening stage, that strain decreases under unloading while it increases under reloading. From Table 13.34, the ratio of strain (ε_u) corresponding to initial unloading in each cycle to strain (ε_r) corresponding to initiation of acoustic emission activity when reloading in the first two cycles is about 1.0, while it declines in the last cycle. The results show that acoustic emission in the softening stage of concrete under uniaxial tensile load can record the maximum strain during the complete damage process. As shown in Table 13.34, it can be indicated that in each loading–unloading cycle, $\varepsilon_r/\varepsilon_u$ is closer to 1.0 than F_r/F_u, which demonstrates that in the process of further unloading–reloading cycles in the concrete axial tensile softening stage, acoustic emission memory effect of strain is better than the stress.

The above characteristics of concrete stress–strain softening stage under uniaxial tensile load may be related to stress state of specimens. The fracture is mode I in the uniaxial tensile damage process of concrete, which is perpendicular to the loading axis. At softening stage, macrocrack has been localized, and crack width will reduce (be closed) when unloading. There is no new crack or dislocation friction similar to the uniaxial tensile damage until the width of a single macrocrack exceeds the previous maximum crack width, with no obvious acoustic emission activity. The results further account for and verify the assumption that concrete present linear elastic characteristics during the damage evolution process under unloading–reloading.

13.4.3.3 AE characteristic parameters and wave spectrum analysis

The time–history scatter diagrams of acoustic emission amplitude and duration time are shown in Figs 13.68 and 13.69. It can be seen that in the softening stage under uniaxial tension load, compared

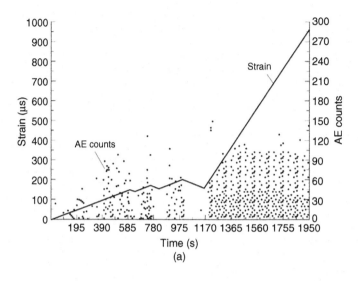

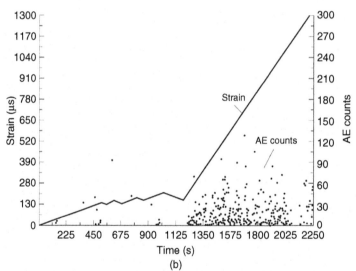

FIGURE 13.67 Time–History Curve of Strain and Acoustic Emission Count in the Concrete Uniaxial Tensile Failure Process

(a) The specimen number: DT03 (b) the specimen number: DT04.

with the prepeak stage, high amplitude (>60 dB) and long duration (>600 s) of the acoustic emission hit, began to produce along with the further propagation of microcracks of concrete specimens, indicating that the microfracture activities release more energy. This clarifies that analysis of AE characteristic parameters of concrete at each loading stage contributes to quantitative research on concrete damage.

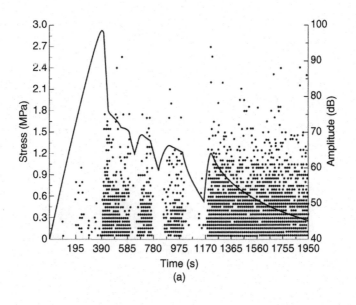

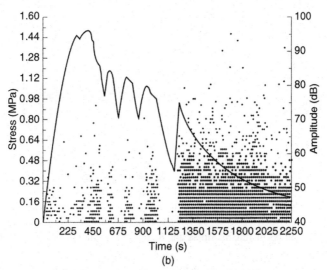

FIGURE 13.68 Time–History Curves of Tensile Stress and Acoustic Emission Amplitude in Concrete Uniaxial Tensile Failure Process

(a) The specimen number: DT03 (b) the specimen number: DT04.

The relationships between AE frequency centroid distribution and amplitude at ascending and softening stage under unloading–loading cycles are shown in Fig. 13.70a and b, respectively. It can be seen that the amplitude distributes in the lower level, and the peak frequency distribution range is between 200 kHz and 500 kHz in ascending stage. At strain softening stage, the range of AE amplitude

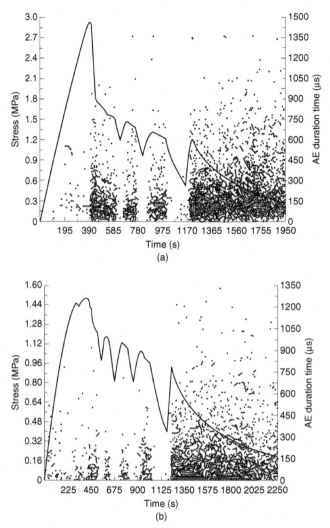

FIGURE 13.69 Time–History Curves of Tensile Stress and Acoustic Emission Duration Time in Concrete Uniaxial Tensile Failure Process

(a) Specimen number: DT03 (b) specimen number: DT04.

distribution is larger, and the AE frequency centroid decreases with the increasing of amplitude, which is similar to the characteristics of concrete specimens under monotonic uniaxial tensile loading.

13.4.3.4 Analysis of AE source location

The distribution region of the AE source can be roughly determined by comparing hit numbers of the AE sensor located at different positions (independent channel localization method). In general, the mechanism of concrete before microcracks localization and peak stress is the intrinsic property of the material, which can be ascribed to the small microcracks compared to the size of concrete specimen

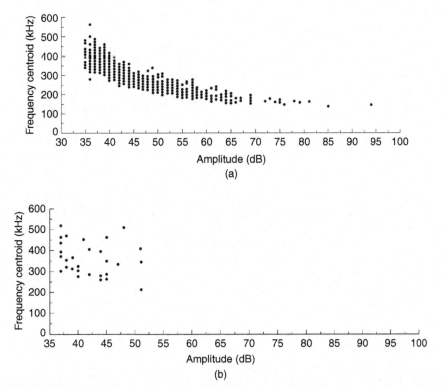

FIGURE 13.70 The Relationship Between Amplitude and AE Frequency Centroid of concrete Under Uniaxial Tensile Load

(a) The distribution of the AE hit at the softening stage (b) the distribution of the AE hit at the rising stage.

and statistically uniform distribution of stress and strain on concrete specimen scale, which can be proved from Fig. 13.71. Figure 13.71 shows the time history curves of tensile stress and cumulative number of hits of AE sensor at different positions. Before uniaxial stress reaches the limit stress (at about before 390 s), cumulative hit curves of AE sensor at different positions are not obviously different, which indicates that at this stage AE activity (microcracks activity) of concrete specimens represents a uniform distribution in the whole specimens.

After concrete crack localization, stress and strain are not statistically uniformly distributed, but obviously different in different positions, which becomes a global parameter, and the corresponding characteristic attributes to structural property. Figure 13.71 confirms that AE sensors near the visible crack receive more AE hit numbers in the stress–strain softening stage. In addition, the microcrack zone and actual position of surface microcracks of concrete specimen are detected by AE source location technology, shown in Fig. 13.72. Acoustic emission events appear in stress–strain softening stage, and it can be seen that most of the AE source locations are coincident with the actual location of cracks. So, microcrack propagation of concrete has obvious localization phenomenon under uniaxial tension loading, which can be reflected through cumulative hit numbers of different position sensors and acoustic emission source location technology. These research results have also been clarified by Zongjin Li (1994).

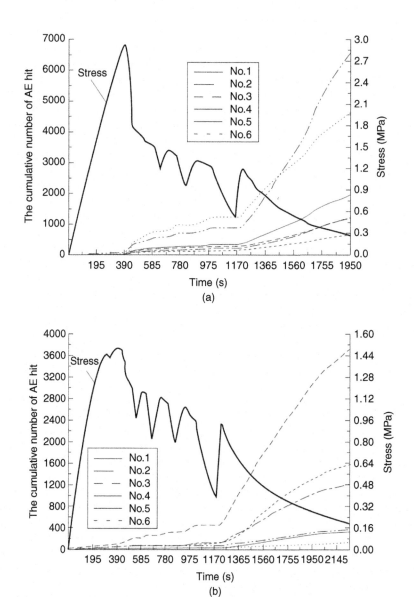

FIGURE 13.71 Time History Curves of Tensile Stress and Cumulative AE Hits of Different Sensors in the Complete Failure Process of Concrete Under Uniaxial Tension

(a) The specimen number: DT03 (b) the specimen number: DT04.

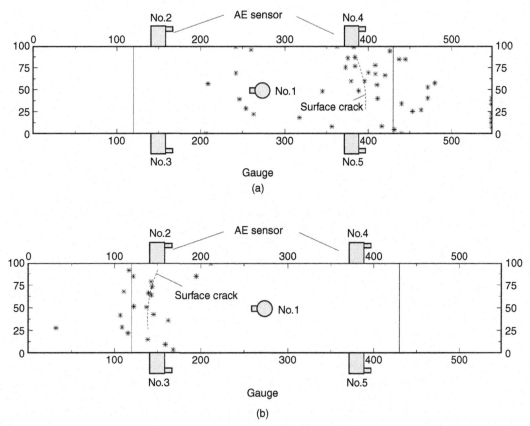

FIGURE 13.72 AE Source Location Diagram of Concrete in Complete Uniaxial Tensile Failure Process (mm)

(a) The specimen number: DT03 (b) the specimen number: DT04.

The stress–strain relationship can be measured by a strain gauge with the short gauge located at the center of the specimen and extensometer with the long gauge shown in Fig. 13.73.

Figure 13.73 shows that the crack zone of concrete is outside the scope of strain gauge, but still within the measuring range of the extensometer. Therefore, the deformation measured by the strain gauge is different from actual cracking in softening stage, reflecting the localized impact of microcracks of concrete specimen under uniaxial tensile load.

In summary, in the concrete damage process under uniaxial tensile load, before microcrack localization, the spatial distribution of microcracks may be different. Actually, the relationship between axial stress and displacement does not change, which reflects the essential equivalent property of concrete material in macro scale. After localized macrocracks appear, the stress–strain relationship of specimen containing localized macrocrack should be measured to reflect the damage characteristics of concrete material in the nonlinear damage stage. Energy parameters obtained by stress–strain relationship at

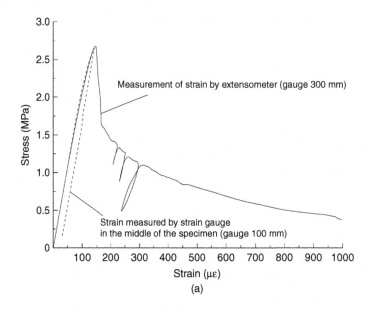

(a)

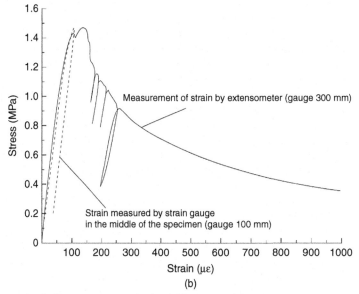

(b)

FIGURE 13.73 Stress–Strain Curves Measured by Strain Gauges and Extensometer

(a) The specimen number: DT03 (b) the specimen number: DT04.

softening stage are related to specimen size and the gauge of strain sensor. Fracture energy determined by load–crack displacement curve can be considered as material characteristic parameter. Therefore, element characteristic length should be taken into account in structure dispersion nonlinear finite element analysis considering material damage evolution laws when the experimental stress–strain curve of concrete materials is applied.

TESTING RESEARCH ON LARGE DAM CONCRETE DYNAMIC-STATIC DAMAGE AND FAILURE BASED ON CT TECHNOLOGY

14

Computed tomography (CT) technology used in materials testing is able to detect the inner changes in materials and structures without any damage and to give out the detected part intuitional image with high-resolution ratio. At present, in studying the inner material structural damage, X-ray CT technology is mainly used in static compression tests of rock specimens, and a series of progress has been made in CT image recognition of compressed rock cracks and the static compressed equipment research and manufacturing (Kawakata et al., 1999; Ueta et al., 2000, 2004). However, in the research on the inner structure damage of nonhomogeneous concrete materials, the application of X-ray CT technology has seldom been seen, and particularly CT test achievements of concrete dynamic tension have never been reported. For this reason, in terms of determining the requirements of large dam concrete material dynamic resisting force in combination with high arch dam engineering earthquake resistance, this chapter introduces the research work on the application of CT scanning technology in probing into the mesocracking process, whose contents include the scanning test principle of application, future description of the medical-use CT machines, and dynamic loading equipment in achievement analysis of the process of concrete static, dynamic, tensile, and compressive damage and failure, concrete subzone failure theory and support vector machine classification of CT images, related fractal theory and fractal dimension computation method, the fractal dimension computation and analysis of concrete un-axial compression CT image under the dynamic loading, the concrete damage evolution equation and constitutive relation establishment on the basis of CT images, the numerical simulation of concrete failure process of damage constitutive relation based on CT images, and the reconstruction of three-dimensional images and demonstration of animation, and so on. Research results indicate that CT tests used to study concrete mesofailure mechanism, check digital concrete computation parameters, and the feasibility of failure model have better application in the future.

14.1 APPLICATION OF CT SCANNING TECHNOLOGY IN CONCRETE MATERIAL TESTS

The purpose of the CT test is to study the concrete mesodamage evolution process, and test results obtain CT digital images of different cross-sections of concrete. Image grayscale or CT number size can reflect the changing process of inner material density. The varying CT image laws in different loading stages through studies can obtain the mesodamage evolution process of concrete materials.

At present, there are two types of CT scanning test equipment: the medical-use CT machine and industrial-use CT machine. The medical-use CT machine is characterized by fast scanning speed and images with high resolution, but a small dosage of ray radiation is difficult to carry out test research on the large-sized test specimens, and at the same, special test-loading installations are required. Industrial-use CT machine with a large dosage of ray radiation can carry out the large-sized specimen research, but its scanning speed is slow with a low-resolution image. Accordingly, at present, there exist many technical problems to be solved in the united application of the mechanics test machine used in direct and conventional materials.

In addition, CT technology must be realized through dynamic scanning, but the scanning process needs a certain time. Although the ongoing improvement made in CT equipment has been likely to make the scanning time shorten to millisecond grade, it is impossible in principle to realize the complete real-time scanning, whereby resulting in the fact that the effect of material dynamic mechanics performance research needs a further in-depth research project.

14.1.1 GENERAL DESCRIPTION OF CT SCANNING IMAGE PRINCIPLE

CT separates the detached faults of tested objects to imaging, thus avoiding having the remaining part interfere and affect the detecting images. In this way, the high-quality images are able to exhibit material components of the tested inner structure and defect conditions.

The working principle of the CT machine is as follows: a ray source and a detecting receiver are fixed on the same scanner frame to carry out the interlocking scanning of the detected subjects. After one scanning action is completed, the frame work turns an angle to carry out the next scanning so that it follows for many, many times in such a way as to collect many groups of data. If the level moving scanning once can obtain 256 data, each time turning $1°$ to scan once, revolving $180°$ can gain $256 \times 180 = 46,080$ data. And after these data are processed comprehensively, some sections of the digital image of the detected object can be obtained. The CT machine mainly consists of a radiation source and detector. The CT machine always takes X-ray as the radiation source, which can run through metal and nonmetal materials. X-rays with different wavelengths may have different piercing or probing-through abilities, while different materials have different absorption capacities for the X-ray with the same wavelength. The higher the material density is, the higher the atom ordinal number among the atom consisting of the matters is, and the stronger the capacity to absorb X-rays is.

The basic principle of CT imaging is this. When radiation emits the rays running through the objects, the ray strength can be attenuated because it is being absorbed by the objects, thus, conforming to the following equation:

$$I = I_0 e^{-\mu X} = I_0 e^{-\mu_m \rho X} = \int_0^{E_{max}} I_0(E) e^{-\int_0^d \mu(E)\mathrm{d}s} \mathrm{d}E \tag{14.1}$$

$$\mu = \mu^m \rho \tag{14.2}$$

where I_0 is ray initial strength; I is ray running through object after strength; μ is absorption coefficient of the object to absorb rays; μ_m is object unit mass absorption coefficient, only related to the incidence wave length; ρ is object density; and X is the length of ray running through.

As to water, $\rho = 1.0$ g/cm^3 so that its absorption coefficient $\mu_w = \mu_m$.

Projection value p is used to record the relative relation between initial strength value I_0 and the attention strength value I running through the objective, and the following equation can be expressed:

$$P = \ln \frac{I_0}{I} = \mu_m \rho X = \sum_{i=1}^{n} \mu_i \rho_i X_i \qquad (14.3)$$

where every section interval X_i producing effects upon the general attenuation depends upon the local attenuation coefficient μ; μ indicates the liner integrate of local attenuation coefficient on this path so as to calculate the attenuation value on every ray from the radiation source to the detector.

CT consists of many integrals exactly measured in fact. Accordingly, in order to obtain the satisfactory image qualities, it is necessary to record enough attenuation integrals or projection values in such a way, one by one within the range of 180°. At present, the surveys or measurements are being carried out usually in the range of 360° in a fan-shape bundle way. Therefore, many intervals and very narrow small data nodes in every projection must be determined so as to guarantee quality and accuracy. In order to calculate $\mu(x, y)$, it is necessary to ensure Nx independent equations obtained from projection surveys and to compute N^2 unknown numbers from $N \times N$ image matrix, of which $N_x = N_p$ (numbers of projections) $\times N_D$ (numbers of data nodes in every projection). When $N_x \geq N^2$ conditions are satisfied, the repeated iterations computation can derive $\mu(x, y)$.

At present, convolution reverse projection method has been adopted to rebuild CT images in carrying out computations. The definition of new function formed through convolution can be as follows:

$$h(x) = f(x)g(x) = \int f(x-t)g(t)\,dt \qquad (14.4)$$

where $f(x)$ is surveyed or measured data; $g(x)$ is convolution core, whose value is dependent on date processing method, data distortion degrees, and accuracy requirements.

CT technology is a reverse problem of projection to rebuild images with the solid mathematics theory serving as the blocking and modern microelectronics and computation technology as the support.

In order for it to be convenient for the direct comparison among CT images, defining CT value (H) is the object relative water attenuation coefficient, that is

$$CT_{value}(H) = \frac{(\mu - \mu_w)}{\mu_w} \times 1000 \qquad (14.5)$$

CT number unit adopts Hu to express in memory the first CT machine inventor in the world, G. Hounsfield, and CT scale is called Hounsfield for the same reason. Water is $H_w = 0$; air is $H_a = -1000$. Water and air CT values are not affected by ray energy so that they are the fixed points on the CT scale. The common material CT value range is from -1000 Hu to $+3095$ Hu, obtaining 4096 (i.e., 2^{12}) different CT values, which correspond with materials to be detected, with every CT unit having 4096 possible densities to display resolutions. CT numbers are used to make the quantitative resolution of each pixel density in an image, being much more accurate than gray scale. Grayscales in the photo are 256 possible resolutions; for instance, after the photo is scanned, the analysis processing of digitalized data is made in a computer, and the accurateness will be much poorer.

14.1.2 A BRIEF INTRODUCTION TO CT SCANNING EQUIPMENT

The CT machine was first successfully developed by engineer Hounsfield of the British EMI Company in 1972. It consists of the following three parts: (1) scanning system – it consists of an X-ray tube, detector, and scanning frame; (2) computation system – it carries out storage and operation of collected information and data; and (3) result output system – it can display the computer-processed and rebuilt image on the display or adopt various kinds of image outputs.

14.1.2.1 Medical-use CT machine

The biggest advantage of the CT scanning method lies in detecting the inner changes in materials and structures without any damages and at the same time with high resolutions. Largely owing to the different structures and functions of the CT machine, it can be divided into five generations in accordance with its development orders, formation, and performances.

The first generation of CT scanning machine adopts the rotating/parallel mode to collect X-ray scanning information used as data in image reconstruction, being the first azimuth irradiation assembly. To begin with, an X-ray bundle within 180° scope makes a synchronous parallel move from every angle. Each scanning needs 3–4 min with slow scanning speed and collection of few data so that the resolution of image reconstruction is poor.

The scanning mode of the second generation of CT scanning machine changes the single-stroke X-ray bundle into the fan-shape ray bundle. Many more detectors in the fan-shape arrangement can replace the single detector. Usually, over 30 detectors are set up so that total scanning time can be shortened to 18 s.

The X-ray tubes and detectors of the third CT scanning machine rotate synchronously. The number of detectors arranged in a fan shape can increase to 300–800 detectors and scanning time can be raised within 5 s, and image quality can obviously be improved.

The fourth generation of CT scanning machine has had only the X-ray tube doing rotary movement, appearing to be ring-arranged; the fixed detectors can increase over 1000 ones, and scanning time can be reduced to 2 s.

The current-use CT machine is the fifth generation of the screw-type of CT scanning machine. Using an electronic gun can shorten the scanning time to high-energy detector within milliseconds. The resolution is improved, and scanning body positions are increased. Apart from cross sections and crown section, it can scan the vector section and slant section. The thickness of the section decreases below 1 mm, and its working principle is shown in Fig. 14.1

Apart from the X-ray scanning machine, CT scanning machines under research and development are single photon emission CT (SPECT), positron emission CT (PET), nuclear magnetic resonance CT (NMR-CT), supersonic CT and microwave CT, and so on, at present.

Nowadays, CT test research on concrete dynamic mechanic performances mainly depends on medical-use the CT machine. The most important problem of application into large dam concrete materials without damage is that loading device scales are subject to scanning space constraints, so that there can be no way to use the large-size loading device in the complete gradation large dam concrete specimen tests. It is even more difficult to carry out three-direction loading to study more mechanics performances of materials under the complex stress conditions. Accordingly, adaptation of direct and conventional material mechanics machines in combination with industrial-use CT machines is likely to be the direction worth being further probed into.

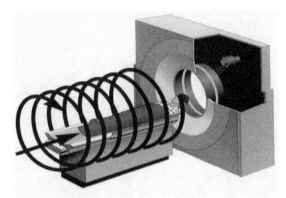

FIGURE 14.1 The Fifth Generation Medical-Use Screw-Type CT Machine

14.1.2.2 Industrial CT machine

Industrial CT technology is the advanced nondestructive detection technology based on the medical-use CT machine developed to gear to the industrial utilization in the 1980s. Detection by the industrial CT machine cannot be subject to test specimen material types, shape structures, and other constraints, whose imaging is intuitive without image overlapping and with high resolution.

The first generation of industrial CT machine adopts the rotary parallel X-ray scanning, single detector to accept mode; the second generation of industrial CT machine adopts the rotary narrow fan-shape light beam scanning, straight linear matrix detector to accept mode; the third generation of industrial CT machine adopts rotary fan-shape light beam scanning, ring linear matrix detector to accept mode; the fourth generation of industrial CT machine adopts the fixed three-dimensional cone X-ray scanning, face array detector to accept mode. The objects to be detected put into the cone-shape light beam rotate one circle, and all the fault images in this regional zone can be obtained, scanning mode rotates by scanning system, the fixed objectives detected are changed into the detected objects rotation, and the scanning system is fixed. This is the main flow scanning mode of present-day industrial CT machines, and the working principle is shown as in Fig. 14.2.

Industrial CT machines can enlarge radiation dosage, increase the penetration thickness, and conduct large-sized test specimen CT tests with high accuracy. In terms of the test requirement of the

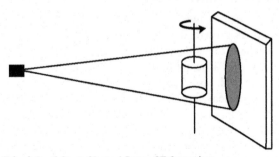

FIGURE 14.2 The Working Principle of Cone-Shaped Beam CT Scanning

FIGURE 14.3 Scheme Consideration Photo of Inner Setting in CT Machine and Testing System in Realignment with Large-Sized Material Testing Machine

complete gradation of large dam concrete dynamic mechanics performance tests, and in matching with the existing large-size dynamic material testing machines, it is adaptable to adopt the high-speed scanning and the scanning system rotary mode so as to shorten the intermittent time caused by scanning in dynamic loading. The scheme consideration of loading equipment supporting post space set in the scanning system is shown in Fig. 14.3. In order to meet the needs of three-direction loading and different testing methods and to avoid loading equipment from blocking the scanning light beam, it is possible to adopt the non-full-fledged rotary fast-speed scanning mode. In order to realize this scheme, it is still necessary to carry out the common research on solving a series of technically difficult projects under the coordination of some departments concerned.

14.2 PORTABLE TYPE DYNAMIC TEST LOADING EQUIPMENT IN REALIGNMENT WITH MEDICAL-USE CT MACHINE

14.2.1 TECHNICAL REQUIREMENT AND STRUCTURE OF TEST LOADING EQUIPMENT

As viewed from the mesomechanics angle, it is necessary to make a real-time observation and measurement of concrete microcrack growth, development, and macrocrack extension process under the action of dynamic loading. Therefore, studying the dynamic failure mechanism of concrete materials needs to carry out the online CT scanning tests of concrete dynamic loading. A new scanning medical-use CT machine with high speed has come into being, whereby creating the prerequisite conditions for carrying out this research project. However, it can only carry out offline CT scanning of test specimens after unloading, so that it is difficult to know the damage cracking process and crack real opening. In order to solve the problem of rock online CT scanning tests developed the static compression test loading equipment in realignment with medical-use CT machine. In order to carry out concrete dynamic loading online CT scanning tests, Dang Faning and Ding Weihua successfully developed the loading equipment that can carry out the dynamic and static tensile and compression tests of concrete specimens and can be in realignment with medical-use CT machines in the important priority research project of great research planning of the National Natural Science Foundation.

In accordance with the requirement of medical-use CT equipment and other utilization environment, this loading equipment must be small in volume and vibration, good in quality, and easy to bring or carry on, and it should work under the voltage of 380 V. This loading equipment must have the damping and noise elimination measures so as to make oil pump noise be less than 50 dB. Aluminum alloy materials for manufacturing compressed silo must satisfy the designed strength requirement in considering the confining pressure and should have the low absorption rate of X-rays so as to guarantee the test specimen CT image is clear.

Loading equipment technical indexes are tensile and pressure maximum outputs of 100 kN, maximum vibration frequency is 5 Hz, and the maximum diameter of test specimen is 150 mm. The pulse waves and sine waves with different frequency and amplitude can be generated; any waves can be given with the iso-loading wave shapes. They should be operated in terms of loadings and displacement controlling ways. And hence, the loading-time curves, the displacement–time curves, and the loading–displacement curves can be output. In the process of loading, the adoption of loading control or displacement control can be converted at any time in terms of designed parameter requirements. When the loading–displacement curve of the test specimen is tested, the loading–control loading is first adopted in general. When the loading reaches 80% of predicted peak strength, the loading can be shifted or converted into the displacement control loading, for it is necessary to be in realignment with medical-use CT machine, and the research and development loading equipment is subject to material quality, size, and quality, so that its pressure silo body rigidity cannot be very large, whereby affecting the testing of the stress–strain complete curve of the material with high strength.

The key to loading equipment is to solve well the problem where the test specimen tensile and pressure will not be subject to eccentric force, and the upper and lower tensile (pressure) transmission rods should center on test specimens. For this reason, the test specimens should be bound in a specially made V-shaped trough to realize.

The two ends of the specimen are bound in a hood-shaped concave trough so as to increase the binding force. Force transducer can be directly connected with the specimens so as to avoid the force transmission rods friction and resistance forces to affect the results of stress measurement.

The pusher piston displacement includes the specimen deformation, and loading system deformation, so that the rigidity test specimen is adopted to test the system deformation quantity, which will be deducted from total deformation quantity so as to obtain the real deformation of the specimen.

Figure 14.4 is the schematic diagram of the loading equipment. Figure 14.5 is the general picture of the material test machine and the whole CT test. Figure 14.6 is the two-way dynamic servo oil cylinder, Fig. 14.7 is the electrohydraulic servo valve, and Fig. 14.8 is the instrument pressure silo. It is a counter force frame when the single axis is stretching, as well as a confining pressure silo when the triaxial test is conducted. Figure 14.9 is the operating platform for the CT machine control system.

Portable type dynamic loading equipment is sponsored by the academician Dang Faning and designed and manufactured by Xi'an University of Technology and Changchun Municipal Chaoyang Testing Instrument Co. Ltd. with joint effort, whose application results indicate that this loading equipment is able to meet the needs of tests, and that this loading equipment is characterized by smart structure, being easy to move, being convenient for specimen installation, smooth operation, and so on.

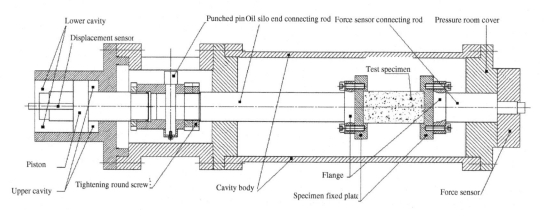

FIGURE 14.4 Structure Schematic Diagram of CT Scanning Used Portable Dynamic Loading Equipment

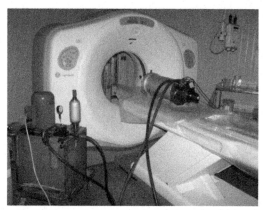

FIGURE 14.5 General Picture of Material Machine and CT Machine

FIGURE 14.6 Two-Way Dynamic Servo Oil Cylinder

FIGURE 14.7 Electro-Hydraulic Servo Value

FIGURE 14.8 Instrument Pressure Silo

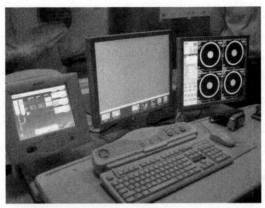

FIGURE 14.9 Operating Platform of CT Machine Control System

14.2.2 SPECIMEN INSTALLATION AND LOAD

The concrete mechanics performance CT test process includes sample preparation, sample installation, loading, CT machine scanning, and so on. Several key procedures, including the main test procedure, are as follows:

1. *Test specimen preparation*: Concrete CT test specimens can be divided into a large test specimen (D = 150 mm, H = 300 mm) and small test specimen (D = 60 mm, H = 120 mm) of two types. Large test specimens are prepared with the standard steel die pouring. The test specimen end part diameter is about 1 mm smaller than its fixed slab concave trough, and hollow parts are filled with bonding admixtures or adhesive materials. Small test specimens are prepared through drilling samples taken from specially poured large block concrete slabs. They are cut off and their end parts are smoothed to become the test specimens.

2. *Specimen binding and centering*: The binding effect is the prerequisite of ensuring tensile test quality. The concave trough with slightly larger diameter than the test specimen sizes should be set up at the binding end of the fixed steel plate of the binding test specimen's two ends. The depths of fixed steel plate concave troughs of big and small test specimen are 16 mm and 8 mm, respectively, so as to increase their side direction binding effect. Through a large number of tests, the research institute of the Chinese Academy of Sciences has been selected, and this glue is able to satisfy the strength requirements.

   The test specimen centering is also a key problem in the process of test specimen installation. For this reason, a special design of the binding fixtures for the test specimen is conducted. As shown in Fig. 14.10, the compression member of the fixture's two ends tightly compress the connecting lever or rod into the fine processing fixture V-shaped trough so as to guarantee the three axial line coincidence of the test specimen, fixed steel plate, and connecting lever.

3. *Installing test specimen*: As shown in Fig. 14.11a, when the test specimen is laid level, the test specimen is in the cantilever arm state. When the main machine end cover is fixed, lateral force will be produced, making it easy to make the test specimen breakdown with very low successful rate. Accordingly, vertical specimen installation is adopted (Fig. 14.11b).

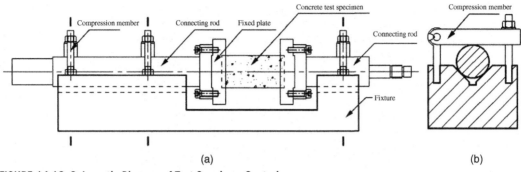

FIGURE 14.10 Schematic Diagram of Test Specimen Centering

(a) Front diagram (b) Cross-section diagram.

FIGURE 14.11 Two Types of Specimen Installation

(a) Vertical installation (b) Level laying specimen installation.

4. *Loading*: When the oil pump starts, and particularly when the oil pump is restarted in the case of tests being interrupted accidentally, the impact force produced is likely to make the test specimen suffer from failure. Therefore, oil pressure and loading controlling values become zero, oil pump is restarted, and pressure rising can be slowly conducted when CT scanning is carried out. Loading should be stopped, but loading should maintain no changes.

14.3 CONCRETE CT RESULTS AND INITIAL ANALYSIS

When CT scanning test is made on the concrete test specimen, the test specimen is, in general, scanned along the height by slicing with minimum scanning layer thickness of 0.8 mm. Chaps observed in CT images are called CT scale or dimension chaps, whose identified minimum width is 0.01 mm magnitude. The basic data of concrete mechanics performance CT tests measures and displays the evolution process of CT scale chaps, including chap growth, extension, healing, branching and breakthrough, and so on.

14.3.1 CT SCANNING TEST OF CONCRETE UNIAXIAL DYNAMIC AND STATIC COMPRESSION DESTRUCTIVE PROCESS

The portable-type loading equipment in realignment with CT machine adopted has carried out the systematic concrete static and dynamic uniaxial compression online CT scanning tests on a Marconi M8000 screw CT machine so as to make an initial study of the concrete static and dynamic destructive mechanism and the laws of dynamic strength variation.

The scanning equipment is a Marconi M8000 screw CT scanning instrument, whose resolution rate is 24-line pairs cm^{-1}; image size is 1024 mm ×1024 mm; imaging speed is the fastest within 0.5 s to scan four layers of CT image. Scanning parameters set in the CT machine are a layer thickness of 2.5 mm; electric voltage of 140 kV; electric current of 200 mA with high resolution; and the filter wave is D mode.

14.3.1.1 Online CT test of grade I concrete uniaxial static compress

1. Specimen conditions

   Grade I concrete is used; specimen size is φ60 × 120 mm. Fine aggregate materials selected from natural borrow area of Mardang hydropower station are natural medium sands. Cement used selected from Yunnan Lijiang Cement Plant is Yongbao 42.5 medium thermal cement. Concrete gradation rate: water–cement ratio is 0.62, sand rate is 45% per unit water use is 170 kg, and per unit cement use is 247 kg. The per unit fine aggregates is 870 kg, and the unit rough aggregates is 1062 kg. Tap water in coincidence with drinking water condition is adopted, and concrete age is 28 days.

2. CT scanning stage and stress–strain curve

   Figure 14.12 is the CONC-17 test specimen stress–strain curve in loading and the CT scanning stage loading curve appears the initial flat section and then embodies the material compacting process steeper section and tends to become flat at last, whereby indicating that the inner part of the material is local expanding capacity, and elastic modulus decreases. Each time after CT scanning loading increases, the curve appears to be steep first and then becomes flat.

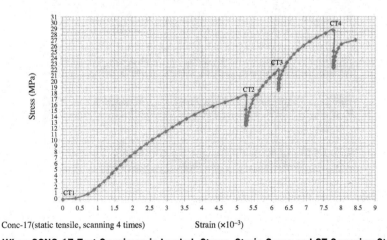

FIGURE 14.12 When CONC-17 Test Specimen is Loaded, Stress–Strain Curve and CT Scanning Stage

3. CT image analysis
   a. *Chap morphology*: Each time five groups are successively scanned, each group includes four-layer profile CT images, and layer thickness is 2.5 mm; therefore, scanning once can give out 20 cross-section CT images in total. The number in the CT image, for instance, "CONC-17-1" indicates CONC-17 test specimen of the first layer image in the first group.

   Selecting the CT image in the first cross-section of each group carries the arrangement of CT images in every section with stress varying sequence, as shown in Fig. 14.13.

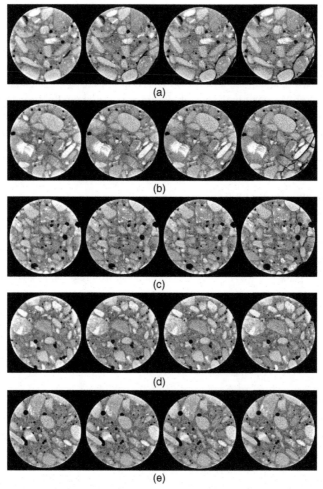

(a)

(b)

(c)

(d)

(e)

FIGURE 14.13 CT Images in the Process of Static Loading

(a) CONC–17-1-1 cross-section CT images with stress changes (b) CONC–17-2-1 cross-section CT images with stress changes (c) CONC–17-3-1 cross-section CT images with stress changes (d) CONC–17-4-1 cross-section CT images with stress changes (e) CONC–17-5-1 cross-section CT images with stress changes.

It can be seen from the chap evolution process of CONC-17 test specimens in different sections with each loading stage in Fig. 14.13 that when the initial scanning is carried out without loading, CT images display the clear aggregate materials, and hollow states within concrete inner parts, and in loading to 17.69 MPa (i.e. the second scanning), CONC-17-1-1 cross section right lower side aggregate boundary edge interface appears to have mesochaps expanding around the aggregates and at the same time emerges to have the crust falling off phenomena. Loading continues, and chap width increases.

When stopping loading scanning, displacement remains unchanged. Binding adhesiveness of the test specimen's two ends, deformation leads to a decrease in stress, and the stress loose phenomenon appears on the macro-loading curve.

When in the fourth scanning, chaps have extended to CONC-17-2-1 and CONC-17-3-1 cross-sections, that is, reaching the middle part of the test specimen, and stress loose phenomenon become more apparent.

One chap is developing in the loading process, and the chap expands bypassing the aggregates, thus displaying that the interface is the relatively weak part. Chap sequence is from the surface layer expanding to the test specimen's inner part. Largely owing to the nonuniformity of the materials, there are no symmetric type chaps to emerge in the tests, but the final failure shape of the test specimen still appears to be cone shaped.

b. *Analysis of average CT numbers of each cross-section A, B, C*: Three CT number statistic zones are selected as shown in Fig. 14.14. The average CT number within the zone is equivalent to all the pixel points within the zone representing average physical density value of concrete elements. Before and after chaps appear, the pixel point average CT number varying laws in each zone display that even in the minimum C zone, CT number variations are able to be reflected. Figure 14.15 shows the curves of A zone average CT number with stress variation in CONC-17, each cross-section under the uniaxial compression conditions. It can be seen from Fig. 14.14 that concrete test specimens, compressive process can be divided into such four stages as dense compactness, expansion capacity, CT magnitude or size chap extension, and failure. The symbol of dense compactness makes the statistics of increase in the zone CT

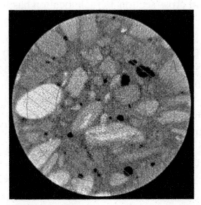

FIGURE 14.14 Schematic Diagram of Statistics Zones

A, big circle statistic zone; B, circular statistic zone; C, small circle statistic zone.

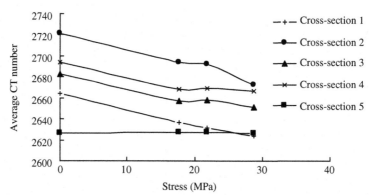

FIGURE 14.15 CONC-17 Each Cross-Section Average CT Number Curves with Stress Changes (Big Circle Statistic Zone)

number average value; the symbol of expansion capacity makes the statistics of decrease in the zone CT number average value; the symbol of CT magnitude chap extension shows the advent of linear or annular low density belt (i.e., dark color strips).

Since there is no CT scanning to be carried out in the dense compact stage without the advent of cracks, Fig. 14.15 does not display the rising stage of average CT numbers. This is because concrete elasticity modulus is very large, but its deformation is very small; the variation in CT numbers at its deformation is very small, and the variation in CT numbers at the dense compact and expansion capacity stages is so small that CT number average value curves are relatively smooth and stable on the whole.

In combination with the analysis of CT images, the 4.5 cross sections of CONC-17 have been in dense compact and expansion capacity so that there has been no advent of CT magnitude chaps. In the stress of 21.92 MPa afterward, cross section 3 appears to have CT magnitude chaps in the right edge. In the stress of 17.69 MPa, cross section 2 has cracks under the right corner, but the main CT magnitude chaps appear after 21.92 MPa; and cross section 1 crack location has been in slow in evolution and run through at last.

In order that CT number varying features can be used to quantify the analysis of the chap evolution process, the cross section CONC-17-1-4 (below 5 mm position of CONC-17-1-2), CONC-17-3-4 (below 5 mm position of CONC-17-3-1), and several key zones of chap positions (see Figs 14.16 and 14.17) are selected to carry out the quantitative analysis of CT number in the specified zones. CT number average value with stress variation source in the specified zones can be seen in Figs 14.18 and 14.19. Since chap growth and expansion are bound to cause the sharp changes in density in the statistic zone, the average CT number variation range in the specified zones in Figs 14.18 and 14.19 is also large, whereby indicating that in small zone statistics including chaps, CT number is sensitive to chap variation. The average CT number of zone 2 in CONC-17-1-4 cross-section in Fig. 14.18 drops faster than in zones 1 and 3, thus illustrating that the cracks in zone 2 appear earlier. CT number average value variation in several specified zones of cross section CONC-17-1-4 in Fig. 14.19 is synchronous, thus illustrating that this chap grows and expands faster and its evolution process is short.

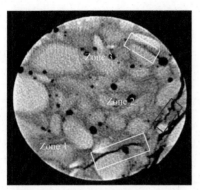

FIGURE 14.16 Cross-Section CONC-17-1-4 Statistic Zone

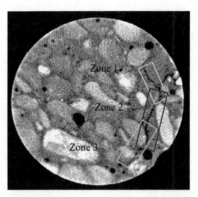

FIGURE 14.17 Cross-Section CONC-17-3-4 Statistic Zone

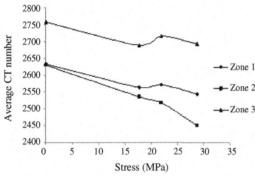

FIGURE 14.18 Cross-Section CONC-17-1-4 Specimen Zone Average CT Number

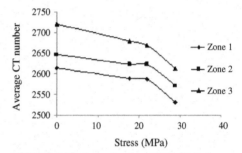

FIGURE 14.19 Cross-Section CONC-17-3-4 Specimen Zone Average CT Number

14.3.1.2 Grade I concrete uniaxial dynamic compression online CT test

1. Loading path and scanning stage

   The test specimen takes grade I concrete cylinder specimen CONC-13. For the simulation, after experiencing seismic alternative loading for a certain time, reaching failure, vibration amplitude is adopted. The predicted destructive loading is 90 kN; loading frequency takes 2 Hz. When CT scanning is carried out, loading must be stopped without unloading. It is necessary to maintain a certain pressure. Loading control mode is adopted for the initial loading. After the second scanning, it is necessary to change into the displacement control mode.

   Forming the cyclic loading load-time process can be expressed as $F = A_i \sin(2\pi ft) + F_0$ (Fig. 14.20), where F_0 is the initial static loading, A_i is amplitude value for each time cycle, f is the vibration frequency. Under the condition of monotone and cycle tensile, concrete stress–strain relation is close to linearity so that the following stress–strain expression formula can be adopted:

$$\sigma(t) = E(t)\varepsilon(t)$$

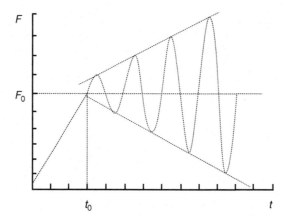

FIGURE 14.20 Schematic Diagram of Cycle Loading Load–Time Process Curve

Strain rate is as follows:

$$\frac{d\varepsilon}{dt} = \frac{1}{E}\left[\frac{d\sigma}{dt} - \frac{\sigma}{E}\frac{dE}{dt}\right] \approx \frac{1}{E}\frac{d\sigma}{dt} = \frac{1}{ES}\frac{dF}{dt} \tag{14.6}$$

where S is cross-sectional area.

Owing to the very large E value, the second item compared with the first item in the bracket of Eq. (14.6) can be omitted. Thus, substituting F into it, we can have the following:

$$\frac{d\varepsilon}{dt} = \frac{1}{ES}\left[2\pi fA\cos(2\pi ft) + \frac{dA}{dt}\sin(2\pi ft)\right] \tag{14.7}$$

where ΔA is the active force increment in each cycle.

In Eq. (14.7), $2\pi fA\cos(2\pi ft)$ plays an important role. This is to say that under the action of variation amplitude cycle loading, the maximum strain rate influence can be divided into two parts: the first part stands for the monotone loading rate maximum value; the second part stands for the rate influence produced by the cycle increasing range. The difference of the two items of rate influence is 90° phase position, whose maximum value does not happen at the same instance, and at the same time, the numerical size of two items are somewhat different, of which the first item plays an important role. Figure 14.21 is the loading time process curves when in loading. Figure 14.22 is the CT images after the five cross-sections experiencing different dynamic loading cycles. Through CT number mathematic analysis of each profile in different stress stages, the concrete mesodestructive mechanism can be studied.

2. CT image analysis
   a. *Chap morphology*: It can be seen from Fig. 14.22 that under the sinusoidal wave dynamic pressurization conditions, the chap distributions and features are obviously different from static loading. In the central part of the test specimen, many positions have yielded cracks at the same time, and chap morphology is relatively smooth and straight.

   After the first loading, cross-section CONC-13-1-2 occurs to have a crack on its left upper part, until it expands to the test specimen destructive stage, so that the crack lasts for a very long time. Chaps meet with the aggregates, stop expansion or change direction, whereby illustrating that the

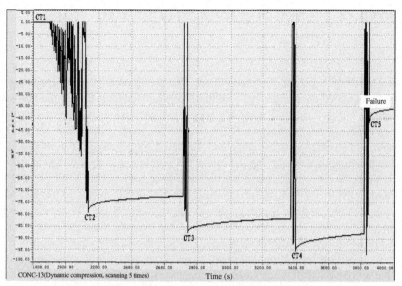

FIGURE 14.21 CONC-13 Loading Time Process Curves (Dynamic Pressure, Scanning for Five Times)

FIGURE 14.22 CT Image in the Process of Dynamic Loading

(a) CONC-14-1-2 cross-section CT images changing with stress (b) CONC-14-2-2 cross-section CT images changing with stress (c) CONC-14-3-2 cross-section CT images changing with stress (d) CONC-14-4-2 cross-section CT images changing with stress (e) CONC-14-5-2 cross-section CT images changing with stress.

aggregates with high strength can inhibit chaps. The evolution laws of the chaps are similar to the previously mentioned ones under the conditions of static loading. Before the last scanning stage, cross-sections 2–2, 3–2, and 4–2 suddenly appear to have obvious running through chaps.

b. *Difference value CT image analysis*: Difference value image is the image formed by the image CT number at the selecting stress stage and the initial stage image CT number to seek for difference solution, thus displaying the image CT number change at different stress stages. Based on the difference value image method, it is necessary to list CONC-13-2-2, CONC-13-3-2, and CONC-13-4-2 cross-section difference value CT images of the second layer of the second, the third, and the fourth groups as shown in Fig. 14.23. Based on every different value image, it is easier to observe and measure chap occurrence, development, and evolution process. The last scope of difference value CT image is characterized obviously by the linear and loop images, thus indicating that many chaps occur, and development and evolution occurs, but among the difference value CT images prior to three stress stages, there are no linear and loop images, thus indicating that this kind of chap is of mutation.

c. *Each cross-section average CT number analysis*: Figure 14.24 shows the curves of CT number average value changing with stress in each cross-section statistic zone of test specimen CONC-13. Statistic zone is the large circle statistic zone A in Fig. 14.14. The basic characteristics of these curves are each cross-section CT number average value in compact expansion capacity stages with smooth variation and same tendency, whereby illustrating that concrete materials

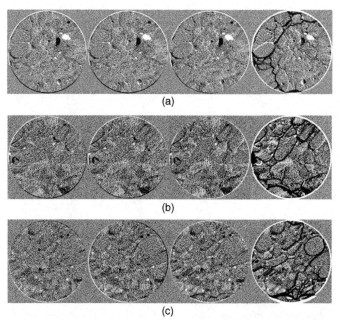

(a)

(b)

(c)

FIGURE 14.23 Difference Value CT Images on Different Cross-Section Changing with Stress

(a) CONC-14-2-2 cross-section difference value CT images changing with stress (b) CONC-14-3-2 cross-section difference value CT images changing with stress (c) CONC-14-4-2 cross-section difference value CT images changing with stress.

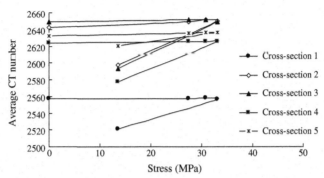

FIGURE 14.24 The Curves of CT Number Average Value with Stress Variation of CONC-13 Each Cross-Section Statistic Zone

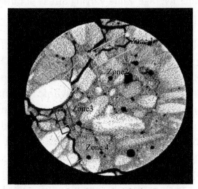

FIGURE 14.25 CONC-13-2-2 Cross-Section Tracking Chap Statistic Zone

are generally in the uniform damage evolution stage. In fact, the left upper part of cross-section CONC-13-1-2 has very early entered into CT magnitude chap evolution stage, but it is only because the chap parts are beyond the statistic zone so that CT number average value prior to failure cannot reflect this change. Therefore, in the last scanning stress stage, the test specimens are damaged, stress drops, and each cross section has produced the obvious running-through chaps in such a way that the test specimens seriously expand transversely and that CT number average value decreases, thus appearing to have the inflexion phenomenon.

Figure 14.26 shows the curves of CT number average value with stress variation of CONC-13-2-2 cross section specified statistic zone in Fig. 14.25. Owing to the specified statistic zone including chap zone, prior to the last scanning (at this time, the test specimen is damaged), CT number average value variation law is the same as that in the large circle statistic zone A. After the specimens' damage, CT number average value drops suddenly in every zone, thus indicating that chap generation is of mutation.

14.3.2 CT SCANNING TESTS OF CONCRETE UNIAXIAL STATIC, DYNAMIC TENSILE FAILURE PROCESS

In a direct tensile test, fracture face is basically parallel with the specimen end face vertical with tensile. In order to obtain the fracture face CT image, it is necessary to adopt the transversal section compacted

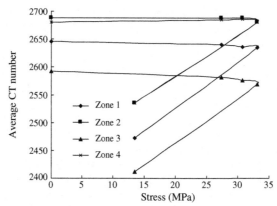

FIGURE 14.26 The Curves of CT Number Average Value Varying Stress of CONC-13-2-2 Cross-Section Specified Statistics Zone

scanning and then to rebuild the longitudinal section CT image, and it is able to obtain concrete chap expansion process image. For this reason, a Philips 16-Row Screw CT machine must be employed to carry out one scanning that can produce 16 CT images of 16 position parts of the specimen transversal section at the same time, whereby uniformly covering the whole specimen's possible destructive range. Scanning once only spends 1 s (one second). The time used by scanning and scanning layer spacing distance can be greatly shortened. Scanning suspension time in the loading process is even shorter so as to make loading curve more approach the real-time tests. This type of CT machine can not only solve the problem of enciphered scanning, but also rebuild the longitudinal section CT images in a CT post-processor system. For this reason, it can give out the transversal section and longitudinal section CT image of the test specimens, being very extremely convenient for observation and measurement of the concrete tensile chap process. The sizes of cross-section scanning images in tests are 512×512 pixels.

Concrete specimens in tests are adopted by using the double-curved arch dam concrete of Dagang-shan hydropower station. Complete adoption of real engineering original materials and mix proportion to pour concrete slabs is drilled with a core for taking samples. Test specimens are the cylinders with a diameter of 60 mm and height of 120 mm.

14.3.2.1 Online CT tests of grade I concrete uniaxial static tensile

1. *Stress–strain curves and CT images*: Figure 14.27 shows the stress–strain curves of three grade I specimens of CONC-042, CONC-043, and CONC-044 in the loading process in the static axial tensile tests. It is found through comparison that the stress–strain curves of the three specimens are very similar. The test repeatability is much better, indicating that the loading system performance is stable. The following only takes CONC-044 specimen as an example to analyze the process of failure.

In the loading process, test specimen CONC-044 in static tensile online CT test, its loading–time curve and displacement–time curve are shown in Fig. 14.28. The test loading–displacement curve converted into the stress–strain curve is shown in Fig. 14-29. Five CT scanning and image analysis time in test are seen in Fig. 14.30. Five longitudinal profiles of three-dimensional rebuilding of scanned images by the CT machine operating system and CT images in different stress stages are shown in Fig. 14.31.

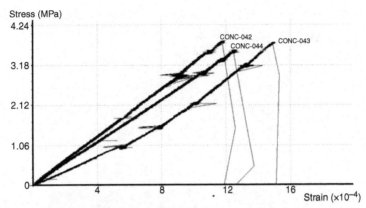

FIGURE 14.27 Stress–Strain Curves of Three Test Specimens

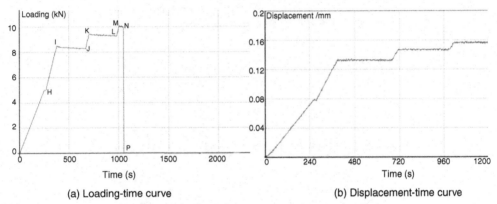

(a) Loading-time curve

(b) Displacement-time curve

FIGURE 14.28 CONC-044 The Loading Process Curve

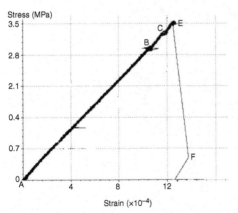

FIGURE 14.29 CONC-044 Stress–Strain Curve

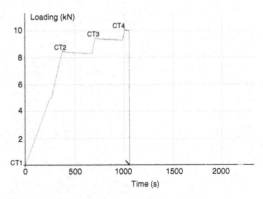

FIGURE 14.30 CONC-044 Scanning Time

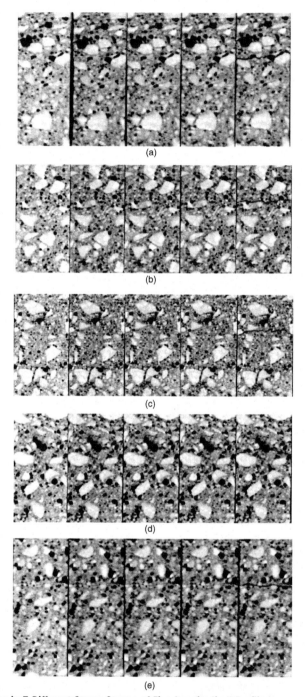

FIGURE 14.31 CT Images in 5 Different Stress Stages of Five Longitudinal Profiles

(a) CT image of one different stress stage of longitudinal profile (b) CT image of two different stress stages of longitudinal profile (c) CT image of three different stress stages of longitudinal profile (d) CT image of four different stress stages of longitudinal profile (e) CT image of five different stress stages of longitudinal profile.

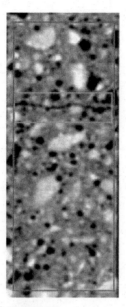

FIGURE 14.32 CONC-044 Test Specimen Statistic Zone

A large rectangle is the whole zone, and the inlaid small rectangle is the fracture zone.

In the intuition analysis of the five longitudinal uniaxial static tensile tests, chaps are vertical to tensile stress so that failure is of mutation in such a way that it is difficult to obtain CT images to reflect the chap expansion process. For this reason, CT number statistic method for the specified statistic zone is adopted, and the statistic zone position is shown in Fig. 14.32.

The five test specimen longitudinal profile locations and CT number average value of chap zone are shown in Figs 14.33 and 14.34. The two figures reflect CT number average value of each layer of the whole zone, and the trend of variance changing with stress are shown in Figs 14.33 and 14.34. The figures reflect CT number average value of each layer of the whole zone and the trend of variance changing with stress. As far as CT number average value is concerned, and in the case of tensile, CT number

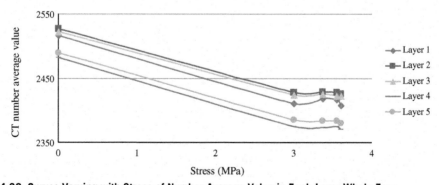

FIGURE 14.33 Curves Varying with Stress of Number Average Value in Each Layer Whole Zone

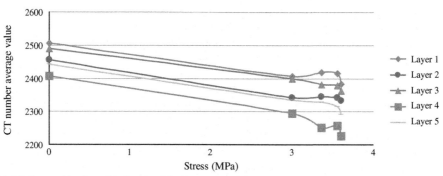

FIGURE 14.34 Curves Varying with Stress of CT Number Variance in Each Layer of Whole Zone

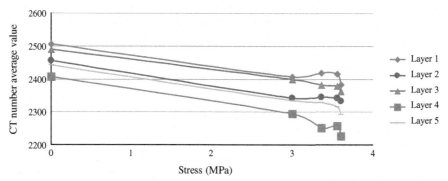

FIGURE 14.35 Curves Varying with Stress of CT Number Average Value of Each Layer Fracture Zone

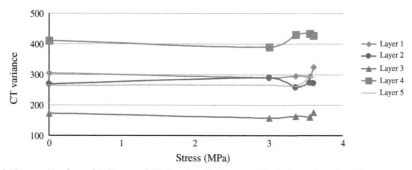

FIGURE 14.36 Curves Varying with Stress of CT Number Variance of Each Layer Fracture Zone

average value in the whole zone drops, while the whole change in variance is smooth, but after the third scanning and when in loading, variance average value increases, thus illustrating that the mesochap Figs 14.35 and 14.36 reflect CT number average value of the fracture zone in each layer surface and the trends of variance changing with stress. As far as CT number average value is concerned and with the last scanning time section, macro-chaps have appeared, and CT number average value at each fracture zone has the trend to drop, and the dropping range is larger than that of the whole CT number average value, whereby indicating that CT number average value of the local fracture zone is more sensitive to damage response. CT number average value of the fifth layer fracture zone in Fig. 14.35 after the third scanning drops obviously, whereby indicating that mesochaps grow from here to expand to other cross sections in sequences. This is the same conclusion as the variance average value judged in Fig. 14.36, whereby confirming mesochap expansion locations in the test specimen's inner parts.

14.3.2.2 Online CT tests of grade I concrete uniaxial dynamic tensile

1. *Loading path and scanning stage*: In the dynamic tensile online CT test, loading–time curves and displacement–time curves of test specimen CONC-055 in the process of loading are shown in Fig. 14.37. The stress–strain curves converted by test loading curves are indicated in Fig. 14.38. Dynamic tensile loading waveforms are different from average value up and down reciprocating

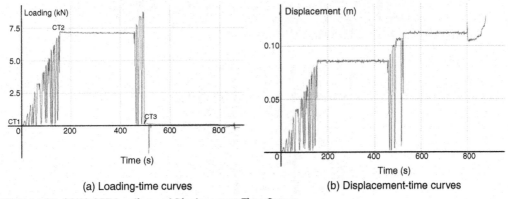

(a) Loading-time curves   (b) Displacement-time curves

FIGURE 14.37 CONC-055 Loading and Displacement Time Curves

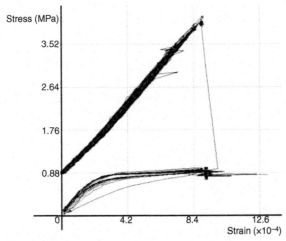

FIGURE 14.38 CONC-055 Stress–Strain Curves

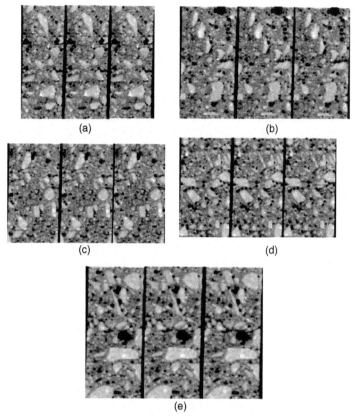

FIGURE 14.39 CT Image of Five Longitudinal Profiles at Different Stress Stages

(a) CT images of longitudinal profile 1 at different stress stages (b) CT images of longitudinal profile 2 at different stress stages (c) CT images of longitudinal profile 3 at different stress stages (d) CT images of longitudinal profile 4 at different stress stages (e) CT images of longitudinal profile five at different stress stages.

vibration waveforms on the horizontal axial line in dynamic compression. Dynamic tensile waveform axial line is the inclined line, and waveforms are not the tensile reciprocating, while the whole process is tensile.

2. *Description of results*: CT machine is used to carry out three-dimensional rebuilding of scanned images CT enciphered images of five longitudinal profiles with different loading as shown in Fig. 14.39.

3. *CT image analysis*: Being similar to the situations of static tensile, the intuition analysis of the mentioned five longitudinal profile CT images indicates that in the uniaxial tensile tests, chaps are vertical to tensile stress. The failure is of mutation so that it is difficult to obtain the CT image of the chap expansion process. For this reason, the specified statistic zone CT statistic method is adopted with the positions of statistic zone shown in Fig. 14.40.

CT number average value zones shown in Fig. 14.40 of the five longitudinal profile and chap zones are given in Figs 14.41–14.44.

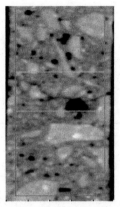

FIGURE 14.40 CONC-055 Test Statistic Zone

A large rectangle is the whole zone and the inlaid small rectangles are the fracture zones.

Figs 14.41 and 14.42 reflect CT number average value and trend of variance changing with stresses of each layer whole zone. When test scanning reaches three times, the test specimens are broken, so that loading is close to zero, but displacement still remains at maximum value. In the case of tensile condition, CT number average value of the whole zone appears to have a decreasing trend, while the variance average value varies smoothly on the whole. But after the second scanning and again loading

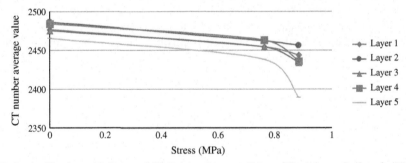

FIGURE 14.41 Curves Varying with Stress of CT Number Average Value of Each Layer Surface in Whole

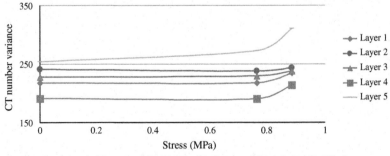

FIGURE 14.42 Curves Varying with Stress of CT Number Variance of Each Layer in Whole

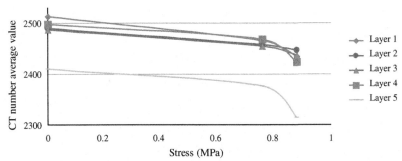

FIGURE 14.43 Curves Varying with Stress of CT Number Average Value at Fracture Zone of Each Layer Surface

scanning, the variance average value increases, thus illustrating that the test specimen inner mesochap activities are enhanced.

Figures 14.43 and 14.44 reflect CT number average value and the trend of variance changing with stress of each layer surface fracture zone. CT number average value indicates that when scanning is carried out in the last stage, macrochaps have emerged, and that CT number average value at each fracture zone has appeared to drop, and the dropped range is larger than that of the whole CT average value, of which CT number average value of the fifth layer fracture zone in Fig. 14.43 has obviously dropped in the third scanning, thus indicating that mesochaps grow and expand from here to other cross sections in sequences. This is the same conclusion as judged by square difference average value in Fig. 14.44, thus confirming the locations of mesochap expansion in the test specimen inner part.

The following important principles and viewpoints are obtained from the concrete CT tests and CT image analysis through grade I dynamic and static tensile conditions:

1. Dagangshan gradation I concrete dynamic and static tensile CT test comparison indicates that CT tests can study the concrete mesofailure process under the tensile conditions, analyze the differences of concrete static and dynamic failure processes under the tensile conditions, and further probe into the concrete dynamic tensile strength change mechanism.

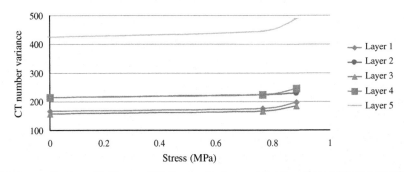

FIGURE 14.44 Curves Varying with Stress of CT Number Variance at Fracture Zone of Each Layer Surface

2. Being the same as other conventional mechanics tests, chaps with the advent of mutations in concrete uniaxial tensile test are hard to obtain the whole stress–strain curves, and there is a certain difficulty to gain the CT image of the chap expansion process. For this reason, it is necessary to improve the test technology to achieve the objective.
3. Dynamic and static tensile CT test comparison indicates that concrete tensile failure only appears to have a macrocrack.

14.3.3 ANALYSIS OF CT TEST RESULTS OF CONCRETE DYNAMIC AND STATIC TENSILE AND COMPRESSION CRACK PROCESS

Based on the applied medical-use CT machine and research and development of dynamic loading equipment in realignment testing conditions, we have successfully carried out online observations and measurements of the concrete inner CT scale chap evolution process and can, intuitively and quantitatively, analyze the test specimen crack occurrence locations, development process, damage degrees, and evolution laws. Concrete CT technology is the effective way to study the concrete inner damage process and failure mechanism.

14.3.3.1 Under the action of static compression loading

Holes in the CT image can display various kinds of initial damage surrounding conditions of concrete, which is likely to form obvious local stress concentration so that most of the chaps develop around these holes to form the local concrete damage evolution. The expanded chap direction is the tensile chap parallel to the main compressive stress direction or the shear chap with a small angle, accompanied with the shear inclined cracks. Accordingly, many parallel cracks with small spacing distances can finally form one passing the main chap, and cubical expansion phenomena may come out.

Chaps in the test specimen appear mostly to have longitudinal distribution with two ends suffering from slight failure, but most parts of the zones in the middle are compressed to break, with full chap development to form rough macrochaps. Test specimen broken face appears to be cone-shaped, caused by the end effects, while the middle part is located in uniaxial compression state without side constraint. Chap developing process is, in general, around the aggregate interface and expands in mortar, and cannot pass through the aggregates with high strength. There are fewer cracking points. The main chaps are first formed in the test specimen. After the main chaps have formed, the stress can shift to the surrounding zones, and finally run through, thus leading to the test specimen failure. After the main chaps have formed, there is still a certain residual strength. It needs a long time for chap formation to test specimen failure. The damaged static pressured specimen shows the obvious softening phenomena so that Type I of splitting damage opening chap or Type II of shearing damage chaps are formed. The damaged surface shapes are shown in Fig. 14.45.

14.3.3.2 In static tensile tests

The initial failure process is similar to static compressed tests. Before the main chaps are formed, it is likely to have many chaps develop at the same time, but at last only one main chap can be formed. The formation of the main chap can promote the release of the surrounding strain energy, and at the same time, owing to the effective tensile small area, its top end can cause the stress concentration on the high degree, whereby inhibiting the further development of other chaps. Therefore, this can greatly distinguish the tensile failure from compressed failure morphology.

FIGURE 14.45 Static Compressed Test Specimen Damaged Surface Shape

FIGURE 14.46 Static Tensile Test Specimen Destructed Surface Shape

In static tensile tests, in general, one horizontal main chap can be formed on the weakest cross section of the whole test specimen so that the broken surface is smooth, and there is only a slight or almost no change in the zones beyond the broken surface. The broken chap developing process is generally around the aggregate interface and expands in mortar and cannot run through the aggregate with high strength. The destroyed or damaged two sections of the test specimen are complete of themselves with the perfect concrete surfaces. The test specimen main chaps formed have broken away very fast, but the residual strength is not obvious. It is a short experiencing time from chap formation to the test specimen failure, appearing to have the brittle abruption, so that Type I of tensile chap is formed, and the destroyed surface shapes are shown in Fig. 14.46.

14.3.3.3 In dynamic compression tests

Comparison of the result between static and dynamic compressed tests shows that the damaged chap characters and distribution of both are apparently different. When dynamic compression suffers from damage, strain speed rate increases, and under the sinusoidal wave form dynamic compression, meso-chap evolution process is obviously accelerated. Chap generation and developing speed are boosted. Chap starting point increases and the damaged area is enlarged. The path of chaps following energy to release develops fastest. Chap form is smooth and straight. Phenomena running through the aggregates are on the increase. Strength gains full development, thus leading to an increase in dynamic strength, while in static compression tests, chaps have plenty of time to follow the trail of weakest structure direction to develop.

14.3.3.4 In dynamic tensile tests

Damage morphology and process are similar to static tensile tests. When loading rate is higher, chaps can run through the aggregates to develop. Accordingly, the higher the rate is, the larger the tendency of running through the aggregates will be. The destroyed two sections of specimens are completely perfect. Concrete surface is perfectly well. The experiencing time from chap formation to the test specimen failure is rather short. However, after the test specimen's main chap is formed, the test specimen cannot break up immediately because of loading instant changes and there being a certain residual strength.

14.4 CONCRETE SUBZONE BREAKING THEORY
14.4.1 EXISTING CONDITIONS AND PROBLEMS OF CONCRETE RESEARCH

The CT method can detect the inner changes in materials and structures with no damage and meanwhile have the higher resolution capacity. The material CT image can reflect the magnitudes of each part of material absorption of X-rays, being a picture of a digital image in nature. Every pixel value is a CT number. The magnitude of CT number on CT image is expressed by gray degrees, which is in positive proportion to the density of corresponding materials. The bright color of CT image expresses the high-density zone of the material while the dark color shows the low-density zone of the material.

Research on mechanics properties of concrete material groups with CT number is characterized by rich information, fineness of test results, and giving out quantitative information for each point. CT number includes the deformation, damage, rupture, and destructive plentiful quantitative information on each medium CT research unit. But at present on the one hand, CT number sizes, average number, variance, CT number statistic frequency number, damage variable, damage evolution equation, and so on, are adopted to observe and analyze material deformation, damage, and mesochap evolution process so that a set of mathematical descriptive methods adaptable and rightful for CT analysis has not been formed. Accordingly, the test specimen obtained by high testing prices for information on each CT resolution point can be averaged as the whole test specimen or some local information so that the valuable quantitative information described by means of the qualitative method wastes CT number resource. On the other hand, owing to being subject to CT resolution, contrasting degrees of gray images, different processing methods for CT images by every author and visual error influence, and so on, the chaps with the same sizes and visual error are influenced, and so on, and the chaps with the same sizes and the states with the same density on CT image will have the different sense perceptions in such a way as to cause research errors by researchers themselves. Internationally, research on material CT without damage has been carried out earlier (Ueta et al., 2000). Also, there has been a vast breakthrough made in the research method. Therefore, nowadays, CT research on the performance of mechanics of concrete materials is still under the stage of CT image observation. Faning (2005) defined the concept of integrity, failure of some spatial points of geo-technique medium, on the basis of which geotechnique medium λ-level perfect region, λ-level damage region, and $(\lambda_1 - \lambda_2)$ cutoff theoretical concept are defined. At the

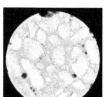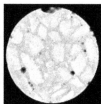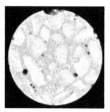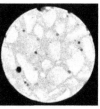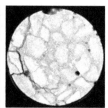

FIGURE 14.47 CT Image of the Fourth Scanning Layer of Cylinder Test Specimen Destructed Posterior to Loading

same time, λ-level perfect region, λ-level damage region, and $(\lambda_1 - \lambda_2)$ cutoff theoretical measurement are suggested. Finally, geotechnique complete space and damage space concepts are given out. Based on CT number defined geotechnique λ-level damage ratio and damage rate, $(\lambda_1 - \lambda_2)$ cutoff theoretical ratio and cutoff theoretical concept, the set theory and measure theory knowledge are used to study the relation among λ-level damage rate and CT number as well as density, whose purpose is to establish a geo-technique CT mechanics quantitative analysis method by using new mathematic means. Relatively to other methods, this method is effective to use CT number including rich information.

Based on the mentioned research thinking (Faning, 2005), this section uses geotechnique failure evolution theory in concrete CT test analysis and probes into or dig CT numbers, including rich information, to the fullest extent. The tests have obtained more images. It is here just to take CT scanning image of the fourth scanning layer posterior to loading failure of C15 concrete cylinder column with Φ60 mm $\times$ 120 mm test specimen as an example (see Fig. 14.47) that this section introduces the concrete subzone failure theory and its application.

14.4.2 BASIC ASSUMPTION OF CONCRETE SUBZONE DAMAGE THEORY

The basic assumptions for concrete subzone damage theory are as follows:

1. The minimum air density within the range of Newtonian mechanics.
2. The maximum density of material matrix without damage.
3. Material matrix and glued joint materials will not produce changes in volume prior to and posterior to subject to force. Changes caused in material volume by microchaps opening and closing.
4. Research base is material CT number so that mesomechanics magnitude is CT resolving unit magnitudes.

14.4.3 DEGREE OF PERFECT AND DEGREE OF FAILURE OF CONCRETE

The research objective is called the whole region using Ω to express:

$$\Omega = \{(x, y, z)|(x, y, z)\} \text{(any point on research object space zone)} \tag{14.8}$$

where "$\|$" express aggregate, "$|$" is the prior equation expressing the aggregate to include the object and "$|$" is the posterior equation to express the objects to satisfy the conditions of this kind of expression, which is often adopted by the following.

CT number uses X-ray computer-aided tomography, using one pixel number to express material corresponding location physical density, and so on, comprehensive information. Nondestructive zone and damage zone of the same kind of materials and rupture zone CT number should distribute in a certain range and exist with maximum value and minimum value. The maximum value of a certain kind of material is recorded as max H, and the minimum value takes the air CT number, that is, -1000. In this way, some material CT number uses its standardized number value of the maximum and minimum, being a definition on Ω field value region [0, 1] function; the definition is as follows:

$$p(x,y,z) = \frac{H(x,y,z)+1000}{\max H +1000} \qquad (14.9)$$

For the material at point (x, y, z) (degree of perfect), it is here that the so-called point (x, y, z) CT number refers to the coordinate point of CT number on CT resolution elements. Accordingly, with the improvement in CT resolution, this small unit can continue to decrease.

Introduction to the concept of perfect degree is based on the following considerations. Any location in the material can be viewed as perfect, but the perfect degrees are different. Material matrix point (the strength highest point) perfect degree is 1, and the complete ruptured zone perfect degree is zero, and the perfect degree in the rest of other zones is a certain number value between 0 and 1.

In corresponding to the perfect degree, the definition is as follows:

$$d(x,y,z) = 1 - p(x,y,z) = \frac{\max H - H(x,y,z)}{\max H + 1000} \qquad (14.10)$$

For the material at point (x, y, z) (degree of damage and fracture), therefore, any location in the material can be considered as having damage and fracture, but their damage and fracture are different in degrees. Material matrix point degree of damage and fracture is zero, and the damage and fracture degree in the complete fracture zone is 1; the damage and fracture degree in the other zones is a certain number value between 0 and 1.

Figure 14.48 is the perfect degree and damage and fracture degree image of the CT image posterior to the fourth scanning layer loading failure in Fig. 14.47. It can be seen that the perfect degrees of each

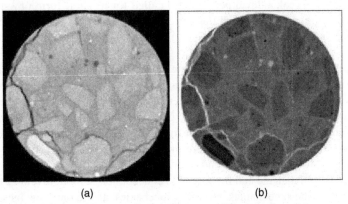

(a)   (b)

FIGURE 14.48 Distribution Diagram of Concrete Perfect Degree and Fracture Degree

(a) Perfect degree image (b) damage and fracture degree distribution diagram.

location on the cross section are different. With the aggregates existing in the zone, the perfect degree is larger, while in the mortar zone, the perfect degree is relatively small and the perfect degree in the chap zone is the minimum, and vice versa; the degree of damage and fracture is the maximum. The general difference value image can reflect CT number variation at each position in the CT image before and after loading in concrete cross section, but cannot well reflect the strength conditions of each zone in the cross section. Figure 14.48 can well make up this shortcoming.

The concrete of perfect degree and damage and fracture degree can reflect the material relative perfect degree and relative damage and fracture degree in that point. Obviously, it is as follows:

$$p(x,y,z) + d(x,y,z) = 1 \tag{14.11}$$

14.4.4 CONCRETE λ-LEVEL PERFECT FIELD AND λ-LEVEL DAMAGE AND FRACTURE FIELD

The perfect degree makes the concept of whether the material is perfect, damaged, or fractured become fuzzy, but people have always been used to applying the distinct physical concept, so that we have introduced the following complete level concept.

Letting $0 \le \lambda \le 1$, and when $p(x_0, y_0, z_0) \ge \lambda$, material at the point (x_0, y_0, z_0) is in the weak λ-level perfection. When $p(x_0, y_0, z_0) > \lambda$, material at the point (x_0, y_0, z_0) is in the strong λ-level perfection. In order to simplify the concept, we have called the weak λ-level perfection. Similarly, when $d(x_0, y_0, z_0) \ge \lambda$, material at the point (x_0, y_0, z_0) can be known as in λ-level damage and fracture.

Material at the point (x_0, y_0, z_0) in λ-level perfect meaning is that its perfect degree is $100\lambda\%$; with the same reason, material at the point (x_0, y_0, z_0) in λ-level damage, and fracture meaning is that its damage and fracture degree is $100\lambda\%$.

Material at the point (x_0, y_0, z_0) in λ-level nonperfect meaning is that its damage and fractures with $(1-\lambda)$ level. With the same reason, material at the point (x_0, y_0, z_0) in λ-level nondamage and fracture meaning is that it is $(1-\lambda)$ level perfection.

Calling aggregate $\{(x,y,z)| \ (x,y,z) \in \Omega, \text{and}, p(x,y,x) \ge \lambda\}$ as material in weak λ-level perfect field and calling aggregate $\{(x,y,z)| \ (x,y,z) \in \Omega, \text{and}, p(x,y,x) \ge \lambda\}$ as material in strong λ-level perfect field later on, the weak λ-level perfect field called for its short form of λ-level perfect field with "p_λ" to express. p_λ expresses that all the perfect degrees are not less than λ value point aggregate, that is, under λ-level, materials are all the perfect parts. Strong λ-level perfect field is expressed with "strong p_λ".

Calling aggregate $\{(x,y,z)| \ (x,y,z) \in \Omega, \text{and}, d(x,y,x) \ge \lambda\}$ as material in the weak λ-level damage and fracture field in its short form of λ-level damage and fracture field as expressed by "d_λ," d_λ can express that all the damage and fracture degrees in material are not less than λ value point aggregate, that is, under λ-level material damage and fracture parts. The strong λ-level damage and fracture field is expressed by using "strong," giving λ different right values, and d_λ can express the damage zone of classical damage mechanics or classical fracture mechanics material chaps. Figure 14.49 is the concrete sample different λ-level perfect field and λ-level damage and fracture field.

By means of the λ-level damage and fracture field concept, it is smooth to realize the natural transition of concrete material mesomechanics into macromechanics, and to unify the mesomechanics damage field with macromechanics fracture field in harmony. λ-level damage and fracture field concept is richer in connotation than the concept of the damage field and the fracture field, and by means of the

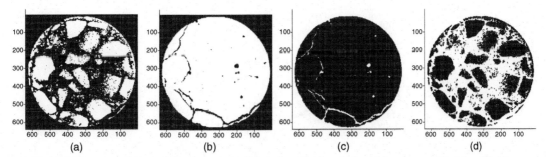

FIGURE 14.49 Concrete Sample Different λ-Level Nonperfect Field and λ-Level Damage and Fracture Field

(a) 0.3 level perfect (b) 0.5 level perfect field (c) 0.5 level damage and fracture field (d) 0.7 level damage and fracture field.

concept of damage and fracture degrees, the quantification and operation are extremely simple, obviously,

$$p_\lambda \cup d_{1-\lambda} = \Omega \tag{14.12}$$

where $\cup$ expresses two aggregates in sum aggregate.

Calling sum aggregate $\Omega - p_\lambda$ as the material λ-level nonperfect field in accounting as $\overline{P_\lambda}$, expressing λ-level perfect field complement aggregate; calling sum aggregate $\Omega - d_\lambda$ as the material λ-level nondamage and nonfracture field in accounting as $\overline{d_\lambda}$, expressing λ-level damage and fracture complement aggregate, of which the upper line is the sum aggregate seeking for complement aggregate operation symbol. Owing to the complement aggregate for the complement aggregate is equal to itself, so that there will be the following equation:

$$p_\lambda = \overline{\overline{p_\lambda}}; \ d_\lambda = \overline{\overline{d_\lambda}} \tag{14.13}$$

With λ value dropping, the investigated chap size is gradually reduced, and zone classified into the damage and fracture field is gradually on the increase so that λ-level damage and fracture field is gradually increasing with λ decrease. Of the process, if as investigated from the angle of perfect field and with λ decrease, λ-level perfect field is also on an increase (see Fig. 14.49). Both are not in contradiction because the problems are in the opposite position. Using a formula to express this can be that when $\lambda_2 < \lambda_1$ the $d_{\lambda_2} \supseteq d_{\lambda_1}$ 1 is established, of which $\supseteq$ is the inclusive relation of the sum aggregate.

14.4.5 CONCRETE λ-LEVEL PERFECT FIELD AND λ-LEVEL DAMAGE AND FRACTURE FIELD MEASURE

In order to describe the magnitude of λ-level perfect field and λ-level damage and fracture field, the measure concept in mathematics is introduced in the following. Sum aggregate (*) measure is expressed by $m(*)$. Usually, this can be interpreted as the product of CT resolution unit number and CT resolution unit length, area or volume; for instance, $m(d_\lambda)$ can be interpreted as λ-level damage and fracture field area.

What measure is used instead of direct use of length, area, or volume to describe is because the object considered here is the sum aggregate and that sum aggregate element is point, and point aggregate

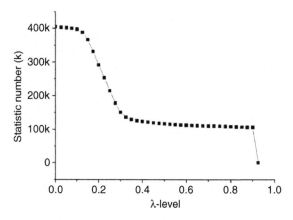

FIGURE 14.50 Curve Varying with λ of Concrete Sample λ-Level Damage and Fracture Field

length, area, or volume cannot exist sometimes. For instance, all the rational numbers comprising sum aggregate is a countable set. All the countable sets are measurable, whose measure is zero, but the set length is immeasurable.

The preceding section has dealt with d_λ with λ varying laws, on the basis of which the mechanics properties of concrete samples and their physical states with $m(d_\lambda)$ varying laws can also be studied. Of process, the direct application of concrete sample CT number $H(x, y, z)$ and perfect degree $p(x, y, z)$ or damage and fracture degree $d(x, y, z)$ can also study concrete sample mechanics properties and its physical states, but great attention should be paid to the fact that $H(x, y, z)$ is defined as the indefinite ternary function in three-dimensional space value field. The perfect degree $p(x, y, z)$ and the damage and fracture degree $d(x, y, z)$ are defined as the ternary function in the three-dimensional space value field [0,1], in which λ-level damage and fracture field d_λ measure is only a monodic function. Therefore, the application of concrete sample λ-level damage and fracture field measure in research has the obvious advantage of making the problems simplified greatly.

Since the so-called point here is, in fact, CT resolution unit, λ-level perfect field p_λ and λ-level damage and fracture field are measurable.

For instance, the mentioned concrete sample λ-level damage and fracture measure $m(d_\lambda)$ with λ varying curve are shown in Fig. 14.50. In Fig. 14.50, the cumulative value of statistic frequency numbers and the product of resolution unit area are the measure degree. It can be seen from Fig. 14.50 that with the gradual reduction from 1 to 0 of λ value, concrete sample λ-level damage and fracture field measure degree $m(d_\lambda)$ has experienced a small sudden growth, gradual growth, and huge growth of three stages, respectively corresponding to the material chap zone, damage and fracture transition zone, and perfect zone. It can be seen from Fig. 14.50 that when $\lambda \geq 0.90$, they are the material chap zone and air unit; when $0.33 \leq \lambda \leq 0.90$, they are the material failure transition zone; while when $\lambda \leq 0.33$, they are the material perfect zone. It can also be seen from Fig. 14.50 that the distribution of concrete samples is dispersive so that the failure transition zone is small.

Based on the research conducted by Guo et al., under the action of different loadings, microholes in rock have experienced the process of being compressed compactness–microchap growth–branch-development–rupture–failure–unloading stages, so that λ-level damage and fracture field on the

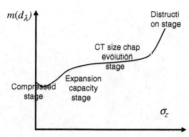

FIGURE 14.51 Curve of Relation Between $m(d_\lambda)$ and Stress

same cross section has experienced a total of four stages as dropping, increasing, level slow increase, and sharp increase with the stress increasing. On the whole, there exist such cases as when $\sigma_2 > \sigma_1$, $d_\lambda\big|_{\sigma_2} \supseteq d_\lambda\big|_{\sigma_1}$ with an increase in outside loading, concrete sample $m(d_\lambda)$ should experience such four sharp increases. This together with the test specimen must experience such four corresponding stages as compact, expansion capacity, CT magnitude evolution, and failure with an increase in outside loading. As shown in Fig. 14.51, it is in fact that there is a relation curve between stress and damage and fracture.

In terms of research, as to any loading state, the magnitude, shape, and position of λ-level damage and fracture field in different cross sections should be different, being caused by the nonuniformity of test specimens.

14.4.6 CONCRETE λ1−λ2 CUTOFF OR INTERCEPT JOINT

Letting $0 \leq \lambda_1 \leq \lambda_2 \leq 1$, the different sets of λ_1-level damage and fracture field and strong λ_2-level damage and fracture field can be defined as concrete material $\lambda_1 - \lambda_2$ intercept joint, accounting as $d_{\lambda_1 - \lambda_2}$. It can be known from definition that

$$d_{\lambda_1 - \lambda_2} = d_{\lambda_1} - \text{strong } d_{\lambda_2} = \left\{(x,y,z)\big|d(x,y,z) \geq \lambda_1\right\} - \left\{(x,y,z)\big|d(x,y,z) > \lambda_2\right\} \tag{14.14}$$

The following two equations are equivalent to $\lambda_1 - \lambda_2$ intercept joint definition:

$$d_{\lambda_1 - \lambda_2} = \left\{(x,y,z)\big|\lambda_1 \leq d(x,y,z) \leq \lambda_2, 0 \leq \lambda_1 \leq \lambda_2 \leq 1\right\} \tag{14.15}$$

$$d_{\lambda_1 - \lambda_2} = \left\{(x,y,z)\big|(x,y,z) \in d_{\lambda_1} \ \& \ (x,y,z) \notin \text{strong } d_{\lambda_2}\right\} \tag{14.16}$$

The $\lambda_1 - \lambda_2$ intercept theoretical physical meaning is the density to be situated among all the material points in a certain range.

In practical engineering works, earth hard layer is a $\lambda_1 - \lambda_2$ intercept joint; soft weak interbedded layer is also an intercept joint. Rock joint is also an intercept joint. Layer joint, chaps, and fracture zones are also intercept joints. The perfect field is an intercept joint; concrete aggregates, mortar, and chaps are all intercept joints. It can be seen from this that the extension of the intercept joint concept is very rich, but its connotation definition is very simple, that is, the whole with the similar densities. In general condition, and as to the same kind of homogeneous materials, the magnitude of density can

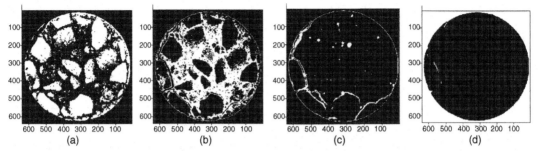

FIGURE 14.52 Concrete Damage and Fracture Field Intercept Joint

(a) 0~0.3 damage and fracture field intercept joint (b) 0.3~0.45 damage and fracture field intercept joint (c) 0.45~0.7 damage and fracture field intercept joint (d) 0.85~1 damage and fracture field intercept joint.

determine material strength so that $\lambda_1 - \lambda_2$ intercept joint is in fact the similar point aggregation of homogeneous materials strength at the time.

Of process, two perfect field difference sets can be used to define the $\lambda_1 - \lambda_2$ intercept joint of concrete materials, accounting as $\lambda_1 - \lambda_2$.

The magnitude of intercept joint on the perfect field and damage and fracture field can also be evaluated with measure, whose measure definition is the same as the definition of intercept set measure. Figure 14.52 shows the concrete damage and fracture field intercept joints. It can be seen that small gravels in the original hole zone and chap zone produced by loading damage, aggregate zone, and mortar zone in the test samples can be clearly recognized

It can be seen from this that the intercept joint concept external extension is not only very rich but also very practicable in nature. Using the intercept joint can make a simple distinction among the perfect field with the fracture field, air, chap zone with mortar zone, compacted zone with matrix zone, and aggregate zone with mortar zone, and so on. In addition, the sum aggregate measure can be used to accurately measure the volume areas and widths of the perfect field, fracture field, air, chap zone, compacted zone, matrix zone, aggregate zone, mortar zone, and so on. The measurement method is very simple. A computer is first used to carry out the statistics of various kinds of three-dimensional, two-dimensional, and one-dimensional numbers of intercept joint points, to be multiplied by resolution unit volume, area, and width, that is all right. These functions of intercept joint diagrams cannot be achieved on the gray degree diagram.

Figure 14.53 shows the concrete sample damage and fracture fields of 0.0–0.1,0.1–0.2,…,0.9–1.0 intercept joint point numbers, to be multiplied by the resolution unit areas, that will be the measure of each intercept joint. The test sample uniformity can be judged by Fig. 14.53. In Fig. 14.53, the head part of the curve is holding high, indicating that the perfect basic body and aggregate occupy the major part of the test sample; the tail parts occur to hold high, illustrating that there exists a bulk of original defects or the advent of fractures (including part of air units) in the test sample. The original defects or fracture measures in concrete test samples are obviously higher than those in sandstone samples. At the same time, it can be seen that the concrete sample obviously consists of three parts: aggregate materials, mortar, and chaps (including the initial defects), of which the density components in the mortar part are complicated but dispersive, and the density components in mortar appear to be successively distributed from the maximum to the minimum.

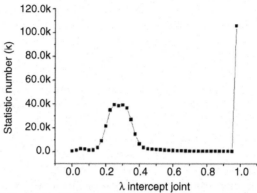

FIGURE 14.53 Statistics Point Numbers of Various Kinds of Intercept Joints of Concrete Samples

With 0.025 length as the intercept joint step length between zones, CT point number statistic values of concrete's every intercept joint at different loading stages can be seen in Fig. 14.54a. Then, with the scanning images as the reference when there is no loading, the comparison is made of the variation in the number of resolution units included in the $\lambda_1 - \lambda_2$ intercept theory of each loading image; with λ as the horizontal coordinate and the number of each intercept theory point relative to difference value of the initial image as longitudinal coordinate, the figure is mapped as shown in Fig. 14.54b. It can be seen from Fig. 14.54 that when λ is in between zones of 0–0175 and 0.425–0.9, each intercept joint zone includes very few pixel point numbers. On this scanning layer, 0–0.175 intercept joint includes the zone with largest density, but this part of the material occupies a very small proportion, while 0.425–0.9 intercept joint includes the chap or hole zone and mortar transition zone. This part also occupies a very small proportion in the figure. In contrast with the original CT image, CT number of chaps or holes is in the range of −600 to +1450. Therefore, selecting 0.425–0.9 intercept joint includes the chap zone or hole zone. When λ_1 is 0.9–1, the number of statistics points sharply becomes large, illustrating that

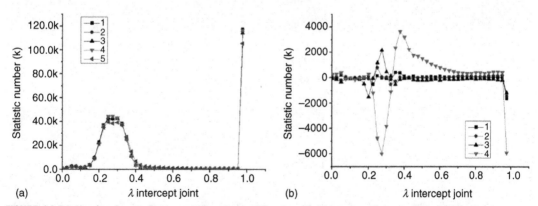

FIGURE 14.54 Varying Curves Between λ Damage and Fracture Field Intercept Joint and Pixel Point Numbers in Each Loading Section

(a) λ intercept joint varying curve of each loading stage (b) λ intercept joint statistics difference value varying curves of each loading stage.

0.9–1 intercept joint includes the large zone scope, but the CT number is small and distributes densely so that there is only the air zone in the CT scanning image. It can be seen from Fig. 14.54b that when λ_1 takes 0.275, λ intercept joint statistics values of each scanning layer cross only one point and the second cross is near 0.325, and there appears to be a peak value between 0.275–0.325 zones. This is because the test specimens have experienced the source of compact first and aggregate density is the biggest, and its variation is much smaller than that of mortar. For this reason, it is considered that the changed zones are the mortar zones. Selecting 0.326–0.425 intercept joint includes the zone being mortar zone and 0–0.325 intercept joint intercept joint includes the zone being aggregate zone.

Using the intercept joint concept can classify concrete components and trace the varying process of loading. To realize the classification of concrete components and to trace the varying laws of the loading process, gray value image cannot be easy to observe and accurate to measure in a general sense, for the human eye resolution capacity is smaller than 256 gray steps so that using $\lambda_1 - \lambda_2$ intercept joint to carry out subzone description of concrete CT image has the obvious advantages.

Figure 14.55 uses a 0.425–0.9 intercept joint to classify chap or hole zones and the varying conditions following the loading process and can observe the slight changes in chap or hole zones in the fourth scanning layer at $\sigma=0.00-20.03$ MPa stage. There suddenly appears to have chaps running through at the stage of $\sigma=20.03-18.74$ MPa. Figure 14.56 uses 0–0.325 intercept joint to classify or partition the aggregate zone, and following the changes in loading process, it can be seen that there is no obvious change in the fourth scanning layer at $\sigma=0.00-18.74$ MPa five stages, illustrating that aggregate density is large and that there is no obvious change in the loading process. Figure 14.57 uses 0.325–0.425 intercept joint to partition the mortar zone and follow the change in loading process. It can

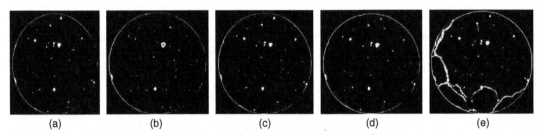

FIGURE 14.55 0.425–0.9 Intercept Joint Extract Diagrams in the Fourth Scanning Layer

(a) $\sigma=0.00$ MPa (b) $\sigma=18.01$ MPa (c) $\sigma=19.69$ MPa (d) $\sigma=20.03$ MPa (e) $\sigma=18.74$ MPa.

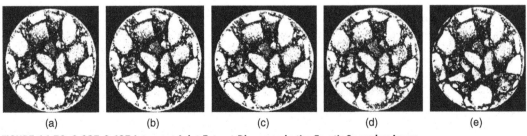

FIGURE 14.56 0.325-0.425 Intercept Joint Extract Diagrams in the Fourth Scanning Layer

(a) $\sigma=0.00$ MPa (b) $\sigma=18.01$ MPa (c) $\sigma=19.69$ MPa (d) $\sigma=20.03$ MPa (e) $\sigma=18.74$ MPa.

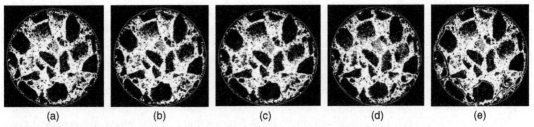

| (a) | (b) | (c) | (d) | (e) |

FIGURE 14.57 0.0–0.325 Intercept Joint Extract Diagrams in the Fourth Scanning Layer

(a) $\sigma = 0.00$ MPa (b) $\sigma = 18.01$ MPa (c) $\sigma = 19.69$ MPa (d) $\sigma = 20.03$ MPa (e) $\sigma = 18.74$ MPa.

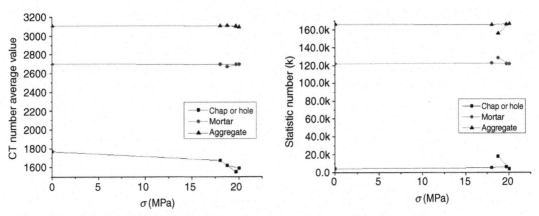

FIGURE 14.58 Curves Varying with Stress of Each Subzones in the Fourth Scanning Layer

(a) Curves varying with stress of CT number average value (b) Curves varying stress of statistics point numbers for each zone.

be seen that there occurs to have more obvious changes in the mortar zone of the fourth scanning layer than in the aggregate zone at $\sigma = 0.00$–20.03 MPa stage. Also, there happens to have obvious variations in part of the mortar zone, lacking inheritance before and after at $\sigma = 20.03$–18.74 MPa stage.

The laws of the mentioned three kinds of zone CT number average value and statistic point numbers varying with stress are seen in Fig. 14.58. It can be seen from Fig. 14.58a that CT number average value in the aggregate zone is the largest, and in the whole loading process, its varying range is the smallest, illustrating that owing to the large density of aggregates, new chap growth or germination, development occurs in the mortar zone, the aggregate failure is not serous, and variation is not obvious in the whole process of loading. CT number average value varying range in the mortar zone is next to that in the aggregate zone, but in the whole process of loading, new chap growth germination, and running-through lead to reduction in CT number average value in the mortar zone. It can be seen from Fig. 14.58b that the statistic point numbers in the aggregate zone after failure are relatively reduced, while the statistic point numbers in the mortar zone and the chap zone or the hole zone increase slightly, of which an increasing range in the chap or hole zone is large relatively. The analysis of its main cause is due to image resolution remaining constant at 640 × 640 pixels, and the test specimen occupying the zone in scanning image occurs to change in the loading process. With the inner magnitudes, chap development and running-through can make more and more pixel points be counted into the chap or hole zones, and at the same time, although the density in the aggregate zone is large, there will occur to have a certain change in it with an increase in loading, whose total density can gradually decrease, whereby resulting in a decrease in the statistic point numbers in the aggregate zone, while there is an increase in statistic point numbers in the mortar zone. The statistic point number in the chap or hole zones have a certain range drop before increasing, whereby indicating that there is a reducing stage in the chap or hole zone area in the loading process. The phenomena are right in proving that concrete material has experienced four stages of compact, expansion capacity, CT magnitude chap extension, and failure in the process from loading to failure.

14.4.7 CONCRETE PERFECT SPACE AND DAMAGE AND FRACTURE

The topologic space concept in point set topology is that if the subset family $\{\mathfrak{R}\}$ of a certain aggregate or set Ω is satisfied, the empty set and universal set belong to the family sets; in this family of sets, the finite element intersects and any multiple element joins or sum aggregate belong to it, and then the family of sets $\{\mathfrak{R}\}$ is called as the topology of this aggregate Ω, and even $(\Omega, \mathfrak{R})$ is called as the topologic space.

The $\{p_\lambda\}$ is used to express λ-level perfect field and strong λ-level perfect field aggregate, and $\{d_\lambda\}$ is used to express all λ-level damage and fracture fields and strong λ-level damage and fracture field aggregate. The $\left\{p_{\lambda_1-\lambda_2}\right\}$, $\left\{d_{\lambda_1-\lambda_2}\right\}$ are used to express all the $\lambda_1 - \lambda_2$ intercept joints and strong intercept joint aggregates. Obviously, the family of subsets serve as the aggregate Ω, $\left\{p_{\lambda_1-\lambda_2}\right\}$, $\left\{d_{\lambda_1-\lambda_2}\right\}$, $\{p_\lambda\}$ and $\{d_\lambda\}$ all satisfy the topology definition so that they are all aggregate Ω topology. Of process, $(\Omega, p_{\lambda_1-\lambda_2})$, $(\Omega, d_{\lambda_1-\lambda_2})$, (Ω, p_λ), and (Ω, d_λ) are all the topologic spaces.

The $(\Omega, p_{\lambda_1-\lambda_2})$, $(\Omega, d_{\lambda_1-\lambda_2})$ are known as the perfect space, damage, and fracture space of concrete materials, respectively. Serving as a special example, (Ω, p_λ) and (Ω, d_λ) are also the perfect space, damage, and fracture space of concrete materials.

The questions to be discussed in the following can be conducted sometimes in the perfect space and sometimes in the damage and fracture space so that attention must be paid to distinguishing the two spaces with different properties.

Since concrete material mechanics mainly concerned with the main questions can be the material strength problem as well as the material damage and fracture process, the concept of concrete material damage and fracture degrees and the λ-level damage and fracture field is inclined to be used in later discussions, that is, to be inclined to discuss the problems in damage and fracture space. It is only particularly necessary that the problem discussion be conducted in the perfect space.

In addition, attention should be paid to λ-level perfect field and λ-level nondamage and nonfracture field; λ-level damage and fracture and λ-level nonperfect field are all the different concepts. λ-Level perfect field and λ-level nonperfect field are the concept of perfect space (Ω, p_λ), while λ-level damage and fracture field and λ-level nondamage and nonfracture field are the concepts of damage and fracture space (Ω, d_λ). "Non" here should be interpreted as the "supplement" in the set operation.

Finally, the important point that must be stressed is that all the strong λ-level perfect field set or all the strong λ-level damage and fracture field set can also conform topology, but all the weak λ-level perfect field set or all the strong λ-level damage and fracture field set cannot conform the topology. The main cause is that any multiclosed sets are likely to be not the closed set any longer.

14.4.8 CONCRETE λ-LEVEL DAMAGE AND FRACTURE RATIO AND λ-LEVEL DAMAGE AND FRACTURE RATE

Knowledge of mathematic measure theory is used to define the concept of concrete material λ-level damage and fracture ratio, measurable space, and measure concepts for reference mathematic measure theory (Faning, 2005).

In the concrete material damage and fracture space $(\Omega,\ d_{\lambda_1-\lambda_2})$, based on λ-level damage and fracture field, we can define:

$$D_e^{\lambda} = \frac{m(d_{\lambda})}{m(\overline{d_{\lambda}})} \qquad (14.17)$$

For the sample λ-level measure damage and fracture ratio, λ-level damage and fracture ratio is its short form. The sample λ-level damage and fracture ratio can be simply interpreted as the test specimen damage and fracture zone and without damage and fracture zone area ratio or volume ratio, in this way, another equation for λ-level damage and fracture ratio can be simply obtained as follows:

$$D_e^{\lambda} = \frac{m(d_{\lambda})}{m(p_{1-\lambda})} \qquad (14.18)$$

where, m (strong $p_{1-\lambda}$) is the strong $(1-\lambda)$-level perfect field in the perfect space $(\Omega, p_{\lambda_1-\lambda_2})$.

Similarly, in concrete material test specimen damage and fracture space $(\Omega,\ d_{\lambda_1-\lambda_2})$, based on λ-level damage and fracture field, we can define:

$$D_n^{\lambda} = \frac{m(d_{\lambda})}{m(\Omega)} \qquad (14.19)$$

For the test specimen λ-level measure damage and fracture rate, λ-level damage and fracture rate is its short form. The specimen λ-level damage and fracture rate can be simply interpreted as the test specimen damage and fracture field and the full field area rate or volume rate.

There can be a comparison among the test sample λ-level damage and fracture ratio, λ-level damage and fracture rate with void ratio in soil mechanics and porosity concept, being soil mechanics void ratio, and porosity concept extension with more popular significance. At the same time, they can serve as the concepts in soil mechanics under the special conditions, being of very obvious physical significance. Relation between λ-level damage and fracture ratio and λ-level damage and fracture rate is the same as the relation between void ratio and porosity, so as to satisfy the following similarly:

$$D_n^{\lambda} = \frac{D_e^{\lambda}}{1+D_e^{\lambda}} \qquad (14.20)$$

$$D_e^{\lambda} = \frac{D_n^{\lambda}}{1-D_n^{\lambda}} \qquad (14.21)$$

With an increase in external loading, the damage and fracture zone continues to enlarge; the D_e^{λ} and D_n^{λ} increase with an increase in external loading. At the same time, with a decrease in λ value in the damage and fracture space, chap magnitude entering into the investigation scope is gradually diminishing and is also gradually expanding. D_e^{λ} and D_n^{λ} are increasing with a decrease in λ value, that is, when $\lambda_2 < \lambda_1$, $D_e^{\lambda_2} \geq D_e^{\lambda_1}$ and $D_n^{\lambda_2} \geq D_n^{\lambda_1}$ are established.

If λ-level damage and fracture field and λ-level damage and fracture field volume can be obtained, $V(d_\lambda)$ and $V(\overline{d_\lambda})$ can be used to express them respectively (d_λ and $\overline{d_\lambda}$ volume are in nonexistence sometimes), and the volume ratio is used to define the concept of λ-level damage and fracture ratio and λ-level damage and fracture rate:

$$D_e^\lambda = \frac{V(d_\lambda)}{V(\overline{d_\lambda})} = \frac{V(d_\lambda)}{V(\text{strong } p_{1-\lambda})} \tag{14.22}$$

$$D_n^\lambda = \frac{V(d_\lambda)}{V(\Omega)} \tag{14.23}$$

They are the volume damage and fracture ratio and volume damage and fracture rate in geotechnical mechanics, respectively.

It can be seen that the volume damage and fracture ratio and volume damage and fracture rate concepts in classical geotechnical mechanics measure damage fracture ratio and measure damage and measure rate concept special examples.

14.4.9 CONCRETE λ1–λ2 INTERCEPT JOINT RATIO AND INTERCEPT JOINT RATE

With the same reason, we can define $\lambda_1 - \lambda_2$ intercept joint ratio and rate.

In concrete material test specimen damage and fracture space (Ω, $d_{\lambda_1-\lambda_2}$) and based on the test specimen $\lambda_1 - \lambda_2$ intercept joint $d_{\lambda_1-\lambda_2}$, we can define:

$$D_e^{\lambda_1-\lambda_2} = \frac{m(d_{\lambda_1-\lambda_2})}{m(\overline{d_{\lambda_1-\lambda_2}})} \tag{14.24}$$

For the test specimen $\lambda_1 - \lambda_2$ intercept joint ratio, we can define:

$$D_n^{\lambda_1-\lambda_2} = \frac{m(d_{\lambda_1-\lambda_2})}{m(\Omega)} \tag{14.25}$$

For the test specimen $\lambda_1 - \lambda_2$ intercept joint rate.

14.4.10 RELATION AMONG CONCRETE λ-LEVEL DAMAGE AND FRACTURE AND CT NUMBER AND DENSITY

It is known from research results by Takashi (1983) that any pixel numerical value in a CT image can be expressed by using CT number H, and that this CT number is the medium for X-ray absorption coefficient linear combination:

$$H = a\mu + b \tag{14.26}$$

where a, b are the constants and μ is the object for X-ray absorption coefficient.

In the test specimen damage and fracture space (Ω, d_λ), it can be known from λ-level damage and fracture rate definition that

$$\mu = (1 - D_n^\lambda)\mu_d + D_n^\lambda \mu_d \tag{14.27}$$

where μ, μ_d, and μ_d are full field, λ-level nondamage and nonfracture field, λ-level damage and fracture field average X-ray line absorption coefficient, respectively.

Letting test specimen unit mass absorption coefficient be μ^m, in general, μ^m is only related with the incident X-ray wave length. Relation formula between object X-ray line absorption coefficient and unit mass of object adsorption coefficient is seen in Eq. (14.2), substituting Eq. (14.2) into Eq. (14.27), we can obtain the following:

$$\rho\mu^m = (1 - D_n^\lambda)\rho_d \mu_d^m + D_n^\lambda \rho_d \mu_d^m \tag{14.28}$$

where $\mu_{\bar{d}}^m$ and μ_d^m are λ-level nondamage and nonfracture field and λ-level damage and fracture field unit mass X-ray line absorption coefficient, respectively.

ρ_d and ρ_d are λ-level nondamage and nonfracture field and λ-level damage and fracture field average density, respectively.

Simplifying Eqs (14.27) and (14.28), we can obtain:

$$\mu = \mu_d + D_n^\lambda(\mu_d - \mu_d) \tag{14.29}$$

$$\rho\mu^m = \rho_{\bar{d}}\mu_{\bar{d}}^m + D_n^\lambda(\rho_d \mu_d^m - \rho_{\bar{d}}\mu_{\bar{d}}^m) \tag{14.30}$$

From Eqs (14.30) and (14.31), we can obtain the theoretical relation formula for λ-level damage and fracture rate and CT number and density in the following:

$$D_n^\lambda = \frac{\mu - \mu_{\bar{d}}}{\mu_d - \mu_{\bar{d}}} = \frac{H - H_{\bar{d}}}{H_d - H_{\bar{d}}} \tag{14.31}$$

$$D_n^\lambda = \frac{\rho\mu^m - \rho_{\bar{d}}\mu_{\bar{d}}^m}{\rho_d \mu_d^m - \rho_{\bar{d}}\mu_{\bar{d}}^m} \tag{14.32}$$

where $H_{\bar{d}}$ and H_d are λ-level nondamage and nonfracture field and λ-level damage and fracture field average CT number, respectively.

Equation (14.32) is the relation formula for λ-level damage and fracture rate for each field average CT number.

14.4.11 CONCRETE DAMAGE AND FRACTURE GENERATED POSITIONS

As to the realistic homogeneous material uniaxial compression test, and as far as some state is concerned, microchap germination position should occur at $d(x,y,z)$ value maximum position or location nearby, and this zone can be expressed by means of the aggregate in the following:

$$\left\{ (x_0, y_0, z_0) \mid d(x_0, y_0, z_0) \geq (d_{max} - \varepsilon) \right\} \tag{14.33}$$

where $d_{max} = \max\{ d(x,y,z) \mid (x,y,z) \in \Omega \}$, ε is any small real number.

As far as some loading process is concerned, microchap germination position should be in the density damage increment ΔD maximum value nearby, that is, the density varying quantity maximum value nearby; this zone can be expressed by means of the aggregate in the following:

$$\left\{ (x_0, y_0, z_0) \middle| \Delta D(x_0, y_0, z_0) \geq (\Delta D_{max} - \varepsilon) \right\} \tag{14.34}$$

where $\Delta D_{max} = \max \left\{ \Delta D(x, y, z) \middle| (x, y, z) \in \Omega \right\}$, ε is any small real number. D is damage variable here, whose definition can be seen in Section 14.7. ΔD is the damage increment.

The preceding text describes the microchap germination positions of the realistic homogeneous material uniaxial compression tests from one state and one process angle. These two positions are the same in spatial distribution. As to the rock materials, this point has been proved by many test results obtained by Yang et al. (1998). However, the common concrete material uniaxial compression tests are difficult to make the loading be vertical to the top surface of the cylinder test specimen strictly without being strictly off-center or eccentric. But it can be achieved owing to the material nonuniformity, and it will result in the test specimen failure as the shear damage at last. Therefore, the positions produced by the failure are worth further studying.

14.4.12 CRITERIONS FOR CONCRETE DAMAGE AND FRACTURE

Whether some points can cause damage and fracture will be judged by the density values and density varying quantity of the point locations. Let some material matrices damage and fracture degree be $d_{(0)}$; as to the specified material, macrochap magnitudes are the fixed numbers, and there are two positive constants ε_f and ε_d, thus making the following:

1. When $d_{(1)}(x, y, z) - d_{(0)} > \varepsilon_f$, the materials at that point location have been damaged.
2. When $\varepsilon_f \geq d_{(1)}(x, y, z) - d_{(0)} \geq \varepsilon_d$, the materials at that point location are being damaged.
3. When $d_{(1)}(x, y, z) - d_{(0)} < \varepsilon_d$, the materials at that point location are in safe state.

Of which, $d_{(1)}(x, y, z)$ is the damage and fracture degree of the test specimen at some points.

If matrices CT number is equal to or close to the maximum CT number max H of the test specimens, and then $d_{(0)} \approx 0$, it is just at this time that the mentioned criteria become:

1. When $d_{(1)}(x, y, z) > \varepsilon_f$, the materials at that point location have been damaged.
2. When $\varepsilon_f \geq d_{(1)}(x, y, z) \geq \varepsilon_d$, the materials that location are being damaged.
3. When $d_{(1)}(x, y, z) < \varepsilon_d$, the materials at that location are in safe state.

14.4.13 CONCRETE SAFETY ZONE, DAMAGING AND FRACTURING ZONE, AND DAMAGED AND FRACTURED ZONE

Based on the mentioned criteria, one test specimen space can be partitioned into safety zone, damaging and fracturing zone, and damaged and fractured zone. Their definitions are as follows:

1. The following set is called the specimen safety zone:

$$S = \left\{ (x, y, z) \middle| d(x, y, z) - d_{(0)} < \varepsilon_d \right\} \tag{14.35}$$

2. The following set is called as the specimen damaging and fracturing zone:

$$D = \left\{ (x, y, z) \middle| \varepsilon_d \leq d(x, y, z) - d_{(0)} \leq \varepsilon_f \right\} \tag{14.36}$$

3. The following set is called as the specimen damaged, and fractured zone:

$$F = \left\{ (x, y, z) \middle| \, d(x, y, z) - d_{(0)} > \varepsilon_f \right\} \tag{14.37}$$

With the same reason, if matrices CT number is equal to or close to the test specimen maximum CT number max H, it is just at this time that the safety zone set S is, in fact, one ε_d-level nondamage and nonfracture field. Damaging and fracturing zone set D is, in fact, one $\varepsilon_d - \varepsilon_f$ intercept joints; damaged and fractured zone set F is one strong ε_f-level damage and fracture field. Indeed, S, D, F are the three different level intercept joints.

It can be seen that ε_f and ε_d are two constants of materials, whose values are determined by the physical properties. Actually, materials are in different physical state limitation value, and they are called as the material damaging and fracturing zone upper limitation, and damaging and fracturing zone lower limitation.

As to concrete materials versus or relative rock materials, nonuniformity is more protruded. Concrete as macroviewed consists of aggregates, mortar, and chap or holes of three parts. If concrete is partitioned, first of all, aggregates and mortar components should be partitioned. Owing to the large density of aggregates, the failure scope in the process of loading is mini. Chaps around the aggregates expand in mortar in general. When chaps and aggregates cross with big angles, chaps can prevent aggregates from expanding forward. This illustrates that viscous surface strength of aggregates with the mortar zone is very low, and that this is the chap producing key zone leading to concrete damage. Therefore, the mortar zone is the priority zone to study the damaging laws of concrete.

14.4.14 CONCRETE DAMAGING AND FRACTURING ZONE LOWER LIMITATION AND UPPER LIMITATION BOUNDARY SURFACE

Using point set definition safety zone, damaging and fracturing zone and damage and fracture zone interface is not a simple curve or curved surface. It is interwoven, and at the same time, there exist some isolated points. Therefore, it is necessary to redefine the boundary surface existing among three zones from the angles of the set.

If the points (x_0, y_0, z_0) any CT magnitude δ neighboring zone exists either safety zone points or damaging and fracturing zone points, these points are the lower limitation boundary points of the damaging and fracturing zone. The set or aggregate consisting of all damaging and fracturing zone lower limitation boundary points is called as the damaging and fracturing lower limitation boundary surface, to be expressed by b_{ε_d}.

With the same reason, if the point (x_0, y_0, z_0) any CT magnitude δ neighboring zone existing either the damage and fracture zone points, or the damaging and fracturing zone or safety zone points, these points are called as the upper limitation boundary points of damaging and fracturing zone. The set or aggregate consisting of all the damaging and fracturing upper limitation boundary points is known as the damaging and fracturing zone upper limitation boundary surface, to be expressed by b_{ε_f}. The damaging and fracturing upper limitation boundary surface is also known as chap boundary surface.

It is here that the so-called point CT magnitude δ neighboring zone magnitude (side length or diameter) should be larger than that of CT resolution unit. It is here that the so-called surface is to maintain agreed with the zone magnitudes habitually, but it is in fact, a point set, having no sense of surface. In addition, the point to say here is the CT magnitude point, having a certain size, but not the point in mathematics.

In addition, attention should be paid to each zone boundary nearby point (x_0, y_0, z_0) any CT magnitude δ neighboring field types is likely to have the following four kinds of situations:

1. Meanwhile including safety zone and damaging and fracturing zone points.
2. Meanwhile including damaging and fracturing and failure zone points.
3. Meanwhile including safety zone and damage and fracture zone points.
4. Meanwhile including safety zone, damaging and fracturing zone and failure and fracture zone points.

In the mentioned definitions, three kinds of conditions of (2), (3), and (4) are finalized into the damaging and fracturing zone upper limitation boundary surface.

14.4.15 CONCRETE WEAKENING–STRENGTHENING NORMS

When some point in concrete material transfers from some stress state σ_{ij} into the stress state $\sigma_{ij} + \Delta\sigma_{ij}$, corresponding to the degree of damage and fracture from $d_{(1)}(x, y, z)$ varying into $d_{(2)}(x, y, z)$, there exist the following:

1. If $d_{(2)}(x, y, z) - d_{(1)}(x, y, z) > 0$, the point is in the weakening state.
2. If $d_{(2)}(x, y, z) - d_{(1)}(x, y, z) < 0$, the point is in the strengthening state.
3. If $d_{(2)}(x, y, z) - d_{(1)}(x, y, z) = 0$, the point is in the neutral state.

It is worth noticing that when material is in triaxial state of being subject to force, the weakening state is not equal to loading increase, and the strengthening state is not equal to loading decrease, and the decrease in external loading can also cause concrete material damage and fracture and weakening. But when materials are in the shearing expansion state, the safety zone continues to transfer to the damage and fracture zone, while the damaging and fracturing zone continues to transfer to the damage and fracture zone.

14.4.16 SUBZONE DESCRIPTION OF CONCRETE CONSTITUTIVE RELATION

In the past, people have established various kinds of constitutive relations, whose common characteristics are based on the materials in different stress states, whereby establishing many stress and strain relation formulas corresponding to various stress states. Concrete materials on the same stress level can cope with many deformation states. Therefore, this constitutive relation must distinguish the materials whether they are the first loading, unloading, reloading, and reunloading times and whether the materials are in the hardening stages or in the softening stages. In this way, making concrete materials constitutive relations become unusually complicated has brought about many difficulties in computation and application.

The following is based on the partition of the above materials in the state zones to establish the subzone damage and fracture constitutive relation of concrete materials.

As to concrete materials safety zone set S, the elastic constitutive relation is adapted as follows:

$$\sigma_e = E\varepsilon_e \tag{14.38}$$

As to the concrete material damaging and fracturing zone set D, the damage and fracture constitutive relation is adopted as follows:

$$\sigma_d = \alpha_1(1-d)E\varepsilon_d = \alpha_1 pE\varepsilon_d \tag{14.39}$$

As to the material damage and fracture zone set F, the force to bear has been very small, and the deformation is hard to recover. Therefore, the plastic constitutive relation is adopted as follows:

$$\upsilon_p = \alpha_2 D_p \varepsilon_p \tag{14.40}$$

where p and d are perfect degree and damage and fracture degree of materials at a certain point; α_1, α_2 are material variability coefficients.

The basic distinction between concrete material subzone damage and fracture constitutive relation and traditional constitutive relation lies not in taking stress state as the base but in establishing the relation based on taking materials in the damage and fracture state as the base, whereby having simple formation and clear physical significance.

14.4.17 CONSTITUTIVE THEORY OF CONCRETE DAMAGE AND FRACTURE SPACE

It is here that the test specimens should be viewed as the formation of safety zone, damaging and fracturing zone, and damage and fracture zone. It can be known from the mentioned contents that the three zones are actually ε_d-level nondamage and nonfracture field, $\varepsilon_d - \varepsilon_f$ intercept joint, and ε_f-level damage and fracture field, respectively, and the three different level intercept joints. Letting their intercept joint rates be $D_n^{0-\varepsilon_d}$, $D_n^{\varepsilon_d-\varepsilon_f}$ and $D_n^{\varepsilon_f-1}$, obviously, $D_n^{0-\varepsilon_d} + D_n^{\varepsilon_d-\varepsilon_f} + D_n^{\varepsilon_f-1} = 1$, can be used as weighting, total average stress, and average strain expression formulas are as follows:

$$\bar{\sigma} = D_n^{0-\varepsilon_d} \bar{\sigma}_e + D_n^{\varepsilon_d-\varepsilon_f} \bar{\sigma}_d + D_n^{\varepsilon_f-1} \bar{\sigma}_p \tag{14.41}$$

$$\bar{\varepsilon} = D_n^{0-\varepsilon_d} \bar{\varepsilon}_e + D_n^{\varepsilon_d-\varepsilon_f} \bar{\varepsilon}_d + D_n^{\varepsilon_f+1} \bar{\varepsilon}_p \tag{14.42}$$

where the labels e, d, and p are safety zone, damaging and fracturing zone, and failure zone.

Total average stress and average strain, and average stress and average strain in three zones can be defined by each formula in the following:

$$\bar{\sigma} = \frac{1}{m(\Omega)} \int_\Omega \sigma \, d\Omega, \quad \bar{\sigma}_e = \frac{1}{m(S)} \int_S \sigma_e \, dS, \quad \bar{\sigma}_d = \frac{1}{m(D)} \int_D \sigma_d \, dD, \quad \bar{\sigma}_p = \frac{1}{m(F)} \int_F \sigma_p \, dF$$

$$\bar{\varepsilon} = \frac{1}{m(\Omega)} \int_\Omega \varepsilon \, d\Omega, \quad \bar{\varepsilon}_e = \frac{1}{m(S)} \int_S \varepsilon_e \, dS, \quad \bar{\varepsilon}_d = \frac{1}{m(D)} \int_D \varepsilon_d \, dD, \quad \bar{\varepsilon}_p = \frac{1}{m(F)} \int_F \varepsilon_p \, dF$$

where $m(*)$ can express the set measure. The integral here is Lebesgue integral.

Defining the local strain coefficients of safety zone, damaging and fracturing zone, and failure zone can be as follows:

$$\beta_e = \frac{\bar{\varepsilon}_e}{\bar{\varepsilon}}, \quad \beta_d = \frac{\bar{\varepsilon}_d}{\bar{\varepsilon}}, \quad \beta_p = \frac{\bar{\varepsilon}_p}{\bar{\varepsilon}}$$

Substituting the partition damage and fracture constitutive relation into Eqn (14.41), we can have following:

$$\begin{aligned}
\bar{\sigma} &= D_n^{0-\varepsilon_d} \bar{\sigma}_e + D_n^{\varepsilon_d-\varepsilon_f} \bar{\sigma}_d + D_n^{\varepsilon_f-1} \bar{\sigma}_p \\
&= D_n^{0-\varepsilon_d} D_e \bar{\varepsilon}_e + D_n^{\varepsilon_d-\varepsilon_f} D_d \bar{\varepsilon}_d + D_n^{\varepsilon_f-1} D_p \bar{\varepsilon}_p \\
&= D_n^{0-\varepsilon_d} \beta_e D_e \bar{\varepsilon} + D_n^{\varepsilon_d-\varepsilon_f} \beta_d D_d \bar{\varepsilon} + D_n^{\varepsilon_f-1} \beta_p D_p \bar{\varepsilon}
\end{aligned} \tag{14.43}$$

$$\text{So,} \bar{\sigma} = \bar{D}\bar{\varepsilon} \tag{14.44}$$

$$\bar{D} = D_n^{0-\varepsilon_d} \beta_e D_e + D_n^{\varepsilon_d - \varepsilon_f} \beta_d D_d + D_n^{\varepsilon_f - 1} \beta_p D_p$$

$$D_e = E$$

$$D_d = \alpha_1(1-d)E = \alpha_1 pE$$

$$D_p = \alpha_2 D_p$$

where D_e, D_d, and D_p are elastic constant matrix, damage matrix plastic matrix in subzone constitutive relation.

If we define $\sigma_e = D_e\,\bar{\varepsilon}, \sigma_d = D_d\bar{\varepsilon}$ and $\sigma_p = D_p\bar{\varepsilon}$, we will have:

$$\bar{\sigma} = D_n^{0-\varepsilon_d} \beta_e \sigma_e + D_n^{\varepsilon_d - \varepsilon_f} \beta_d \sigma_d + D_n^{\varepsilon_f - 1} \beta_p \sigma_p \tag{14.45}$$

Equation (14.45) is the basic equation of the material damage and fracture space theory, where σ_e, σ_d, and σ_p are no longer average stress.

If the damaging and fracturing zone will not be taken into account, the zone is converged into the safety zone or the damage and fracture zone, and material damage and fracture space theory can be degraded into two types of material phase damage and fracture theory. It is here that the research theory of Zhujiang (2003) is used for reference.

14.4.18 SUMMARY

This section points out that the current concrete material CT number mechanics research is still in CT image observation and research stage and defines the concepts concerning the perfect degree, damage and fracture degree, λ-level perfect field, λ-level damage and fracture, $\lambda_1 - \lambda_2$ intercept joint as well as λ-level perfect field, λ-level damage and fracture and $\lambda_1 - \lambda_2$ intercept joint measure, and so on, on one point in concrete material medium space, and gives out the damage and fracture space concept, and studies their some basic properties. And then this section defines λ-level damage and fracture ratio and rate, the concepts of $\lambda_1 - \lambda_2$ intercept joint ratio and rate based on CT number and studies the relation between λ-level damage and fracture rate and CT number and density. The application of knowledge of set theory, measure theory and Lebesgue integral can partition or divide concrete materials into the safety zone, damaging and fracturing zone, and damage and fracture zone, and give out the definition of material weakening and strengthening norms, and establish the constitutive relation of concrete material partition damage and fracture and further infer the basic equations of damage and fracture spatial theory so as to lay a solid foundation for the application of modern mathematic quantitative analysis of CT mechanics properties of concrete materials and for establishment of damage and fracture spatial mechanics. These theories are, at the same time, adaptable to macromechanics and mesomechanics as well as successive medium mechanics and multiporosity medium mechanics or fracture medium mechanics.

14.5 CLASSIFICATION OF SUPPORT VECTOR MACHINE FOR CONCRETE CT IMAGES

Support vector machine (SVM) is a kind of machine learning method established on the basics of statistics learning theory. SVM aims at seeking the existing information optimum solution, but not only for the specimen or sample numbers; it is tending to seeking the optimum solution at infinity under the conditions of limited samples. At present, SVM is widely used in the geotechnical engineering field, such as the classification of adjoining rock stability, and prediction of side-slope stability, and so on. It is here that support vector machine is used to recognize or interpret the aggregates, mortar, fracture, or holes in concrete CT images so as to carry out research on concrete mesomechanics.

Vapnik completely suggested the support vector machine-learning algorithm in the book "The Nature of Statistical Learning Theory," published in 1995. The complete mathematical description of support vector machine can refer to related reference literatures. It is here that a brief description of the support vector machine algorithm is given out.

14.5.1 OPTIMUM CLASSIFICATION FACES

Support vector machine theory is suggested from the optimum classification face under the linear separable situations. Figure 14.59 shows the two types of linear separable situations. In the Figure 14.59, hollow dots and solid round dots express two types of training specimens, and H is the classification line. H1, H2 are the nearest dots to the classification line and parallel to the classification straight line of specimens of each type separately. The distance between H1 and H2 is called classification margins. The so-called optimum classification line requires that not only can two types be separated accurately and correctly, but also the classification margins must be great. If it is extended to the high dimensional space, the optimum separating line will become the optimum classification face. Supposing that the training specimens are (x_i, y_i), $i = 1, 2, \ldots\ldots n$, $x \in Rd$, $y \in (-1, +1)$, where x is the output vector, y is the input vector belonging to the type, and n is the number of specimens. If this vector set can be separated by the super-plane $wx + b = 0$ accurately and correctly, and the distance (i.e., margin) between the nearest vector from the super-plane and the super-plane is the maximum, this vector set is separated

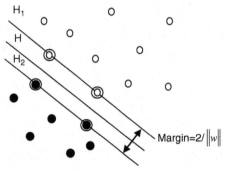

FIGURE 14.59 Optimum Classification Face Support Vector

by this optimal super-plane (the largest margin super-plane). D type space linear discriminate is $g(x) = w^T x + w_0$, the linear separable conditions can be generalized as follows:

$$\begin{cases} w \cdot x_i + b \geq +1 & \text{for} \quad y_i = +1 \\ w \cdot x_i + b \leq -1 & \text{for} \quad y_i = +1 \end{cases} \tag{14.46}$$

It is at this time that classification face equation is $wx + b = 0$; the decision-taking function is $f(x) = sgn(wx + b)$; the classification margin is $2/\|w\|$.

If the classification margin needs to be the largest, $\|w\|$ must be made the largest. If all the specimens need to be separated rightly, the following constraint conditions must be satisfied:

$$y_i[w \cdot x_i + b] - 1 \geq 0, i = 1, 2, \cdots n \tag{14.47}$$

In Eq. (14.47), the equal symbol established is called support vector. For instance, the round circle is used to label the dots in Fig. 14.59.

As to the mentioned constraint optimal problem, the problem can be converted into the following constraint-face problem through the introduction of Lagrange multiplier $\alpha_i \geq 0$:

$$W(w, b, a) = \frac{1}{2} \|w\|^2 - \sum_{i=1}^{n} a_i \left[y_i \left(w^2 x_i + b \right) - 1 \right] \tag{14.48}$$

However, the problem has become the minimum value for w, b and a_i: to seek the solution to Lagrange function. Equation (14.48) versus w and b respectively to seek for partial different, let them equal to zero, the original problem can be converted into the following simple dual problem.

In the case of constraint conditions, there will be the following:

$$\frac{\partial W}{\partial b} = 0 \Rightarrow \sum_{i=1}^{n} y_i a_i = 0 \quad (a_i \geq 0, i = 1, 2, \ldots, n) \tag{14.49}$$

Under this equation, it is for a_i to seek for the maximum value of the following function:

$$W(a) = \sum_{i=1}^{n} a_i - \frac{1}{2} \sum_{i=1}^{n} \sum_{j=1}^{n} y_i y_j (x_i^T x_j) a_i a_j \tag{14.50}$$

If a_i^* is the optimum solution, there will be the following:

$$\frac{\partial W}{\partial w} = 0 \Rightarrow w* = \sum_{i=1}^{n} a_i^* y_i x_i \tag{14.51}$$

The weight coefficient vector of the best classification face is the linear combination of training specimen vector.

This is a quadratic function to seek for the best problem under the inequality constraint, there exists only one solution, and in terms of the Kuhn–Tucker conditions, this optimum problem solution must satisfy the following:

$$a_i \left[y_i \left(w^T x_i + b \right) - 1 \right] = 0, i = 1, 2, \ldots, n \tag{14.52}$$

Therefore, as for most specimens a_i^* being 0, taking value will not be 0 a_i^* corresponding to Eq. (14.52) equal symbol established specimen, that is, support vector; they are usually only a few parts of all the specimens.

After seeking for the solution to the mentioned problem, the best classification function obtained is as follows:

$$f(x_i) = \text{sgn}\left\{(w^T x_i) - b\right\} = \text{sgn}\left\{\sum_{i=1}^{n} a_i y_i (w^T x_i) - b\right\} \tag{14.53}$$

where, sgn( ) is symbol function; since nonsupport vector corresponding to a_i is 0, summation in Eq. (14.53) is, in fact, carried out only for support vector; b is a classification threshold value, which can be obtained by means of any one support vector, or through two types of any one pair of support vector in taking the mid-value to obtain.

14.5.2 BROAD SENSE OPTIONAL CLASSIFICATION FACE

The optional classification face is discussed under the prerequisite of linear separation, but in the case of nonseparation, that is, some specimens cannot satisfy Eq. (14.47) conditions, one loose item $\xi_i \geq 0$ can be increased in the conditions, thus becoming the following:

$$y_i (w^T x_i + b) - 1 + \xi_i \geq 0 \tag{14.54}$$

As to being small enough, we can only make:

$$F_\sigma (\xi) = \sum_{i=1}^{n} \xi_i^\sigma \tag{14.55}$$

Being the minimum, thus rendering the number of misclassification of specimens to become the minimum, in corresponding to linear classification situations, the classification margin can be made to be the maximum. In the case of linear nonseparable situations, the constraint conditions can be introduced into:

$$\|w\|^2 \leq c_k \tag{14.56}$$

Which is to ensure the classification margin is not too small. Under the condition of constraint Eqs (14.54) and (14.56), it is necessary to seek the extremely small value of Eq. (14.55) to obtain the optional classification face under the condition of nonlinearity, which is called as the broad sense of optional classification face. In order to be convenient for computation, it is necessary to take $\sigma = 1$ MPa.

In order to render computation to be further simplified, the broad sense optimal classification face problem can further evolve into the conditional Eq. (14.54) constraint to seek for function extremely small value:

$$\phi(w, \xi) = \frac{1}{2}(w^T w) + C\left(\sum_{i=1}^{n} \xi_i\right) \tag{14.57}$$

where C is some specified constant, it actually plays a role in controlling over the penalty degrees for the right or wrong specimen classification, so as to achieve the compromise between the ratio of wrong classification of specimens and complexity of the algorithm.

When the optimal classification face for finding a solution is used, the same method can be used to find a solution to this optimal problem, so as to obtain the same one quadratic function extreme value problem, whose results are as completely the same as the Eqs (14.44)–(14.53) under the separable situations, but it is only conditional Eq. (14.49) that becomes as follows:

$$0 \le a_i \le C, i = 1, 2, \ldots, n \tag{14.58}$$

The optimal and broad sense optimal classification function Eq. (14.50) obtained from the earlier discussion only contains a support vector inner product operation (x, x_i) of the real classification specimens and training specimens. It can be seen that to solve the optimal linear classification problem in a special space, one only needs to carry out the inner product operation in this space. Since many practical model identification problems are not linearly separated in the definition space, a nonlinear change can be adopted and then input into imaging one characteristic space z, of which $z = \varphi(x)$. In the characteristic space z, data are linearly separable so that one linear sort function can be found in z. It corresponds to the input space nonlinear sort function, and requiring to seek a solution to the optimal sort function at this movement involves the computing of inner product $(z_i \cdot z_j) = \sum_{k=1}^{m} \varphi_k(x_i)\varphi_k(x_i)$ directly seeking the solution cannot only know the nonlinear imaging φ forms but also calculate the quantity, which appears to have an exponential growth with an increase in characteristic space dimensional number, thus making solution-seeking more difficult and even unable to get the solution. This is because the optimal problem and sort function involve the inner product computation (z_i, z_j). Based on Hilbert–Schmidt principle, it is able to find the kernel function $K(x_i, x_i)$ to satisfy Mercer conditions, thus making $K(x_i, x_i) = z_i \cdot z_j$ optimal function become:

$$Q(a) = \sum_{i=1}^{n} a_i - \frac{1}{2} \sum_{i,j=1}^{n} a_i a_j y_i y_j K(x_i, x_j) \tag{14.59}$$

The corresponding sort function also becomes:

$$f(x) = \mathrm{sgn}\left[\sum_{i=1}^{n} a_i y_i K(x_i \cdot x_j) + b \right] \tag{14.60}$$

This is the support vector machine to solve the problem of nonlinear classification, whose form and structure are similar to a neural network. It is easy to see that the basic thinking of support vector machine can be generalized. First of all, the nonlinear exchange should be input into the space to image one-dimensional space and then in this high-dimensional space, the optimal linear classification face can be obtained. The nonlinear exchange is realized through the appropriate definition of the inner product function (kernel function). Accordingly, selection of different forms of kernel functions can generate the different kinds of support vector machines.

14.5.3 CLASSIFICATION METHOD OF SUPPORT VECTOR MACHINE IMAGE

1. Support vector machine image classification

   To use support vector machines to carry out image classification, basic thinking is through the extraction of one or many characteristics from the selected specimen points in the images to train the SVM classifier or sorter, and then the pixel dots in the waiting classification images are classified by the well-trained classifier. The concrete steps are as follows:

a. In the images, the zones for extracting objectives should be selected, and then the characteristics of these specimen dots are extracted as training specimen zones. As to the image classification, the trained specimens may select specimen dot RGB components, gray degrees, average values, and so on.

b. It is necessary to select the suitable kernel function and parameter classifiers. The frequently used kernel functions are as follows:

   – Polynomial kernel function is as follows:

$$K(x_j, x_i) = \left[\left(x_j^T, x_i \right) + 1 \right]^d$$

   – Gouss radial base kernel function is as follows:

$$K\left(x_j, x_i \right) = \exp\left\{ -\frac{\left| x_j - x_i \right|^2}{\sigma^2} \right\}$$

   – Sigmoid kernel function is as follows:

$$K\left(x_j, x_i \right) = \tanh[v(x_j^T x_i) + c]$$

c. The well-trained SVM classifier is used to carry out the classification of classifying specimen set.

d. The classification results are separated with different colors.

2. Classification of multitypes of problems:

Since the types of mesostructures in concrete CT image are not two classes but many, this needs to deal with the multitypes of problems. There are two kinds of main methods for support vector machine to deal with the multitypes of problems:

a. *One-to-one method*: In general, in IV class classification, it is likely to build up all the possible class II classifier in class II, it needs to build up $n(n-1)/2$ classifiers. After all the waiting specimens are classified through by the classifiers, where the specimen belongs will be decided through voting.

b. *One pair multimethod*: Class I and the other classes are used to make judgment and classification. Its realization strategy is to aim at N class classification of problems to build up n pieces of class II SVM classifiers, the $i(1 \le i \le n)$ piece of SVM classifiers will take the training specimens in the i class as the positive training specimens, while the other training specimens are used as the negative training specimens. The waiting classifying specimens are classified through all the classifiers afterward, from which one specimen can be found out; that is the result of classification.

14.5.4 CLASSIFICATION OF CONCRETE COMPONENT SUPPORT VECTOR MACHINE

Since more images can be obtained from the tests, this section takes Fig. 14.60 as an example to introduce the application of support vector machine in the analysis of CT image concrete components.

One-pair of multiclassification method is adopted in this part (Yanwen, 2007). CT number of aggregates, mortar, and holes in concrete CT image varies greatly. Aggregates are the brighter zones in the images. CT numbers are, in general, between 2501~3600. Mortar CT numbers are, in general,

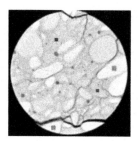

FIGURE 14.60 Section of Specimen Zone

between 2001~2500. Hole CT numbers are, in general, below 2000. CT numbers have great differences so that pixel dot CT numbers are taken as training specimens. First of all, the part of zones with the representatives can be selected to serve as the specimen zone; pixel dots selected total 964 dots, of which 147 dots are selected in the hole zone, 436 dots are selected in the aggregate zone, and 381 dots are in the mortar zone. These 964 pixel dots CT numbers can be used as the training specimens, and at the same time, three classifiers are constructed. In order to train suitable classifiers, it is necessary to select the appropriate kernel function and penalty factor C. Different parameters can have the direct effect upon the learning efficiency of support vector machine, but the support vector machine method cannot give out the selection method of various kinds of parameters. It has been found through the comparisons that the better kernel function is the polynomial kernel function $K(x_j, x_i) = \left[\left(x_j^T, x_i\right) + 1\right]^d$. Parameters are determined in the past experiences, to select $d = 3, C = 11$, is able to obtain images as shown in Fig. 14.61. Here, we only take the first layer cross-section results of classification as an example as shown in Table 14.1.

Result analysis of support vector machine:

It can be seen from Table 14.1 that after 436 aggregate specimens are classified through three classifiers, 429 specimens are correctly classified (excluding Mi Xing classified specimens); 370 specimens from 381 mortar specimens are correctly classified through three classifiers; 140 specimens from 147 hole specimens are correctly classified through three classifiers, whereby illustrating the high correctness of classification results.

Aggregates, mortar, chaps, or hole zones and their distribution positions can be clearly seen from Fig. 14.61. This is because there are large zones with high density in aggregates. In contrast with the analysis of chaps or hole zone variation in SVM image under each stress condition, it can be found that the magnitude of chaps or hole zones and their distribution continue to change with the variation in stress. This zone in $\sigma = 18.01$ MPa being in opposite to $\sigma = 0.00$ MPa becomes small, while after $\sigma = 19.69$ MPa, this zone gradually expends, whereby demonstrating that in the loading process, concrete test specimens have experienced the process of compact first, capacity expansion afterward, CT magnitude chap expansion and finally reaching failure.

Accordingly, it can be seen that after SVM analysis, each zone obtained has obvious subzones, clear chaps or hole zones, aggregate zones, and mortar zone distribution, coinciding with test results. Therefore, the advantages of SVM analysis lies in the fact that not only can it well reflect chap or hole zones, aggregate zone and mortar zone, thus making the final classification results have that there are

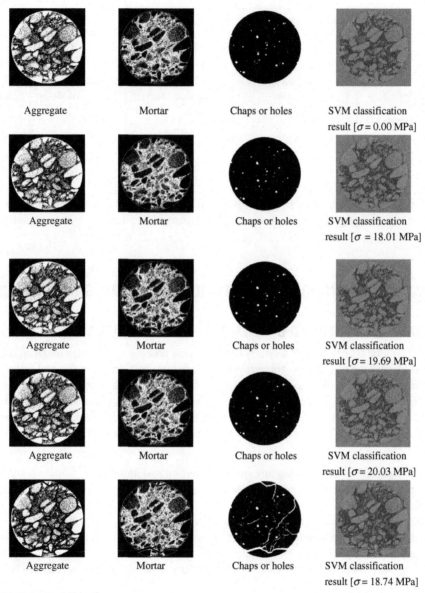

FIGURE 14.61 SVM Classification

Table 14.1 SVM Classification Result

	Classifier No.			
Training Specimens	Classifier 1 (Separated Aggregate)	Classifier 2 (Separated Mortar)	Classifier 3 (Separated Holes)	Identification Rate (%)
Aggregate specimens (436)	429	1	6	98.4
Mortar specimens (381)	11	370	0	97.1
Hole specimens (147)	2	5	140	95.2

only three different numbers, being favorable for each zone to carry out statistical analysis, studying chap and hole zones changing laws with loading, and meanwhile, it provides a quantitative feasible way to measure chaps and hole zones.

14.6 FRACTAL DIMENSION COMPUTATION AND ANALYSIS OF CONCRETE CI IMAGE

14.6.1 FRACTAL THEORY AND DIMENSIONAL COMPUTATION METHOD

In this section, fractal theory is used to study the evolution process of concrete material damage and fracture so as to open up a new vista for studying the concrete material damage and failure process. The Hausdorff–Besicovitch dimensional number is strictly larger than the topologic dimensional set called a fractal. Later on, he revised this definition to be "whose components with some kind of manner to be similar to the integrity physique called fractal." But, at present, fractal has no reliable definition. Wu Zheng (2002) pointed out that if F is of the following character set, it can be called as fractal:

1. F has a fine structure, that is, having any small proportion in details.
2. F is so very irregular that its integrity and locality cannot be described by means of traditional geometric languages.
3. F is usually of some similar figures, and it is likely to be approximate or statistics.
4. In general, F fractal dimensional number (with some manner to define it) is larger than topologic dimensional number.
5. In an appropriate case, F can be produced by a very simple method, and it is likely to be produced by the interaction.

In these definitions, self-similar figure is an important characteristic of fractal. The so-called self-similar figure refers to the similarity between the system total situation and parts, this part and that part. In Euclidean space, point, line, surface, and spherical surface correspond to zero, one, two, and three dimensions, respectively. Also, the higher dimensional space can be introduced, but they are the integers, while the dimensionality in the fractal can be used to express the degrees of irregularity of the fractal set, whereby expanding the dimensionality from the integer to the fraction as viewed from the measure angle, making a breakthrough to the boundary of the common topologic set dimensionality as the integer. In a certain sense, the magnitude of fractal dimensionality is a kind of measurement of the object irregularity. Therefore, dimensionality that is not the integer becomes the fractal's second characteristics.

In the fractal, people have carried out in-depth studies and understanding of the dimensionality concept and suggested new concepts concerning dimensionality. Heping (1996) pointed out that there are many similar dimensionalities: capacity dimensionality, Hausdorff dimensionality, information dimensionality, Lyapunov dimensionality, spectrum dimensionality, topology dimensionality, generalized extension function, and cartridge dimensionality method, and so on. In practical application, the different definition methods of dimensionality are adopted with an aim at the different research objectives (Heping, 1996). For instance, Hausdorff dimensionality mathematic expression equation is as follows:

$$D = \lim \frac{InN(\delta)}{In(1/\delta)} \qquad (14.61)$$

In Hausdorff dimensionality definition, consideration is only given to the needed δ covering pieces number $N(\delta)$, but not to every coverage U_i including fractal set elements. Letting P_i express fractal set elements belonging to coverage U_i probability, the information dimensionality D_i should be as follows:

$$D_i = \lim_{\delta \to 0} \frac{\sum_{i=1}^{N} P_i \ln P_i}{\ln \delta} \qquad (14.62)$$

In the case of equal probability $P_i = 1/N(\delta)$ situations, information dimensionality is equal to Hausdorff dimensionality. Sometimes, information dimensionality is also called as information content dimensionality.

Since (Xie Heping, 1996) first used the fractal quantity to analyze metal material fracture surface, scholars at home and abroad have done much work on research and probing into concrete fracture fractal dimensionality. Mesodefects (such as holes, microfractures, and so on) special distribution, chap distribution after concrete macrobreakages, and broken blocks are characterized by strong fractals. Also, the fractal dimensionality can rightly characterize the situations of concrete damage and fracture CT gray image fractal dimensionality are Peleg Model, Pentland based on the FBM model, box accounting method, and the difference cartridge model suggested by Chaudhuri. Box accounting method and differential box accounting methods are introduced respectively in the following.

In box accounting method, $r \times r$ cartridges are used to form the grid coverings, to compute nonempty cartridge quantity, accounting as $N_i(r)$. If the sizes of box are reduced, the obtained quantity of nonempty boxes must be enlarged. The minimum grid of this method is subject to the minimum pixel of images. In order to reduce errors, the grids with different sizes should cover the geometric figures with the sizes. The scales of this method without changing nature only exist in finite scope (in the straight line part measure range); it can be defined as follows:

$$\ln N(r) = \ln a - D \ln r \qquad (14.63)$$

where r is measuring unit size, a is one constant, N is number of nonempty cartridges boxes, and D is this irregular zone fractal dimensionality.

Differential box accounting method (Wu Zheng, 2002) is built up on the basis of Eq. (14.63). As to the given area of $M \times M$ image, supposing that it has been decomposed into an integer between 1 and $m/2$, and at the same time, $r = s/M$. $M \times M$ image can be seen as a three-dimensional space, (x, y) denotes the pixel dot plane position, and the third dimension can be used to express the gray value of the pixel dot. In this way, if the image plane (x, y) is divided into several $s \times s$ grids, it can be considered that there will be a series of $s \times s \times s'$ small box on every grid, s' expresses single box height, and the image gray degree grade number is G, the s' value taking should satisfy $G/s' = M/s$. Supposing that the highest value of gray degree and the lowest value of gray degree in the (i, j) grids of the image fall into the L and the K small boxes in correspondence, respectively, $N(r)$ distribution $n_r (i, j)$ in the (i, j) grids can be as follows:

$$n_r(i, j) = l - k + 1 \tag{14.64}$$

As to all the grids' $n_r (i, j)$ summation, there is the following equation:

$$N(r) = \sum_{i,j} n_r(i, j) \tag{14.65}$$

Being corresponding to as variation, r occurs to have changes in the close following, and the least squares method can be adopted to fit out $\lg N(r) \sim \lg(1/r)$ gradient so as to derive the gradient, that is, corresponding to the fractal dimensionality value D.

Owing to the large range of CT image gray degree, the box counting method cannot obtain the accurate dimensional results so that a kind of differential box counting method suggested by B.B. Chaudhuri can be adopted.

14.6.2 FRACTAL DIMENSIONALITY COMPUTATION AND ANALYSIS OF CONCRETE UNIAXIAL COMPRESSION CT IMAGES

Based on the mentioned method, MATLAB language is adopted to compile the computer program; VB language is used to compile the users' interface and the fractal dimensionality computation program has been developed. The compiled differential box counting computation program is used to select CONC10-3 concrete CT images of scanning cross sections in static compression tests at five stress stages, to compute the fractal dimensionality of each image with the results shown in Fig. 14.62, of which the fitting straight inclined rate is the image fractal dimensionality under the different stresses. The computed results of fractal dimensionality computation of complete stress stages of CONC10-1, CONC10-3, CONC10-5 three scanning sections are shown in Table 14.2. The relation between fractal dimensionality and strain is indicated in Fig. 14.63.

It can be seen from Fig. 14.63 that each fractal dimensionality in damage and fracture evolution image can satisfy fractal equations with fractal characteristics, thus indicating that fractal dimensionality can serve to characterize material damage and fracture state parameter, and that damage and fracture fractal dimensionality evolution laws can reflect material damage and fracture evolution laws.

Fractal dimensionality is related with surface morphologic range value changes. The higher is the fractal dimensionality, the richer the surface details are. It can be known from Table 14.2 and Fig. 14.63

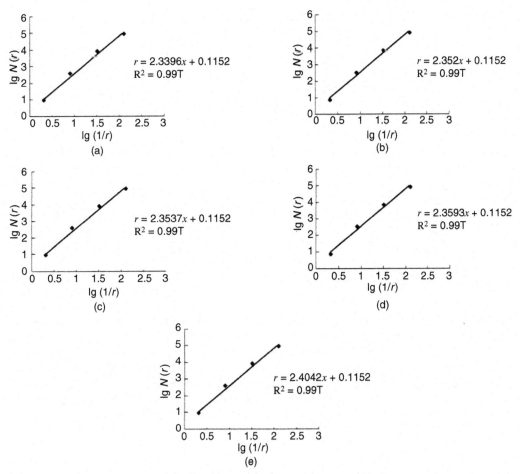

FIGURE 14.62 lg *N(r)* ~ lg (1/*r*) Diagrams of Parts of Scanning Cross Section CONC10-3 Under Loading

(a) δ= 0 MPa straight line fitting results (b) δ = 24.59 MPa straight line fitting results (c) δ = 34.89 MPa straight line fitting results (d) δ = 33.61 MPa straight line fitting results (e) δ = 12.71 MPa straight line fitting results.

that chap evolution source appears on the whole to have the tendency of dimensionality increase, but in the middle, there are some ups and downs. For instance, when initial scanning, cross section 1 fractal dimensionality is $D = 2.3063$. Cross section 3 fractal dimensionality is $D = 2.3396$, and cross section 5 fractal dimensionality is $D = 2.3244$. When initial loading, fractal dimensionality has a certain increase. When loading increases to 24.59 MPa, the fractal dimensionality of the three cross-sections has some decrease; when loading is between 24.59–27.24 MPa, there is not much change in fractal dimensionality, but it is still in the tendency of increasing. When stress reaches the peak value loading of 36.08 MPa, fractal dimensionality appears to have a small fast increase range, cross section 1 fractal

Table 14.2 Fractal Dimensionality Computation Results Under Each Loading

σ (MPa)	Strain (%)	Scanning Section 1 Dimensionality	Scanning Section 3 Dimensionality	Scanning Section 5 Dimensionality
0	0	2.3063	2.3396	2.3244
21.23	0.55	2.3111	2.3513	2.3263
24.59	0.66	2.3115	2.352	2.3220
27.24	0.75	2.3108	2.3459	2.3276
33.19	0.958	2.3113	2.3495	2.3294
34.89	1.041	2.3116	2.3537	2.3295
36.08	1.149	2.3138	2.3518	2.3291
33.61	1.25	2.3155	2.3593	2.3320
32.22	1.375	2.3186	2.374	2.3328
12.71	1.417	2.3253	2.4048	2.3477

dimensionality is $D = 2.3138$, cross section 3 fractal dimensionality is $D = 2.3518$, cross section 5 fractal dimensionality is $D = 2.3291$. In comparison with the initial fractal dimensionalities, they increase 0.33, 0.52, and 0.22%, respectively. When after loading gradually unloads to 33.61 MPa, concrete materials enter into the residual deformation stage, fractal dimensionality appears to have a sudden increase, so that cross section 1 fractal dimensionality is $D = 2.3253$; cross section 3 fractal dimensionality is $D = 2.4048$, and cross section 5 fractal dimensionality is $D = 2.3477$; the maximum increase ranges are 0.82, 2.78, and 1%, respectively, whereby indicating that the chaps on test specimens have well developed and formed the macrochaps.

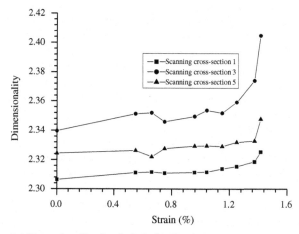

FIGURE 14.63 Image Fractal Dimensionality–Strain Relation Curves

Through the analysis, it shows that in the case of initial loading action, initial chap expansion has high irregular degrees (because of microchap accumulation and expansion in the main in the initial stage). At this time, fractal dimensionality is on the increase. With an increase in loading, concrete test specimens appear to have the closed tendency, accompanied by energy drop in this process, whereby causing a decrease in fractal drop. With an increase in loading, concrete test specimens appear to have a capacity expansion process, and fractal dimensionality appears to have a stable successive increase; after the peak stress passes and enters the residual deformation stage, microchaps rapidly expand, run through and gradually form the main chaps, and concrete test specimens' geometric shapes are becoming more and more broken so that fractal dimensionality reaches the maximum. It can be seen from this that the variation of fractal dimensionality in the loading process can embody the loading changing process of concrete test specimens.

14.6.3 FRACTAL DIMENSIONALITY COMPUTATION AND ANALYSIS OF CONCRETE UNIAXIAL COMPRESSION CT IMAGE UNDER DYNAMIC LOADING ACTION

In order to study the fractal characteristics of concrete materials under the dynamic loading conditions, it is here that CT images of scanning cross section CONC14.3 in dynamic tensile tests are selected to carry out fractal dimensionality computation as shown in Fig. 14.64, and contrasts are made with fractal dimensionality of concrete CT images under static action. The results of CT image fractal dimensionality computation of each loading step scanning cross section CONC14.1, CONC14.3, and

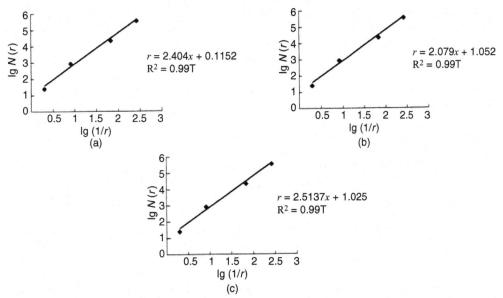

FIGURE 14.64 lg $N(r)$ ~ lg (1/r) Diagrams of Scanning Cross-Section CONC-14.3 Part Under Loading

(a) $\delta = 0$ MPa image straight line fitting results (b) $\delta = 30.79$ MPa image straight line fitting results (c) $\delta = 30.79$ MPa image straight line fitting results.

CONC14.5 are listed in Table 14.3, and relations between fractal dimensionality and stress are indicated in Fig. 14.64.

It can be seen from Table 14.3 and Fig. 14.65 that under the dynamic compression action, fractal dimensionality appears to have similar laws with the actions of static compression. When in initial scanning, fractal dimensionality of cross section 1 is $D = 2.3999$, fractal dimensionality of cross section 3 is $D = 2.404$, fractal dimensionality of cross section 5 is $D = 2.4031$. And in the initial loading, fractal dimensionality has a certain increase. When loading increases to 30.79 MPa, fractal dimensionality of three scanning cross sections has a certain reduction; when loading reaches 33.13 MPa, fractal dimensionality has an apparent inflection point, appearing to have a successive increasing trend or tendency so that fractal dimensionality of cross-section 1 is $D = 2.4023$, fractal dimensionality of cross section 3 is $D = 2.4086$, and fractal dimensionality of cross section 5 is $D = 2.4044$. In comparison with the initial fractal dimensionality, there is an increase of 0.1, 0.2, and 0.24%, respectively. When loading is unloaded to 14.5 MPa, and concrete materials enter the residual deformation stage, fractal dimensionality has a rapid increase so that fractal dimensionality of cross section 1 is $D = 2.5069$, fractal dimensionality of cross section 3 is $D = 2.5137$, fractal dimensionality of cross section 5 is $D = 2.51$;

Table 14.3 Computation Results of Fractal Dimensionality Under Each Step Loading

σ (MPa)	Scanning Cross-Section 1	Scanning Cross-Section 3	Scanning Cross-Section 5
	Dimensionality	Dimensionality	Dimensionality
0.00	2.3999	2.404	2.4031
27.37	2.4001	2.4086	2.4036
30.79	2.3974	2.4079	2.4043
33.13	2.4023	2.4086	2.4044
14.5	2.5069	2.5137	2.51

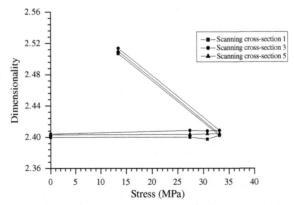

FIGURE 14.65 Fractal Dimensionality–Stress Relation Curves of Each Scanning Cross Section

and the maximum increasing range of fractal dimensionality is 4.5, 4.6, and 4%, respectively, whereby indicating that the breakage degree of test specimen geometry has been very high and that macrochaps have been formed.

Through this analysis, it can be indicated that under the action of initial loading, the initial microchaps and holes densely compressed have caused an increase in concrete density, but fractal dimensionality of each cross section has some decrease, and at this time, concrete strength has been improved. With an increase in loading, fractal dimensionality of each cross section appears to have an increase on the basis of decrease in dense compression, thus showing that the test specimens appear to have a process of capacity expansion. After the peak value strength is over, and entering the residual deformation stage, fractal dimensionality reaches the maximum, but the process of fractal dimensionality reaching the maximum is obviously more rapid than static compression action so that there is a mutation process, whereby illustrating that under the situation of dynamic compression, chap extension and running-through have a very strong sharp or sudden occurrence, and the test specimens are developing from capacity expansion stage to failure within the shortest possible time. This is in agreement with the conclusion obtained from the mentioned contents manipulated through the application of the different image processing technologies.

14.6.4 COMPARISON OF CONCRETE FRACTAL CHARACTERISTICS UNDER STATIC AND DYNAMIC ACTIONS

Through the analysis of concrete fractal characteristics under the static and dynamic loading actions, the following conclusions can be detained:

1. *Common points*: Chap evolution process on the whole appears to have the fractal dimensionality increase trend, but there are risings and fallings in the middle. They have experienced the initial fractal dimensionality increase → fractal dimensionality drop → when in peak value loading, fractal dimensionality rapid increase → the residual deformation stage at which fractal dimensionality reaches the maximum different stages. This indicates that fractal dimensionality variation can also better reflect concrete material damage and fracture evolution laws, that is, dense compactness, capacity expansion, microchap extension running through final damage, and fracture process of failure.

2. *Different points*: It can be seen that after the dense compression process appears, concrete fractal dimensionality under the dynamic compression action is obviously higher than fractal dimensionality under the static compression action. Particularly when reaching peak value stress, fractal dimensionality under the dynamic compression action suddenly appears to have an inflection point with a sharp increase range. After the peak value stress is over, entering the residual deformation stage, fractal dimensionality under the dynamic compression condition has an increasing range larger than that of fractal dimensionality under the static compression action. It can be seen from this that the curve variation of fractal dimensionality of concrete material under the static compression action is rather flat and the curve incline rate change is smaller, while fractal dimensionality curve variation of concrete materials under the dynamic compression action is rather steep and the curve incline rate variation is larger.

These comparisons have also proved the behaviors of concrete material failure under the dynamic compression action, that is, more points of fracture starting in materials when in failure, and more locations produce the fractures at the same time. Once fractures appear or occur they will expand and run through very fast, with a certain sharp or sudden occurrence and large destructive energy.

14.7 CONCRETE DAMAGE EVOLUTION EQUATION AND CONSTITUTIVE RELATION BASED ON CT IMAGES

The balance equation and geometry equation of elastic mechanics, plastic mechanics, and damage mechanics are the same. The distinction is only the constitutive equation. Concrete materials are neither elastic nor plastic. Damage mechanics is used to describe the mechanics behavior of concrete, being more agreeable to the real conditions. The application of damage mechanics needs to establish damage variables. The traditional damage variable definition is the ratio value of an area of material damage part and total area. The definition is simple and physical significance is clear. But in practical utilization, the damaged area of the material can be by no means measured so that the macromeasurable other mechanics indexes are used to replace them. CT test is a kind of damage-free detection method to display the means for the people to measure the sizes of inner damage zone within materials. Therefore, the following is based on the evolution equations and constitutive models. Therefore, the damage variables are established on the basis of concrete material CT tests so as to further deduce the damage evolution process and damage constitutive relation and to lay a solid foundation for the analysis of materials subject to force.

14.7.1 GEOTECHNICAL DAMAGE EQUATION AND CONSTITUTIVE MODEL BASED ON CT TESTS

Yang Genshe and Xie Dingyi et al. (1996, 1998) first used CT number special distribution laws to carry out research on the evolution laws of rock damage and suggested CT number damage variables:

$$D = \frac{1}{m_0^2}[1 - \frac{E(\rho)}{\rho_0}] = -\frac{1}{m_0^2}\frac{\Delta\rho}{\rho_0} \tag{14.66}$$

Where $\rho0$ is the test specimen matrix material density, $E(\rho)$ is the test specimen material density average value at some time, and M_0 is CT machine spatial resolution.

Jianxi et al. (2000) introduced the initial damage influence factor and closed influence coefficient, revised the damage variables, and suggested a damage evolution equation under the rock uniaxial compression test conditions:

$$D = \frac{\alpha_e}{m_0^2}(1 - \frac{1000 + H_{rm}}{1000 + \alpha_c H_{rm0}}) \tag{14.67}$$

Where α_e, α_c are the initial damage influence factor and closed influence coefficient, m_0 is CT machine spatial resolution, H_{rm0} is rock initial state CT number, and H_{rm} is rock CT number at a certain stress stage.

Ding Weihua et al. (2003) suggested the concept of rock density damage increment and deduced the relation equation between the increment and rock mass strain as follows:

$$\Delta D = \frac{\rho_i - \rho_0}{\rho_0} = \frac{H_i - H_0}{1000 + H_0} \tag{14.68}$$

$$\varepsilon_v = \frac{\Delta D}{1 + \Delta D} = \frac{H_0 - H_i}{1000 + H_i} \qquad (14.69)$$

where ρ_0 and ρ_i are the density of test specimen initial and any stress state; H_0, H_i are CT number at initial state and any stress state at a certain point or in any zone; ε_v is rock mass strain.

Lu Zhaihua and Chen Zhenghan et al. (2002) have suggested the concept of expansive soil fracture damage increment and deduced the relation equation between expansive soil and accumulated drying shrinkage strain as follows:

$$D = \frac{H_0 - H}{H_i - H_f} \qquad (14.70)$$

$$D = e^{A\varepsilon_v^Z} \qquad (14.71)$$

where H_0, H_i, and H_f are CT number of a certain point or any zone of the initial state, any zone of the initial state, any stress state and complete damage state; ε_v is body strain; A is coefficient reflecting strong and weak expansion; Z is coefficient reflecting curve shape.

Hao et al. (2002). have suggested a rock mass damage variable based on CT joint, under the concept of CT number reducing rate, defined damage evolution rate in the following:

$$D = -\frac{1}{m_0^2} \frac{\rho_w - \rho_r}{\rho_r} \frac{H_r - H}{H_r} \qquad (14.72)$$

$$\frac{dD}{d\sigma} = -\frac{1}{m_0^2} \frac{\rho_w - \rho_r}{\rho_r H_r} \frac{dH}{d\sigma} \qquad (14.73)$$

where ρ_r and ρ_w are the test specimen matrix material and water density, respectively; m_0 is CT machine spatial resolution; H_r is matrix material CT number; H_i is the brittle material and fracture water whole CT number at some time.

Dang Faning and Yin Xiaotao et al. (2005) have defined a statistic damage variable based on the set theory and measure theory knowledge and in combining with the digital performances provided by CT test result as follows:

$$D = 1 - \frac{\frac{1}{n}\sum_{i=1}^{n} H_i}{\frac{1}{m}\sum_{j=1}^{m} H_{j0}} = 1 - \frac{H_{rmi}}{H_{rm0}} \qquad (14.74)$$

where m is in initial scanning CT unit number of safety zone, n is total sum of damaging and fracturing zone and failure zone CT unit numbers, H_i is CT unit is CT number on CT unit in damaging and fracturing zone or failure zone under a certain grade of loading, and H_{j0} is CT unit CT number in the case of damage-free state.

Chen Shili (2007) has established the damage evolution equation and constitutive equation for cement stone mesofracture process, and the established damage variable is as follows:

$$D = D_0 + m\varepsilon^n \qquad (14.75)$$

$$D_0 = \frac{1}{1-k_0}(1-k_0\frac{\rho_w}{\rho_0})-1$$

$$m = \left[\frac{1}{1-k_0}(1-k_0\frac{\rho_w}{\rho_0})-\frac{\rho_w}{\rho_0}\right]C$$

where D_0 is the initial damage variable of cement stone material, and m is the damage coefficient of cement stone material.

The established damage constitutive equation is as follows:

$$\sigma = \begin{cases} \sigma_A(\varepsilon/\varepsilon_A)^2, \varepsilon \leq \varepsilon_A \\ \sigma_A + E[(1-D_0)(\varepsilon-\varepsilon_A)-m(\varepsilon-\varepsilon_A)^{n+1}], \varepsilon \geq \varepsilon_A \end{cases} \quad (14.76)$$

where σ_A and ε_A are the separation points of microfracture compacted dense stage and elastic compacted dense stage.

14.7.2 CONCRETE STATISTIC DAMAGE EVOLUTION EQUATION BASED ON CT TESTS

In Section 14.4, the concrete materials in some certain loading stages are divided into the safety zone, damage and fracture zone, and failure zone. Based on the image damage and fracture zone, the adoption of porosity can effectively reflect the evolution process of the test specimen damage and fracture and quantitatively describe the degree of damage and fracture. Therefore, by analogue of the damage mechanics, the adoption of test specimen cross section porosity or chap area (measure) and total area ratio value can define the damage variable:

$$D = \frac{A_i}{A_0} = \frac{n_i}{n} \quad (14.77)$$

where A_0 is the initial image total area (image all pixel dot number n), and A_i is the porosity or fracture area at each loading stage (the porosity or fracture pixel dot number n_i).

Damage variable defined in such a way seems to be very similar to the traditional definition, but before and after the material damage, their total quantity remains unchanged. The failure area expressed by the CT number is unlikely to be equal to the initial total area, whereby leading to that the maximum is far smaller than 1. For this reason, it is suggested that the following equation be used to express damage variable more rationally:

$$D = \frac{A_i}{A_f} = \frac{n_i}{n_f} \quad (14.78)$$

Where A_f is the maximum value of the test specimen porosity or fracture area when loading continues to damage stage (the maximum value n_f of porosity or fracture pixel dot number).

In this way, when in the initial state, $D = 0$ represents the damage-free specimen at this time; when in failure state, $D = 1$ stands for the formation of macrochaps, and the test specimens have occurred to fail. In the process of analysis of test results, it is only necessary to compute porosity or fracture area of every larger position at the different stages (porosity or fracture pixel dot number n_i) and that can the degree of damage and fracture of the test specimens be quantitatively analyzed with the results shown in Table 14.4 and Fig. 14.66.

Equation (14.78) can be used to compute the damage variables in the damage and fracture process of concrete. It can be seen that with an increase in stress, the damage variable D appears to have a nonlinear increase, whereby indicating that concrete mesodamage and fracture evolution is an irreversible nonlinear accumulative process. It can be known from Fig. 14.66 that the initial damage variable is not equal to zero. The main cause is that there exist the initial damages of microchaps and holes in concrete materials and when loading reaches 27.24 MPa, the damage variable has some reduction. This is because the part of microchaps and holes are compacted densely, and this stage is the damage weakening stage. When loading is 27.24–33.19 MPa, the damage variable will have a stable development trend, whereby demonstrating that under the loading action, the inner holes and microchaps in the test specimens begin to expand, pores increase, and porosity is improved in the failure zone. This is the stage of damage and fracture beginning to evolve and to develop steadily. When loading is 36.08 MPa, reaching peak value loading, it can be seen from Fig. 14.66 that the damage variable has a sudden and rapid increase trend and that there is a sudden increase in pores and large increase margin in porosity in the damage and fracture zone, and also there is a sudden change in the curve inclined rate, thus indicating that the inner microchaps that rapidly fuse and run through is the fastest stage of damage evolution. Accordingly, this stage is the forewarning of concrete damage. When entering the unloading stage, loading decreases to 33.61 MPa afterward, the damage variable increasing range reaches the maximum, the porosity in the failure zone reaches the maximum, and the curve inclined rate becomes steeper. This illustrates that the inner chaps of the test specimens have run through; the chaps of length and width have further increased so as to form the macrochaps.

Table 14.4 Statistic Damage Variable of Each Scanning Section

σ (MPa)	Strain (%)	Scanning Cross Section 1		Scanning Cross Section 3		Scanning Cross Section 5	
		Porosity or Fracture Pixel Dot Number	D	Porosity or Fracture Pixel Dot Number	D	Porosity or Fracture Pixel Dot Number	D
0	0	3135	0.1209	6254	0.2411	2955	0.1139
21.23	0.55	3144	0.1212	7450	0.2872	3781	0.1458
24.59	0.66	3066	0.1182	7429	0.2864	3756	0.1448
27.24	0.75	2523	0.0973	5884	0.2269	3224	0.1243
33.19	0.958	2582	0.0996	6320	0.2437	2852	0.1100
34.89	1.041	2925	0.1128	6382	0.2461	2882	0.1111
36.08	1.149	3606	0.1390	7620	0.2938	4273	0.1648
33.61	1.25	3500	0.1349	8229	0.3173	4780	0.1843
32.22	1.375	3392	0.1308	10428	0.4021	5801	0.2237
12.71	1.417	5252	0.2025	25936	1.0000	11747	0.4529

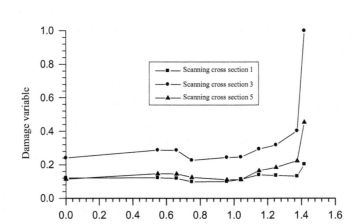

FIGURE 14.66 Curves of Damage Variable and Strain Relations

From this analysis, it is known that the relation curves of damage variable and strain can reflect that concrete test specimens have experienced the earliest beginning damage and damage weakening, damage stable development, and damage having sudden growth to complete failure process and reflected the basic behaviors of concrete damage evolution.

It is necessary to fit the relation curves of the damage variable and strain so as to observe the obtained damage variable and strain curves appearing to have the exponent function relation so that it is suitable to adopt the exponent function to carry out the fitting, and the fitting function should be as follows:

$$D = A(e^{B\varepsilon})$$ (14.79)

where A and B are parameters and ε is strain.

Through the fitting, $A = 0.11392$, $B = 1.533$, we can obtain the damage evolution equation in the following:

$$D = 0.11392 \, (e^{1.533\varepsilon})$$ (14.80)

14.7.3 CONCRETE AVERAGE CT NUMBER DAMAGE EVOLUTION EQUATION BASED ON CT TESTS

First of all, it is necessary to separate the whole stress–strain process curve of concrete macrofailure into stages or sections so as to establish a more rational concrete damage evolution equation. The following defines all the average values of scanning cross section CT numbers, which are the CT number average value of concrete test specimens.

Taking CONC-10 test specimen loading process as the example to carry out the analysis, the test specimen damage process can be divided into five sections so as to establish the damage evolution equations. CT number stages in Fig. 14.67 are labeled as OA, AB, BC, CD, DE, which are corresponding to the stress–strain curve stages in Fig. 14.68.

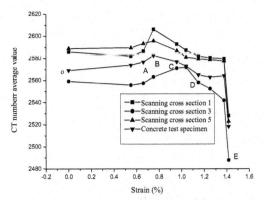

FIGURE 14.67 Curves Between CT Number Average Value and Strain Relation in Each Scanning Cross Section

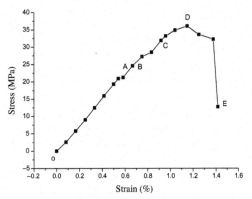

FIGURE 14.68 Curves of Stress–Strain Relation of CONC-10 Concrete Test Specimen

The stress in the first stage is 0–24.59 MPa, being the damage weakening stage OA; CT number in this stage has a slight rising on the basis of the initial damage. Concrete density increases largely and the strength is improved, that is, strain is 0–0.66%; the stress in the second stage is 24.59–27.24 MPa, being the elastic stage AB. There is not much variation in CT number in this stage, and also, CT observed image has not much change. In this stage, there is not much change in mesodamage and fracture of the test specimens. Accordingly, concrete materials are in the elastic deformation stage, that is, strain is 0.66–0.75%. The stress in the third stage is 27.24–33.19 MPa, being damage and fracture beginning to evolve and stable development stage BC; it is just at this time that the test specimen average CT number begins to reduce, showing that the mesodamage begins to evolve, and this stage is a capacity expansion stage so that the test specimen density is reduced and the strength decreases, that is, strain is 0.75–0.958%; the stress in the fourth stage is 33.19–36.08 MPa, being damage development stage CD with the acceleration; the stress reaches the peak value loading, CT number of each layer decreases at the rapid rate so that this stage is the forewarning stage of failure, that is, the fastest stage of damage evolution speed. It is just at this time that the test specimen density decreases by a wide margin, whose mesodirect performances are the sharp drops of CT numbers. Each scanning cross-section appears to have the significant capacity expansion, that is, strain is 0.958–1.149%. The stress in the fifth stage is from 36.08 MPa unloading to 33.61 MPa, being the softening stage DE after the peak. It is just at this time that the test specimens enter the residual deformation stage. CT number drops sharply, and the test specimen expands its capacity at a fast speed. Chaps within the cross section are further fused and run through to form macrochaps, that is, strain is 1.25–1.417%.

14.7.3.1 Establishing damage variable

Based on the thought of mesostatistic damage mechanics, the damage variable is established as follows:

$$D = \frac{H_{0m} - H_i}{H_{0m} - H_{fm}} \tag{14.81}$$

$$H_{0m} = \frac{1}{n}\sum_{i=1}^{n} H_0$$

$$H_{fm} = \frac{1}{n}\sum_{i=1}^{n} H_f$$

where H_{0m} is CT number average value of the initial state of concrete test specimen; H_i is CT number average value of concrete test specimen at a certain stress stage; n is scanning larger number; H_0 is CT number average value of any scanning cross-section at the initial state; H_{fm} is CT number average value of the broken test specimen; H_f is CT number average value of any broken scanning cross-section.

It is stipulated that the densely compressed CT number (i.e., the maximum value H_{0m} of each scanning step) serves as the initial CT value, while the process prior to this can be viewed as the damage-free stage, that is, damage variable $D = 0$, and when CT number average value is over H_{0m}, its damage variable $D = 0$ is considered.

When after peak softening stage, there will be a rapid drop in stress and a sharp decrease in CT number so that the test specimen expands its capacity rapidly, and CT number average value of the ending point of the test specimen at the softening stage is used as CT number when concrete fractures, that is, damage variable $D = 1$, whereby indicating that the concrete test specimen macrofracture has completely run through and lost the bearing capacity, and when CT number average value at each scanning cross section is less than H_{fm}, its damage variable $D \leq 1$ is considered.

Figure 14.69 is the curve relation between damage variable and strain for each scanning cross section, of which five scanning cross sections are considered as having no damage to occur prior to their dense compression, that is, damage variable is $D = 0$, and the part CT number in scanning cross-section 1 and scanning cross-section 5 is over H_{0m}, whose damage variable is also considered as $D = 0$. If damage variable is $D = 1$, this indicates that the macrofracture of concrete test specimens after the peak value loading has run through, and the test specimens have completely damaged and lost their bearing capacity.

14.7.3.2 Establishment of damage evolution equation

Since the first stage is the damage weakening stage OA, the CT number in this stage rises slightly on the basis of the initial damage, while there are very few changes in the CT number in the second stage

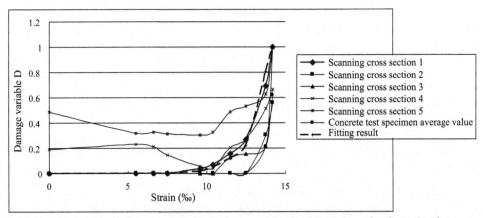

FIGURE 14.69 Relation Curves Between Damage Variable and Strain of Each Scanning Cross Section

AB, and so there are not much changes in mesodamage, so that the first and the second stage in the stress–strain curves are converged into one stage, and this stage $D = 0$ is considered so that there is no occurrence of damage. The third and fourth stages can be converged into the damage and fracture development stage; when in the softening stage after the peak value, there can be a rapid drop in stress and a sharp decrease in CT number; the test specimens expand their capacity at a fast rate and lose their bearing capacity at least so that $D = 1$ is considered in this stage, and the test specimens are completely destroyed. In this way, concrete test specimens have experienced the following four stages from beginning to load to the final failure:

1. *Linear elastic stage*: It is considered that damage fracture will not begin to develop, $D = 0$.
2. *Damage and fracture developing stage*: It is essential to fit the relation curves between damage variable and strain to discover that damage variable and strain approximate to become the polynomial function so that the polynomial function is adopted to fit the test curves, as indicated in Fig. 14.69. The fitting function is as follows:

$$D(\varepsilon) = a\varepsilon^3 + b\varepsilon^2 + c\varepsilon + d \qquad (14.82)$$

where ε is strain; a, b, c, d are the test constants.

Through fitting, the damage evolution equation is obtained as follows:

$$D(\varepsilon) = 0.0012\varepsilon^3 - 0.0234\varepsilon^2 + 0.1446\varepsilon - 0.2745 \qquad (14.83)$$

3. *Softening stage after the peak*: The relation curves between damage variable and strain are fitted so as to obtain the following:

$$D(\varepsilon) = a\varepsilon + b$$

Through fitting, we can obtain the damage evolution equation as follows:

$$D(\varepsilon) = (77.95\varepsilon - 937.5515)/167 \qquad (14.84)$$

4. *Failure stage*: It is just at this time that the materials are considered to have been completely destroyed, $D = 1$, so that the damage evolution equation is obtained at last as follows:

$$D = \begin{cases} 0 & 0 \leq 1000 \ \varepsilon \leq 7.5 \\ 0.0012 \ \varepsilon^3 - 0.0234 \ \varepsilon^2 + 0.1446 \ \varepsilon - 0.2745 & 7.5 < 1000 \ \varepsilon \leq 12.5 \\ (77.95\varepsilon - 937.5515)/167 & 12.5 < 1000 \ \varepsilon \leq 14.17 \\ 1 & 1000 \ \varepsilon > 14.17 \end{cases} \qquad (14.85)$$

14.7.4 CONCRETE FRACTURE DIMENSIONALITY DAMAGE EVOLUTION EQUATION BASED ON CT TESTS

14.7.4.1 Establishment of mesostatistics damage mode

Rocks are viewed as a microelement mass containing several different defects. Since the inner structure of rock materials is extremely nonuniform containing different defects, each microelement is of different strength. Weibull (1939) first used the probability method to describe the nonuniformity of materials and held that accurate investigation or survey is impossible to destroy the strength, but in the case of a given stress level, the occurrence of damage probability can be defined. It is suggested that the power function law with threshold value be used to describe the strength extreme value distribution laws. This distribution is the Weibull distribution in statistics. It can play an important role in magnitude of effect and strength theory.

Krajcinovic et al. (1981) combined the successive damage theory with statistics strength theory in an organic way to suggest a statistics damage model. He abstracted the test specimens into N pieces of finely long pole sets, whose constitutive relation is as follows:

$$\sigma = E\varepsilon(1 - \frac{n}{N}) \qquad (14.86)$$

$$D = \frac{n}{N}$$

where n is the number of broken poles.

When N is very large, Krajcinovic held that Weibull distribution can be used to replace Eqn (14.86). Though the finely long pole combination is used to simulate the complete test specimen behaviors in lack of theoretical seriousness, Krajcinovic's model combines statistics theory with damage theory to study the rock brittle material constitutive theory, whereby opening up a new vista for this research, on the basis of which, rock mesostatistics damage theory is suggested, and also the rock mesostatistics damage model is established.

Concrete can serve as a kind of sub-brittle material, into which a bulk of apparent microchaps and holes are filled so that the initial damages, microchaps, and so on, are formed, expanded, connected, or run through so as to produce an apparent effect upon the material rigidity, strength, and many properties, whereby leading to the gradual deterioration of material properties or nature till the failure. Therefore, any one-dimensional element is taken from concrete, whose strength is bound to subject to some certain statistics laws. Supposing that the one-dimensional strength $\phi(f)$ is subject to be the following formula:

$$\varphi(f) = \frac{m}{a} \left(\frac{f}{a} \right)^{m-1} e^{-(f/a)^m} \qquad (14.87)$$

$$\frac{dD}{df} = \varphi(f)$$

(14.88)

where a is magnitude parameters, m is the parameters to reflect defects distributing into material conditions; and f is the one-dimensional element strength with random distribution variable.

The following equation can be deduced from Eqs (14.87) and (14.88):

$$D = 1 - e^{-(f/a)^m}$$

(14.89)

Equation (14.89) is the statistics law to which the concrete material damage variable D and one-dimensional element strength should be subject. That is based on the one-dimensional element strength distribution function in concrete material to satisfy the Weibull damage evolution equation under the assumed conditions, that is, concrete mesostatistics damage model.

On the one hand, in terms of damage variable concept, the damage variable D is a measure for material damage degree, while damage degree is more or less related with each one-dimensional element including defects, which can directly affect one-dimensional element strength. On the other hand, much more fractal researches indicate that the defects distribution in concrete materials are of approximate characteristics so that the fractal dimensionality can be used to characterize the defect distribution, while the parameter m in Eq. (14.87) can reflect the defect distribution conditions in rock materials. For this reason, Bai Shengguang (1996), Wen Shiyou (2002), and Ming et al. (2005) have established the relation between rock fractal dimensionality and the damage variable. The following Eq. (14.90) is used in concrete material mesodamage to obtain concrete mesostatistics damage evolution laws.

In the damage and fracture analysis of rocks, concrete, pottery, and porcelain or ceramics, brittle and sub-brittle materials, the strain produced by stress is often used to measure the damage variable (Ming et al., 2005); therefore, Eq. (14.89) can be rewritten or revised as follows:

$$D = 1 - e^{-(\varepsilon/a)^m}$$

(14.90)

From Eq. (14.90), we can define the following equation:

$$\lg\left(\lg\frac{1}{1-D}\right) = m\lg\frac{1}{a} + m\lg\varepsilon$$

Letting $\lg(1/[1-D]) = \bar{D}$, we can obtain the following equation:

$$\lg\bar{D} = m\lg\frac{1}{a} + m\lg\varepsilon$$

(14.91)

Owing to $\mathrm{grad}_D(\bar{D}) > 0$, $\bar{D}$ and D have the same varying trend so that $\bar{D}$ and D physical implications are the same physical quantity to describe the degree of material damage. Based on the dimensional concept in fractal theory, it is known that m is the fractal dimensionality of D set. Therefore, based on the previous discussions, m is an unchangeable measure to describe the complicated geometric problem in concrete material defects distribution situation, that is, the fractal dimensionality D_f from which we can obtain:

$$D_f = m$$

(14.92)

Substituting Eq. (14.92) into Eq. (14.90), we can obtain:

$$D = \left(1 - e^{-(\varepsilon/a)^{D_f}}\right) + \xi \tag{14.93}$$

Up to now, the damage evolution model containing fractal dimensionality D_f to express the initial defects of concrete materials has been obtained.

The mentioned equations have established relations between the fractal dimensionality and damage variable in concrete materials. Since there exist microchaps, holes, bubbles, and other initial damages in concrete materials, there exist no damages in true significance. There will be some errors in replacing the fractal dimensionality in complete damage free with the approximate fractal dimensionality in the initial damage. For this reason, the authors have introduced the initial damage coefficient ξ, and ξ take the average value of the difference value between the different strain values corresponding to fractal dimensionality and the initial fractal dimensionality.

14.7.4.2 Determination of parameter a in mesostatistics damage model

The following damage constitutive relation is obtained from the strain equivalence principle:

$$\sigma = E_0 \left(e^{-(\varepsilon/a)^{D_f}} - \xi\right)\varepsilon \tag{14.94}$$

D_f value is fixed. The fractal dimensionality is taken when the materials are damaged at the initial time. E_0 takes the initial elastic modulus; a takes its value at the macrostrain average value nearby in such a way that the stress–strain curves with different a values can be obtained, and in comparison with tests, the stress–strain in Fig. 14.70. Curve value increases with an increase in a, but a variation cannot change the peak value prior linear deformation curve. Therefore, the parameter a in concrete damage statistics model can reflect the mechanics properties of concrete deformation; parameter a to the nonlinear deformation part affecting the curve is clear, with an obvious effect upon peak value posterior curve in particular. The curve morphology can be changed, and the comprehensive figure can select $a = 1.5$.

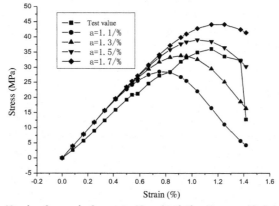

FIGURE 14.70 Parameter a Varying Curves in Concrete MesoStatistics Damage Model

14.7.4.3 Establishment of damage variables

Table 14.5 is the damage variable statistics of each scanning cross section.

Equation (14.93) is used to compute the relation between damage variable and strain of concrete damage and fracture process (see Fig. 14.71). It can be seen from Fig. 14.71 that with an increase in strain, damage variable D appears to be nonlinearly growing, again illustrating that the concrete mesodamage and fracture process is an irreversible nonlinear accumulative process. It can be known from Fig. 14.71 that the initial damage variable is not equal to zero. The main reason is that there exist the initial damages such as microchaps and holes in concrete materials, and when stress reaches

Table 14.5 Damage Variable Statistics of Each Scanning Cross Section

σ (MPa)	Strain (%)	Scanning Cross Section 1		Scanning Cross Section 3		Scanning Cross Section 5	
		Fractal Dimensionality	Damage Variable	Fractal Dimensionality	Damage Variable	Fractal Dimensionality	Damage Variable
0	0	2.3063	0.00496	2.3396	0.01661	2.3244	0.00363
21.23	0.55	2.3111	0.09796	2.3513	0.10961	2.3263	0.09563
24.59	0.66	2.3115	0.09696	2.352	0.10861	2.3220	0.09363
27.24	0.75	2.3108	0.21496	2.3459	0.22661	2.3276	0.21363
33.19	0.958	2.3113	0.30496	2.3495	0.31661	2.3294	0.30363
34.89	1.041	2.3116	0.35496	2.3537	0.36661	2.3295	0.35363
36.08	1.149	2.3138	0.42496	2.3518	0.43661	2.3291	0.40363
33.61	1.25	2.3155	0.41496	2.3593	0.42661	2.332	0.42363
32.22	1.375	2.3186	0.56496	2.374	0.57661	2.3328	0.52363
12.71	1.417	2.3253	0.58496	2.4048	0.59661	2.3477	0.60163

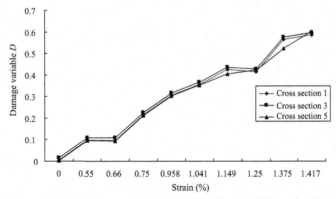

FIGURE 14.71 Relation Curves Between Damage Variable and Strain of Each Scanning Cross Section

21.33~24.59 MPa (i.e., strain is 0.55%~0.66%), it can be seen that the damage variable appears to have a certain reducing trend; this is because parts of microchaps and holes are densely compressed and chaps have the closed trend, and this stage is the damage weakening stage. When stress is 24.59~36.08 MPa (i.e., strain is 0.66%~1.149%), there is a stable increasing trend in damage variable, demonstrating that under the loading action, the inner holes and microchaps within the test specimens begin to expand, and there are plenty of fine joints on the concrete test specimen surface so that geometric shapes become more and more broken. This stage is the damage stable development and acceleration stage. When entering the unloading stage, loading decreases to 32.22 MPa (i.e., strain is 1.375%), the range of damage variable reaches the maximum, so that curves become very steep. When loading decreases to 12.71 MPa (i.e., strain is 1.417%), the damage variable reaches the maximum, showing that the inner chaps within the test specimen have run through so that chap length and width have further enlarged, thus forming macrochaps to cause the unstable failure to the test specimens.

It can be known from the analysis that these curves can reflect that concrete test specimens have experienced the whole process of initial damage, damage weakening, damage stable development, and failure, and can better describe the evolution behaviors of the full concrete material damage development process.

14.7.4.4 Establishment of damage evolution equation

A polynomial function is adopted to fit the relation curves between damage variable and strain:

$$D = A\varepsilon^2 + B\varepsilon \tag{14.95}$$

where A and B are parameters and ε is strain.

Through fitting $A = 0.0013$, $B = 0.0475$, related coefficient $R^2 = 0.9678$ the damage evolution equation can be as follows:

$$D = 0.0013\varepsilon^2 + 0.0475\varepsilon \tag{14.96}$$

14.7.4.5 Establishment of damage constitutive relation

It is here that the fractional damage evolution equation based on CT number variation is selected to establish the constitutive relation:

1. When stress reaches 0~27.24 MPa (strain is 0.075%), there is no damage occurrence in $D = 0$, it can be considered that each point is of elasticity so that an elastic constitutive relation can be adopted. The corresponding stress, strain, and original connecting inclined rate on the stress–strain curves at this time can replace the elastic modulus at the initial damage.

The constitutive equation can take the linear elasticity $\overline{\sigma} = E\varepsilon$; through fitting, we can obtain the following equation from Fig. 14.72:

$$\sigma = -0.4587 + 48.72\varepsilon \tag{14.97}$$

We can obtain the elasticity modulus $E_0 = 4.73$ GPa for the initial damage.

2. When $\sigma = 27.24~33.61$ MPa (strain is 0.75%~1.25%) and in terms of Lemaitre strain equivalence principle, the damage constitutive equation is as follows:

$$\sigma = E_0(1 - (0.0012\varepsilon^3 - 0.0234\varepsilon^2 + 0.1446\varepsilon - 0.2745))\varepsilon \tag{14.98}$$

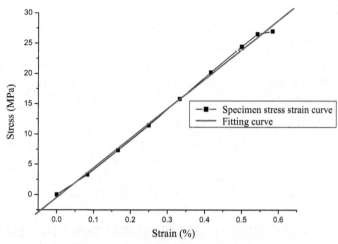

FIGURE 14.72 Comparison of Test Value with Computation Value

3. When $\sigma = 33.61 \sim 12.71$ MPa (strain is $1.25\% \sim 1.417\%$), the damage constitutive equation is as follows:

$$\sigma = E_0(1 - [\{77.95\varepsilon / 167\} - \{937.5515 / 167\}])\varepsilon \tag{14.99}$$

In order to test the rationality of the established damage evolution equation and constitutive equation, the 19 stress points are substituted into the equation to carry out testing values in comparison with theoretical computation values, with the results shown in Fig. 14.73. It can be seen from Fig. 14.73 that the stress–strain curves obtained from tests are found to be basic in agreement with the stress–strain

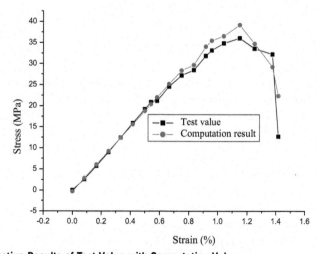

FIGURE 14.73 Comparative Results of Test Value with Computation Value

curves obtained from theoretical computation, whereby testing that the computation value of the damage variable is better fitted with the whole testing value.

14.8 CONCRETE 3-DIMENSIONAL MESOMECHANICS ANALYSIS BASED ON CT TESTS

There are two points for the concrete CT test objective: one is to study the concrete mesodamage mechanism; another is to apply the CT test results to production practice directly. At present, the main methods of application of concrete CT test results to production practice are the application of CT test results to building a concrete numerical computation model and the application of damage constitutive numerical computation and application of CT test results to checking numerical computation results.

14.8.1 THREE KINDS OF CONCRETE 3-DIMENSIONAL COMPUTATION MODELS

14.8.1.1 Concrete three-dimensional stochastic aggregate digital computation model

Concrete can be viewed as a compound material body consisting of aggregates, mortars, and adherence surface. The Monte Carlo method is used to produce aggregate positions stochastically and to establish concrete three-dimensional stochastic distribution mathematical model. The whole computation area is classified into the finite element units. The projection model method and the spatial distribution positions of various kinds of materials can judge which are the aggregate units and which are the mortar units and adherence surface.

The profile division technique is adopted to divide the adherence surface for many times so as to achieve the fine effect of adherence surface units to form a three-dimensional finite element grid model of three-phase materials. The so-called unit multiprofile division technique refers to the second division of adherence surface unit into the smaller size unit after the first unit division. After the fineness of adherence surface unit, the material classification should be again carried out, until the adherence surface units and pore defect unit sizes meet the needs of requirements.

Based on the elastic damage constitutive relation, the double polygonal line damage evolution model is adopted to describe the degradation process of elastic natures of each mesophase materials, and the elastic modulus depreciation degree is used to reflect the degree of damage and fracture of concrete specimens in the loading process, and to carry out the numerical simulation computation of cylinder test specimens. The comparison from two aspects of concrete damage image and loading–displacement curve diagram is made to evaluate the differences between the numerical simulation results of the concrete damage process and CT test results so as to further improve the mesocomputation model and each phase of material computation parameters. Finally, the actual computation and analysis of a large-scale volume of concrete dynamics behaviors can reveal the dynamic damage mechanism.

14.8.1.2 Concrete three-dimensional contact surface digital computation model

There are existing problems in the concrete three-dimensional stochastic aggregate digital computation model. The large thickness of the interface affects the reliability of computation results, while fine division unit can produce a vast number of units being subject to computer computation capacity constraints. The life and death unit method is adopted to eliminate the destroyed units. In the model, a hole zone is formed but not a fracture zone. After the units are classified, the aggregate volume ratio is reduced prior to classification in computation. In order to solve these problems, concrete should be viewed as a two-phase compound material body consisting of aggregates and mortars. In the software ANSYS, the concrete cylinder test specimen should be built up first, and Monte Carlo method is used to produce a stochastic aggregate sphere, and then the cylinder and aggregate sphere can form the test specimens, respectively, to carry out BOOLEAN operation in such a way as to form aggregate position being the hollow concrete test specimen. Again, the program is used to read in the previous generating sphere position coordinate, to produce aggregate sphere for the second time. SOLID45 unit is selected to carry out the whole grid classification. The aggregate projection grid method is adopted to judge the unit material attribute. TARGE170, CONTA174 unit types are selected to add the contact unit to in between the aggregate unit and mortar unit (see Figs 14.74 and 14.75). The contact surface unit is used to simulate the combined interfaces of all kinds of materials in better solving all the mentioned problems when the old models are used in computation. The revised model can be called as the concrete three-dimensional contact digital computation model.

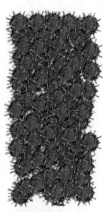

FIGURE 14.74 Concrete Aggregate Contact Unit Diagram

FIGURE 14.75 Concrete Aggregate Contact Unit Cross Section Diagram

The program realization method of contact surface digital computation model in ANSYS software is as follows:

```
ALLSEL, ALL
FLST, 5, 206, 5, ORDE, 2
FITEM, 5, 7
FITEM, 5, -212
ASEL, S, , , P51X
APLOT
NSLA, S, 1
NPLOT
r, 1
type, 3
esurf, all
ALLSEL, ALL
FLST, 5, 206, 5, ORDE, 2
FITEM, 5, 215
FITEM, 5, -420
ASEL, S, , , P51X
APLOT
NSLA, S, 1
NPLOT
r, 1
type, 2
esurf, all
```

14.8.1.3 Concrete three-dimensional rebuilding digital computation model

In terms of CT test results, image three-dimensional rebuilding technique is used to generate the concrete test specimen three-dimension mesomechanics model, called the concrete three-dimensional rebuilding digital computation model.

In terms of concrete CT plane graph information, the program compiled and written by means of APDL language for concrete three-dimensional mesostructure can carry out three-dimensional rebuilding in the large-scale commercial finite element software ANSYS. In the rebuilding process, there are no such geometric elements as point, line, plane, body, and so on, to be reproduced, while CT plane graph information is used to directly generate nodes, to link nodes to generate units in such a way as to avoid the advent of phenomena of having no way to profile the grids. The generated units are of regular shapes and rational ratio of length and width. Through building the numerical model closer to the actual concrete mesostructures, the finite element method can be used to further study concrete mechanics behaviors based on the mesostructures.

The three-dimensional rebuilding method and steps are as follows:

1. *CT image format conversion, reading, and writing*: The original CT images are stored by adopting DICOM standard. The images stored with DICOM standard read into MATLAB are not easy to

be processed so that it is necessary to carry out the image format conversion; that is, the original CT images must be converted to 256 color BMP images.

2. *Image enhancement*: Image enhancement based on some kind of needs can protrude some information in the images by man, whereby inhibiting or eliminating other information processing ware. In this way, the input images will have better image qualities, being favorable for their analysis and interpretation. It is here that aggregates and mortars information in images should be particularly protruded. The morphology wave filter (alternatively opening and closing operation) is adopted for filtration of the small interference zone with high or low gray degrees in images. The threshold value separation method is used to distinguish the aggregate zone from the mortar zone.

3. *Read-in CT images into MATLAB*: In MATLAB, the imread (two-dimensional image address, bmp) command is used to read-in CT images with continuous-BMP form by consequence. Images of every layer must be saved with two-dimensional matrix 345×345 format. Array element value is the pixel attributive value corresponding to CT images; that is, after threshold splitting processing, the aggregate zone array element value is 1 and that in the mortar zone and background zone is zero.

4. *Generating element connection or node*: APDL language is used to compile the program, and the pixel attribute value matrix in MATLAB is read-in into ANSYS array element; the array element name is MSG.

5. *ANSYS generating connection or node according to layers*: The coordinate information in RMSG is read-in, APDL command, that is, N. INUM, X, Y, Z is used to generate the first layer node, and then APDL command, that is, IVGEN, ITIME, INC, NODEI, MODE2, NINC, DX, DY, DZ SPACE is used to copy nodes of the rest of every layer. The separation distance between the layers is the scanning image.

6. *Connecting the corresponding nodes to generate element*: ANSYS with self-contained SOLID45 unit, using APDL command, that is, E, i, j, k, l, m, n, o, p. The four nodes of every layer totaling eight nodes can generate one SCLID45 unit or element. The unit generated after the program in operation is the regular rectangle. The ratio of length, width, and height is about 3:3:8. Accordingly, regular unit shape, relative realistic length, and width ratio can provide a guarantee for the accuracy of following-up numerical computation.

7. *Making discrimination of eight-node attributes of every element or unit, being endowed with mechanics parameters*: strong getting command in APDL is used to extract node numbering, and to extract the corresponding node attribute value from RMSG array element, so as to compile the program for discriminating node attribute. If eight nodes in the same unit are the aggregate nodes, this unit can be endowed with mechanics parameters; if eight nodes are the mortar nodes, this unit can be endowed with mortar material mechanics parameters. The other situations are endowed with interface material parameters.

Figures 14.76 and 14.77 are the concrete three-dimensional finite element model and displacement and stress computed results generated by using the continuous 290 CT images. It can be seen from this that in comparison with the computed results by random aggregate digital computation model, the results computed by the three-dimensional rebuilding model are finer, closer to real concrete behaviors, and better displacement continuity. The advantages of this method are that it is different from the method of concrete mesobuilding model based on the stochastic or random sampling principle. It is suggested that the building model method should be based on CT image information three-dimensional

FIGURE 14.76 Finite Element Three-Dimensional Rebuilding Model in ANSYS(a) Test specimen 3-D rebuilding in ANSYS (b) Aggregate 3-D rebuilding in ANSYS (c) Mortar 3-D rebuilding in ANSYS.

finite element. The aggregate unit or element position is generated in terms of real test specimen CT images, whereby making the finite element model be closer to real concrete structure and the accuracy of numerical simulation be further improved. As a result, the computation results can be compared with test results one by one.

14.8.2 APPLICATION OF DAMAGE CONSTITUTIVE RELATION BASED ON CT IMAGES TO CARRY OUT CONCRETE DAMAGE PROCESS NUMERICAL SIMULATION

This section studies the application of damage constitutive relation based on CT image to numerical computation.

14.8.2.1 Computation conditions

The random mathematic method is used to generate aggregate particle sizes and special position and to build the same numerical model with CT tests (diameter is 60 mm, height is 120 mm); aggregate grading is grade I. The model test specimens are classified into units based on projection model method, and the material types of each unit are discriminated. The multiprofile division technique is adopted to divide the interface and at last, the finite element grid model for concrete three-phase materials is shown in Fig. 14.78.

It is generally considered that concrete stress–strain curve nonlinearity mainly subjects to force afterward; the microunit continues to damage causing microchaps to germinate and expand so that the simple mesomechanics model can be used to simulate the complicated macrofracture process. The linear elastic constitutive relation is adopted in computation. Material parameters of each item in concrete are selected in terms of mechanics test (see Table 14.6). The material damage process is carried out in terms of the following damage model:

$$D = \begin{cases} 0 & 0 \leq \varepsilon_{max} \leq \varepsilon_1 \\ 0.0012\varepsilon_{max}^3 - 0.0234\varepsilon_{max}^2 + 0.1446\varepsilon_{max} - 0.2745 & \varepsilon_1 < \varepsilon_{max} \leq \varepsilon_2 \\ (77.95\varepsilon_{max} - 937.5515)/167 & \varepsilon_2 < \varepsilon_{max} \leq \varepsilon_3 \\ 1 & \varepsilon_{max} > \varepsilon_3 \end{cases}$$

where ε_1, ε_2, and ε_3 are selected in terms of material damage and fracture model Eq. (14.85) of CT test results; after loading, the unit largest tensile stress is ε_{max}. When $\varepsilon_{max} \pi \varepsilon_1$, this unit is considered to have no occurrence of damage and fracture. When $\varepsilon_1 \pi \varepsilon_{max} \leq \varepsilon_2$, this unit is in the first damage stage.

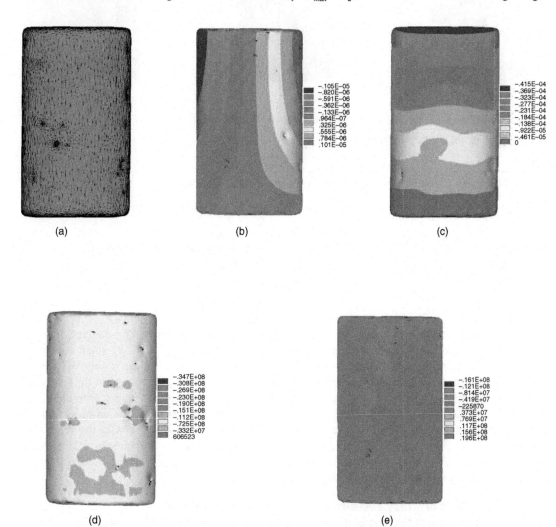

FIGURE 14.77 The Finite Element Three-Dimensional Rebuilding Grids and Computation Results in ANSYS

(a) 3-D finite element grids (b) specimen horizontal displacement (c) test specimen vertical displacement (d) test specimen horizontal stress (e) test specimen vertical stress.

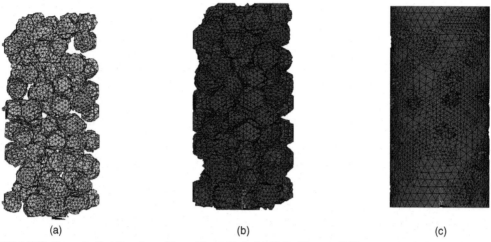

(a)   (b)   (c)

FIGURE 14.78 Concrete Test Specimen Three-Phase Material Finite Element Grids

(a) Aggregate unit (b) interface unit (c) mortar unit.

Table 14.6 Each Phase Component Material Parameters of Concrete			
Materials	**Elastic Modulus (Pa)**	**Poisson's Ratio**	**Tensile Strength (MPa)**
Aggregate	5.8731×1010	0.2407	9.25
Solidified cement mortar	1.7458×1010	0.1960	2.78
Adherence interface	1.3967×1010	0.2000	1.56

When $\varepsilon_2 \pi \varepsilon_{max} \leq \varepsilon_3$, this unit is considered to be in the second damage stage. When $\varepsilon_{max} \phi \varepsilon_3$, this unit is considered to have occurrence of damage. If there are the units occurring to have damage, fracture, and failure, the rigidities of these elements are multiplied by less than one or a very small coefficient to make them degrade the mentioned process and should be repeated until loading is finished. In computation, loading is taken as a uniform distribution force applied to the upper surface of the test specimen. Loadings applied to the test specimens should keep in agreement with CT test, with nine steps of completing loadings in total. The loading values after loadings in every step should be as follows: the first step is 21.23 MPa, the second step is 24.59 MPa, the third step is 27.24 MPa, the fourth step is 33.19 MPa, the fifth step is 34.89 MPa, the sixth step is 36.08 MPa, the seventh step is 33.61 MPa, the eighth step is 32.22 MPa, and the ninth step is 12.71 MPa. The results of damage and fracture of concrete cylinder specimen units are shown in Figs 14.79 and 14.80.

14.8.2.2 Analysis of stress and displacement computation results

It can be seen from Fig. 14.80 that in the first step there are damages occurring in the interface units, and in the second and third steps loading, there are damages occurring in more and more interfaces, but they are in the first damage and fracture stage (i.e., $\varepsilon_1 < \varepsilon_{max} \leq \varepsilon_2$). With the gradual increase in loading, the damaged mortar units are increasing, and the interface units are gradually developing toward

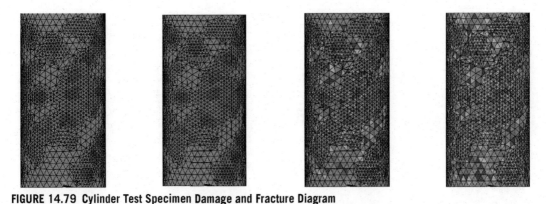

FIGURE 14.79 Cylinder Test Specimen Damage and Fracture Diagram

Step 1 damage; step 3 damage; step 6 damage; step 8 damage.

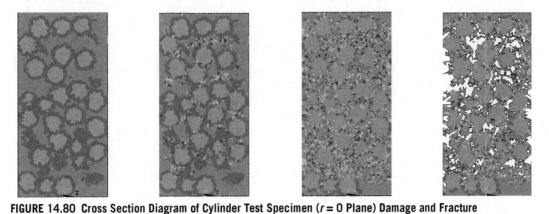

FIGURE 14.80 Cross Section Diagram of Cylinder Test Specimen ($r = 0$ Plane) Damage and Fracture

Step 1 damage cross section; Step 3 damage cross section; Step 6 damage cross section; Step 8 damage cross section.

the second damaged stage (i.e., $\varepsilon_2 < \varepsilon_{max} \leq \varepsilon_3$). In the seventh step loading, the loading reaches its peak value, and the number of the interface unit damage and mortar unit damage reaches the maximum, but the number of mortar units in the second damage stage increases remarkably. At this time, there are only a few aggregate units occurring to have damages. In the follow-up, unloading phenomena appear, and the number of damaged interface and mortar units continues to increase to form macrochaps, and the test specimens begin to be out of stable damage. Accordingly, numerical simulation results are found to be similar to CT test process.

Through step loading computation, each loading step displacement cross section diagram and stress cross section diagram. The displacement cross section diagram and stress cross section diagram of test specimens in the first, third, sixth, and eighth loading steps are shown in Fig. 14.81~fig.14.84.

It can be seen from Fig. 14.81 that when loading is smaller, the displacement distribution and uniform body are similar, and the displacement appears to be in continuous distribution with better regulations. The horizontal displacement of each cross section is in symmetry with the right and left sides, near the

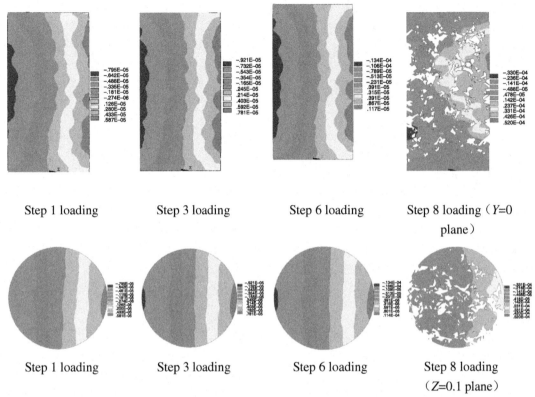

FIGURE 14.81 Horizontal Displacement Cross Section of Each Loading Step of Cylinder Test Specimen

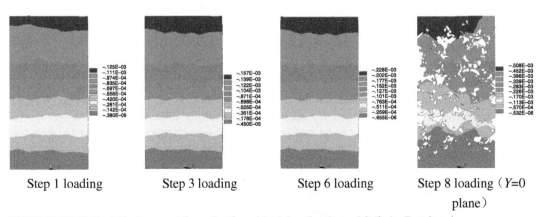

FIGURE 14.82 Vertical Displacement Cross Section of Each Loading Step of Cylinder Test Specimen

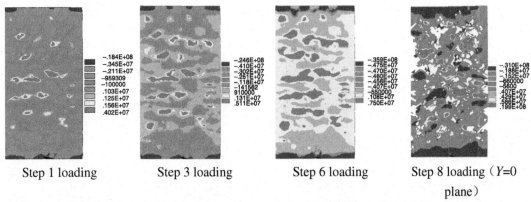

Step 1 loading   Step 3 loading   Step 6 loading   Step 8 loading（*Y*=0 plane）

FIGURE 14.83 Horizontal Stress Cross Section of Each Loading Step of Cylinder Specimen

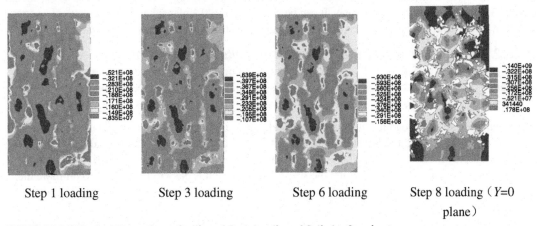

Step 1 loading   Step 3 loading   Step 6 loading   Step 8 loading（*Y*=0 plane）

FIGURE 14.84 Vertical Stress Cross Section of Each Loading of Cylinder Specimen

middle, is gradually decreased, whereby indicating that under the uniaxial compression action, the middle part of the test specimen expands outward. After the sixth loading step, the horizontal displacement still remains in symmetry with the right and left sides, but its value iso-line is of inclined phenomenon, whereby illustrating that under loading action, the test specimens occur to have failure, and the horizontal distribution regulations begin to become poor.

It can be seen from Fig. 14.82 that the vertical displacement continuous variation of the concrete test specimen cross section after each loading distributes regularly. Vertical displacement appears to have the trend of dropping from the upper to the lower. The maximum vertical displacement occurs at the top of the cross section. Being subject to the aggregate influence in test specimens, the displacement distribution contour line appears to be wave-shaped. In the case of large loading, the displacement distribution trend remains to have no changes, but not as good as regular in the case of low stress. It can be seen from Fig. 14.82 that in the case of initial loading, vertical displacement maximum value occurs at the right side of the cross section. With an increase in loading, vertical displacement maximum value occurs at the left side of the cross section so that the trend continues to decrease from the left side to the right side basically.

It can be seen from Fig. 14.83 that there can be a very great difference between the stress distribution diagram of concrete test specimens and uniform body. After each loading step, concrete test specimen cross section stress distribution is extremely uneven with the upper and lower ends larger than that in the middle. There is a discontinuous tensile stress zone in the middle. This is caused by the upper and lower two ends boundary displacement constraints. It can be seen from $Z = 0.1$ plane diagram that the horizontal stress distribution is not very uniform. The stress in the destroyed zone changes obviously, fully reflecting nonuniformity of concrete material distribution.

It can be seen from Fig. 14.84 that vertical stress distribution of concrete test specimen is also very nonuniform. In the initial stage of loading, test specimen vertical stress is large at both ends but small in the middle. In the case of smaller loading, the aggregate bearing vertical stress is large, and the vertical stress of horizontally-arranged two aggregate interfaces and in the mortar zone is smaller. With an increase in loading, the vertical stress shared by the aggregate transfers into the surrounding mortar body surrounding aggregates suffers from damages. The stress shared by the stress enlarges again. The stress on the test specimen's upper and lower ends changes obviously. It can be seen from $Z = 0.1$ plane diagram that the vertical stress is larger, and the surrounding stress distribution is small, and stress in the destroyed zone varies obviously.

14.8.2.3 Comparison of numerical simulation results with CT test results

Three concrete test specimen CT test results corresponding to the mentioned numerical test loading steps are indicated in Fig. 14.85. The following understandings can be obtained through the comparison with numerical tests as follows:

1. In the early stage of loading, stress is small, and part of the units begins to be damaged. These damaged units are mainly located in the contact interface of aggregates and mortars because the concrete's weak location is there. The damaged units have led to stress redistribution of the whole test specimens, around which the stress concentration is easy to form so as to promote the surrounding units to occur to have the tensile stress concentration to generate tensile damage. With an increase in the external loading, a vast bulk of unit tensile damages can take place between the interfaces and the mortars. When surpassing loading peak values (including peak value), the test specimens have entered the residual strength deformation stage, chaps begin to branch, elongate and later on, a large bulk of microchaps run through, whereby forming one or several main chaps that can be recognized with eyes intuitionally. It just at this time that the test specimens happen to be out of stability and be damaged.
2. Concrete chaps generate in the interface layer between aggregates and mortars, whereby illustrating that the aggregates with higher strength can block or prevent microchap expansion and in addition, that CT test results are found to be in good concert with numerical simulation results.
3. The kinds of damage morphology of concrete cylinder test specimens are different, with each test specimen location, expansion direction, and destroyed zone being different. Microchaps germinate in the interface layer, and expand along the aggregate surrounding nonuniform weakest zones, whereby demonstrating that the concrete inner structure random distribution has an important effect upon the morphological failure.
4. CT test results are found to be in agreement with the results chap germination, expansion process, or process of numerical simulation. Therefore, as viewed from statistics sense, the damage laws are the same.

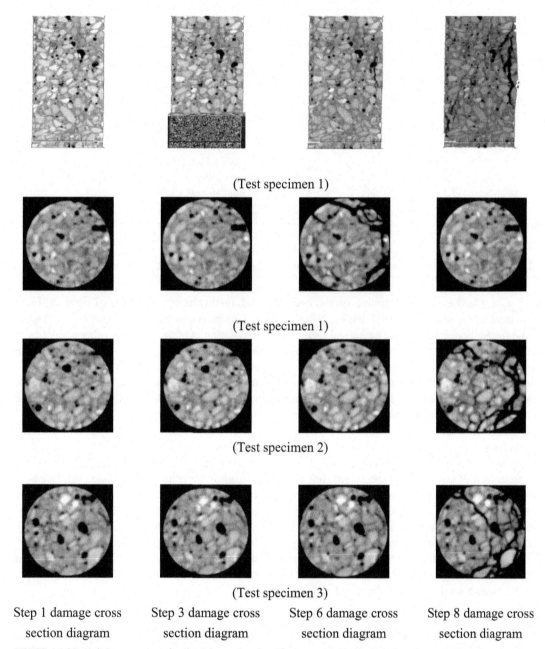

(Test specimen 1)

(Test specimen 1)

(Test specimen 2)

(Test specimen 3)

| Step 1 damage cross | Step 3 damage cross | Step 6 damage cross | Step 8 damage cross |
| section diagram | section diagram | section diagram | section diagram |

FIGURE 14.85 Unit Damage Longitudinal Cross Section Diagram of Cylinder Test Specimen at Each Loading Step

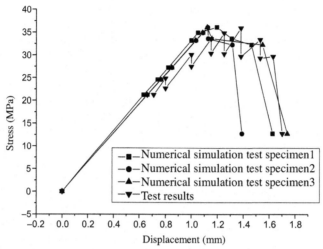

FIGURE 14.86 Loading–Displacement Curves Contrast of Different Numerical Test Specimens

Figure 14.86 is the comparative results between loading–displacement curves of three different numerical test specimens and CT loading–displacement curves. It can be seen that with Fig. (14.85) as the foundation, carrying out numerical simulation computation is of the following characteristics.

As to numerical model test specimen loading–displacement curves, its linear elastic section is the basic straight line while the loading–displacement curves of mechanics test concrete specimens are the curves in the case of small stress. This is because the inner structure and each component properties of real concrete test specimens are not fully reflected in the numerical model, leading to the fact that the curves of numerical model loading–displacement curves tend to become realistic.

Mechanics test results are found to be very close to the results of numerical simulation as a whole, particularly being in better coincidence with curve peak and peak value previous zones, whereby illustrating that the results of the numerical test simulation are reliable and can be used to test the mechanics properties of concrete test specimens, but peak value loading corresponding damage displacement differences are so great that the difference of three numerical test specimen simulation results, and mechanics test results corresponding to the damaged displacement are 18.4, 18.7, and 14.3%, respectively, whereby showing that there exist some certain errors in numerical tests so that it is felt essential to carry out further improvement of numerical tests.

As viewed from the results of numerical simulations for many times, concrete mesodamage and fracture constitutive relation built on the basis of CT tests is feasible for implementing numerical simulation research on the concrete damaged process.

14.9 THREE-DIMENSIONAL CARTOON DEMONSTRATION OF CONCRETE LOADING DAMAGE PROCESS

Concrete CT probe result displays with stratified CT image format. Owing to the CT image with high resolution, it is easy and convenient to observe and analyze the concrete structure information of every cross section. But, chaps after concrete failure are the spatial three-dimensional bodies to analyze chap expansion laws, and quantitatively to describe chap three-dimensional morphology so as to build up the chap evolution model under various kinds of conditions in such a way that it is necessary to observe chap morphology intuitively from different angles. This needs to carry out the visual rebuilding of three-dimensional chap spatial distribution.

In a frame of concrete CT image, chap morphology and development process within the cross section can be clearly displayed in terms of low-density tape of image features. As far as the vertical direction expansion is concerned, the method adopted in general is to make a horizontal arrangement of CT images in different stress stage of cross-section or vertically to arrange them together by relying on observation and imaginations to obtain the chap spatial shape concept and concrete recognition. Owing to many cross sections, it is, in fact, difficult to imagine the morphological and expansive process of chaps in the vertical direction, and it is very difficult for chap expansive and aggregate inhibiting action to for the whole complete image in three-dimensional space. For this reason, it is necessary to rebuild the chap three-dimensional images of each stress stage in terms of cross-section CT images to weaken the structure information of the other parts, and to protrude the image features of displaying each chap position.

The objective of three-dimensional visual chaps within the concrete CT image is that based on the cross section CT image, first of all, it is necessary to determine chap concrete images in this layer, and then a special software is adopted to rebuild the three-dimensional images of concrete test specimens to form cartoons at last, so as to study and analyze the chap spatial morphological behaviors and aggregate intersecting relation.

Software, such as 3ds max, Adobe Premiere, and Adobe Photoshop are used to rebuild concrete chap three-dimensional image cartoons, the manufacturing process involving technology package such as previous data processing, model preparation, cartoon settings, and outputs building-up of five major parts:

1. *Prestage data processing*: The prestage data processing steps are as follows. First of all, based on CT physical principle and image processing method, it is necessary to determine the chap zones within concrete CT images; the next, CT image data obtained are introduced into Photoshop, and the contrast degrees of each piece of CT photo or picture data are regulated in order. Belaying tools are used to select the chaps and aggregate outline shapes in detail so as to form selection zones i, and then in the selection zones, working paths are formed so as to lead the working paths to Illustrator and stored as AI format. Each piece of image data can be processed in order, and stored in chap AI format information and aggregate AI format information.

2. *Model preparation*: Steps for model preparation are as follows. To begin with, a cylinder (12 cm high, diameter is 6 cm) with the same size of the test specimen should be set up in 3ds max 6.0, and one is duplicated in the original place, being named as A and B. Again, the AI format chap information and aggregate information are introduced into 3ds max 6.0 in order, every chap data are edited by using Edit Spline commanded strictly in terms of real sizes and positions in orders, and using Extrude to tensile a certain length makes 20 slice information arranged in order, and length is 12 cm. Finally, Edit Spline command is used to edit every aggregate data in order, and Extrude is adopted to tensile to the real height and to regulate the sizes and outer morphological states so as to make them be more natural and more real. Also, HSDS command is used to carry out the smooth treatment of aggregate model, and the Optimize command is used to implement the optimal treatment of the aggregate model.

3. *Material quality regulation*: Material quality regulation process is as follows. One concrete image-sticking material quality is given to prepare cylinder A. Normal command is used to make the normal line inversion, and one white line frame material quality is given to the prepared cylinder B. One red transparent material quality (transparency degree is 70) is given to the chaps of the first layer and the first reciprocal layer. Again, one red transparent material quality (transparency degree is 60) is given to the chaps of the second layer and the second reciprocal layer, and later on, when each layer is downward, transparency is reduced by 10 in succession; one granite image-sticking material quality (transparency is 80) is given to the aggregate materials of the topmost layer and the most reciprocal bottom layer. Again, one granite image-sticking material quality (transparency is 70) is given to the aggregate materials of the second layer and the second reciprocal layer, and later on, when each layer is downward, transparency is reduced by 10 in succession.

4. *Cartoon setting*: Along the cylinder outline, one smooth screw curve line 01 is drawn to serve as the camera cartoon pathway (Fig. 14.87). The camera is adjusted to the right angle to enter the Motion panel to carry out the operation of Parameters/Assign Controller/position: position XYZ/Path Constraint, to pick up the well-watered motion path. It is just at this time that the camera moves along the path starting point to the path terminal point in terms of setting time.

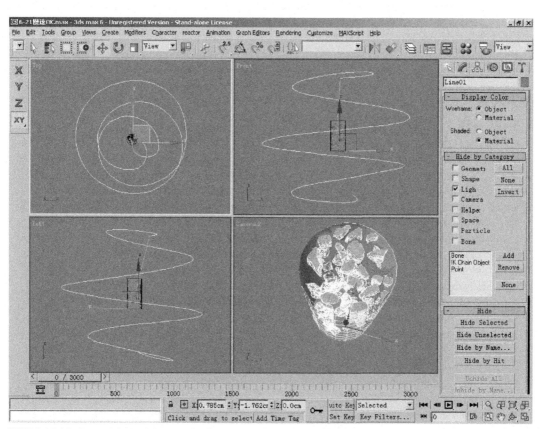

FIGURE 14.87 Cartoon Path of Cartoon Setting

FIGURE 14.88 Damage Chaps of Three-Dimensional Images of Static Loading

FIGURE 14.89 Damage Chaps of Three-Dimensional Images of Dynamic Loading

5. *Output synthesis*: This cartoon setting time length is 2400 picture frequencies (30 picture frequencies/s), that is, the camera moves from the path starting point to the path terminal point with the uniform velocity within 2400 picture frequencies. The setting sizes of this cartoon can be 800 × 600 pixels in the scope of 0–2400 picture frequencies. Saved cartoon is in AVI format and to implement the output. The production of AVI segment with the corresponding audio materials can be introduced into Adobe Premiere 6.0 to implement the video editing and rearrangement and output.

Figures 14.88 and 14.89 give out the intercept frames of different visual image states of three-dimensional cartoons of concrete state and dynamic loading damage chaps picture. The chap spatial morphological states effectively observed from each angle can be realized through three-dimensional chap cartoon picture displays damage process under the concrete static and dynamic compression conditions so as to deepen the recognition of chap morphology and to be convenient for research on the concrete damage and fracture processor.

RESEARCH ON NUMERICAL ANALYSIS OF FULL GRADATION LARGE DAM CONCRETE DYNAMIC BEHAVIORS

15

In order to explain and confirm the macrotest results of full-gradation large dam concrete dynamic behaviors, to deepen the understanding of inner-structure damage and failure mechanism of concrete materials, and at the same time to probe into the new thinking of digital concrete research, the authors have carried out research on the numerical analysis of the dynamic behaviors of full-gradation large dam concrete.

Based on the engineering background of high arch dam aseismic safety evaluation of large dam concrete dynamic resisting force requirements, the research contents of present-day full-gradation large dam concrete dynamic behavior numerical analysis mainly concentrate on two aspects: (1) to explain and confirm the prestatic loading upon large dam concrete dynamic resisting force enhancement behaviors; and (2) to study the damage morphology and failure mechanism of large dam concrete under the dynamic–static actions.

In order to realize the numerical test of concrete dynamic behaviors, it is necessary to establish the numerical damage model of large dam concrete and the failure process. In the process of research, a series of key technical difficult problems need to be solved. For instance, in order to satisfy the practical mix proportion of large dam concrete coarse aggregates, input as high as 70% of aggregates can be the first difficult subject for carrying out concrete three-dimensional mesomechanics analysis. For this reason, the authors suggest the random aggregate placeholder reject method. In addition, the large magnitude of concrete test specimen is to find the solution to the large-scale mesonumerical problem. In order to reduce the large-scale computation freedom, the authors have studied the composite medium double-dimension analysis theory and suggested the equivalent mechanics parameter solution method of the composite medium consisting of small-dimension coarse aggregates and mortars, and at the same time, the authors have studied and developed the parallel computation software for implementing large dam concrete dynamic performance nonlinear numerical analysis based on the PFEPG platform. This chapter introduces research achievements of these difficult problems, while some problems still require in-depth studies.

15.1 MESOMECHANISM NUMERICAL METHOD OF FULL-GRADE LARGE DAM CONCRETE

While considering large dam concrete gradation and uniformity of each phase medium, and based on the concrete material basic test materials, the concrete mesodamage mechanics numerical method carries out numerical simulation and analyzes and discusses the inner-damage process and mechanism under the different operation conditions. This is to establish the relationship among the defects and performances of various kinds of concrete material as well as their macromechanics behaviors in meso and microstructure.

The concrete mesomechanics method views concrete as a three-phase nonhomogeneous composite maternal consisting of coarse aggregates, hardened cement colloid, and the interface adhesive belt. This method needs classifying the units in concrete material mesolayers and establishes the appropriate mesostructure model. In considering the differences of the mechanics behaviors of aggregate unit, hardened cement mortar units, and interface unit material, and with the simple failure norms or damage model to reflect the unit rigidity degradation, the numerical method is used to simulate the concrete test specimen fracture expansion process and failure morphology to reflect test specimen damage and fracture mechanism (Huaifa et al., 2004).

15.1.1 THREE-DIMENSIONAL RANDOM AGGREGATE MODEL

The concrete mesomechanics method in the initial stage adopts the random aggregate model and concrete mesoanalysis is only restricted to two-dimensional plane problems. As to gravels and pebbles, spherical or perfectly round aggregates, their particles can be assumed as spherical ones. used spherical aggregates to have equal probability distribution in the test specimens and any size circular tangential face without probability holding superiority, and established the relations of inner aggregate gradation and contents of concrete test specimen space and the inner cut-off section with cutting aggregate area – this is the so-called Walraven formula. In terms of the Walraven formula, a two-dimensional random circular aggregate model is generated, on the basis of which, and in considering the real shapes of common macadam, Huaifa et al. (2006) suggested the method of generating a concrete random convex polygon aggregate model. This method is based on the circular aggregate model generated in terms of the Walraven formula. The circular aggregate inner-connected with polygons can be outward extended with convex, until the circular aggregate model occupies the area of the concrete's inner cut-off sections, that is, the polygon random aggregate model with the same area as the circular aggregate area is generated. The convex polygon aggregate model generated by this method can not only reflect aggregate practical morphology, but also maintain the aggregate real gradation and contents.

With the progress in computation technology and deepening research, the people naturally expect that the aggregate model established can be possibly near the aggregate morphology so as to make concrete mesomechanics analysis operate in three-dimensional space. Therefore, this book deals with the technology to generate the three-dimensional random aggregate model.

In terms of real gradation, it is necessary to find out coarse aggregate particle sizes, quality ratio of each particle size group, and concrete containing coarse aggregate qualities in order to determine the number of particles of aggregate of various types. The random spherical aggregate model is the initial research achievement of the three-dimensional random aggregate model study. In order to determine three-dimensional random aggregate distribution, first of all, it is essential to determine the special scope in which aggregates are, and then within the concrete test specimen space. Monte Carlo method is used to determine the sphere position of aggregates randomly. Every aggregate sphere determination needs four variables: sphere diameter and sphere center coordinate values x, y, z. Owing to the determination of sphere diameter, it is here that only the sphere center coordinate values are determined, which will be enough. Accordingly, the larger aggregate sphere center coordinate is determined first.

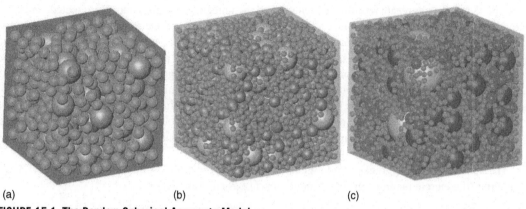

(a)   (b)   (c)

FIGURE 15.1 The Random Spherical Aggregate Model

(a) Grade 2 gradation (aggregate contents 47%), (b) Grade 3 gradation (aggregate contents 59%), (c) Grade 4 gradation (aggregate contents 66%).

Each time, three random variables can be produced, and when the positions of larger aggregates are determined afterward, the positions of smaller aggregates are determined in such a way by following the circulations in order until the particles of all the aggregates are generated. In order to guarantee that the spheres generated are mutually independent and there will be no occurrence of position overlap or intersection, it is required that the distance between two aggregate sphere centers should be over the sum of the radius of two spheres, and at the same time, it must be guaranteed that the spheres generated must be within the decided scope. Figure 15.1 is the distribution model for concrete random aggregates with two-stage gradation, three-stage gradation, and four-stage gradation.

15.1.1.1 Placeholder reject method

Research on concrete mesomechanics, sizes, shapes, and distribution of aggregate materials has a direct effect upon concrete mechanics behaviors. In order to reflect the nonuniformity of concrete mesostructure and to carry out the mesomechanics numerical simulation, people have suggested many mesomechanics models for studying concrete damage and failure processes such as the grid structure model, random mechanics behavior model, and random aggregate model. Prior to this, the numerical simulation research on concrete mesomechanics was mainly restricted to the plane problem. Despite some three-dimensional aggregate model research, the research basically stayed at the "convex type" polygonal face aggregate random input algorithms, and aggregate contents are low without considering the real aggregate gradation. For instance, the content of big aggregate concrete in large dams is as high as 60–70%. For this reason, in accordance with the real gradation aggregate percentage, the practical positioning of random aggregate particles is the main difficulty in building-up the model. Aggregate input algorithm with high efficiency is the prerequisite and foundation for carrying out concrete mesomechanics analysis.

Random aggregate inputs can be directly completed within the concrete test specimen zone. Particularly when aggregate contents are high, it is difficult to achieve the objective. Huaifa et al. (2005) suggested the "occupied zone reject method" or called as the "placeholder reject method." This method is able to effectively complete the inputs of high-content aggregates in certain zones. The so-called occupied zone reject method refers to the profile of the test specimen zone into cubic units with less than the minimum aggregate radium step length, and the units near the boundary are rejected, and then Monte Carlo method is used to select a certain unit at random from the inner units with equivalent probability. Again, in this unit zone, the coordinate points are selected at random and with equivalent probability, and it is necessary to discriminate whether some kinds of particle aggregates with the point as the sphere center have been introduced into the already-decided aggregates; the rest of the units are selected or else the aggregate positions can be determined. If some unit has come to the maximum distance less than the sum of its radius with the minimum aggregate radius to the aggregate sphere center, the unit can be rejected. When the rest of the aggregate positions are determined, these rejected units will not be selected. This method is used to raise the two-dimensional random aggregate generating program efficiency.

The sphere random aggregate model-generating program is a three-dimensional model F90, with gradation III concrete random aggregate model serving as an example:

1. Input data: 3D3.dat

0.30D0 0.30D0 0.30D0		/meso-profile spatial length, width & height	
50	0.060D0	/ large aggregate particle number, diameter D=0.06m）	
300	0.030D0	/ medium aggregate particle number, diameter D=0.03m）	
2400	0.015D0	/ small aggregate particle number, diameter（D=0.015m）	

2. Output data: 3D3AGG.dat

2750		/aggregate particle total number		
Aggregate number		Circle center coordinate		aggregate particle radius
	（x	y	z）	
1	.4043E-01	.1151E+00	.1730E+00	.3000E-01
2	.1946E+00	.1858E+00	.4981E-01	.3000E-01
3	.2515E+00	.2008E+00	.1762E+00	.3000E-01
......				
149	.2820E+00	.6447E-01	.2498E+00	.1500E-01
150	.1816E+00	.1835E+00	.1840E+00	.1500E-01
......				
2749	.1160E+00	.2912E+00	.2181E+00	.7500E-02
2750	.1761E+00	.6819E-01	.4259E-01	.7500E-02

The aggregate distribution model is drawn in terms of output data (Fig. 15.1b). The main steps of the aggregate input procedure main steps are given next, as shown in Fig. 15.2.

15.1.1.2 Random convex polyhedron aggregate model
Node intrusion into convex spatial judgment norms:

Based on the spatial vector mixture product principle, it is necessary to judge whether some node is in the convex space or not. The vector mixture product will produce one volume scalar quantity from this, with spatial volume as the scale. It is essential to set up some node to check whether it intrudes into the convex space serving as the judgment norm. Figure 15.3 indicates the convex ABCDEF. Supposing

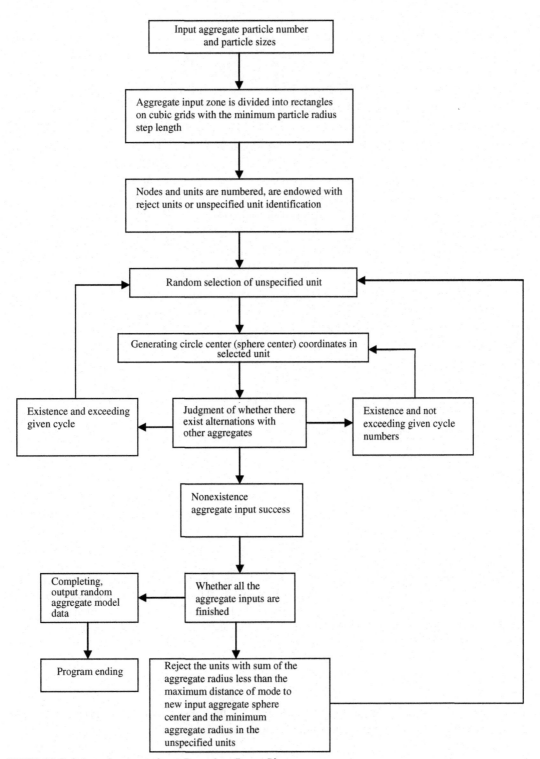

FIGURE 15.2 Sphere Aggregate Input Procedure Frame Diagram

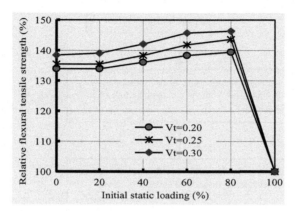

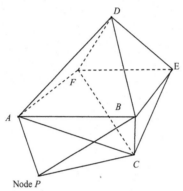

FIGURE 15.3 Relation Between Nodes and Spatial Convex

that the following node coordinate is $P(x, y, z,)$, $A(x_1, y_1, z_1)$, $C(x_2, y_2, z_2)$, $B(x_3, y_3, z_3)$. The tetrahedron *PACB* volume is as follows:

$$V = \frac{1}{6} \begin{vmatrix} x & y & z & 1 \\ x_1 & y_1 & z_1 & 1 \\ x_2 & y_2 & z_2 & 1 \\ x_3 & y_3 & z_3 & 1 \end{vmatrix} \tag{15.1}$$

Node *P* in the external with the triangle *ACB* face can surround a tetrahedron, whose volume is a positive value. If *P* is in the inner convex, it, together with the triangle *ACB* face, can surround a tetrahedron, whose volume is negative value. Obviously, when *P* is in the convex boundary, this tetrahedron volume is zero. For this reason, it is necessary to define node *P* whether it is in the inner convex judgment norm in the following:

$$\begin{cases} P \notin \Omega & V_P > 0 \\ P \in \Gamma_\Omega & V_P = 0 \\ P \in \Omega & V_P < 0 \end{cases} \tag{15.2}$$

where Ω is the spatial zone surrounded by the convex, Γ_Ω is the convex boundary surface, and V_P is the tetrahedron volume surrounded by node *P* together with the face element.

The face element top node of the convex triangle seen from the external into the internal is arranged in an anticlockwise way.

Intersection monitoring of two triangles in three-dimensional space:

In the extending process of the spatial convex polyhedron, it is necessary to judge whether two convex polyhedrons intersect. But in the final analysis, it is to carry out the intermonitoring of two triangles in the three-dimensional space. In the grid profile separation, to judge whether the interface element intersects with the aggregates can also use the intermonitoring of two triangles in the three-dimensional space; this is an important basic mathematical operation.

As indicated in Fig. 15.4 Δabc is located on the m plane, and Δdef is located on the n plane; the two-triangle intermonitoring algorithm is as follows: as to Δabc, it is necessary to test Δdef whether it intersects, and it is only to test the three sides Δdef to prove whether they pass through Δabc that is enough, with the de side serving as an example:

1. To find Δabc normal vector M first.
2. Separately to find the direction vector **da** from e point to a point as well as the direction vector **ea** from e point to a point.
3. Separately to find the direction vector **da**, **ea** with M point product, that is, $DA = da$, $EA = ea$ M. If DA and EA are the same number, this indicates that the two points d and e are located on the two sides of the Δabc plane; obviously, they will not intersect with Δabc.
4. If DA and EA are the different numbers or one of them is zero, this illustrates that de will pass through the Δabc plane. At this time, it is necessary to make further judgment and to find out the intersection point of the de side and Δabc plane. If the intersection point is located on the Δabc plane (or on the side), this shows that Δabc and Δdef intersect with each other. If DA, EA are zero at the same time, this means that de side and Δabc are coplanar. But they are processed in terms of nonintersection, and the further judgment can be made through the neighboring triangle; Δdef has only one side that intersects with Δabc, and two triangles intersect with each other.
5. As to the situations indicated by Fig. 15.4, the three sides Δdef cannot intersect with Δabc, but two triangles can intersect with each other. At this time, in terms of the mentioned procedures, it is necessary to rejudge whether the three sides Δabc can intersect with Δdef.

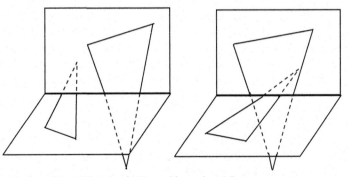

FIGURE 15.4 Intermonitoring of Two Triangles in Three-Dimensional Space

6. In the mentioned repeated cycle calculation process, there is only one side of one triangle to intersect with another triangle; it can be out of this cycle to obtain the conclusion of triangle intersection. If all the situations are not satisfied, two triangles cannot intersect with each other.

In terms of the mentioned generating spherical aggregate distribution, to generate the convex aggregate algorithm is as follows:

1. The node numbers can be generated on the spherical aggregate particle surface at random (larger than four, but less than some certain controlled number N, here taking 15); first of all, the four different octant nodes located in the space must be generated to form the tetrahedron aggregate base.
2. Adding to a new node, it is necessary to allow some new-insertion nodes into the aggregate spherical center with the distance slightly larger than the sphere radius. One inner node within the formed convex belongs to two sides of some convex side surface. This side surface is visible so that all the visible surfaces of this node must be found.
3. These new nodes can be connected with each top node of these visible surface triangles to form a new convex side surface.
4. Data structure must be adjusted so as to eliminate the mentioned visible triangle side surfaces.
5. Return to the Step (2), until all the nodes complete the join or participation.
6. To output the top node code and coordinate value of each triangle side surface of polyhedron aggregates.

The new joined node and one node in the inner convex aggregate belong to two kinds of algorithms of two sides of the triangle plane. The first kind of algorithm forms the tetrahedron volume positive and negative value judgment in terms of node and triangle, whose principal part is seen in Eqs (15.1) and (15.2). The second kind of algorithm finds the two-node connection line and intersection nodes of the triangle plane. If the intersection nodes are located in the middle of nodes between the joined nodes and convex aggregate inner nodes, this shows that the two nodes belong to the plane two sides, respectively, or else they belong to the plane same sides.

The program rough frame diagram (Huaifa et al., 2006) is shown in Fig. 15.5.

The third gradation random sphere aggregate distribution model based on Fig. 15.1b can generate the random convex aggregate model as indicated in Fig. 15.6.

It is worth noticing that if the inner convex aggregate connects with the original sphere aggregates, the sphere model can become a convex polyhedron model. Thus, aggregate contents of the generated polyhedron model cannot reach the aggregate contents of the original sphere aggregates unless many nodes are inserted into the aggregate sphere surface in such a way that the generated polyhedrons are made to be closer to the original sphere aggregate shape. Accordingly, under the condition of satisfying the non-interintrusion of aggregates in Step (2), more new inserted nodes are allowed into the aggregate sphere center distance being slightly larger than the radius of the original sphere aggregates. In this way, it can not only make the newly generated polyhedron aggregate contents reach the requirements of the original sphere aggregate contents, but also basically reflect the polyhedron morphology of the real aggregates.

Particle size continuous distribution random aggregate model:

In the generated aggregate models, particle sizes have basically taken the average value of each gradation particle range. For instance, in gradation 3 concrete, small stone or gravel particle size range is 5–20 mm, small stone or gravel particle size is 15 mm; medium stone or gravel particle size range is 20–40 mm; medium stone or gravel particle size is 30 mm; big stone or gravel particle size range is 40–80 mm; big stone or gravel particle size is 60 mm. Each particle size gradation should take

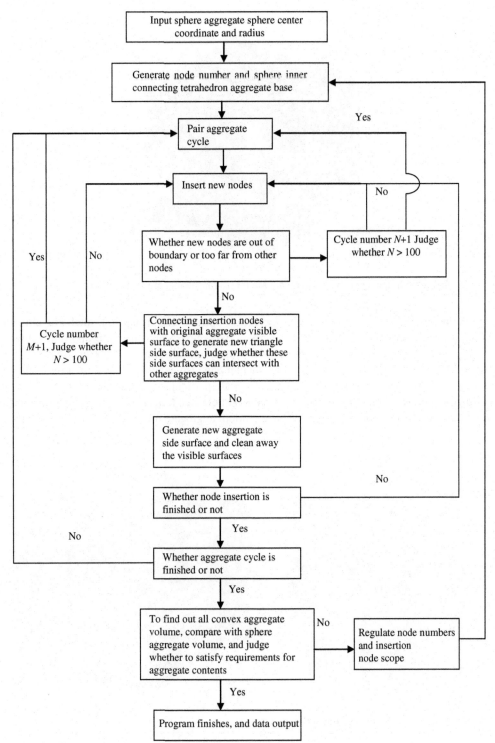

FIGURE 15.5 Convex Polyhedron Aggregate Model Generating Program Diagram

FIGURE 15.6 Gradation 3 Random Convex Aggregate Model (Li Yongzhang, Ma Huaifa et al., 2006)

one fixed value, and then, in accordance with each grade particle size proportion, small gravel:medium gravel:big gravel proportions (3:3:4), and concrete coarse aggregate quality proportion (59%) can get small gravel 2400 grains, medium gravel 300 grains, and big gravel 50 grains, respectively. In fact, the random aggregate input method in this chapter can also generate the continuous particle size random aggregate model. In accordance with each grade particle size aggregate proportion and corresponding to aggregate contents, grade rate in each particle size range can select particle magnitudes or sizes can generate the continuous particle size random aggregate mode. Figures 15.7–15.9 give out

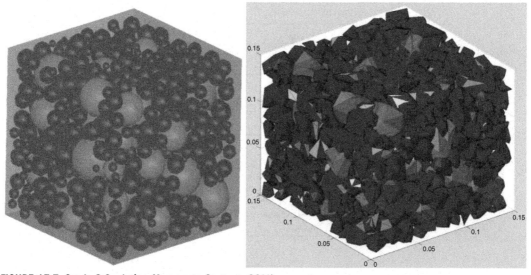

FIGURE 15.7 Grade 2 Gradation (Aggregate Contents 60%)

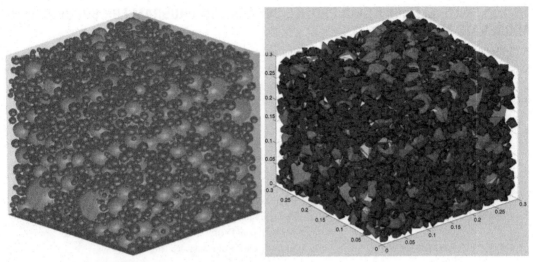

FIGURE 15.8 Grade 3 Gradation (Aggregate Contents 65%)

grades 2, 3, and 4 gradation sphere and their gradation contents corresponding to polyhedron aggregate models.

Aggregate particle size distribution in terms of actual gradation aggregates can take the continuous particle sizes so that grade 2 gradation aggregate input contents can reach 60%, grade 3 gradation aggregate input contents reach 65%, and grade 4 gradation aggregate input contents over 70%, whereby, completely reaching the requirement of full-gradation aggregate actual contents. But this can bring about difficulties to element profile separation. If the small gravels with particle sizes below 10 mm are involved in design, it will lead to finding the freedom degrees with a large scale so as to make it difficult to realize mesoanalysis.

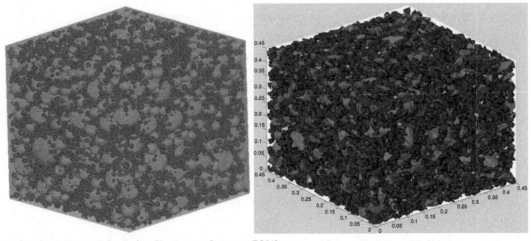

FIGURE 15.9 Grade 4 Gradation (Aggregate Contents 70%)

15.1.2 FINITE ELEMENT GRID PROFILE SEPARATION PROGRAM FOR SPHERICAL AGGREGATE MODEL

A similar method for circle aggregate profile separation is adopted to implement spherical aggregate model profile separation. Structure tetrahedron or hexahedron grids are generated in the selected concrete test specimen spatial zones, and then the generated random spherical aggregate model is projected into these grids. In terms of aggregate model location, mesoelement or unit is endowed with attribute.

Program name: 3PGRID.F90.

1. Input data: This program has two input data blocks. One block is 3D3AGG.dat, that is, three-dimensional aggregate model, f90 output data; another block is 3DFN.dat, whose data are:

8	\ element type, 8 express 8-node hexahedron element or unit.
0.0D0 0. 0D0 0.D0 0.3D0 0. 3D0 0.3D0	\ profile separation zone left down & right upper corner node coordinate.
0.00375D0 0. 00375D0 0.00375D0	\ grids on x, y and z direction step length.

2. Output data : 3DXY.dat and 3DPNUM.dat.

3DXY.dat format is as follows:

531441   Node coordinate

Node No	(x,	y,	z)
1	.000000E+00	.000000E+00	.000000E+00
2	.000000E+00	.000000E+00	.375000E-02
3	.000000E+00	.000000E+00	.750000E-02
......			
531439	.300000E+00	.300000E+00	.292500E+00
531440	.300000E+00	.300000E+00	.296250E+00
531441	.300000E+00	.300000E+00	.300000E+00

512000   /unit total number

Element	unit type	unit node No.							Material type
	8 node unit								
1	8	1 6562 82 6643 2 6563 83 6644							20
2	8	2 6563 83 6644 3 6564 84 6645							20
......									
512000	8	524798 531359 524879 531440 524799 531360 524880 531441							20

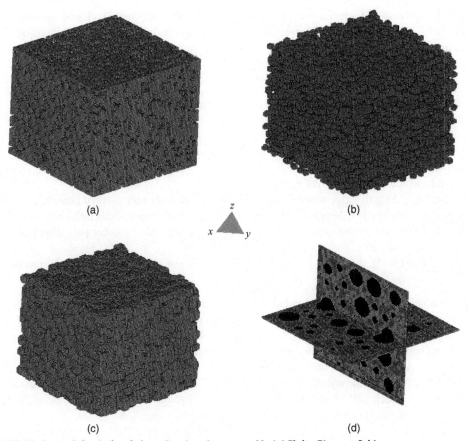

FIGURE 15.10 Grade 3 Gradation Sphere Random Aggregate Model Finite Element Grids

(a) Solidified cement mortar unit, (b) Aggregate unit, (c) Interface unit, (d) Symmetric tangent plane grid diagram.

In accordance with node coordinate code 3DXY.dat and unit number 3DPNUM.dat, GID software is used to output as Fig. 15.10 indicates the mesoaggregate structure grid diagram.

15.1.3 MESOFINITE ELEMENT PROFILE SEPARATION OF CONVEX POLYHEDRON AGGREGATE MODEL

Convex polyhedron aggregate model profile separation is relatively complicated. The following two kinds of unit grids are adopted to implement the analysis: one is the regular eight-node hexahedron unit, and the other is the tetrahedron unit. The tetrahedral unit encrypts the hexahedron unit. Every hexahedron unit can be divided into five tetrahedron units. The discrimination algorithm of the two kinds of unit attribute is the same. Taking the tetrahedron unit as an example illustrates the discrimination algorithm of the mesounit attribute.

To begin with, the test specimen space can be profiled into regular finite element grids with eight-node hexahedrons, whose grid sizes are less than one-third of the minimum aggregate diameter. For instance, the concrete test specimen is 300 mm × 300 mm × 300 mm, and grid spacing of three coordinate axis direction can be 3.75 mm ($D_{min}/4$). Every hexahedron unit may be classified into five tetrahedron units, and then, the generated random aggregate model space should be superimposed on the finite element grid space. In terms of designed algorithm, the computer can automatically recognize each unit attribute, whose concrete algorithm is as follows:

1. All the tetrahedron units are first assumed as the mortar units.
2. Each random aggregate is cycled.
3. Each tetrahedron unit is cycled.
4. If the unit is mortar unit, it will enter the next step, or else, return to Step (3).
5. Each note of every unit must be cycled.
6. Each said surface of aggregate is cycled.
7. Based on Eq. (15.6), to judge whether nodes are in the inner or outer parts of convex aggregates.
8. Returning to Step (5), if four nodes are in the inner part of aggregates, this unit should be defined as the aggregate unit; if four nodes belong to aggregate inner, side, or outer part. This unit can be defined as interface unit.
9. Returning to Step (2), if the unit is neither aggregate unit nor interface unit, this unit can be defined as mortar unit.

When the unit attribute is discriminated in terms of the mentioned method, the situations as indicated in Fig. 15.11 might be missed; that is, nodes of each unit are beyond the aggregate, but aggregates are still likely to intrude into the unit. In order to prevent the situation from occurring, it is necessary to check the two surface intrusions in three-dimensional space; that is, the aggregate triangle side surface and the unit side surface must be checked to find out whether there are intrusions. It is at this time that all mortar units defined by the algorithm should have a second cycle screening. It is only to discriminate whether the line of each side of aggregate triangle passes through the mortar unit side surface. If the line passes through, this unit should be redefined as the interface unit.

The grid profile separation program frame diagram is shown in Fig. 15.12. The application of meso-finite element profile separation program to the profile separation view is shown in Fig. 15.13.

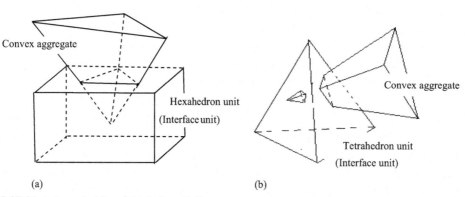

(a)   (b)

FIGURE 15.11 Judgment of Special Interface Unit

(a) Aggregate intrusion into hexahedron unit, (b) Aggregate intrusion into tetrahedron unit.

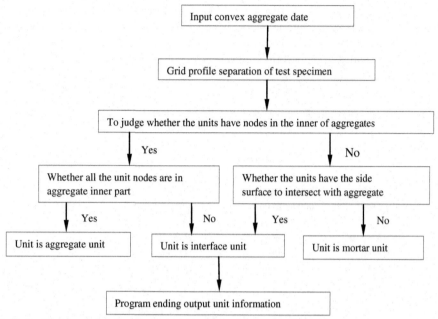

FIGURE 15.12 Profile Separation Program Frame Diagram of Random Aggregate Model

FIGURE 15.13 Mesounit Profile Separation of Random Convex Aggregate Model

GID before and after processing software has a powerful finite element with prior processing and posterior processing function, having the interactive operation interface. If the output date of the mesoprofile separation program is output in terms of *.flavia.msh format, the concrete mesounit graphic patterns can be output under GID platform posterior processing environment.

*.flavia.msh format can be as follows:

Mesh "aec8" Dimension 3 Elemtype Hexahedra Nnode 8   \ dimensionality; unit type
Coordinates   \8-node unit coordinate

\ node No.;		node coordinate	
1	-.228125E+00	.000000E+00	.000000E+00
2	-.228125E+00	.000000E+00	.187500E-01
3	-.228125E+00	.000000E+00	.375000E-01
......			
24255	.321875E+00	.150000E+00	.112500E+00
24256	.321875E+00	.150000E+00	.131250E+00
24257	.321875E+00	.150000E+00	.150000E+00

End coordinates
Elements

unit No.;			8-node unit node code;					unit material type
1	1	82	91	10	2	83	92	11	1
2	2	83	92	11	3	84	93	12	1
3	3	84	93	12	4	85	94	13	2
......								
20990	24164	24245	24254	24173	24165	24246	24255	24174	3
20991	24165	24246	24255	24174	24166	24247	24256	24175	4
20992	24166	24247	24256	24175	24167	24248	24257	24176	4

End elements
Mesh "aew4" Dimension 3 Elemtype Tetrahedra Nnode 4
Coordinates   \ node unit coordinate(concrete node coordinate values are the same as above, to be omitted, but coordinate labeled words can not be omitted)
End coordinates
Elements

note No;	4-node unit node code;				Unit material type
1	406	424	423	325	1
2	406	407	424	325	3
3	325	423	334	424	2
......					
8959	23834	23851	23852	23933	2
8960	23834	23852	23835	23933	4

End elements

If there are other types of units, the rest may be inferred. The units of different material types can be independently coded but coordinates should be unified in label.

After the mentioned data are prepared, CID interface is started to select the postprocessing. Open *flavia.msh, wait for input data, and later on the unit view can be seen in the screen window, and then, other operations can be carried out. For instance, the display visual angle can be changed, the sizes of graphic patterns or the unit material types are displayed in layers, the color of graphic patterns and profile displays can be changed with *.flavia res. GID prior and postprocessing software interface can be used to give out all the required physical quantity cloud graph contour maps and node displacement time process curves, and so on, in the whole time process. Also, the model can be sliced, and the result on the interface can be output.

15.2 DAMAGE AND FAILURE NUMERICAL SIMULATION FINITE ELEMENT EQUATION OF CONCRETE TEST SPECIMENS

Through mesomechanics numerical simulation, it is necessary to consider larger dam concrete gradation and nonuniformity behaviors of each phase medium. Based on basic testing data of concrete materials and under the various operational situations, it is essential to carry out numerical simulation tests, and analyze and discuss its inner failure and mechanism so as to establish relations among various defects and behavior nonuniformity as well as macromechanics of concrete meso and microstructures. Under the conditions of rational computation model and enough accuracy of each phase concrete material behavior, numerical analysis may replace part of tests and can avoid the objective constraints and artificial effect upon test results under the testing conditions.

Concrete mesomechanics concentrates concrete materials, damage mechanics, computation mechanics, and computer programming technology into one body of newly emerged interdisciplinary discipline. Research on concrete mesomechanics method can provide a new approach or means for studying concrete dynamic behavior and open up a new vista for high arch dam seismic motion safety evaluation. Research contents are mainly in the following aspects: (1) to study concrete and each phase material strain rate effect enhancement mechanism; (2) to study concrete mesoeach-phase material constructive relations and the evolution laws; (3) to establish the static and dynamic equations of simulating concrete test specimen damage and failure physical tests so as to carry out numerical calculation method; and (4) to study mesofinite element grid or mesh profile separation method. The following introduces the established basic equations for carrying out concrete test specimen damage and failure nonlinear numerical analysis.

15.2.1 CONCRETE MATERIAL STRAIN RATE EFFECT ENHANCEMENT RELATIONS

Strain rate effect (rate effect for its short form) is the solid material basic behavior. Research on concrete material dynamic behavior indicates the following: the rate effort of nonuniform materials is more apparent than that of uniform materials; concrete strength is highly sensitive to the loading process; when static and dynamic loading to the same loading level, the test specimen will yield the different damage accumulation, while the strain rate effect is subject to the influence of material damage rate; concrete elastic modulus increases with strain rate enlargement, but there is no apparent increase in strength; concrete Poisson's ratio has no obvious change with an increase in strain. Based on some test analysis and some related research literature, the relations among tensile strength, elastic modulus, and tensile (or compression) strain rates are as follows:

$$\begin{cases} f_t(\dot{\varepsilon}) = H_t f_{ts} \\ E(\dot{\varepsilon}) = H_E E_s \end{cases} \tag{15.3}$$

where f_{ts} is concrete material static tensile strength, E_S is static elastic modulus, and H_t is concrete tensile strength enhancement coefficient being the ratio value between the dynamic and static tensile strengths or called the relative tensile strength under some strain rate.

H_E is elastic modulus enhancement coefficient, being the ratio value between the dynamic elastic modulus and the static elastic modulus under some strain rate.

$f_t(\dot{\varepsilon})$ and $E(\dot{\varepsilon})$ express concrete dynamic tensile strength and dynamic elastic modulus, respectively, being the tensile (or compression) strain rate $\dot{\varepsilon}$ function H_t and H_E strain rate $\dot{\varepsilon}$ relation (Huaifa et al., 2005) can be expressed as follows:

$$\begin{cases} H_t = \exp\{[A_t(\lg|\dot{\varepsilon}|+B_t)]^{C_t}\} \\ H_E = \exp\{[A_E(\lg|\dot{\varepsilon}|+B_E)]^{C_E}\} \end{cases} \tag{15.4}$$

where A_t, B_t, C_t are strength enhancement parameters; A_E, B_E, C_E are elastic modulus enhancement parameters.

A_t, A_E express the tensile strength and elastic modulus to the degree of sensitive of strain rate respectively.

B_t, B_E reflect the magnitude of the minimum strain rate produced by the tensile strength and elastic modulus, respectively.

C_t, C_E reflect the tensile strength and elastic modulus to the sensitivity of high strain rate, respectively, displaying the magnitude of curvature of concrete enhancement curves. Owing to the different inner structures of different concretes, their enhancement coefficients are also different. Therefore, the concrete problems can be obtained in terms of the corresponding concrete materials test results.

If strain rate is less than $10^{-6}\,\mathrm{s}^{-1}$, the tensile strength and elastic modulus of concrete materials will not produce the rate effect so that B_t and B_E are equal to σ. When A_E and A_t take 0.12 and 0.20, respectively, the index parameter C_t and C_E take 2.0, the curve can be drawn with schematic enhancement shapes, as seen in Fig. 15.14.

If static strength and static elastic modulus of some concrete are known, the dynamic tensile strength and dynamic elastic modulus in the case of certain strain rate can be easily obtained through Eqs (15.3) and (15.4), whereby avoiding considering such concrete details as the concrete humidity, gradation, and other factors affecting concrete rate effect. This kind of method is a very practical method adopted in the case of present-day research level.

As far as the physical sense of rate effect enhancement parameters in Eq. (15.4) are concerned, concrete damage delayed behavior discussions by Huaifa et al. (2008, 2010) can be referred to.

15.2.2 CONCRETE CONSTITUTIVE RELATION AND DAMAGE EVOLUTION MODEL

When the microdefects in concrete materials are evolving, materials will be bound to produce worsening in different degrees in such a way that there must exist some kind of relation between the macrodamage variable used to describe material damages and micro (meso) defects in materials. But it is

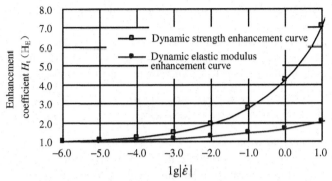

FIGURE 15.14 Relation Curves of Concrete Tensile Strength (Elastic Modulus) Enhancement Coefficient and Strain Rate

difficult to realize the connection of micro (meso) relation so that at the present, there has been no good approach. The effective stress concept in the existing damage models is adopted to realize this kind of connection.

The damages caused by the concrete inner defects (microfractures) can be qualitatively accounted for in a macroconcept. In the beginning stage of loading, microfractures are in the uniform distribution states and in every independent fracture surrounding, there is a stress-releasing zone, that is, damage and fracture zone. When loading continues, the stress-releasing zone can increase, and the independent fractures begin to connect with each other until they form macrofractures to cause local failure.

When the concrete test specimen is in the uniaxial tensile, the test specimen fracture is vertical to the loading direction, and along one main fracture tensile failure is formed, that is, along with the maximum tensile deformation direction, tensile fracture is caused. When uniaxial compression is carried out, the test specimen fractures appear to have many parallel fractures; that is, the openings develop along with the loading direction. This is mainly due to the fact that being subject to compression, freedom surfaces expand outward to produce the tensile deformation, from which it can be considered that concrete subject to force to produce deformation and damaged fracture is controlled by the tensile deformation.

Owing to concrete in tensile and compression stress state, there exist some differences in microfracture development behaviors so that damage variables are different. Stress states can be expressed by two parts of sphere stress and deviation stress tensile. When concrete occurs to have tension crack failure, the sphere stress component part is a positive value; sphere stress, together with deviation stress component, can cause expansions of microchaps. Vice versa, when concrete occurs to have shear pressure and crushing types of failure, sphere stress component is negative value, whereby making microchaps tend to close without damage occurrence, and the damage and fracture are mainly caused by deviation stress. In this process, the key is rationally to select damage and fracture evolution laws, and the parameter values are determined by tests. Concrete is a kind of sub-brittle material. After loading, stress–strain curves appear to have macrononlinearity that originate from the microchap germination and expansion and is not caused by plastic deformation. Owing to the damage and fracture within the inner concrete, actually nondamaged and fractured equivalent resistance volume definition being able to bear loading is V_a, the damaged and fractured zone volume is V_d, and total volume (or in name) is V. Obviously, in the process of loading, damaged zone V_d increases, while V_a decreases. Supposing that V_a

part is subordinated to the linear plastic constitutive relation, the corresponding stress is called effective stress $\bar{\sigma}_{ij}$.

The constitutive relation of isotropic plastic damage mechanics is adopted to describe the mechanics behavior of concrete materials. Scalar damage variable $d = V_d/V$ ($0 \leq d \leq 1$) is introduced, and again with $V = V_d + V_n$, we can obtain $1 - d = V_n/V$, whereby the effective stress can be define as follows:

$$\bar{\sigma}_{ij} = \sigma_{ij} / (1 - d)$$
$$\sigma_{ij} = (1 - d)\bar{\sigma}_{ij}$$

(15.5)

where σ_{ij} is stress in name (Cauchy stress).

In terms of Lamete strain equivalent principle, the damaged material name stress σ can be expressed through effective stress in the nondamage material ε; without considering damage effect upon Poisson's ratio, there will be the following:

$$E = E_0(1 - d)$$

(15.6)

where E_0 is the initial elastic modulus, and E is the elastic modulus after damage.

Considering when the unit maximum tensile stress σ_{max} reaches the given limit value f_t, this unit begins to occur to have tensile damage and fracture. In the case of uniaxial tensile state, the damage variables conform to double polygonal line evolution laws as indicated in Fig. 15.15 (Huaifa et al., 2004, 2005); there will be the following:

$$d = \begin{cases} 0 & \varepsilon_{max} < \varepsilon_0 \\ 1 - \dfrac{\eta - \lambda}{\eta - 1}\dfrac{\varepsilon_0}{\varepsilon_{max}} + \dfrac{1 - \lambda}{\eta - 1} & \varepsilon_0 < \varepsilon_{max} \leq \varepsilon_r \\ 1 - \lambda\dfrac{\varepsilon_0}{\varepsilon_{max}} & \varepsilon_r < \varepsilon_{max} \leq \varepsilon_u \\ 1 & \varepsilon_{max} < \varepsilon_u \end{cases}$$

(15.7)

In Eq. (15.7) and Fig. 15.15, f_t is the concrete material tensile strength; f_{tr} is the damaged unit tensile residual strength, $f_{tr} = \lambda f_t$ ($0 < \lambda \leq 1$), where λ is the residual strength coefficient; ε_0 is the unit stress reaching the tensile strength of the main tensile strain; ε_r is the strain corresponding to tensile residual strength, $\varepsilon_r = \eta \varepsilon_0$, of which η is the strain coefficient corresponding to the tensile residual strength; as

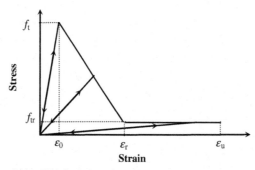

FIGURE 15.15 Double Polygonal Line Elastic Damage and Fracture Model

to concrete, it is $1 < \eta \leq 5$ in general. The limit tensile strain is $\varepsilon_u = \xi \varepsilon_0$ ($\xi > \eta$), where ξ is limit strain coefficient, ε_{max} is maximum value of tensile strain in the unit loading history. Under the complicated stress condition, the tensile strain can be replaced by the equal effect.

Tensile strain $\bar{\varepsilon}$, that is $\bar{\varepsilon} = \sqrt{\sum_{i=1}^{3} < \varepsilon_i >^2}$. Accordingly, the equal effect stress $\bar{\sigma} = \sqrt{\sum_{i=1}^{3} < \sigma_i >^2}$

is used to judge whether the tensile stress reaches the limit tensile strength, where $< \varepsilon_i > = \frac{1}{2}(\varepsilon_i + | \varepsilon_i |)$

and $< \sigma_i > = \frac{1}{2}(\sigma_i + | \sigma_i |)$, ε_i and σ_i are the main strain and the main stress, respectively.

15.2.3 VIRTUAL WORK EQUATION IN STATIC–DYNAMIC SYSTEM

15.2.3.1 Increment form of static balance equation

At t time, physical force f_i^t in V zone, loading $\bar{T}_i^{(t)}$ on the stress boundary S_u, is displacement. $\bar{u}_i^{(t)}$ on the displacement boundary S_u are known, and stress $\sigma_{ij}^{(t)}$ and $\varepsilon_{ij}^{(t)}$ are derived, $t + \Delta t$ time loading and displacement have one increment:

$$
\begin{aligned}
f_i^{(t+\Delta t)} &= f_i^{(t)} + \Delta f_i && \in V \\
\bar{T}_i^{(t+\Delta t)} &= \bar{T}_i^{(t)} + \Delta \bar{T}_i && \in S_\sigma \\
\bar{u}_i^{(t+\Delta t)} &= \bar{u}_i^{(t)} + \Delta \bar{u}_i && \in S_u
\end{aligned}
\tag{15.8}
$$

Hence, at $t + \Delta t$ displacement, strain and stress are as follows:

$$
\begin{aligned}
u_i^{(t+\Delta t)} &= u_i^{(t)} + \Delta u_i \\
\varepsilon_{ij}^{(t+\Delta t)} &= \varepsilon_{ij}^{(t)} + \Delta \varepsilon_{ij} \\
\sigma_{ij}^{(t+\Delta t)} &= \sigma_{ij}^{(t)} + \Delta \sigma_{ij}
\end{aligned}
\tag{15.9}
$$

1. $t + \Delta t$ time balance equation can be as follows:

$$
\sigma_{ij,j}^{(t+\Delta t)} + f_i^{(t+\Delta t)} = 0 \qquad \in V,
$$

This equation changes its forms as

$$
\sigma_{ij,j}^{(t)} + \Delta \sigma_{ij,j} + f_i^{(t)} + \Delta f_i = 0,
\tag{15.10}
$$

2. Strain and displacement relation can be as follows:

$$
\varepsilon_{ij}^{(t+\Delta t)} = \varepsilon_{ij}^{(t)} + \Delta \varepsilon_{ij} = \frac{1}{2}\left[u_{i,j}^{(t)} + u_{j,i}^{(t)} \right] + \frac{1}{2}\left[\Delta u_{i,j}^{(t)} + \Delta u_{j,i}^{(t)} \right] \qquad \in V
\tag{15.11}
$$

3. On the stress boundary S_σ, there will be as follows:

$$
\sigma_{ij}^{(t)} n_j = \bar{T}_i^{(t)}; \qquad \Delta \sigma_{ij}^{(t)} n_j = \Delta \bar{T}_i^{(t)}
\tag{15.12}
$$

where n_j is the three-direction cosine of boundary external normal line.

4. In the displacement boundary S_u, there will be as follows:

$$
u_i^{(t)} = \bar{u}_i^{(t)}, \qquad \Delta u_i^{(t)} = \Delta \bar{u}_i^{(t)}
\tag{15.13}
$$

5. Physical equation is as follows:

$$\Delta\sigma_{ij} = D_{ijkl}^{(\tau)}\Delta\varepsilon_{kl} \qquad (t \le \tau \le t + \Delta t) \tag{15.14}$$

$$[D_{ijlk}^{(\tau)}] = \frac{E_0[1 - d(\tau)]}{(1+v)(1-2v)}\begin{pmatrix} 1-v & v & v & & & \\ v & 1-v & v & & 0 & \\ v & v & 1-v & & & \\ & & & 1-2v & & \\ & 0 & & & 1-2v & \\ & & & & & 1-2v \end{pmatrix} \tag{15.15a}$$

Letting

$$[C_{ijlk}] = \begin{pmatrix} 1-v & v & v & & & \\ v & 1-v & v & & 0 & \\ v & v & 1-v & & & \\ & & & 1-2v & & \\ & 0 & & & 1-2v & \\ & & & & & 1-2v \end{pmatrix}$$

Therefore

$$[D_{ijlk}^{(\tau)}] = \frac{E_0[1 - d(\tau)]}{(1+v)(1-2v)}[C_{ijlk}] \tag{15.15b}$$

where v is Poisson's ratio.

15.2.3.2 Virtual work form of static balance

Virtual strain $\delta\Delta\varepsilon_{ij} = (1/2)\delta\left[\Delta u_{i,j}^{(t)} + \Delta u_{j,i}^{(t)}\right] \in V$, virtual displacement $\delta(\Delta u_i) = 0 \in S_u$.

In terms of the principle of virtual work, work done by the balance force system in corresponding to meeting the needs of virtual displacement of deformation coordinating conditions or virtual strain should satisfy the virtual work equation (Huaifa et al., 2005, 2007); there will be the following:

$$\iiint_V \sigma_{ij}^{(t)} \, \delta\Delta\varepsilon_{ij} \, dV - \iiint_V f_i^{(t)} \, \delta\Delta u_i \, dV - \iint_{S\sigma} \overline{T}_i^{(t)} \, \delta\Delta u_i \, dS = 0 \quad (t \text{ time}) \tag{15.16}$$

$$\iiint_V \sigma_{ij}^{(t+\Delta t)} \, \delta\Delta\varepsilon_{ij} \, dV - \iiint_V f_i^{(t+\Delta t)} \, \delta\Delta u_i \, dV - \iint_{S\sigma} \overline{T}_i^{(t+\Delta t)} \, \delta\Delta u_i \, dS = 0 \quad (t + \Delta t \text{ time}) \tag{15.17}$$

Equation (15.17) minus Eq. (15.16), we can obtain:

$$\iiint_V \left[\sigma_{ij}^{(t+\Delta t)} - \sigma_{ij}^{(t)}\right]\delta\Delta\varepsilon_{ij} \, dV - \iiint_V \Delta f_i \, \delta\Delta u_i \, dV - \iint_{S\sigma} \Delta\overline{T}_i \, \delta\Delta u_i \, dS = 0 \tag{15.18}$$

In terms of constitutive relation Eq. (15.14), we can obtain:

$$\begin{aligned} \sigma_{ij}^{(t+\Delta t)} &= D_{ijkl}^{(\tau)}\varepsilon_{kl}^{(t+\Delta t)} = \left[D_{ijkl}^{(t)} + \Delta D_{ijkl}^{(\tau)}\right]\varepsilon_{kl}^{(t+\Delta t)} = D_{ijkl}^{(t)}\varepsilon_{kl}^{(t+\Delta t)} + \Delta D_{ijkl}^{(\tau)}\varepsilon_{kl}^{(t+\Delta t)} \\ &= D_{ijkl}^{(t)}\varepsilon_{kl}^{(t)} + D_{ijkl}^{(t)}\Delta\varepsilon_{kl} + \Delta D_{ijkl}^{(\tau)}\varepsilon_{kl}^{(t+\Delta t)} \end{aligned} \tag{15.19}$$

$$\sigma_{ij}^{(t)} = D_{ijkl}^{(t)}\varepsilon_{kl}^{(t)} \tag{15.20}$$

Substituting Eqs (15.19) and (15.20) into Eq. (15.18), we can obtain:

$$\iiint_V D_{ijkl}^{(t)} \Delta\varepsilon_{kl}\, \delta\Delta\varepsilon_{ij}\, dV = \iiint_V \Delta f_i^{(t)}\, \delta\Delta u_i\, dV + \iint_{S\,\sigma} \Delta\bar{T}_i^{(t)}\, \delta\Delta u_i\, dS - \iiint_V \Delta D_{ijkl}^{(\tau)} \varepsilon_{kl}^{(t+\Delta t)}\, \delta\Delta\varepsilon_{ij}\, dV$$

The $\varepsilon_{kl}^{(t+\Delta t)}$ of the third item of the equal symbol to the right side of the preceding equation can be rewritten as the $\varepsilon_{kl}^{(\tau)}$, so as to express the middle process value in the iteration process. We can obtain:

$$\iiint_V D_{ijkl}^{(t)} \Delta\varepsilon_{kl}\, \delta\Delta\varepsilon_{ij}\, dV = \iiint_V \Delta f_i^{(t)}\, \delta\Delta u_i\, dV$$
$$+ \iint_{S\,\sigma} \Delta\bar{T}_i^{(t)}\, \delta\Delta u_i\, dS - \iiint_V \Delta D_{ijkl}^{(\tau)} \varepsilon_{kl}^{(\tau)}\, \delta\Delta\varepsilon_{ij}\, dV \tag{15.21}$$

$$\left[\Delta D_{ijlk}^{(\tau)}\right] = \frac{-E_0 \Delta d(\varepsilon)}{(1+v)(1-2v)}\left[C_{ijlk}\right] \tag{15.22}$$

Damage and fracture parameter increment:

$$\Delta d(\varepsilon) = d(\varepsilon_{t+\Delta t}^\tau) - d(\varepsilon_t) \tag{15.23}$$

15.2.4 PRESTATIC LOADING DYNAMIC EQUATION

15.2.4.1 Prestatic loading dynamic increment equation

After a certain static loading is applied to the system, the dynamic loading is applied. First of all, it is necessary to carry out static loading computation and to consider that $u_i^s = u_i^s(t=0)$ is the static displacement of applying dynamic prior to a time $(t=0)$, which will remain to have no change in the process of the postdynamic loading. Total displacement is $u_i^T = u_i^s(t=0) + u_i$, and the damage and fracture parameter is computed in terms of total displacement, being similar to static derivation.

1. In V zone, there is the following:

$$\sigma_{ij,j}^{(t+\Delta t)} + f_i^{(t+\Delta t)} + \sigma_{ij,j}^{s(t+\Delta t)} + f_i^s = \rho\ddot{u}_i^{(t+\Delta t)} + \mu\dot{u}_i^{(t+\Delta t)} \tag{15.24}$$

where ρ is quality density; μ is damping coefficient; $\ddot{u}_i$ and $\dot{u}_i$ are for dynamic displacement u_i second derivative and first derivative; that is the i direction and speed.

Particular attention should be given to the prestatic loading to produce the displacement (or strain) that remains to have no change. Owing to the material weakening produced by further loading the static stress bearing static displacement at the previous loading step will occur to have a shift, that is, $\sigma_{ij}^{s(t+\Delta t)}$ will change in the process of further loading, thereby being identified in relation with time.

As to the variable in $t + \Delta t$ and t time, the increment relation can be expressed as follows:

$$\sigma_{ij}^{(t+\Delta t)} = \sigma_{ij}^{(t)} + \Delta\sigma_{ij}; \quad f_i^{(t+\Delta t)} = f_i^{(t)} + \Delta f_i$$
$$\dot{u}_{ij}^{(t+\Delta t)} = \dot{u}_{ij}^{(t)} + \Delta\dot{u}_{ij}; \quad \ddot{u}_i^{(t+\Delta t)} = \ddot{u}_i^{(t)} + \Delta\ddot{u}_i$$

$$\varepsilon_{ij}^{(t)} = \frac{1}{2}\left[u_{i,j}^{(t)} + u_{j,i}^{(t)}\right]; \qquad \Delta\varepsilon_{ij} = \frac{1}{2}\left[\Delta u_{i,j} + \Delta u_{j,i}\right]$$

$$\varepsilon_{ij}^{(t+\Delta t)} = \varepsilon_{ij}^{(t)} + \Delta\varepsilon_{ij} = \frac{1}{2}\left[u_{i,j}^{(t)} + u_{j,i}^{(t)}\right] + \frac{1}{2}\left[\Delta u_{i,j} + \Delta u_{j,i}\right]$$

2. Stress, displacement boundary conditions and physical equation are the same as Eqs (15.12)–(15.14). But Eq. (15.14) elastic matrix is related with enhancement parameter $H_E(\dot{\varepsilon})$, and it can be expressed as follows:

$$\left[D_{ijlk}^{(\tau)} \right] = \frac{E_0 H_E(\dot{\varepsilon})[1 - D(\varepsilon)]}{(1+v)(1-2v)} \left[C_{ijlk} \right] \tag{15.25}$$

15.2.4.2 Virtual work of finite element increment equation

In terms of Galerkin method, the area weak integral solution form can be obtained through dynamic Eq. (15.24) and stress boundary conditions Eq. (15.12) at $t + \Delta t$ time (Huaifa et al., 2005, 2007):

$$\iiint_V [\sigma_{ij,j}^{(t+\Delta t)} + f_i^{(t+\Delta t)} + \sigma_{ij,j}^{s(t+\Delta t)} + f_i^s - \rho\ddot{u}_i^{(t+\Delta t)} - \mu\dot{u}_i^{(t+\Delta t)}]\delta\Delta u_i dV$$
$$- \iint_{S\sigma} [(\sigma_{ij}^{(t+\Delta t)} + \sigma_{ij}^{s(t+\Delta t)})n_j - (\overline{T}_i^{(t+\Delta t)} + \overline{T}_i^s)]\delta\Delta u_i\, dS = 0 \tag{15.26}$$

$\iiint_V [\sigma_{ij,j}^{(t+\Delta t)} + \sigma_{ij,j}^{s(t+\Delta t)}]\delta\Delta u_i dV$ in Eq. (15.26) can be deformed to obtain:

$$\iint_{S\sigma}\left[\sigma_{ij}^{(t+\Delta t)} + \sigma_{ij}^{s(t+\Delta t)}\right]\delta\Delta u_i n_j\, dS - \iiint_V\left[\sigma_{ij}^{(t+\Delta t)} + \sigma_{ij}^{s(t+\Delta t)}\right]\delta\Delta u_{i,j} dV.$$

Owing to $\sigma_{ij}\delta\Delta u_{i,j} = \sigma_{ij}\delta\Delta\varepsilon_{ij}$, there will be the following:

$$\iiint_V [\sigma_{ij,j}^{(t+\Delta t)} + \sigma_{ij,j}^{s(t+\Delta t)}]\delta\Delta u_i dV$$
$$= \iint_{S\sigma} [\sigma_{ij}^{(t+\Delta t)} + \sigma_{ij}^{s(t+\Delta t)}]\delta\Delta u_i n_j\, dS - \iiint_V [\sigma_{ij}^{(t+\Delta t)} + \sigma_{ij}^{s(t+\Delta t)}]\delta\Delta\varepsilon_{ij} dV$$

Substituting this equation into Eq. (15.26), we can obtain:

$$\iiint_V [\sigma_{ij}^{(t+\Delta t)}\delta\Delta\varepsilon_{ij} + \sigma_{ij}^{s(t+\Delta t)}\delta\Delta\varepsilon_{ij} + \rho\ddot{u}_i^{(t+\Delta t)}\delta\Delta u_i + \mu\dot{u}_i^{(t+\Delta t)}\delta\Delta u_i]dV$$
$$= \iiint_V [f_i^{(t+\Delta t)} + f_i^s]\delta\Delta u_i\, dV + \iint_{S\sigma} [\overline{T}_i^{(t+\Delta t)} + \overline{T}_i^s]\delta\Delta u_i\, dS \tag{15.27}$$

With the same reason, virtual work equation at t time is obtained as follows:

$$\iiint_V [\sigma_{ij}^{(t)}\delta\Delta\varepsilon_{ij} + \sigma_{ij}^{s(t)}\delta\Delta\varepsilon_{ij} + \rho\ddot{u}_i^{(t)}\delta\Delta u_i + \mu\dot{u}_i^{(t)}\delta\Delta u_i]dV$$
$$= \iiint_V [f_i^{(t)} + f_i^s]\delta\Delta u_i\, dV + \iint_{S\sigma} [\overline{T}_i^{(t)} + \overline{T}_i^s]\delta\Delta u_i\, dS \tag{15.28}$$

Equation (15.27) minus Eq. (15.28), we can obtain:

$$\iiint_V [(\sigma_{ij}^{(t+\Delta t)} - \sigma_{ij}^{(t)})\delta\Delta\varepsilon_{ij} + (\sigma_{ij}^{s(t+\Delta t)} - \sigma_{ij}^{s(t)})\delta\Delta\varepsilon_{ij} + \rho\Delta\ddot{u}_i^{(t)}\delta\Delta u_i + \mu\Delta\dot{u}_i^{(t)}\delta\Delta u_i]dV$$
$$= \iiint_V \Delta f_i^{(t)}\,\delta\Delta u_i\, dV + \iint_{S\sigma} \Delta\overline{T}_i^{(t)}\,\delta\Delta u_i\, dS \tag{15.29}$$

It can be derived from the physical equation:

$$\sigma_{ij}^{(t+\Delta t)} = D_{ijkl}^{(\tau)}\,\varepsilon_{kl}^{(t+\Delta t)} = [D_{ijkl}^{(t)} + \Delta D_{ijkl}^{(\tau)}]\varepsilon_{kl}^{(t+\Delta t)} = D_{ijkl}^{(t)}\,\varepsilon_{kl}^{(t)} + D_{ijkl}^{(t)}\,\Delta\varepsilon_{kl} + \Delta D_{ijkl}^{(\tau)}\,\varepsilon_{kl}^{(t+\Delta t)}$$

$$\sigma_{ij}^{(t)} = D_{ijkl}^{(t)}\,\varepsilon_{kl}^{(t)}$$

$$\sigma_{ij}^{s(t+\Delta t)} - \sigma_{ij}^{s(t)} = [D_{ijkl}^{s(t)} + \Delta D_{ijkl}^{s(\tau)}]\varepsilon_{kl}^{s} - D_{ijkl}^{s(t)}\,\varepsilon_{kl}^{(s)} = \Delta D_{ijkl}^{s(\tau)}\,\varepsilon_{kl}^{s}$$

Substituting the three preceding equations into Eq. (15.29), we can obtain:

$$\iiint_{V}[D_{ijkl}^{(t)}\,\Delta\varepsilon_{kl}\delta\Delta\varepsilon_{ij} + \rho\Delta\ddot{u}_{i}^{(t)}\delta\Delta u_{i} + \mu\Delta\dot{u}_{i}^{(t)}\delta\Delta u_{i}]\mathrm{d}V = \iiint_{V}\Delta f_{i}^{(t)}\,\delta\Delta u_{i}\,\mathrm{d}V$$
$$+ \iint_{S\sigma}\Delta\bar{T}_{i}^{(t)}\,\delta\Delta u_{i}\,\mathrm{d}S - \iiint_{V}\Delta D_{ijkl}^{(\tau)}\,\varepsilon_{kl}^{(t+\Delta t)}\delta\Delta\varepsilon_{ij}\mathrm{d}V - \iiint_{V}\Delta D_{ijkl}^{s(\tau)}\,\varepsilon_{kl}^{s}\delta\Delta\varepsilon_{ij}\mathrm{d}V \qquad (15.30)$$

In iteration process $\varepsilon_{kl}^{(t+\Delta t)}$ be $\varepsilon_{kl}^{(\tau)}$, Eq. (15.30) can be written as the following:

$$\iiint_{V}[D_{ijkl}^{(t)}\,\Delta\varepsilon_{kl}\delta\Delta\varepsilon_{ij} + \rho\Delta\ddot{u}_{i}^{(t)}\delta\Delta u_{i} + \mu\Delta\dot{u}_{i}^{(t)}\delta\Delta u_{i}]\mathrm{d}V = \iiint_{V}\Delta f_{i}^{(t)}\,\delta\Delta u_{i}\,\mathrm{d}V$$
$$+ \iint_{S\sigma}\Delta\bar{T}_{i}^{(t)}\,\delta\Delta u_{i}\,\mathrm{d}S - \iiint_{V}\Delta D_{ijkl}^{(\tau)}\,\varepsilon_{kl}^{(\tau)}\delta\Delta\varepsilon_{ij}\mathrm{d}V - \iiint_{V}\Delta D_{ijkl}^{s(\tau)}\,\varepsilon_{kl}^{s}\delta\Delta\varepsilon_{ij}\mathrm{d}V \qquad (15.31)$$

where

$$[\Delta D_{ijlk}^{(\tau)}] = \frac{-E_{0}H_{E}(\dot{\varepsilon})\Delta d(\varepsilon)}{(1+v)(1-2v)}[C_{ijlk}] \qquad (15.32)$$

$$[\Delta D_{ijlk}^{s(\tau)}] = \frac{-E_{0}\Delta d(\varepsilon)}{(1+v)(1-2v)}[C_{ijlk}] \qquad (15.33)$$

Damage and fracture parameter is as follows:

$$\Delta d(\varepsilon) = d(\varepsilon_{t+\Delta t}^{\tau}) - d(\varepsilon_{t}) \qquad (15.34)$$

15.3 **CONCRETE NUMERICAL SIMULATION TEST FEPG DOCUMENT**

The previous section has given out the description of the finite element equation virtual work form of concrete test specimen damage and fracture process under static and dynamic action. The concrete mesofinite element mechanics analysis problem is a kind of special nonlinear dynamic problem. The next step is to compile and write a VDE and NFE scenario document in accordance with the finite element grammar rules set in FEPG subject and in terms of weak solution form and corresponding steps. In addition, the concrete mesofinite element mechanics analysis problem is also a multiphysical field coupling problem, including displacement field, stress field, and damage variable field. Different physical fields are in correspondence with different partial differential equations. In order to reflect the coupling relations among various kinds of physical fields, it is necessary to have prepared a GCN document to describe the relations among physical fields of various kinds and organization computation flow and to realize GCN and PDE connection GIO (or MDI) documents, and functions of these documents are introduced next.

15.3.1 GIO COMMAND STREAM DOCUMENT

The project title is stdy, whose GIO document is named as stdy GIO; its filling content consists of two-paragraph information introduced in Chapter 8. The first paragraph information of GIO document gives out the PDE (or VDE) document name of each physical field after decoupling; the second paragraph information gives out the system unit type of information and coordinate system in the solution zone.

1. Stdy. GIO document:

stdya	\displacement field corresponding unit computation document is stdya. VDE
stdyb	\ stress field corresponding unit The computation document is stdyb. VDE.
	\null one line
#elemtype q4g2	\unit type,4-node quadrangle unit, The second order Gauss's integral
2dxy z	\2-dimensional right angel coordinate system

2. stdy. MDI document:

2dxy y	
#a 3 2 u v	\physical field name is a;3 initial values, 2 unknown quantities
VDE stdya q4g2 1q4g2	\from stdya.PDE generating aeq4g2.for and aelq4g2.for
# b 0 4 sxx syy sxy did	\physical field name is b; O initial value; 4 unknown quantities.
VDE stdyb q4g2 1q4g2	\from stdyb.PDE generating;beq4g2.for and I bleq4g2.for

15.3.2 GCN COMMAND DOCUMENT

The GCN document gives out the coupling patterns among every physical field, signal field problem, algorithms, and solution flow. GCN document name is the same with GIO document name, and with GCN as the document expansion name, that is, the problem GCN document name is stdy GCN. The filling contents consist of a two-paragraph information: the first paragraph information is the algorithm paragraph, giving out each physical field algorithm after decouplings; the second paragraph is command stream, giving out the multifield coupling computation flow.

Defi	\key words
a stdya &	\a field solution displacement,
	corresponding a field algorithm document
	\Is stdy. NFE; compressing generated document
	without rename. Output document is unoda
	or else is unoda
b stdyb a	\b field solution stress coupling with a field.
	Corresponding b field document is
	Stdyb NEF ,output document is unodb.
	\null one line expressing definition end
	The following via command steam to organize
	computation stream.
startilu a	\corresponding pre-processing conjugate gradient
	method without storage single stiff manner
	displacement field initial
if exist stop del stop	
:1	
bft a	\side value computation
if exist end del end	
:2	
solvilu a	\pre-treatment conjugate gradient method
	solution displacement
if not exist end goto 2	\control non-linear iteration
stress b	\least square method solution stress field
post a	\in terms of time step output result
if not exist stop goto 1	\control time cycle

15.3.3 **NFE ALGORITHMS DOCUMENT**

NFE algorithm document is used in the GCN system to generate unit computation algorithm E element program and postprocessing computation U element program. It mainly consist of defi, coef, equation, and solution information paragraph.

15.3.3.1 *Displacement algorithm stdya NFE document*

In terms of dynamic, static mixture solution steps (Huaifa et al., 2008), FEPG finite element language is used to compile and write the following solution displacement:

Defi	\key words
stif s	\stiffness matrix s
mass m	\mass matrix m
damp c	\damping matrix c
load f	\load vector f
type w	\wave motion equation
mdty l	\mass matrix adopting concentrated form

step 0   \non-existing previous time unit stiffness matrix

                                    null one line

                                    \expressing information section end

coef u1 du us   \transfer variables to unit computation:u1 for the

                                    upper time total displacement , u^t du is this time k

                                    iteration step displacement increment $\Delta u_k^{t+\Delta t}$.

\null one line   \us is static displacement u^s vector

Equation   \forming algebra equation group

vect u1,v1,w1,du,us   \define vector u^t ,velocity $\dot{u}^t$,acceleration

                                    $\ddot{u}^t, \Delta u_k^{t+\Delta t}, u^s$

read(s,unod) u1,v1,w1,us,du   \read-in the upper iteration result: $u^t, \dot{u}^t,$

                                    $\ddot{u}^t, u^s, \Delta u_k^{t+\Delta t}$

matrix = [s]+[m]*a0+[m]*a1*ealph+[s]*a1*ebeta   \Compute generalized stiffness matrix

$$[K]=[s]+a_0[m]+a_1[c]$$

\Null one line

orc=[f]+[m]*[v1]*a2+[m]*[w1]*a3

   +[m]*[v1]*a4*ealph+[s]*[v1]*a4*ebeta

   +[m]*[w1]*a5*ealph+[s]*[w1]*a5*ebeta   \Compute generalized loading

\Null one line

solution u   \Solution displacement

vect u, v, w, u1, v1, w1, du, ue, us   \define vector $\Delta u_{k+1}^{t+\Delta t}, \dot{u}^{t+\Delta t}, \ddot{u}^{t+\Delta t}, u^t,$

                                    $\dot{u}^t, \ddot{u}^t, \Delta u_k^{t+\Delta t}$,ue, u^s ,u2

$c6 open（99, file='elem.it', form='formatted', status='unknown')

$c6 read（99, *） itn

Read (s,unod) u1,v1,w1,us,du   \Read-in the upper iteration step

                                    $u^t, \dot{u}^t, \ddot{u}^t, u^s, \Delta u_k^{t+\Delta t}$

[ue]=[u]-[du]   \Find iteration error

                                    $\Delta u_{k+1}^{t+\Delta t} - \Delta u_k^{t+\Delta t}$

$c6 err=0.0

%nod

%dof

err=err+[ue]**2

\Find error $\Delta u^{t+\Delta t}_{k+1} - \Delta u^{t+\Delta t}_{k}$
range number

%dof

\To node every freedom cycle

%nod

\To node cycle

$c6 if ((err.lt.ferrd) .or. (itn.ge.itfd)) then

\if reach displacement convergence criterion or surpass the maximum time ,end iteration , to obtain t+Δ t time results.

$c6 open (11, file='end', status='new')
$c6 close (11)

[u1]=[u1]+[u]

\$t + \Delta t$ time total displacement

$$u^{t+\Delta t} = u^t + \Delta u^{t+\Delta t}$$

[w]=[w1]+[u]*a0-[v1]*a2-[w1]*a3

\ $\ddot{u}^{t+\Delta t} = \ddot{u}^t + a_0 \, \Delta u^{t+\Delta t} - a_2 \, \dot{u}^t$

$- a_3 \, \ddot{u}^t$

[w]=[w1]+[u]*a0-[v1]*a2-[w1]*a3   \$\ddot{u}^{t+\Delta t} = \ddot{u}^t + a_0 \, \Delta u^{t+\Delta t} - a_2 \, \dot{u}^t - a_3 \, \ddot{u}^t$

[v]=[v1]+[u]*a1-[v1]*a4-[w1]*a5   \$\dot{u}^{t+\Delta t} = \dot{u}^t + a_1 \, \Delta u^{t+\Delta t} - a_4 \, \dot{u}^t - a_5 \, \ddot{u}^t$

write (s, unod) u1, v, w, us, u   \$u^{t+\Delta t}, \ \dot{u}^{t+\Delta t}, \ \ddot{u}^{t+\Delta t}, \ u^s, \ \Delta u_0^{(t+\Delta t)+\Delta t}$

$c6 itn=1   \Iteration time itn endowed with initial value 1
$c6 else

write (s, unod) u1, v1, w1, us, u   \$u^t, \ \dot{u}^t, \ \ddot{u}^t, \ u^s, \ \Delta u_{k+1}^{(t+\Delta t)}$

$c6 itn=itn+1
$c6 end if
$c6 if (time.le.ectrltime) then

\ ectrltime is static to dynamic computation conversion control time parameter

[us]=[u1]

\renew static displacement until starting dynamic computation

$c6 end if
$c6 rewind (99)
$c6 write (99, *) itn

\Keeping current iteration times

$c6 close (99)

\Null one line

Fortran

\Inserting fortran program

@begin

\program begins to join public block, to be used in parallel computation

```
 common/ctrparam/ectrltime, eign1, eign2, ezit1, ezit2, ferrd, itfd
@subfort
 open (77, file='pinput.txt', form='formatted', status='old')
 read (77, *)  eblen, ebwid, ebhei, eign1, eign2, ezit1, ezit2
 read (77, *)  ebfor, ebfor1, ectrltime
 read (77, *)  ebcon, ferrd, itfd
 read (77, *)  npfs
 close(77)                    \read-in parameter, to be deleted in parallel computation, to change
                                transfer by bfts public block read-in
a0=0.0                       \static computation
a1=0.0
a2=0.0
a3=0.0
a4=0.0
a5=0.0
Else                         \domestic computation
Delta =0.5                   \ Newmark   method interpolating coefficient
aa=0.25* (0.5+ delta) **2
a0=1./ (dt*dt*aa)
a1= delta / (aa*dt)
a2=1./ (aa*dt)
a3=1./ (2.0*aa)
a4= delta /aa
a5=dt/2.0* (delta /aa-2.0)
ealph=2.0* (ezit1*eign2-ezit2*eign1) *eign1*eign2   \computing damping coefficient
ealph=ealph/ (eign2*eign2-eign1*eign1)
ebeta=2.0* (ezit1*eign1-ezit2*eign2)
ebeta = ebeta/ (eign1*eign1-eign2*eign2)
 end if
\                           null one line
```

15.3.3.2 Stress algorithm stdyb NEF document

After obtaining displacement, least squares method is used to find stress. For this, it is necessary to define function $F(\sigma) = \iiint_V [\sigma - \sigma_0(u)]^2 dV$, where $\sigma_0(u)$ expresses stress obtained from known deformation, σ finds stress after smoothness. Take $F(a)$ extreme value in terms of least squares method. Letting $\delta F(a) = 0$, there will be:

$$\iiint_V \sigma \, \delta\sigma \, dV = \iiint_V \sigma_0(u) \delta\sigma \, dV$$

(15.35)

The stress algorithm is:

```
defi                \key word
stif s              \stiffness matrix s
mass m              \mass matrix m
load f              \loading vector f
type e              \oval type equation
mdty l              \l or d, 1 is centered mass matrix. d is distribution matrix
step 0              \non-existing previous time unit stiffness matrix.
                    \null one line, express information section end


coef 1a 2a          \to define coupling variable, transfer non-linear equation coefficient. Here
                    is the upper step displacement component, corresponding to VDE or PDE
                    document
                    \null one line
equation
var 1a 2a 3a 4a 5a 6a 7a 8a 9a 10a        \ to define sealer variable
read (s, unod) 1a 2a 3a 4a 5a 6a 7a 8a 9a 10a      \Read-in unod document
matrix = [s]              \stiffness matrix. There is zero
l,  mass = [m]            \l indicates centered muss matrix
forc=[f]                 \equation right end item
                         \null one line
solution w               \to find node stress and damage parameter after smoothness
vect w                   \to define vector
write (s,unodb) w        \to write b field solution in unodo document
                         \null one line

fort
@kdgof
       kdgof=4           \freedom degree in agreement with one solution end variable
                         in stress computation VDE document
                         \null one line
End
```

15.3.4 SCENARIO DOCUMENT OF DIFFERENT EQUATIONS

In accordance with FEPG grammar rules, the differential equation expression formula is described with the VDE document. This document can generate stiffness matrix, unit mass matrix, unit damping matrix, unit loading vector, unit cub-programs, and so on. In an isoparametric element, the unit stiffness matrix, mass matrix, damping matrix, and unit loading vector are obtained via Gauss' integration integral. Therefore, in the unit VDE document, to find the elastic modulus, damage and fracture parameters on Gauss's integral nodes is the key to compile and write this document.

15.3.4.1 Unit displacement computation VDE document

Concrete test specimen dynamic numerical solution can be realized using two physical fields, one of which is the displacement field. The plane problem has two variables to satisfy the virtual work equation given out by Eq. (15.13), from which VDE document is compiled and written, giving the name stdya VDE.

```
defi                            \the following only labeled in chapter 8 without the advent of statement
disp u v                        \unknown displacement function name
vect u u v
coor x y
coef ut vt dut dvt us vs        \non-linear coefficient corresponding to displacement algorithm
styda.NFE
func exx eyy exy                \to define stain function
vect ev exx eyy
vect ep exy
 vect evt exxt eyyt
vect ept exyt
vect devt dexxt deyyt
vect dept dexyt
vect evs exxs eyys
vect eps exys
vect fv fxx fyy
vect fp fxy
vect dfv dfxx dfyy
vect dfp dfxy
shap 1% 2%
gaus 3%
$c6 common/c1/epsm (1000000),eps0(1000000), gpstr(6,1000000)   \FORTRAN public block
$c6 common/c2/enhance (2000000),epstm(2000000)
$c6 common/c3/dmgd(2000000),dmgd0(2000000)
$c6 common/ctrparam/ ectrltime, eign1, eign2,ezit1,ezit2,ferrd,itfd
                               \transfer control parameter
array etn(3),etnm(3),ets(3),etsm(3)              \Definition number group
mate pe ft pv fu fv rou yita elamda ekesi at bt ct ae be ce act bct cct ace bce cce\
3.2D10 4.0D6 0.167 0.0 -2.4D4 2.4D3 3.0 0.1 10.0 0.20 6.0 2.0 0.17 6.0 2.0
                                                        \material parameter
mass q rou
damp q 1.0
vect f fu fv
matrix sm 2 2
1.0   pv
pv    1.0

$c6 fact=pe/(1.+pv)/(1.-pv)              \ $ C6 indicating 6 null after insert
                                FORTRAN program
$c6 shear=(1.-pv)/2.0
$c6 if (time.gt.1.5*dt) fv=0.0

Func                    \noticing FEPG specifying Gauss node cycle of    this type unit after
                                Func or stif calculation

exx=+[u/x]

eyy=+[v/y]
```

exy=+[u/y]+[v/x]

stif
$c6 nengs=nelem*ngaus
$c6 if ((time.lt.1.5*dt).and.(num.eq.1).and.(igaus.eq.1)) then
\correlative variables are endowed with initial values.
$c6   do i=1,nengs
$c6   epsm(i)=0.0d0
$c6   eps0(i)=0.0d0
$c6   epstm(i)=0.0d0
$c6   epstm(nengs+i)=0.0d0
$c6   enhance (i)=1.0d0
$c6   enhance (nengs+i)=1.0d0
$c6   dmgd(i)=1.0d0
$c6   dmgd(nengs+i)=1.0d0
$c6   dmgd0(i)=1.0d0
$c6   dmgd0 (nengs+i)=1.0d0
$c6   do j=1,3
$c6   gpstr(j,i)=0.0d0
$c6   end do
$c6   end do
$c6 end if
$c6 factd0=dmgd((num-1)*ngaus+igaus)
$c6 facttb=dmgd(nengs+(num-1)*ngaus+igaus)
$c6 if (time .le.ectrltime) then
$c6   exxs=0.0
$c6   eyys=0.0
$c6   exys=0.0
$c6 else
$cv   exxs=+{us/x}   \ $ CV expresses insertion including known function upon
                              the derivative of coordinate variable, at this time, brace
                              "{}"is used to express,
$cv eyys=+{vs/y}   \ but not PORTRAN language source program to compute
                              Strain   .E.q.{vs/y}indicates   to   find   known   quantity

                              corresponding to $\dfrac{\partial v_s}{\partial y}$

$cv exys=+{us/y}+{vs/x}
$c6 end if
$c6 eht=enhance((num-1)*ngaus+igaus)
$c6 ehe=enhance(nengs+(num-1)*ngaus+igaus)
$c6 ftd=ft*eht
$cv exxt=+{ut/x}
$cv eyyt=+{vt/y}

```
$cv exyt=+{ut/y}+{vt/x}
$cv dexxt=+{dut/x}
$cv deyyt=+{dvt/y}
$cv dexyt=+{dut/y}+{dvt/x}
$c6 factd=dmgd0((num-1)*ngaus+igaus)
$c6 if (time.le.ectrltime) then
$cv   dfv_i=+factd*sm_i_j*devt_j*fact+(factd-factd0)*sm_i_j*evt_j*fact
$cv   dfp_i=+factd*shear*dept_i*fact+(factd-factd0)*shear*ept_i*fact
$c6 else
$c6   fdh0=factd0*ehe
$c6   fdh=factd*ehe
$cv   dfv_i=+fdh*sm_i_j*devt_j*fact+(fdh-fdh0)*sm_i_j*(evt_j-evs_j)*fact\
                              \compute stress component increment
          +(factd0-facttb)*sm_i_j*evs_j*fact
$cv dfp_i=+fdh*shear*dept_ i*fact+(fdh-fdh0)*shear*(ept_ i-eps_ i)*fact\   \"\"indicating
                              ongoing line
          +(factd0-facttb)*shear*eps_i*fact
$c6 end if
$cv ets_i=dfv_i
$c6 ets(3)=dfxy
$c6 do j=1,3
$c6    gpstr(j+3,(num-1)*ngaus+igaus)=ets(j)
$c6 end do
$c6 do j=1,3
$c6    ets(j)=gpstr(j,(num-1)*ngaus+igaus)
$cc          +gpstr(j+3,(num-1)*ngaus+igaus)
$c6 end do
$cv etn(1)=+{ut/x}+{dut/x}
$cv etn(2)=+{vt/y}+{dvt/y}
$cv etn(3)=+{ut/y}*0.5+{vt/x}*0.5+{dut/y}*0.5+{dvt/x}*0.5
$c6 call mstress3(3,etn,etnm)                       \computing main stress

$c6    eigstrn=0.5*sqrt((abs(etnm(2))+etnm(2))**2
$cc            +(abs(etnm(3))+etnm(3))**2)
$c6 if (eigstrn.gt.epsm((num-1)*ngaus+igaus)) then
$c6   epsm((num-1)*ngaus+igaus)=eigstrn
$c6 end if
$c6 if(eps0((num-1)*ngaus+igaus).lt.1.0d-15) then
$c6    call mstress3 (3,ets, etsm)
$c6    eigstrs=0.5*sqrt((abs(etsm(2))+etsm(2))**2
$cc            + (abs (etsm(3))+etsm(3))**2)
$c6    if(eigstrs.gt.ftd) then                \judging where new unit are damaged or not
$c6       eps0 ((num-1)*ngaus+igaus)=eigstrn*ftd/eigstrs
$c6    end if
```

```
$c6 end if
$c6 if (eps0((num-1)*ngaus+igaus).ge.1.0d-15) then
$c6     epn0=eps0((num-1)*ngaus+igaus)
$c6     epnm=epsm((num-1)*ngaus+igaus)
$c6     epnr=epn0*yita
$c6     epnu=epn0*ekesi
$c6     if(epnm.le.epn0) then
$c6         edd=0.0
$c6     elseif((epnm.gt.epn0).and.(epnm.le.epnr)) then     \renew damaged variables
$c6         edd=1.0-epn0*(yita-elamda)/(yita-1.0)/epnm
$c6         edd=edd+(1.0-elamda)/(yita-1.0)
$c6     elseif ((epnm.gt.epnr).and.(epnm.le.epnu)) then
$c6         edd=1.0-elamda*epn0/epnm
$c6     elseif (epnm.gt.epnu) then
$c6         edd=1.0-1.0D-5
$c6     end if
$c6     factd=(1.0-edd)
$c6 else
$c6     factd=1.0d0
$c6 end if
$c6 if (time. le. ectrltime) then
$c6     etfcts=1.0
$c6     etfctd=0.0
$c6 else
$c6 etfcts=0.0
$c6 etfctd=1.0
$c6 end if
$c6 dmgd0((num-1)*ngaus+igaus)=factd
$c6 dmgd0(nengs+(num-1)*ngaus+igaus)=factd0
$cv exxt=+{ut/x}+{dut/x}
$cv eyyt=+{vt/y}+{dvt/y}
$cv exyt=+{ut/y}+{vt/x}+{dut/y}+{dvt/x}
$c6 eff1=factd0*ehe*fact
$c6 eff2=etfcts*fact
$c6 eff3=etfctd*fact
dist=+[ev_i;ev_j]*sm_i_j*eff1+[ep_i;ep_i]*shear*eff1     \computing stiffness matrix
load=+[u_i]*f_i                                          \computing loading increment
    -[ev_i]*evt_j*sm_i_j*(factd-factd0)*eff2
    -[ep_i]*ept_i*shear*(factd-factd0)*eff2
    -[ev_i]*(evt_j-evs_j)*sm_i_j*(factd-factd0)*ehe*eff3
    -[ep_i]*(ept_i-eps_i)*shear*(factd-factd0)*ehe*eff3
    -[ev_i]*evs_j*sm_i_j*(factd0-facttb)*eff3
    -[ep_i]*eps_i*shear*(factd0-facttb)*eff3

End
```

15.3.4.2 Unit stress computation VDE document

After obtaining the displacement field, stress distribution and damage variables can be obtained by the displacement field through physical equations and damage–fracture evolution relations. VDE document takes the name of stdyb VDE.

```
defi
disp sxx,syy,sxy,did            \unknown stress component and damage function name
coor x,y
coef ut vt                      \transfer the precious step displacement component
hap 1% 2%
gaus 3%
mass q 1.0                           \to define unit mass matrix to be decided by unit
                                      shape .1.0 are coefficient corresponding to equation
```

$$\text{left end item } \int_V \sigma\delta\sigma dV$$

```
$c6 common/c1/epsm(1000000),eps0(1000000),gpstr(6,1000000)
$c6 common/c2/enhance(2000000),epstm(2000000)
$c6 common/c3/dmgd(2000000),dmgd0(2000000)
$c6 common/ctrparam/ectrltime, eign1, eign2,ezit1, ezit2,ferrd,itfd
array etn(3),etnm(3),ets(3),etsm(3)
mate pe ft pv fu fv rou yita elamda ekesi at bt ct ae be ce act bct cct ace bce cce \
3.2D10 4.0D6 0.167 0.0 -2.4D4 2.4D3 3.0 0.1 10.0 0.20 6.0 2.0 0.17 6.0 2.0
load fxx fyy fxy eid                 \to define unit load matrix corresponding to equation
```

$$\text{right item } \iiint_V \sigma_0 (\varepsilon)\delta\sigma dV$$

```
vect fv fxx fyy fxy


stif
$c6 nengs=nelem*ngaus
$c6 if ((num.eq.1).and.(igaus.eq.1)) then         \correlative variables are endowed with
                                                    initial values.
$c6    do i=1,nengs*2
$c6    dmgd(i)=dmgd0(i)
$c6    end do
$c6 end if
$cv etn(1)=+{ut/x}
$cv etn(2)=+{vt/y}
$cv etn(3)=+{ut/y}*0.5+{vt/x}*0.5
$c6 call mstress3 (3,etn,etnm)
$c6    eigstrnt=0.5*sqrt((abs(etnm(2))+etnm(2))**2   \computing equal effect tensile strain.
$cc +(abs(etnm(3))+etnm(3))**2)
```

```
$c6  eigstrnc=0.5*sqrt((abs(etnm(2))-etnm(2))**2        \computing equal  effect compression
                                                            strain
$cc                +(abs(etnm(3))-etnm(3))**2)
$c6 if (time.le.ectrltime) then
$c6     eht=1.0
$c6     ehe=1.0
$c6 else
$c6     do j=1,3
$c6        ets(j)=gpstr(j,(num-1)*ngaus+igaus)
$c6     end do
$c6     call mstress3 (3,ets, etsm)
$c6     if (etsm(1).gt. 0.0D0) then          \computing strength enhancement coefficient.
$c6        elgstn=abs(eigstrnt-epstm((num-1)*ngaus+igaus))/dt
$c6        eth=LOG10(elgstn)
$c6        if(eth.gt.-bt) then
$c6           edt=(at*(eth+bt))**ct
$c6           eht=exp(edt)
$c6        else
$c6           eht=1.0
$c6        end if
$c6        if (eth.gt.-be) then    \computing elastic modulus enhancement coefficient
$c6           ede=(ae*(eth+be))**ce
$c6           ehe=exp(ede)
$c6        else
$c6           ehe=1.0
$c6        end if
$c6     else
$c6        elgstn=abs (eigstrnc-epstm(nengs+(num-1)*ngaus+igaus))/dt
$c6        eth=LOG10 (elgstn)
$c6        if (eth.gt.-bct) then
$c6           edt=(act*(eth+bct))**cct
$c6           eht=exp(edt)
$c6        else
$c6           eht=1.0
$c6        end if
$c6        if (eth.gt.-bce) then
$c6           ede=(ace*(eth+bce))**cce
$c6           ehe=exp(ede)
$c6        else
$c6           ehe=1.0
$c6        end if
$c6     end if
$c6 end if
$c6 epstm((num-1)*ngaus+igaus)=eigstrnt
```

```
$c6 epstm (nengs+(num-1)*ngaus+igaus)=eigstrnc
$c6 enhance((num-1)*ngaus+igaus)=eht
$c6 enhance(nengs+(num-1)*ngaus+igaus)=ehe
$c6 if (eps0 ((num-1)*ngaus + igaus).gt.1.0d-15) then
$c6    epn0=eps0((num-1)*ngaus+igaus)
$c6    epnm=epsm((num-1)*ngaus+igaus)
$c6    epnr=epn0*yita
$c6    epnu=epn0*ekesi
$c6     if (epnm.le.epn0) then
$c6     eid=0.0
$c6     elseif ((epnm.gt.epn0).and.(epnm.le.epnr)) then
$c6     eid=1.0-epn0*(yita-elamda)/(yita-1.0)/epnm
$c6     eid=eid+(1.0-elamda)/(yita-1.0)
$c6     elseif ((epnm.gt.epnr).and.(epnm.le.epnu)) then
$c6     eid=1.0-elamda*epn0/epnm
$c6     elseif (epnm.gt.epnu) then
$c6     eid=1.0
$c6     end if
$c6 else
$c6     eid=0.0
$c6 end if
$c6 do j=1,3
$c6     ets(j)=gpstr(j,(num-1)*ngaus+igaus)
$cc            +gpstr(j+3,(num-1)*ngaus+igaus)
$c6     gpstr(j,(num-1)*ngaus+igaus)=ets(j)
$c6 end do
$cv fv_i=ets_i
dist=+[sxx;sxx]*0.0    \to define unit stiffness   matrix the left end item in equation takes
                            centered mass matrix form
                       \Null one line , express by mass line but not adopting rigidity matrix
                            so that rigidity item here is zero.

End
```

Based on the mentioned VDE document and algorithms (NFE, GCN, and GIO), the FEPG system is used to generate the finite element series source code program, and then, with an aim at concrete problems, and preproc operation entering into FEPG, GID can do preparatory work for processing. After having the data ready, the finite element computation can be carried out, and finally enters the postprocessing of computation results to analyze the computation results. In the following, the finite element program generated based on FEPG can be introduced, and the numerical simulation tests for dynamic flexural tensile failure process of concrete test specimens.

15.4 STRAIN RATE EFFECTS UPON CONCRETE DAM DYNAMIC FLEXURAL STRENGTH

15.4.1 NUMERICAL COMPUTATION MODEL

Material damage evolution can make the mechanics indexes of its strength, elastic modulus, etc., decrease. Loading with different strain rates can form the different mesobehavior damage process, whereby rendering material macromechanics behavior index to display deterioration or worseness to some extent, and making materials have different mechanics properties under the different strain rates. That is, there exist correlative relations between material damage evolution behaviors and strain rates, while material inner damage strain rate correlation properties are just the inherent causes of strain rate effects (Yon et al., 1992). Starting from concrete material basic strain rate enhancement relation formulas (15.3) and (15.4), and based on concrete test specimen three-element point breakage resistance tests, we can carry out the numerical analysis of concrete dynamic load flexural tensile destructive mechanism and probe into the accumulative relations among concrete dynamic strength, strain rate effects, and damage and fracture.

15.4.1.1 Parameter taking values

In order to probe into the general laws of concrete rate effect upon its dynamic flexural tensile strength, as well as the cause of improvement of concrete test specimen flexural resistance capacity under a certain prestatic loading action, and in the following computation, the considerations are, for the time being, not given to difference of rate effects produced by the tensile, compression, and strain rate upon concrete material tensile strength and elastic modulus, and it is considered that when strain rate is less than 10^{-6} s^{-1}, concrete material tensile strength and elastic modulus will not produce rate effects, that is, the enhancement parameters B_t and B_E in Eq. (15.4) are equal to σ, A_E and A_t take 0.12 and 0.20, respectively, and the indexes C_t and C_E can take 2.0.

Concrete material elastic modulus is 28.00 GPa, Poisson's ratio is 0.2, and tensile strength is 2.80 MPa, and the bulk density is 24.00 kN/m^3.

In the following computation, the model parameter in Eq. (15.7) λ takes 0.1, η takes 3, and ξ takes 10; when dynamic computation is carried out, concrete tensile strength, elastic modulus, and the unit strain rate correlations are computed in terms of Eqs (15.3) and (15.4), respectively.

15.4.1.2 Computation model

The test specimen magnitude and unit profile separation are shown in Fig. 15.16, and there are 2745 nodes and 5350 units in total. Static loading plus loading step length can take 0.25 kN, and static strength f_t obtained via computation is 3.20 MPa, and corresponding static limit loading is 210.400 kN. In the process of dynamic loading computation, time step length is 0.001 s. In the following stress–displacement curves, the node name strength in the lower verge of concrete test specimen ($\sigma = PL/bh^2$, P is loading; L is support gap distance; b is the thickness of the test specimen; h is the height of the test specimen.) corresponds to the nodes of vertical displacement of the test specimen in the upper verge. Abscissa is the nondimensional displacement after the vertical displacement v on the test specimen verge midpoint divided by the test specimen height. In the process of computation, the loading is increased gradually in terms of specified loading step length. The solution to node displacement and stress are found in accordance with the incremental unit; the test specimen occurs to have a large deformation. When the specimen is out of stability, the end of stress–displacement process is close to

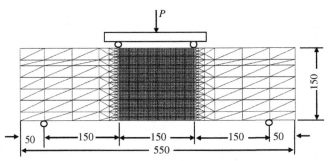

FIGURE 15.16 Schematic Diagram of Flexural Tensile Test and Unit Profile Separation (unit: mm)

the horizontal line, so that the test specimen limit stress taken will be the stress–displacement curves end peak value.

15.4.2 NUMERICAL COMPUTATION ANALYSIS

15.4.2.1 Effects of loading rate upon dynamic flexural strength

Taking elastic modulus enhancement parameter A_E of 0.12 and strength enhancement parameter A_t of 0.20 for example, loading rate taking 168, 200, 250, 300 kN/s to carry out computation to obtain the stress–displacement curves as shown in Fig. 15.17. Dynamic strength corresponding to the mentioned loading rates are 3.92, 4.03, 4.23, and 4.36 MPa, respectively. With an increase in loading rate, dynamic strength increase and dynamic elastic modulus also increase slightly.

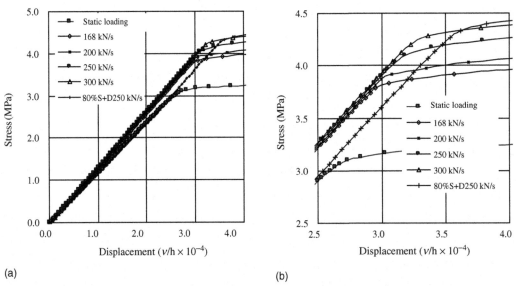

FIGURE 15.17 Effects of Different Loading Rates Upon Dynamic Elastic Modulus and Dynamic Strength

(a) Stress–displacement curve, (b) stress–displacement local magnified curves.

Research on damage mechanics theory and related testing observations indicate that owing to static loading acting for a long time, the inner damages in materials mainly display microchap expansion, linkage and running through, and then leading to failure, while in the short duration, and under the action of high-speed impacting load the inner damages in materials mainly distance a large bulk of microhole and microchap generations or produces, but they have not enough time to expand or extend. When there are microholes and microchaps spreading over a certain cross-section, this cross-section can completely break or repute so that the material dynamic loading strength is, in general, higher than the static loading strength, and with an increase in loading rate, the loading strength increases.

Testing research has found that applying dynamic loading in terms of a certain loading rate can obtain the different dynamic flexural tensile strength (static–dynamic comprehensive flexural tensile strength) in the case of different prestatic loading level. Under a certain static loading level, the static–dynamic comprehensive flexural tensile strength is higher than the dynamic flexural tensile strength without initial static loading. As shown in Fig. 15.17, after adding the prestatic loading to 80% of static limit loading, and reapplying the impacting load with the rate of 250 kN/s, the static–dynamic comprehensive flexural tensile strength has surpassed the dynamic loading when the loading rate of pure dynamic loading is 300 kN/s. This kind of physical phenomena illustrate that the dynamic flexural tensile strength is related with not only loading rate but also with static loading.

In the following, it is worth discussing that under the conditions of different preloading the concrete material strength and elastic modulus enhancement parameters can affect test specimen flexural tensile strength and probing into the accumulation relation among concrete dynamic strength and strain rate effect as well as damages and fractures. The loading rate and dynamic load can take 168 kN/s.

15.4.2.2 Effect of initial prestatic loading upon dynamic flexural tensile strength

Elastic modulus enhancement parameter A_E takes 0.12, and strength enhancement parameter A_t takes 0.20, thus, pure dynamic loading can be obtained via computation.

Applying prestatic loading to static limit loading of 20, 40, 60, and 80%, the test specimen stress–displacement curves are shown in Fig. 15.18. It can be seen from Fig. 15.18 that pure dynamic loading and several prestatic loading stress–displacement curves are close to parallel before they are out of stability and failure; that is, dynamic elastic modulus differences are not great but they are all larger than the static modulus. Figure 15.18 gives out pure dynamic loading and several prestatic loading flexural tensile strength. The relative values of pure dynamic flexural tensile strength is 122.5%, and the relative values of the prestatic loading level are 20, 40, 60, and 80%, corresponding to static–dynamic comprehensive flexural tensile strength (the ratio between static–dynamic comprehensive flexural tensile and static flexural tensile strength is f_{dt}/f_{st}) of 124.1, 125.7, 128.3, and 131.3%, respectively. These computation results and testing results have displayed the same trend that static–dynamic comprehensive flexural tensile strength change with prestatic loading levels.

These computation results indicate that static–dynamic comprehensive flexural tensile strength under different initial prestatic loading is larger than the pure dynamic loading strength, and that dynamic loading strength increases with an increase in prestatic loading, when at 80%, reaching the peak value, and then begins to drop, just as Fig. 15.19 shows. It can be seen from Fig. 15.20 that when static loading is applied to 2.63 MPa, some units begin to destroy, the whole dynamic loading is applied to 3.65 MPa, the damage units begin to generate. This illustrates that the damages produced by the static loading

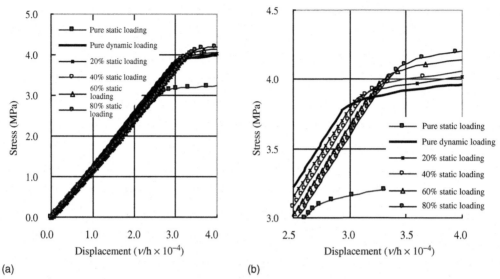

(a)   (b)

FIGURE 15.18 Effect of Prestatic Loading Upon Dynamic Flexural Tensile Strength and Dynamic Elastic Modulus ($A_E = 0.12$, $A_t = 0.20$)

(a) Stress–displacement, (b) stress–displacement local magnification.

with the same loading levels are larger than the damages caused by dynamic loading. In the process of dynamic loading prior to loss of stability, the same loading level and loading rate can produce. The different quantities of failure units are because of different initial prestatic loading levels, whereby indicating the very great correlation between dynamic strength and prestatic loading. This kind of phenomena can be accounted for; when there is prestatic loading, static displacement stress is loose so as to make loading increase, strain enlarge, and strain rate raise. At the same time, rate effect makes the unit strength raise, and stiffness enlarge. These factors interact with each other to produce the phenomena that dynamic strength is raised with an increase in the prestatic loading. Damage and fracture can make

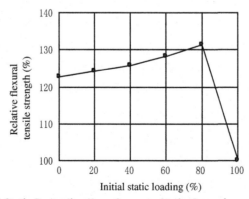

FIGURE 15.19 Effect of Initial Static Preloading Upon Concrete Static–Dynamic

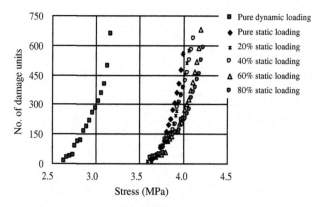

FIGURE 15.20 Different Loading Levels Producing the Damage

the unit stiffness weaken and strain rate increase, while rate effect makes the unit strength and stiffness rise. On the contrary, strain rate improvement accelerates further worsening of the unit. When the enhancement is dominant, dynamic strength is upgraded; when worsening is dominant, dynamic strength drops. In Fig. 15.19, when prestatic loading level is prior to 80%, the enhancement is dominant, but posterior being over 80%, dynamic strength drops sharply. Obviously, the dynamic flexural tensile strength is relevant with not only loading rate but also with strain rate and damage accumulation.

15.4.2.3 Effect of concrete strain rate sensitivity upon dynamic flexural tensile strength

In order to make a further investigation into the effect of strain rate upon test specimen flexural tensile damage process and dynamic flexural tensile strength, it is necessary to carry out strain rate sensitivity analysis of concrete dynamic flexural tensile strength and strain rate enhancement laws of flexural tensile strength and strain rate enhancement laws of static–dynamic comprehensive flexural tensile strength; that is, in terms of a certain law, taking different elastic modulus enhancement parameters and different tensile strength enhance parameters, it is important to compute concrete dynamic flexural tensile strength and static–dynamic comprehensive flexural tensile strength and to probe into their varying laws.

H_t and H_E affect dynamic flexural tensile strength. When (A_E, A_t) takes (0.12,0.16), (0.14,0.18), and (0.16,0.12), respectively, the corresponding dynamic strength is computed, and when the prestatic loading are 20, 40, 60, and 80% of static limit loadings, respectively, the series of corresponding results are shown in Fig. 15.21. It can be seen from Fig. 15.21 that with the enhancement parameters A_E, A_t, that is, dynamic strength, dynamic elastic modulus increases the sensitivity to strain rate, concrete dynamic flexural tensile strength also increases from 114.1, 116.9, to 122.5%. In addition, with an increase in the parameter values, the different initial static loading strengths are also increasing; for instance, the corresponding to 80% initial static loading dynamic strength increases from 117.3, 122.9, to 125.7%.

As viewed generally, with an increase in dynamic strength and dynamic elastic modulus to strain rate sensitivity, dynamic strength rises; and static–dynamic comprehensive flexural tensile strength subject to the prestatic loading level influence functions are enhanced. In addition, with an increase in enhancement parameters A_t, A_E, static–dynamic comprehensive flexural tensile strength peak value corresponding to the prestatic loading level is somewhat different, whereby indicating that static–dynamic comprehensive flexural tensile strength is the common results functioned by the rate sensitivity enhancement and damage worsening.

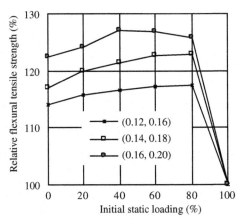

FIGURE 15.21 Effect of Rate Sensitivity Upon Static–Dynamic Comprehension Flexural Tensile Strength

When H_E takes a fixed value, H_t affects dynamic flexural tensile strength. Elastic modulus enhancement parameter always takes 0.12. Strength enhancement parameter A_t takes 0.16 and 0.20. It is necessary to compute dynamic strength and prestatic loading of 20, 40, 60, and 80% of static limit loading in static–dynamic comprehensive flexural static strength as shown in Fig. 15.22. Dynamic strength enlarges with an increase in A_t, and static–dynamic comprehensive flexural tensile strength for the initial prestatic loading sensitivity is also raised. When $A_t = 0.16$, dynamic strength relative value is 114.1%; when $A_t = 0.18$, dynamic strength relative value is 118.3%; when $A_t = 0.20$, dynamic strength relative value is 122.5%. Obviously, dynamic strength and the static–dynamic comprehensive flexural tensile strength corresponding to the same prestatic loading are raised with an increase in the strength enhancement parameter A_t. That is, the concrete test specimen flexural tensile strength is improved as the concrete material strength rate sensitivity is enhanced. Figure 15.23 gives out the stress–displacement curves under several kinds of strength enhancement parameters. It can be seen from Fig. 15.23 that

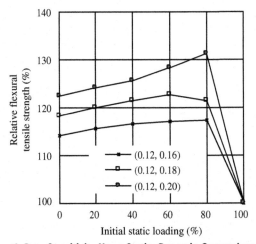

FIGURE 15.22 Effects of Strength Rate Sensitivity Upon Static–Dynamic Comprehension Flexural Tensile Strength

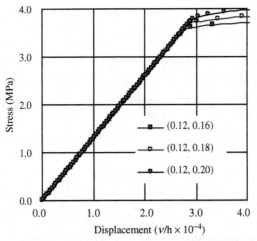

FIGURE 15.23 Effects of Strength Rate Sensitivity Upon Dynamic Flexural Tensile Strength

under the conditions of there being no changes in elastic modulus enhancement parameters, the gradients of these curves remain to have no changes basically; that is, the test specimen resistant flexural stiffness remains unchanged basically, and it is only the dynamic strength that is raised as the strength enhancement parameters are enlarged.

When H_t takes a fixed value, H_E affects dynamic flexural tensile strength. Strength enhancement parameter A_t takes 0.20, remaining unchanged; elastic modulus enhancement parameter A_E can take 0.12, 0.14, and 0.16, respectively. It is important to compute strength and the prestatic loading of static limit loading of 20, 40, 60, and 80% of static–dynamic comprehensive flexural tensile strength as shown in Fig. 15.24. Figure 15.24 indicates that in the case of having no changes in strength enhancement parameters, as the elastic modulus sensitivity is enhanced, the effect of the prestatic loading level upon

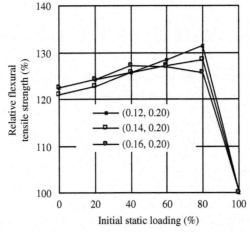

FIGURE 15.24 Effects of Elastic Modulus Sensitivity Upon Static–Dynamic Comprehension Flexural Tensile Strength

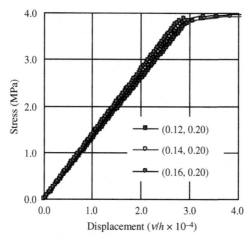

FIGURE 15.25 Effects of Elastic Modulus Rate Sensitivity Upon Dynamic Flexural Tensile Strength

the static–dynamic comprehensive flexural strength is decreased, while the stress–displacement curves in Fig. 15.25 become steep with an increase in A_E; that is, the flexural stiffness increases. This can be accounted for from the angle of concrete material physical properties. When strength remains unchanged, under the action of the same loading (static loading + dynamic loading), the test specimen with smaller stiffness deforms greatly, strain is also larger, and rate effect is also obvious. Static–dynamic comprehensive flexural tensile strength improves because of rate effect enhancement. From this it can be deduced that materials with smaller stiffness but bigger strength can have their static–dynamic comprehensive flexural tensile strength effect upon the sensitivity of the prestatic loading level.

Through stress–displacement curves in Figs 15.24 and 15.25, it is easy to see that dynamic strength is mainly decided by concrete material strength rate sensitivity, while dynamic elastic modulus is mainly subject to the effect or influence of concrete material elastic modulus rate sensitivity.

15.5 PHYSICAL SIGNIFICANCE OF CONCRETE MATERIAL ENHANCEMENT PARAMETERS

Inertia force is an important factor to produce concrete rate effect. When concrete material is subject to dynamic action, inertia force produces some particles to form the delay functions from stress state converted into strain state, whereby making these particle place damage delayed. Apart from inner friction force, some unknown factors by the people can also produce this delayed effect, and these factors to produce the delayed effect can produce the functions with different natures in the case of different strain rates and damage levels. It is generally considered that under the high strain rate level, inertia force plays an important role in rate effect, and under the low strain rate level, viscosity plays an important part in rate effect. The dynamics equation includes the inertia force item and dumping item, and introduces the strain rate to enhance concrete material strength and elastic modulus so that it is worthwhile to further discuss parameter value taking and physical significance in concrete material dynamic strength, elastic modulus, and strain rate equations. It is worth noticing that the structure dumping ratio

Table 15.1 Mechanics Characteristics of Each Phase Component Material of Concrete

Materials	Elastic Modulus (GPa)	Poisson's Ratio	Tensile Strength (MPa)	Bulk Density (kN·m^{-3})
Aggregates	50.00	0.200	6.00	27.00
Solidified cement	40.00	0.160	4.50	21.00
Viscosity interface	30.00	0.150	3.50	24.00
Concrete	32.00	0.167	4.00	24.00

Table 15.2 Concrete Damage and Enhancement Parameter

	Residual Strength Coefficient	Residual Strain	Limit Strength Coefficient	Strength Enhancement Parameter			Elastic Modulus Enhancement Parameter		
	λ	η	ξ	A_t	B_t	C_t	A_E	B_E	C_E
Tensile	0.10	3.00	10.00	0.20	6.00	2.00	0.17	6.00	2.00
Compression	0.10	3.00	10.00	0.12	6.00	2.00	0.10	6.00	2.00

adopted in the dynamic equation is a constant and that there are no considerations to give to the micro-chap inner frictions in the process of concrete test specimens, and to capillary water and pore water in concrete materials as well as to viscosity effects produced by material viscosity.

Taking three-element point flexural test results of grade 3 gradation concrete test specimens numerical simulation, we can discuss the inertia force item, dumping item, and the effect of enhanced parameter taking value upon calculation results and analyze the physical significance of enhanced parameters.

Each phase component material mechanics behavior parameter taking values of each concrete group can be seen in Table 15.1. Concrete damage and enhanced parameters are listed in Table 15.2. In considering the differences of tensile and compressed strain rate sensitivity, the compressed strain rate enhanced coefficient takes the value smaller than the tensile strain rate enhanced coefficient; static loading step length is 0.6 kN; impact loading is 600 kN/s; time step length takes 0.001 s. The computation model is seen in Fig. 15.26. After applying 20, 40, 60, and 80% of prestatic loading to pare

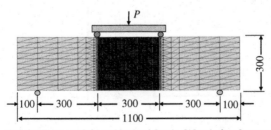

FIGURE 15.26 Classification of Finite Element Grid or Mesh of Grade 3 Gradation Concrete Test Specimens (unit: mm)

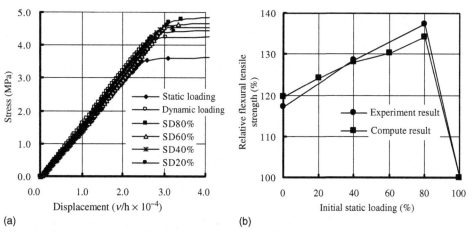

(a)   (b)

FIGURE 15.27 Considering Damping, Inertia Force, and Introduction to Enhanced Parameters

(a) Stress–displacement (b) effect of prestatic loading on dynamics enhanced coefficients.

dynamic loading by order and again applying dynamic loading, numerical tests are carried out. This numerical test is mainly to examine Eqs (15.3) and (15.4) enhanced coefficients, as well as the effects of damping and inertia force upon numerical and computation results. For this reason, in the process of dynamic loading, the following three kinds of schemes will be carried out, respectively.

1. Considering damping, dynamic tensile strength, and dynamic elastic modulus are computed in accordance with enhanced coefficients given out in Table 15.2 and Eqs (15.3) and (15.4). Stress–displacement curves are obtained via computation as shown in Fig. 15.27a. The effect of prestatic loading upon concrete enhancement coefficient is shown in Fig. 15.27b. Figure 15.27b gives out that dynamic enhancement coefficient curves varying with the prestatic loading level at the same time. It can be seen that computation results are in good agreement with test results.
2. Only considering damping without introducing Eqs (15.3) and (15.4) enhanced coefficients. The computation results are shown in Fig. 15.28.

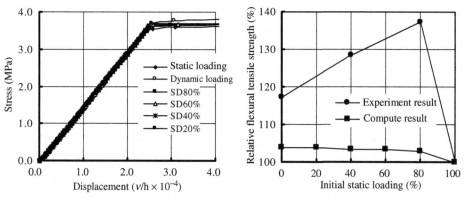

FIGURE 15.28 Considering Damping and Inertia Force, Not Introducing Enhanced Parameters

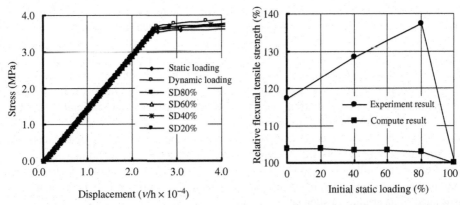

FIGURE 15.29 Considering Inertia Force, Not Introducing Enhanced Parameters, Without Accounting Damping

3. Neither considering damping nor introducing Eqs (15.3) and (15.4) enhanced coefficients, but obtaining computation results as indicated in Fig. 15.29.

In stress–displacement curves, it is necessary to take the name stress of concrete test specimen lower verge mid-node ($\sigma = PL/bh^2$, P is loading, L is support gap distance, b is the thickness of test specimen, and h is the height of test specimen) corresponding to vertical displacement (v/h) of the test specimen upper stress when stress–displacement is out of stability. And later, the displacement appearing in stress–displacement curves also has this connotation. "SD" in the figure expresses reloading of dynamic loading after loading of static loading. For instance, SD80% indicates that 80% of prestatic loading is applied and that dynamic loading is added again.

It has been found though that the computed results by the mentioned three schemes that Rayleigh structure damping introduced into the numerical model has hardly influenced the computation results and that the dynamic enhanced coefficient produced by inertia force has only been 1.04. Obviously, without counting enhanced parameters, the numerical simulation has displayed that there has hardly been any dynamic loading phenomena, and even no enhancement effect of prestatic loading upon the dynamic resistance flexural strength. It can be judged from test results that apart from inertia force and Rayleigh damping force, there have been other factors to produce rate enhancement functions, such as capillary water or free water Stefan effect, thermal activation of concrete material in itself, and phonon damping functions and other factors. Concrete rate effects produced by these factors can be reflected in the relations among strain rate, dynamic elastic modulus, dynamic strength, and so on. The enhanced parameter $H_t(\dot{\varepsilon})$ and $H_E(\dot{\varepsilon})$ curves can be obtained by fitting the results of the specific material simple dynamic loading test and observation and measurement, with details affecting factors neglected.

15.6 EFFECTS OF CONCRETE MATERIAL NONUNIFORMITY UPON DYNAMIC FLEXURAL STRENGTH

15.6.1 RANDOM AGGREGATE PARAMETER MODEL (RAPM) 1

Strain rate effect is the basic property of solid materials while strain rate effect of nonuniform materials is more obvious than that of uniform materials. Concrete strength is highly sensitive to the process of loading. When static and dynamic loading reaches the same load level, damage accumulation produced by test specimens is different, while strain rate effect is subject to material damage influence. Nonuniformity is the essential property of concrete material, and it is just this kind of nonlinear property of concrete macromechanics behavior. Accordingly, in order to better and completely reflect that the nonuniformity of concrete mesostructure can affect its macromechanics behavior, it is necessary to carry out the detailed research on concrete aggregate magnitudes, aggregate gradation, and mesoeach-phase mechanics parameter random distribution characteristics from the mesostratum simulation and analysis of concrete dynamic flexural tensile failure process.

Concrete mesostructure differences and nonuniformity depend on the aggregate magnitudes and gradation. Amount of cement used, water–cement ratio, solidification strength, maintenance and repair conditions, environmental humidity and concrete evaporation, and other factors, whose aggregate, solidification mortar, and interfacial mechanics parameters are the random numbers in the test specimen space distributed of a certain statistics law.

In order to reflect the nonuniformity of concrete mesostructure in a more complete way, the random aggregate and random parameter model will be adopted in the following computation, and within the two-dimensional test specimen cross section. Walraven formula converted from Fulei three-dimensional aggregate gradation curves into two-dimensional aggregate gradation curves can be used to determine the number of aggregate particles as well as the unit attribute. It is considered that concrete and its mesoeach-phase material tensile strength and elastic modulus are the random parameters, following the logarithmic normal distribution. This is the following suggested random aggregate and random parameter model.

The random aggregate and random parameter model assumes that concrete and its mesoeach-phase-material tensile strength and elastic modulus are the random parameter $p_{Ln}(x)$ and follow a logarithmic normal distribution:

$$p_{Ln}(x) = \begin{cases} \dfrac{1}{\sqrt{2\pi}\sigma x}\exp\left[-\dfrac{(\ln x - \mu)^2}{2\sigma^2}\right] & x > 0 \\ 0 & x \leq 0 \end{cases} \tag{15.36}$$

Here, $-\infty < \mu < \infty, \sigma > 0$, and then $\ln x$ satisfies the normal distribution $N(\mu,\sigma)$, Random variable x average value is $\mu_x = \exp(\mu + \frac{1}{2}\sigma^2)$, and variance is $\sigma_x^2 = \mu_x^2[\exp(\sigma^2) - 1]$.

Monte Carlo method is still used to generate each phase material mechanics property parameter. Supposing that each phase material mechanics property parameter statistic average value μ_x and variability coefficient V_x ($V_x^2 = [\sigma_x^2 / \mu_x^2]$) are known, then $\sigma^2 = \ln(1 + V_x^2)$, $\mu = \ln\mu_x - \frac{\sigma^2}{2}$.

Two uniform random numbers u_n and u_{n+1}, generated in (0.1) range and two interindependent random numbers y_n, y_{n+1} are obtained by Eq. (15.37) in coincidence with standard normal distribution $N(0,1)$.

$$\begin{cases} y_n = \sqrt{-2\ln u_n}\,\cos(2\pi u_{n+1}) \\ y_{n+1} = \sqrt{-2\ln u_n}\,\sin(2\pi u_{n+1}) \end{cases} \tag{15.37}$$

from the change formula:

$$\begin{cases} \ln x_n = y_n\sigma + \mu \\ \ln x_{n+1} = y_{n+1}\sigma + \mu \end{cases} \tag{15.38}$$

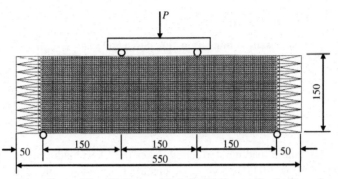

FIGURE 15.30 Concrete Beam Flexural Tensile Test and Unit Profile Separation (unit: mm)

Obtaining $\ln x_n$ and $\ln x_{n+1}$ can satisfy $N(\mu, \sigma)$ normal distribution. Equation (15.36) can find each concrete phase logarithmic normal distribution material parameter as follows:

$$\begin{cases} x_n = \exp(y_n \sigma + \mu) \\ x_{n+1} = \exp(y_{n+1} \sigma + \mu) \end{cases} \tag{15.39}$$

15.6.2 MECHANICS PARAMETER DISCRETENESS AFFECTING DYNAMIC FLEXURAL TENSILE STRENGTH

Numerical model: First of all, 150 mm × 150 mm × 550 mm wet-sieved concrete test specimens are divided in profile into finite element meshes (or grids) as shown in Fig. 15.30. Spacing between two supports should be divided into triangle units in detail.

The transverse and longitudinal magnitudes are 15/4 mm. (i.e., one-fourth of small particle size), 5025 nodes in total, and 9780 units. Mechanics property parameters of each concrete phase component materials take values in terms of Table 15.3. Concrete damage and enhanced parameters are listed in Table 15.4. The three groups of variability coefficients (V_E, V_t) take (0.00,0.00), (0.10,0.10), and

Table 15.3 Mechanics Property Parameters of Each Concrete Phase Component Materials						
	Elastic Modulus			**Tensile Strength**		
Materials	**Mean Value (GPa)**	**Variability Coefficient (V_E)**	**Poisson's Ratio**	**Mean Value (MPa)**	**Variable Coefficient (V_t)**	**Bulk Density (kN•m⁻³)**
Aggregate	50.00	0.00~0.20	0.20	6.00	0.00~0.20	27.00
Solidified cement mortar	25.00	0.00~0.20	0.20	3.50	0.00~0.20	21.00
Viscous interface	21.00	0.00~0.20	0.20	2.70	0.00~0.20	24.00
Concrete	28.00	0.00~0.20	0.20	3.00	0.00~0.20	24.00

Table 15.4 Concrete Damage and Enhanced Parameters

Residual Strength Coefficient	Residual Strain Coefficient	Limit Strain Coefficient	Strength Enhanced Coefficient			Elastic Modulus Enhanced Coefficient		
λ	η	ξ	A_t	B_t	C_t	A_E	B_E	C_E
0.1	3.0	10.0	0.2	6.0	2.0	0.14	6.0	1.5

(0.10,0.20), respectively. Static loading step length takes 0.25 kN; loading rate of impacting load is 150 kN/s; time step length takes 0.001 s, whose computation results are shown in Fig. 15.31.

Discrete coefficients of three groups, test specimen static flexural tensile strength are 3.40, 2.90, and 2.50 MPa in their order; dynamic flexural tensile strengths are 4.60, 3.55, and 3.11 MPa in their order. The enhanced coefficients corresponding to dynamic flexural tensile coefficients (the ratio between static–dynamic comprehensive flexural tensile strength is f_{dt}/f_{st}) are 119.3, 122.4, and 124.4%, in their order. As the concrete elastic modulus and tensile strength discreteness increase, static flexural tensile strength and dynamic flexural tensile strength decrease. This is in agreement with the basic laws. But it can be found through Fig. 15.31 that with an increase in discreteness, the sensitivity of static–dynamic comprehensive flexural tensile strength to the prestatic loading is enhanced, and that the enhanced coefficients of dynamic flexural tensile strength are raised. As the unit stiffness and the strength discreteness are enlarging, damaged units under the same loading level are increasing. Damage makes the unit stiffness worsen and strain rate enlarge, while strain rate effect makes the unit strength be raised and stiffness be enhanced. The raising of strain rate can also further accelerate the worsening of unit damage. The joint actions of enhancement and weakening produce the phenomena as shown in Fig. 15.31, that is, when enhancement is dominant, dynamic strength rises, and when damage worsening is dominant, dynamic strength drops so that the prestatic loading level reaches 80%, and after, dynamic strength drops sharply.

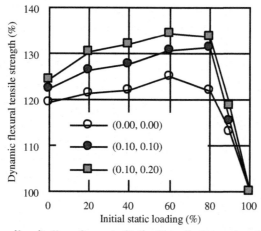

FIGURE 15.31 Effects of Nonuniformity Upon Concrete Static–Dynamic Comprehensive Flexural Tensile Strength

15.6.3 MESOANALYSIS OF WET-SIEVED CONCRETE TEST SPECIMENS

In mesostrata sequence, concrete consists of coarse aggregates, cement and hydrates, pores and aggregates, and cement mortar viscous interface, and so on, mesocach. In considering aggregate gradation and concrete mesoeach-phase mechanics property differences, and in terms of random aggregate and random parameters, a kind of model is generated with aggregate distribution to project on Fig. 15.30 finite element meshes, forming a mesofinite element mesh model as shown in Fig. 15.32. In this random aggregate model, mid-aggregates are 18 grains, particle size is 30 mm; small aggregates are 107 grains, particle size is 15 mm; aggregate units are 2702 pieces, solidified cement mortar body units are 3142 pieces, interface units are 3756 pieces, and concrete units beyond two supports are 180 pieces. Mechanics property parameter values of various types are listed in Tables 15.3 and 15.4. The stress–displacement curves obtained from this mesomodel computation, when in static loading, are found to be in agreement with the stress–displacement curves from the mentioned macroanalysis of variability coefficient (V_E, V_t) being (0.10, 0.10). Static loading strengths are all 2.90 MPa, but the dynamic flexural tensile strength enhanced coefficients are different to the initial prestatic loading level variation, as shown in Fig. 15.33. This is because there exist differences in concrete mesoeach-phase material

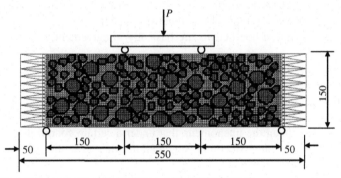

FIGURE 15.32 Wet-Sieved Concrete MesoFinite Element Model (unit: mm)

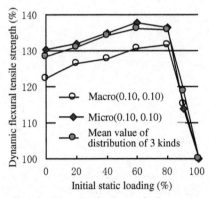

FIGURE 15.33 Effects of MesoNonuniformity Upon Static–Dynamic Comprehensive Pull and Flexural Strength

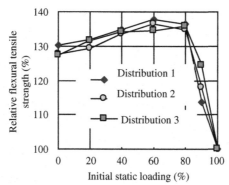

FIGURE 15.34 Three Kinds of Aggregate Random Distribution Static–Dynamic Comprehensive Flexural Tensile Strength

mechanics properties, whereby making dynamic, flexural tensile strength enhanced coefficients increase the initial prestatic loading sensitivity. Figure 15.34 gives out three kinds of aggregate random distribution computation results with the same statistic performances. Static loading strength average value is 2.92 MPa. It can be seen from Fig. 15.35 that the three kinds of aggregate distribution dynamic flexural tensile strength enhanced coefficients are basically in agreement with the changing trend of prestatic loading level.

These computation cases illustrate that the larger the material parameter discreteness is, the larger the dynamic flexural tensile strength enhanced coefficient is, and the larger the effect degree of initial prestatic loading upon the static–dynamic comprehensive strength will be. Usually, the complete graduation concrete is of more apparent mesostructure nonuniformity than that of wet-sieved concrete. Accordingly, it is easy to explain the phenomena that under a certain condition, the dynamic flexural tensile strength enhanced coefficients of complete gradation concrete are higher than the dynamic flexural tensile strength enhanced parameters.

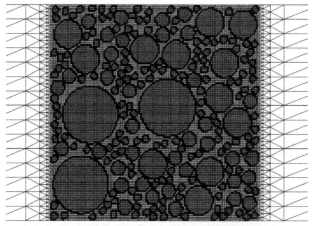

FIGURE 15.35 The Complete Gradation Concrete Aggregate Grains Distribution and Unit Profile Separation

15.6.4 EFFECT OF CONCRETE GRADATION UPON DYNAMIC FLEXURAL TENSILE STRENGTH

The mentioned computation analysis has clearly indicated that the nonuniformity of concrete mesomechanics property has an obvious effect upon the dynamic flexural tensile strength of concrete test specimens. Test results have shown that the complete gradation concrete static–dynamic flexural tensile strength is smaller than that of the wet-sieved concrete corresponding to static–dynamic flexural tensile strength, and the growth coefficient of the complete gradation concrete dynamic flexural tensile strength is larger than that of the wet-sieved concrete dynamic flexural tensile strength growth coefficient. The complete gradation concrete with large aggregate and more viscous interface has more protruded the mesostructure and mechanics property nonuniformity than the wet-sieved concrete does. In the following, it is worth further carrying out mesoflexural tensile analysis of the complete graduation concrete test specimens.

The complete gradation concrete flexural tensile tests are carried out still in terms of the three-element point method. Test specimen magnitude is 450 mm × 450 mm × 1700 mm. Mesoprofile separation zone should take the span mid-width of 450 mm zone; the rest of the part units should take macrouniform complete gradation concrete mechanics property parameters. Random aggregate model generating the complete gradation concrete aggregate particle distribution and the unit profile separation are shown in Fig. 15.35, of which there are the particularly three large grains with a particle size of 120 mm, 12 large aggregate grains with a particle size of 60 mm, 37 midaggregate grains with a particle size of 30 mm, and 167 small aggregate grains with a particle size of 15 mm. Also, there are 13,505 aggregate units, 7086 solidification cement units, 8209 viscous interface units, and 990 macroconcrete units, with 15,049 nodes and 29,790 units in total.

In order to compare the computation results between the complete gradation concrete test specimens and wet-sieved concrete test specimens, the test specimens with two kinds of magnitudes are made to macrolinear elastic plane in the loading process, and then there is $P_f/P_s = (L_f/L_s)^2(E_f/E_s)$, where P_f, P_s; $\{\hat{a}_{ijhk}\}$, L_s; E_f, E_s are the complete gradation concrete test specimens and the wet-sieved concrete test specimens, respectively. When $E_f = E_s$, $L_f/L_s = 3$, $P_f = 9P_s$ and then, $\Delta P_f = 9\Delta P_s$ that is, the complete gradation test specimen static loading step length, and dynamic loading rate are nine times as much as the corresponding values of the wet-sieved test specimens. In this section, static loading step length of the complete gradation concrete should take 2.25 kN; impact loading rate is 1350 kN/s; time step length still takes 0.001 s.

Concrete and its mesoeach-phase material elastic modulus and tensile strength should conform to logarithmic normal distribution. The complete gradation concrete test specimen mesoeach-phase statistic property parameters still take values in terms of Tables 15.3 and 15.4, of which variability coefficients (V_E, V_t) are (0.10, 0.10). Test specimen stress–displacement curves are shown in Fig. 15.36. Static loading strength is 2.77 MPa, in pure dynamic loading, dynamic flexural tensile strength enhanced coefficient is 1.32, and applying prestatic loading to static limit loading of 20, 40, 60, and 80%, corresponding to dynamic flexural tensile strength enhanced coefficients are 1.34, 1.36, 1.38, and 1.40, respectively. In order to be convenient for the comparison of concrete gradation affecting dynamic strength, the dynamic flexural tensile strength enhanced coefficients along with the prestatic loading level change curves of the complete gradation concrete model, and the curves indicated in Fig. 15.33 are set in Fig. 15.37 to carry out the comparative analysis. Obviously, when considering mesodifferences, the complete gradation concrete dynamic, flexural tensile strength enhanced coefficients are higher than the wet-sieved concrete dynamic flexural tensile strength

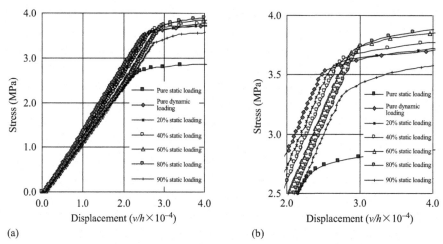

FIGURE 15.36 Complete Gradation Concrete Beam Stress–Strain Curves Under Prestatic Loading Action

(a) Stress–displacement curve, (b) Stress-displacement local.

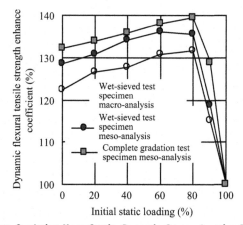

FIGURE 15.37 Effect of Concrete Gradation Upon Static–Dynamic Comprehensive Flexural Tensile Strength

enhanced coefficients, while the wet-sieved concrete dynamic flexural tensile strength enhanced coefficients are higher than the wet-sieved concrete dynamic flexural tensile enhanced coefficients without considering mesodifferences; at the same time, dynamic flexural tensile strength enhanced coefficients have enhanced the sensitivity to the prestatic loading levels in sequences. This conclusion has been confirmed in some relevant researches.

15.6.5 EFFECT OF SELECTION OF MESOPROFILE SEPARATION ZONES UPON COMPUTATION RESULTS

Owing to the limitation of computer memory capacity and computation speed, sometimes it will have to take a part of the test specimen to carry out mesocomputation, and the rest of the test specimens are treated as the macrohomogeneous or uniform concrete materials. Apart from taking span mid-width of 450 mm zone in Section 15.6.4 to carry out mesoanalysis, it is again essential to take a span mid-width of 375 mm and 525 mm (which can be divided by the minimum aggregate particle sizes) zones to carry out the mesonumerical computation of the complete gradation concrete test specimen, whose computation results are indicated in Fig. 15.38. With an enlargement of zone width, mesononuniformity has been enhanced, dynamic flexural tensile strength enhanced coefficients have been raised, and the sensitivity of dynamic flexural tensile strength enhanced coefficient to the prestatic loading level is also enhanced. This is in agreement with the conclusion of the earlier concrete mesostructure to dynamic flexural tensile strength influence. However, it can be seen from Fig. 15.38 that the result of taking span mid-width zones of 450 mm and 525 mm to carry out mesocomputation are close to real conditions.

15.6.6 CONCRETE TEST SPECIMEN STATIC–DYNAMIC COMPREHENSIVE FLEXURAL TENSILE MESODAMAGE MECHANISM

The complete gradation concrete test specimen magnitude is 450 mm × 450 mm × 1700 mm. Meso profile separation zone takes span mid-width of 375 mm zone, and the rest of the units can take the macrohomogenous complete gradation concrete mechanics property parameters. As seen in Fig. 15.39, of which there are three special large aggregate grains with particle size of 120 mm, 10 large aggregate grains with particle size of 60 mm, 29 midaggregate grains with particle size of 30 mm, and 139 small aggregate grains with particle size of 15 mm. There are 11,924 units, 5270 solidification cement mortar units, 6806 viscous interface units, and 836 macroconcrete units, totaling 12,545 nodes and 24,836 units. Mesoeach-phase unit elastic modulus and tensile strength should conform to the logarithmic normal

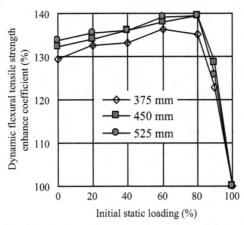

FIGURE 15.38 Comparison of Different MesoZone Numerical Computation Result

distribution, and the complete gradation concrete test specimen mesoeach-phase statistic property parameters are shown in Table 15.5.

In terms of three-element point loading method, static loading step length takes 1.00 kN; impacting loading acceleration rate is 1000 kN/S; time step length takes 0.001 s. Viscous interface tensile strength variability coefficients V_t take 0.20, 0.25, and 0.30, respectively, the rest of the parameters take the values given in Table 15.5. Static flexural tensile strength obtained from calculation are 2.67, 2.55, and 2.43 MPa, respectively, and dynamic flexural tensile strength are 3.57, 3.45, and 3.36 MPa, respectively, whose corresponding dynamic strength growing coefficients are 1.34, 1.35, and 1.38, respectively. Figure 15.40 gives the complete gradation test specimen variability coefficients of interface strength of three kinds as well as the corresponding static–dynamic comprehensive flexural tensile strengths in the case of different initial prestatic loading functions. Obviously, concrete static–dynamic strength decreases as the viscous interface tensile strength discreteness increases. At the same time, as the interface strength variability coefficient increases, the rest of the specimen static–dynamic comprehensive flexural tensile strength to the sensitivity of the initial prestatic loading level increase.

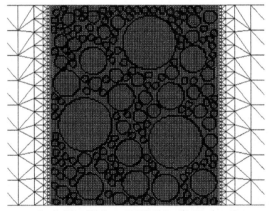

FIGURE 15.39 Concrete Aggregate Grain Distribution and Unit Profile Separations

Table 15.5 Complete Gradation Concrete Test Specimen Each-Phase Component Material Parameters

| Materials | Elastic Modulus | | Poisson's Ratio | Tensile Strength | | Bulk Density (kN·m⁻³) |
	Mean Value (GPa)	Variability Coefficient (V_E)		Mean Value (MPa)	Variability Coefficient (V_t)	
Aggregate	50.00	0.10	0.20	5.00	0.15	27.00
Solidification cement mortar	25.00	0.10	0.20	3.50	0.15	21.00
Viscous interface	22.00	0.10	0.20	3.00	0.20~0.30	24.00
Concrete	28.00	0.00	0.20	2.80	0.00	24.00

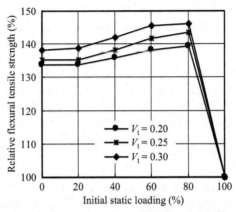

FIGURE 15.40 Effect of Interface Strength Discreteness Upon Static–Dynamic Comprehensive Flexural Tensile Strength

Concrete dynamic flexural tensile strength is higher than that of static flexural tensile strength. It can be considered that the high dynamic flexural tensile strength obtained from the computation of concrete mesononuniformity can be viewed as the dynamic flexural tensile strength obtained from macrouniform materials. The dynamic strength-growing coefficient of the complete gradation concrete test specimen is higher than the wet-sieved concrete test specimen dynamic strength growing coefficient. Under the action of static loading and dynamic loading, these mechanics property differences macrodisplayed are tested in the previous computations.

Taking the interface variability coefficient $V_t = 0.30$ as an example, it is worth further analyzing the concrete mesodamage and failure mechanism to produce the phenomena.

Figure 15.41 gives the static loading and the number of failure units produced by the dynamic loading under the different prestatic loading levels. When stress reaches 2.3 MPa, there are about 140 units

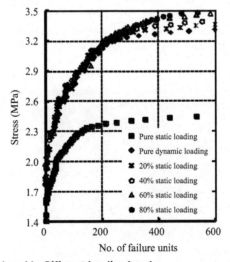

FIGURE 15.41 Failure Units Produced by Different Loading Levels

to produce damage and failures under the static loading actions, while the impact loading under the same loading level produces only about 30 units of damages and failures; that is, under the same loading level, the damaged area produced by static loading is larger than the accumulated area produced by the dynamic loading.

At the same time, it is noticed that before stress reaches 3.2 MPa, under the same loading level, the number of damage and failure units by pure dynamic loading and dynamic action with the prestatic loading are basically the same. After the stress level is over 3.2 MPa, the quantity of destructive units is affected greatly by the prestatic loading level. Dynamic loading needed by the same quantity of unit failure increases with the prestatic loading raised. Under the same loading (the prestatic loading dynamic loading) level, and owing to the existence of previous prestatic loading, the damage accumulation is larger than the damage accumulation produced by pure dynamic loading (noninitial static loading) conditions, and there are more microchaps. It can be seen from this that the previous situation is that nonuniformity of concrete mesostructure can be more obvious than the latter situation so that rate effect is macrodisplayed as the static–dynamic comprehensive flexural tensile strength being higher than pure dynamic flexural tensile strength.

As viewed from lost stability when fracture occurs, with the crack distribution situations, concrete mesoinner chap yielding and expansion patterns are different under the action of static and dynamic loading. As indicated in Fig. 15.42, apart from only a few scattered units damaged under the static loading, there is only one macrofracture connected through more destroyed units, while under the dynamic actions, there are many damaged units to run through to form fractures, as shown in Fig. 15.43. The main cause that produces this phenomenon (Zhaixia et al., 1994) lies in the fact that static loading can function for a long time so that the damages within the inner materials mainly display a microfracture expansion, linkage, and running through, and then lead to failure. While in a short time, under high-speed impact loading actions, deformation and stress can transmit in a wave pattern. Concrete mesostructures exist on the interface surface (for instance, viscous interface between aggregates and mortars, and the interfaces produced by mesoeach-phase material inner mechanics performance differences, and so on). Therefore, the larger the nonuniformity is, the more the interfaces are. Wave motion on these interfaces can produce reflections and refraction, whereby delaying energy spreading and stress shifting so that the inner material parts mainly display a large number of microholes and microchap generation, but there is not enough time for them to expand. At the same time, these microholes and microchaps have lowered the mass density where they are located so as to reduce the wave speed, becoming

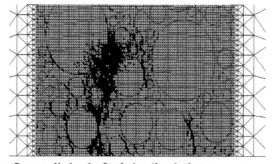

FIGURE 15.42 Beam Fracture Process Under the Static Loading Action

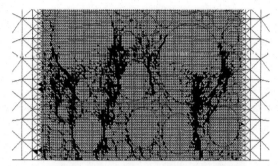

FIGURE 15.43 Beam Fracture Process Under the Dynamic Loading Action After Prestatic Loading

another reason to cause energy expansion and stress shift and delay so that the general dynamic loading (static–dynamic integration) strength is higher than static strength. These microholes and microchaps develop to a certain degree so that the test specimens will break and lose stability. In meso and microstrata, there are larger aggregate particle sizes, more grains, and more viscous interfaces of complete gradation concrete than those of the wet-sieved concrete so that nonuniformity is more obvious, while these mesostructure differences can lead to macro-mechanics property differences. Accordingly, the effect produced by concrete mesostructure nonuniformity upon its dynamic loading flexural tensile strength cannot be neglected, thus indicating the importance of research on the concrete microdamage and failure mechanism.

15.6.7 EFFECT OF AGGREGATE MORPHOLOGY UPON CONCRETE FLEXURAL TENSILE STRENGTH

Spherical aggregate model and round aggregate models generated based on Walraven formula are able to close pebbles and gravel and spherical or perfectly round aggregates, but they are different from the general macadam aggregates in morphology. The random round aggregate, sphere aggregate, polygon, and polyhedron aggregate models are adopted to carry out the mesonumerical simulation of complete gradation large dam concrete test specimen flexural tensile damage process, to contrast the analysis of the effect of different aggregate morphologies upon computation results and to discuss the effectiveness of using sphere aggregates or round aggregates to simplify actual macadam aggregates (Houqun et al., 2007).

15.6.7.1 Model parameters taking values

The main mechanics property parameters of aggregate units, solidification cement mortar unit, and viscous interface units are listed in Table 15.6. Concrete damage and enhanced parameters are seen in Table 15.7. The double polygonal line elastic damage evolution model is adopted to describe each phase material constitutive relation.

15.6.7.2 Numerical test comparison of different gradation concrete mesomodels

1. Computation analysis of complete gradation concrete test specimen flexural breakage
   Complete gradation concrete test specimen size (DL/T5150-2001) is 450 mm × 450 mm × 1700 mm as shown in Fig. 15.44. Mesoprofile separation takes a span mid-width of 450 mm

Table 15.6 Each Concrete Phase Component Material Mechanics Property Parameters

Material	Elastic Modulus (GPa)	Poisson's Ratio	Tensile Strength (MPa)	Bulk Density (kN•m^{-3})
Aggregate	50.00	0.20	6.00	27.00
Solidification	25.00	0.20	3.50	21.00
Viscous interface	22.00	0.20	2.50	24.00
Concrete	28.00	0.20	3.00	24.00

Table 15.7 Concrete Damage and Enhanced Parameters

Residual Strength Coefficient	Residual Strain Coefficient	Limit Strain Coefficient	Strength Enhanced Parameter			Elastic Modulus Enhanced Parameter		
λ	η	ξ	A_t	B_t	C_t	A_E	B_E	C_E
0.1	3.0	10.0	0.15	6.0	2.5	0.15	6.0	2.5

zone; the rest of the part units take the macrohomogenous complete concrete mechanics property parameters. In random aggregate model, there are three special large grains with a particle size of 120 mm, 12 large aggregate grains with a particle size of 60 mm, 37 mid-aggregate grains with a particle size of 30 mm, and 167 small aggregate grains with a particle size of 15 mm. The mesoprofile separation of the two kinds of aggregate models is shown in Fig. 15.45. Aggregate gradation and cross section area of models of two kinds are the same. Static loading step length takes 0.25 kN. Impact loading rate is 1000 kN/s. Power computation time step length still takes 0.001 s. Test specimens are loaded in terms of three-element point loading pattern so that static

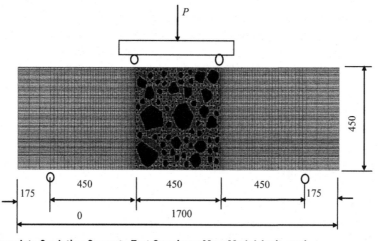

FIGURE 15.44 Complete Gradation Concrete Test Specimen MesoModel (unit: mm)

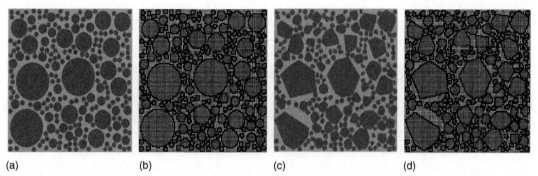

(a)   (b)   (c)   (d)

FIGURE 15.45 Complete Gradation Concrete Aggregate Grain Distribution and Unit Profile Separation (Huaifa et al., 2006)

(a) Circular aggregate, (b) Circular aggregate mesh, (c) Convex polygon aggregates, (d) Convex polygon aggregate mesh.

loading breakage tests and dynamic loading breakage tests are carried out, respectively. In considering the effect of prestatic upon dynamic strength, the first applying static loading to the limit loading is 20, 40, 60, and 80%, and the impact loading is applied again, and then the united loading numerical test analysis can be carried out.

Figure 15.46 a gives the reaction curves of static–dynamic loading limit flexural tensile strength with the prestatic loading level variation. Being in different computation results from wet-sieved test specimens, static limit strength (f_{st}) obtained from the round aggregate model is 2.38 MPa; the static limit strength obtained from the polygon aggregate model is 2.23 MPa. Figure 15.47b is the relation curves between dynamic flexural tensile strength enhanced coefficients (the ratio between static–dynamic comprehensive flexural tensile strength and static flexural tensile strength is (f_{dt}/f_{st}) along with the change in the prestatic loading level. Owing to an increase in the interface

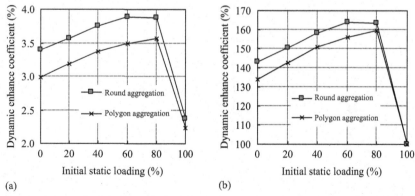

(a)   (b)

FIGURE 15.46 Result Comparison of Two Kinds of MesoAggregate Model Computation of Complete Gradation Concrete

(a) Static-dynamic loading limit flexural tensile strength varying with prestatic loading level relation curves,
(b) Dynamic flexural tensile strength enhancement coefficient varying with prestatic loading level relation curves.

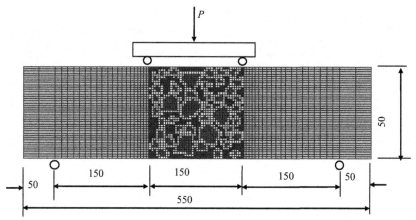

FIGURE 15.47 Wet-Sieved Concrete MesoFinite Element Model (unit: mm)

units, the polygon aggregate side angle stress concentration effect is more obvious than that of round aggregates. Therefore, flexural tensile strength calculated from the polygon aggregate model is lower than that of round aggregates.

2. Wet-sieved concrete test specimen flexural tensile damage calculation analysis

a. *Two-dimensional random aggregate model*: The wet-sieved concrete test specimen sizes are 150 mm × 150 mm × 550 mm. Applying loading is in terms of three-element point method; mesoprofile separation of random aggregate model is indicated in Fig. 15.48. After a certain

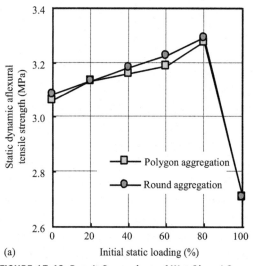

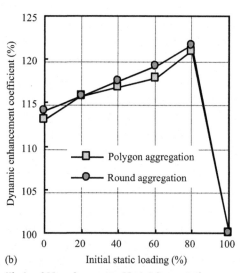

FIGURE 15.48 Result Comparison of Wet-Sieved Concrete Two Kinds of MesoAggregate Model Computation

(a) Relation curves of static–dynamic flexural tensile strength varying with prestatic level variation, (b) Relation curves of dynamic flexural tensile strength enhanced coefficient with prestatic level damage.

static loading is applied, the impacting loading is applied, the united loading tests analysis is carried out, of which static loading step length takes 0.1 kN; the impacting loading of applying loading rate is 150 kN/s; time step length takes 0.001 s. The rest of the parameters take values in terms of Tables 15.6 and 15.7, whose computation results are shown in Fig. 15.48.
Figure 15.48a is the relation curves of dynamic flexural tensile strength varying with prestatic loading levels. Figure 15.48b is dynamic flexural tensile strength enhanced coefficient varying with prestatic loading level relation curves. Static limit strength obtained from the round aggregate model and polygon aggregate model is 2.71 MPa. But dynamic flexural tensile strength is slightly different from the static–dynamic comprehensive strength.

b. *Three-dimensional random aggregate model*: Span mid-width is 150 mm zone, adopting the random sphere aggregate model and random convex polygon aggregate model, and the rest of the positions are processed in terms of macroconcrete materials. Grade 2 gradation aggregate medium and small stones: mid-stone is 55%:45%. The number of grains of various aggregates are determined in terms of the ratio of coarse aggregate occupying concrete mass. Within the span mid-cube of 150 mm × 150 mm × 150 mm, there are 45 grains of mid-aggregates with particle size of 30 mm, 438 grains of small aggregates with particle size of 15 mm, and coarse aggregate mass ratio is 47%, as shown in Fig. 15.49. The spatial mesh profile separation is shown in Fig. 15.50. There are 76,810 nodes and 11,988 units in total. Concrete mesoeach-phase component material mechanics parameters and loading step length taking values are the same as those of the two-dimensional model.

Stress–displacement curves obtained from computation are shown in Fig. 15.51. It can be seen from the Fig. 15.52 that the limit flexural tensile stress computed from sphere aggregate model and convex polygon aggregate model is closer to each other. Static flexural tensile strength is about 2.70 MPa. Dynamic flexural tensile strength is 3.10 MPa. The stress–displacement under the corresponding static loading or dynamic loading action obtained by adopting two kinds of aggregate models is basically overlapped.

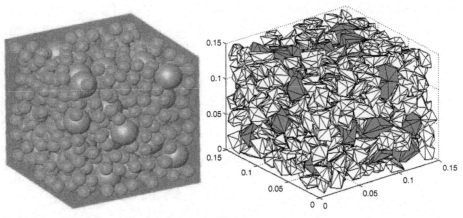

FIGURE 15.49 Three-Dimensional Random Aggregate Model

(a) Sphere aggregate model, (b) Convex aggregate model.

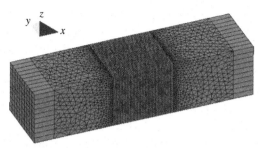

FIGURE 15.50 Spatial Mesh Profile Separation

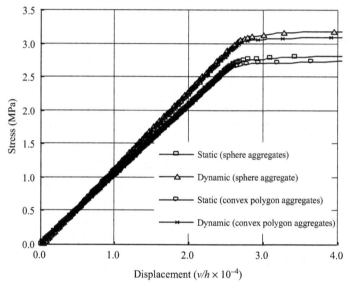

FIGURE 15.51 Stress–Displacement Curves

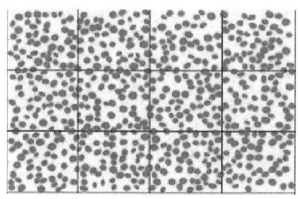

FIGURE 15.52 Statistics Zone Ω of Grains Distribution with the Same Particle Sizes

The mentioned three-dimensional mesoflexural tensile numerical test of wet-sieved concrete test specimens are realized on the present mainstream PC machine. Every loading process will take about a week's time so that work on computation is extremely large, while it is difficult to realize the mesonumerical simulation of grade 3–4 multigradation large dam concrete test specimens. However, the testing results indicate that in the large dam concrete flexural tensile tests, damages are the main fractures starting from the weakest mid, coarse aggregate, and cement mortar interfaces and to develop toward cement mortar media gradually. It is only in higher-grade concrete when aggregates are broken artificially or their textures are very poor. The mid and coarse aggregates possibly fracture, but usually, fine aggregates with particle sizes of 5–20 mm and their interfaces seldom break.

In recent years, as the compound materials have developed, great progress has been made in the prediction of mesomechanics property parameters. Being subject to the revelation of research achievements in this respect, and based on the testing results of large dam concrete material tests, it is considered that if the fine aggregates and cement mortars are considered as a kind of new double-phase compound material in the double-size stratum, on the basis of obtaining the equivalent mechanics parameters, the minimum particle sizes in large-dam concrete mesomechanics model can increase by 20–40 mm so that the number of aggregate grains in the test model will be reduced greatly, or the major obstacles in the analysis of three-dimensional mesomechanics of the multigradation large dam concrete can be overcome.

In the existing research achievements obtained from the prediction of multiphase compound material equivalent mechanics property parameters, the equivalent inclusion theory by Eshelby (1957,1959) is relatively mature. It finalizes the double-phase compound material equivalent elastic modulus computation into the average problem, that is, under the uniform boundary conditions, the average stress–strain relation in continuous base mass-phase with discrete phase inclusion sandwiched can be predicted. This is based on the infinite large base mass phase that there exists only separated inclusion mass, and the solution is also obtained in terms of linear elastic body. It is here that many more inter-reactions between the inclusion gaps cannot be accounted for, and the spatial distribution randomization in each inclusion and the spatial distribution randomization of mechanics property parameters themselves cannot be accounted for either, while research on the compound material multimagnitude analysis method for grain random distribution can provide the theoretical basis for the fine aggregates and cement mortars in the large dam concrete material mesomechanic model compounded into a kind of equivalent new medium.

15.7 MULTISCALE ALGORITHM TO PREDICT CONCRETE MATERIAL PARAMETER METHOD

15.7.1 MULTISCALE ALGORITHM BASIC THEORY

The multiscale algorithm to predict compound material mechanics property parameters is based on the double-size cycle recurrence. Double-scale compound material consists of basic mass and grain within the same magnitude range. The whole zone structure Ω can be logically decomposed into a series of sizes far larger than the largest size of including grains, serving as statistics window single cellular body set εQ^s. Grain probability distribution in every single cellular body is the same. For this reason, this is equal to the structure grain probability distribution. Letting ω^s be the statistic window inner grain random parameter distribution, sample $\omega^s \in P$ (sample space), as shown in Fig. 15.52. The sample set $\omega = \{\omega^s / x \in \varepsilon Q^s \subset \Omega\}$

Mechanics parameters in double-scale analysis method for two-phase materials can be carried out though the cycle random distribution, of which mechanics parameters of one sample can be expressed as follows:

$$a_{ijhk}\left(\frac{x}{\varepsilon},\omega^s\right) = \begin{cases} a^1_{ijhk} & x \in e_{i_1} \\ a^2_{ijhk} & x \in \varepsilon Q^s - \bigcup\limits_{i=1}^{N} e_{i_1} \end{cases} \tag{15.40}$$

where εQ^s is statistic window zone; e_{i_1} is in εQ^s of i_1 oval sphere grains; N is the maximum number of grains in static window; a^1_{ijhk}, a^2_{ijhk} are the positive constant in given range, and right normal, symmetric, expecting that uniformed value has a boundary in statistic zone Ω.

$a_{ijhk}\left([x/\varepsilon],\omega^S\right)$ is a measurable random variable set, and at same time, supposing that Ω zone is of the same size grain small cycle random distribution compound material zone, that is, $\Omega = \bigcup_{(\omega^s,t\in Z)} \varepsilon(Q^s + t)$, (Z is the integer vector space), as seen in Fig. 15.49. In order to obtain the equivalent mechanics parameters with the small cycle grain random distribution compound material, the following elastic problems can be considered:

$$\begin{cases} \dfrac{\partial}{\partial x_j}\left[a^\varepsilon_{ijhk}(x,\omega)\dfrac{1}{2}\left(\dfrac{\partial u^\varepsilon_h(x,\omega)}{\partial x_k} + \dfrac{\partial u^\varepsilon_k(x,\omega)}{\partial x_h}\right)\right] = f_i(x) & x \in \varepsilon Q^s \subset \Omega \\ u^\varepsilon(x,\omega) = u_0(x) & x \in \partial\Omega \end{cases} \tag{15.41}$$

where $u^\varepsilon(x,\omega) = [u^\varepsilon_1(x,\omega), u^\varepsilon_2(x,\omega), \ldots, u^\varepsilon_n(x,\omega)]$, $x = [x_1, x_2, \ldots, x_n]$, $i, j, h, k = 1, 2, \ldots, n$.

The structure method is used to set up double-scale formula on every single cellular εQ^s to compute equivalent mechanics parameter $\hat{a}_{ijhk}$ (expecting the uniformed coefficient). First of all, it is necessary to assume that there exists a group of mechanics parameter $\hat{a}_{ijhk}$ in the whole zone Ω to satisfy oval conditions, and vector value function $u^0(x)$ is the solution to the following problem:

$$\begin{cases} \hat{a}_{ijhk}\dfrac{1}{2}\dfrac{\partial}{\partial x_j}\left(\dfrac{\partial u^0_h(x)}{\partial x_k} + \dfrac{\partial u^0_k(x)}{\partial x_h}\right) = f_i(x), & i=1,\cdots,n, \quad x \in \Omega \\ u^0(x) = 0, & x \in \partial\Omega \end{cases} \tag{15.42}$$

Here, $u^0(x)$ is called the excepted uniformed solution on Ω. $\{\hat{a}_{ijhk}\}$ is the expecting uniformed coefficient in Ω with symmetric and right normal natures.

As to every ω^s sample, Eq. (15.41) has only one solution $u^\varepsilon(x, \omega)$. From Eq. (15.41), displacement $u^\varepsilon(x, \omega)$ and stress are known to rely on Ω and on every single cellular εQ^s microdistribution so that displacement can be expressed as $u^\varepsilon(x, \omega) = u^\varepsilon(x, \xi, \omega)$, where x expresses the whole structure performances and behaviors as the following $\xi = x / \varepsilon$, $\xi = [\xi_1, \xi_2, \ldots, \xi_n]$, to express the local coordinate and random structure influence in Q^s.

In order to obtain random grain distribution compound material mechanics property parameters double magnitude formula, it can be assumed that $u^\varepsilon(x, \omega)$ will have the following open formula:

$$u^\varepsilon(x,\omega) = u^0(x) + \varepsilon N_{\alpha_1}(\xi,\omega^s)\frac{\partial u^0(x)}{\partial x_{\alpha_1}} + \varepsilon^2 N_{\alpha_1\alpha_2}(\xi,\omega^s)\frac{\partial^2 u^0(x)}{\partial x_{\alpha_1}\partial x_{\alpha_2}}$$
$$+ \varepsilon^3 P_1(x,\xi,\omega P) \quad x \in \varepsilon Q^s \subset \Omega, \quad \xi = \frac{x}{\varepsilon} - \left[\frac{x}{\varepsilon}\right] \in Q^s \tag{15.43}$$

where $N_{\alpha_1}(\xi,\omega^s)$, $N_{\alpha_1\alpha_2}(\xi,\omega^s)$ $(\alpha_1,\alpha_2=1,...,n)$ is the function matrix. On Q^s, it can only be necessary to define $N_{\alpha_1 m}(\xi,\omega^s)$, $u^\varepsilon(x,\omega)$ to satisfy Eq. (15.42) boundary conditions and as to any ω^s on every Q^s, there will be a boundary equation:

$$\begin{cases} \dfrac{\partial}{\partial\xi_j}\left[a_{ijhk}(\xi,\omega^s)\dfrac{1}{2}\left(\dfrac{\partial N_{\alpha_1 hm}(\xi,\omega^s)}{\partial\xi_k}+\dfrac{\partial N_{\alpha_1 km}(\xi,\omega^s)}{\partial\xi_h}\right)\right]=-\dfrac{\partial a_{ij\alpha_1 m}(\xi,\omega^s)}{\partial\xi_j} & \xi\in Q^s \\[4mm] N_{\alpha_1 m}(\xi,\omega^s)=0 & \xi\in\partial Q^s \end{cases} \tag{15.44}$$

There exists only one solution to the preceding equations. Owing to ε from 1~2 orders, power rising is arranged, the right side of the equation is independent of ε; therefore, the left side ε coefficient is equal to $f1(x)$, there will be the following:

$$a_{ijhk}(\xi,\omega^s)\frac{1}{2}\frac{\partial}{\partial x_j}\left(\frac{\partial u_h^0(x)}{\partial x_k}+\frac{\partial u_k^0(x)}{\partial x_h}\right)+a_{ijhk}(\xi,\omega^s)\frac{1}{2}\left(\frac{\partial N_{\alpha_1 hm}(\xi,\omega^s)}{\partial\xi_k}+\frac{\partial N_{\alpha km_1}(\xi,\omega^s)}{\partial\xi_h}\right)\frac{\partial^2 u_m^0(x)}{\partial x_{\alpha_1}\partial x_j}$$
$$+\frac{1}{2}\frac{\partial}{\partial\xi_j}(a_{ijhk}(\xi,\omega^s)N_{\alpha_1 hm}(\xi,\omega^s))\frac{\partial^2 u_m^0(x)}{\partial x_{\alpha_1}\partial x_k}+\frac{1}{2}\frac{\partial}{\partial\xi_j}(a_{ijhk}(\xi,\omega^s)N_{\alpha_1 km}(\xi,\omega^s))\frac{\partial^2 u_m^0(x)}{\partial x_{\alpha_1}\partial x_h}$$
$$+\frac{\partial}{\partial\xi_j}\left[a_{ijhk}(\xi,\omega^s)\frac{1}{2}\left(\frac{\partial N_{\alpha_1\alpha_2 hm}(\xi,\omega^s)}{\partial\xi_k}+\frac{\partial N_{\alpha_1\alpha_2 hm}(\xi,\omega^s)}{\partial\xi_h}\right)\right]\frac{\partial^2 u_m^0(x)}{\partial x_{\alpha_1}\partial x_{\alpha_2}}$$
$$=f_i(x)\qquad i=1,2,\ldots,n \tag{15.45}$$

Owing to $\{a_{ijhk}(x,\omega)\}$ symmetry and right normal natures, the material parameters on every single Q^s are independent of each other; therefore, there will be the following:

$$E\left[\int_{Q^s}\left\{\hat{a}_{i\alpha_2 m\alpha_1}-a_{i\alpha_2 m\alpha_1}(\xi,\omega^s)-a_{i\alpha_2 hk}(\xi,\omega^s)\frac{\partial N_{\alpha_1 hm}(\xi,\omega^s)}{\partial\xi_k}-\frac{\partial}{\partial\xi_j}\left(a_{ijh\alpha_2}(\xi,\omega^s)N^{\alpha_1 hm}(\xi,\omega^s)\right)\right\}d\xi\right]=0 \tag{15.46}$$

Equation (15.46) describes the material properties of a single cellular. Owing to εQ^s single cellular in which random parameters obey the smooth random distribution, in the sense of statistics, the integral on εQ^s expects to be equal to zero. In considering Eq. (15.44) boundary conditions, there will be the following:

$$\int_{Q^s}\frac{\partial}{\partial\xi_j}\left[a_{ijh\alpha_2}(\xi,\omega^s)N_{\alpha_1 hm}(\xi,\omega^s)\right]d\xi=0 \tag{15.47}$$

Letting:

$$\hat{a}_{ijhk}(\omega^s)=\int_{Q^s}\left[a_{ijhk}(\xi,\omega^s)+a_{ijpq}(\xi,\omega^s)\frac{1}{2}\left(\frac{\partial N_{hkp}(\xi,\omega^s)}{\partial\xi_q}+\frac{\partial N_{hkq}(\xi,\omega^s)}{\partial\xi_p}\right)\right]d\xi \tag{15.48}$$

If $\{\hat{a}_{ijhk}(\omega^s)\}$ expected value exists, we can obtain M several sample expected uniformed parameters as follows:

$$\hat{a}_{ijhk} = \frac{\sum_{s=1}^{M} \hat{a}_{ijhk}(\omega^s)}{M}, \qquad M \to +\infty \tag{15.49}$$

15.7.2 MULTISCALE ALGORITHM BASED ON DOUBLE-SCALE METHOD

The random distribution compound materials with multiscale grains have been mentioned previously, whose expecting uniformed parameters (equivalent parameters) can be computed by means of classical finite element method. But, since there are too many grains in the materials, and when the numerical method is used to carry out the simulation, it is difficult to generate grains and meshes or grids. In order to reduce the complexity of computer simulation and computation of expecting unformed behavior parameters, the grain magnitude can be divided into m scales according to the circulation of grains with multiscales. Then, it is important to determine each grain distribution under the current scale, and to generate the whole material stage by stage; the double-scale method is used to compute the final expecting uniformed performance parameters in a recursive way.

In the following, it is worth introducing the solution steps of the application of the double-scale algorithm to compute the equivalent mechanics behavior parameters with multiscale grain compound materials. It is here that there is only one kind of aggregate compound material serving as an example. First of all, it is necessary to assume that the mechanics parameter of the base mass material is $\{a_{ijhk}\}$, and the grain-enhanced material mechanics parameter is $\{a'_{ijhk}\}$.

1. Computing the expecting uniformed parameter $\hat{a}_{ijhk}(\varepsilon^m)$ under the M scale:

   a. Statistic law based on M scale grains, $\forall \omega^s$, with grain sample is generated, whose mechanics behavior parameter $\{a_{ijhk}([x/\varepsilon^m], \omega^s)\}$ can be defined as follows :

   $$a_{ijhk}\left(\frac{x}{\varepsilon^m}, \omega^s\right) = \begin{cases} a_{ijhk}, & x \in \varepsilon^m \hat{Q}^s(\varepsilon^m) \\ a'_{ijhk}, & x \in \varepsilon^m \tilde{Q}^s(\varepsilon^m) \end{cases} \tag{15.50}$$

   where $\varepsilon^m \hat{Q}^s(\varepsilon^m)$ is the base mass material; $\varepsilon^m \hat{Q}^s(\varepsilon^m)$ is grain enhanced material.

   b. From material coefficient $\{a_{ijhk}([x/\varepsilon^m], \omega^s)\}$, and through finite element solution to Eq. (15.48), we can obtain $N_{\alpha_1 m}(\xi^m, \omega^s)$, and then, through Eq. (15.48) this sample uniformed coefficient $\{a_{ijhk}([x/\varepsilon^m], \omega^s)\}$ can be obtained.

   c. As for $\omega^s \in P$, $s = 1, 2, \cdots M^m$ (1) and (2) to repeat M^m times, several M^m uniformed coefficients $\{\hat{a}_{ijhk}([x/\varepsilon^m], \omega^s)\}$ $(s = 1, 2, \cdots M^m)$ can be obtained in such a way that the expecting uniformed coefficient with the grain scale smaller than ε^m can be predicted.

2. If $r = m-1, m-2, \cdots, 1$, every scale uniformed coefficient $\{\hat{a}_{ijhk}(\varepsilon^r)\}$ will be obtained in terms of the recursive method.

   a. The same as earlier, statistics law based on the r scale grain distribution performances, the grain parameter distribution patterns of curious types can be obtained, and then the computer is used to simulate sample ω^s, whereby obtaining material mechanics behavior parameters and is a sample of random variable $\{a_{ijhk}([x / \varepsilon^r], \omega^s)\}$.

$$a_{ijhk}(\tfrac{x}{\varepsilon^r}, \omega^s) = \begin{cases} a_{ijhk}, & x \in \varepsilon^r \widehat{Q}^s(\varepsilon^r) \\ a'_{ijhk}, & x \in \varepsilon^r \widetilde{Q}^s(\varepsilon^r) \end{cases}$$

where $\varepsilon r \widehat{Q}^s(\varepsilon^r)$ is base mass material; $\varepsilon^r \widetilde{Q}^s(\varepsilon^r)$ is grain enhanced material.

b. In terms of material coefficient $\{a_{ijhk}([x / \varepsilon^r], \omega^s)\}$ and through the finite element solution to Eq. (15.45) variation form, $N_{\alpha_1 m}(\xi^r, \omega^s)$ can be obtained, and then, through Eq. (15.49), the uniformed coefficient $\{\widehat{a}_{ijhk}([x / \varepsilon^r], \omega^s)\}$ for this sample can be obtained.

c. As for $\omega^s \in P$, $(s = 1, 2, \cdots M^r, 1 \text{ and } 2)$, to repeat M^r times, several M^r uniformed coefficient $\{\widehat{a}_{ijhk}([x / \varepsilon^r], \omega^s)\}$ $(s = 1, 2, \cdots M^r)$ can be obtained in such a way that the expecting uniformed coefficient with grain scale less than ε^r.

3. Finally, the uniformed coefficient $\{\widehat{a}_{ijhk}(\varepsilon^1)\}$ is on Ω expecting uniformed coefficient (equivalent mechanics property parameter).

15.7.3 APPLICATION OF MULTISCALE ALGORITHM TO COMPLETE GRADATION CONCRETE MESOPARAMETER PREDICTION

In the complete gradation of large dam concrete test specimens, the mentioned method can be adopted to synthesize fine aggregates with cement mortars as the equivalent base mass materials so as to calculate the equivalent elastic modulus. Concrete flexural tensile test specimens are seen in Fig. 15.53. In combination within Xiaowan arch dam engineering works, the mix proportion of grade 3 gradation aggregates, coarse, medium, and fine aggregate magnitude ranges are 40~80, 20–40, 5–20 mm, respectively, whose mass ratios are 4:3:3. Aggregates, cement mortars, and both interface stratum elastic modulus take 55.5, 26, and 25 GPa, respectively. Poisson's ratio is 0.16, 0.22, and 0.16, respectively. The thickness of the interface between two grains is less than 1/5 of its radius. Taking the width of the test specimen midpart is 40 cm zone, serving as the mesomechanics analysis zone. Fifty random samples are taken for small aggregates so as to generate meso zones containing small, medium,

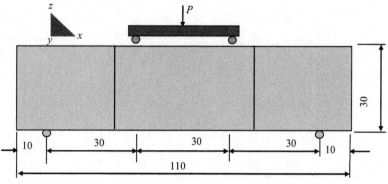

FIGURE 15.53 Grade 3 Gradation Concrete Test Specimen MesoModel (unit: cm)

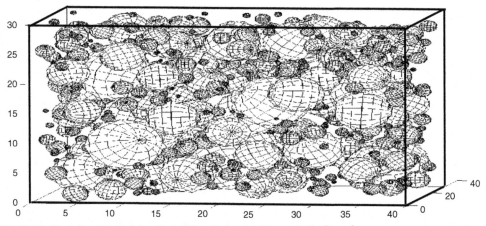

FIGURE 15.54 Meso-Zone Containing Part of Small, Coarse Aggregate (unit: cm)

coarse aggregates, as shown in Fig. 15.54. The mesozone containing medium and coarse aggregates is shown in Fig. 15.55. The statistics window taking small aggregates in terms of proportions is shown in Fig. 15.56.

Figures 15.57 and 15.58 give out the statistics mean elastic modulus and Poisson's ratio in different times, being 27.6 GPa and 0.21, respectively.

In fact, aggregate shapes in large dam concrete are very irregular, and usually distribute in different directions and closer to the changing oval sphere body in a given range. But in order to simplify computation, it is usual to take the spherical body, whereby making the grain distributive geometric variables change from 10 to 4. In order to test this approximate influence, the mentioned theory and program are used to compute the equivalents elastic modulus and Poisson's ratio approximate to spherical,

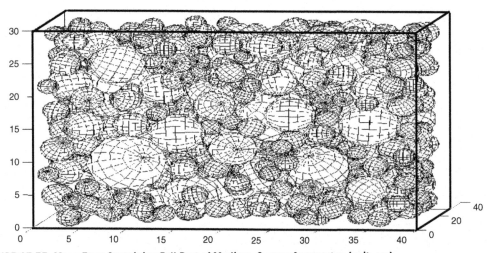

FIGURE 15.55 Meso-Zone Containing Full Part of Medium, Coarse Aggregates (unit: cm)

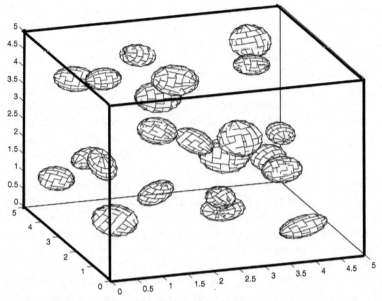

FIGURE 15.56 Statistics Window Containing Small Aggregates (unit: cm)

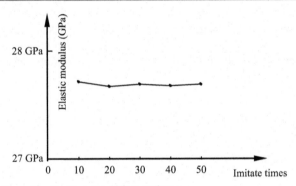

FIGURE 15.57 Statistics Mean Elastic Modulus Different Times

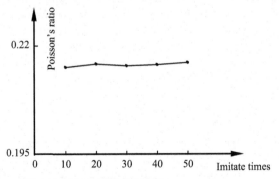

FIGURE 15.58 Statistics Mean Poisson Ratio in Different Times

new compound base mass, whose values are 28.5 GPa and 0.20, respectively. They are only 3% or so different from the results computed in terms of spherical shapes. It can be considered that in the analysis of concrete mesomechanics, aggregates are approximately simplified as spherical shapes, which can be accepted from the angle of engineering, and these parameters become small after the number of samples is over 10 so that taking 10 samples is enough.

It should be pointed out that the division scale analysis theory or the equivalent inclusion theory by Eshelby (1957, 1959) can only predict the elastic compound elastic modulus and Poisson's ratio equivalent values, but cannot give out the equivalent post and new compound base mass tensile strength – one of key parameters. However, the division scale method tests and proves in theory that it is feasible for the predictable performance parameter and new equivalent base mass materials synthesized by random distribution aggregates and solidification cement mortars.

As viewed from the practical engineering use, the coarse aggregate with small particle size and cement mortar can pour into the test specimens in terms of the corresponding proportion in such a way that the equivalent elastic modulus and tensile strength can be directly measured. Also, the equivalent elastic modulus and tensile strength with their axle centers being subject to tensile strength can be obtained through 50 mm × 50 mm × 150 mm virtual test specimens consisting of three statistics windows. For this reason, the mentioned double-scale method is used to generate the fine aggregate statistics windows from which nine virtual test specimens are obtained, which are analyzed through ANSYS software to obtain nine equivalent elastic modulus of new compound base mass. They are 27.65, 28.11, 28.17, 28.79, 28.21, 28.20, 27.91, 28.17, and 28.59 GPa, respectively, whose average value is 28.2 GPa. They are found to be very close in agreement with the results obtained by means of double-scale theory analysis. Therefore, in the mesomechanics analysis of large dam concrete, it is easy to obtain the equivalent elastic modulus and tensile strength of new compound base mass through fine aggregate random distribution virtual test specimens, whereby avoiding the mentioned complicated double-scale theoretical analysis work.

15.7.4 MESOFLEXURAL TENSILE NUMERICAL SIMULATION OF COMPLETE GRADATION CONCRETE TEST SPECIMEN

15.7.4.1 Small aggregate melting

If in terms of previous doing, the mechanics property parameters of compound base mass materials consisting of coarse aggregates with small particle sizes and cement mortars are obtained. In concrete mesomechanics analysis, this kind of compound base mass material can be used to replace the original solidification cement mortar base. In this way, concrete in three-phase mesoscale compound material consists of coarse aggregates with larger particle sizes (removing the coarse aggregates with the minimum particle sizes), compound base mass material, and both viscous interface. In mesofinite element analysis of concrete, the unit sizes can be controlled by the minimum aggregate of the upper aggregate particle sizes, whereby achieving the purpose of decreasing finding a solution to freedom degrees.

Grade 3 concrete coarse aggregates of small stone, medium stone, and large stone with ratios of 3:3:4 can serve as the example. Also, small aggregate mass occupies 18% (30% × 60%) of total mass proportion of test specimens. Taking small test specimen magnitude is 45 mm × 45 mm × 135 mm and within this small test specimen, there are 25 grains with a particle size of 15 mm. Aggregates, mortars, and viscous interface elastic modulus take 50, 35, and 31 GPa, respectively.

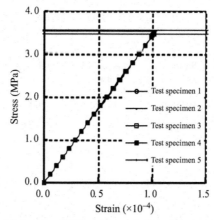

FIGURE 15.59 Stress–Strain Special Curves of Different Sample Test Specimens

Tensile strength takes 6, 4, and 3.75 MPa, respectively, axial static tension mesonumeric tests for five test specimens are carried out, whose stress–strain curves are shown in Fig. 15.59. The stress–strain curves obtained from random distribution test specimens of five kinds of aggregates are basically overlapped. It can be seen from computation results that the comprehensives and solidification cement mortars are subject to less effect of larger interface, but mainly controlled by aggregate elastic modulus, whose mean value is 34.38 GPa, while tensile strength is subject to the large influence of larger interface strength, with mean value of 3.50 MPa. Such grade 3 large test specimen mesozone can be viewed as the zone consisting of large aggregates, small aggregates, and solidification cement mortars compound mass as well as medium aggregates and the compound mass viscous interface zones. In considering the nonuniformity effect, whether this uniformity (small aggregate melting) is suitable to concrete dynamic performance is worth further discussing.

15.7.4.2 Mesoflexural tensile numerical tests of grade 3 concrete test specimens

Complete gradation test specimen size is 300 mm × 300 mm × 1100 mm. Taking span medium width as 300 mm zone to carry out mesoprofile separation, and the rest of the units take macrohomogeneous concrete mechanics performance parameters. There are 51 large aggregate grains with particle size of 60 mm, 305 medium aggregate grains with particle size of 30 mm, and 2450 small aggregate grains with particle size of 15 mm. Each concrete phase component material mechanics property parameters take values in terms of Table 15.8. Concrete damages and enhanced parameters are shown in Table 15.9. After small aggregates are melted into solidified cement mortars, the elastic modulus is 34.38 GPa and tensile strength is 3.50 MPa. Supposing that aggregate and the compound material viscous strength can still adopt the interface strength in Table 15.5.

Mesohexahedron unit side length is 7.5 mm (medium aggregate particle size 1/4). Mesh or grid from fine to coarse can be transferred in terms of the ways set in Fig. 15.2. There are 74,499 nodes and 97,841 units in total in the solution zone. Static loading step length is 0.25 kN, dynamic loading rate is 600 kN/s and time step length is 0.001 s.

Table 15.8 Concrete Each Phase Component Material Mechanics Property Parameters

Materials	Elastic Modulus		Poisson's Ratio	Tensile Strength		Bulk Density (kN·m⁻³)
	Mean Value (GPa)	Variability Coefficient		Mean Value (MPa)	Variability Coefficient	
Aggregates	50.00	0.05	0.20	6.00	0.05	27.00
Solidification	35.00	0.05	0.20	4.00	0.05	21.00
Viscous interface	31.00	0.05	0.20	3.75	0.05	24.00
Concrete	31.50	0.05	0.20	3.78	0.05	24.00

Table 15.9 Concrete Damage and Enhanced Parameters

Residual Strength Coefficient	Residual Strain Coefficient	Limit Strain Coefficient	Strength Enhanced Parameters			Elastic Modulus Enhanced Parameters		
λ	η	ξ	A_t	B_t	C_t	A_E	B_E	C_E
0.10	3.00	10.00	0.15	6.00	2.00	0.10	6.00	2.00

Stress–displacement curves obtained through computation are shown in Fig. 15.60. Static limit flexural tensile strength is about 3.49 MPa, and dynamic limit flexural tensile strength is 4.34 MPa. In the research report of full-gradation large dam concrete dynamic property, the mean value of static flexural tensile strength of four test specimens obtained through test measurements is 3.58 MPa; the mean value of dynamic flexural tensile strength of three test specimens is 4.20 MPa. Accordingly, the computation values are found to be in basic agreement with the test values.

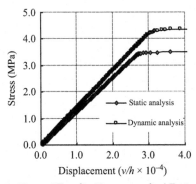

FIGURE 15.60 Full Gradation Concrete Flexural Tensile Mesonumerical Tests

15.8 RESEARCH ON CONCRETE MESOANALYSIS AND PARALLEL COMPUTATION BASED ON PFEPG

When the random aggregate and random parameter model is used to carry out numerical simulation of concrete material, first of all, it is necessary to carry out the finite element profile separation of test specimens in mesostratum. In considering aggregate units, solidification cement mortar units, and the differences in interface unit material mechanics properties, the simple failure standard or damage model are reused to reflect the deterioration of unit stiffness or rigidity. Then it is essential to simulate the test specimen crack extension process or process and breakage morphology, since the relations between loading and deformation to which concrete test specimen is subject to display as the nonlinearity so that it is necessary to carry out the unit classification in such a way that memory and computation quantity have become the bottleneck problem of concrete, particularly for full-gradation concrete three-dimensional mesonumerical simulation.

In order to realize three-dimensional mesomechanics numerical computation, there are two groups of difficult problems to be solved: The first is the input problem of three-dimensional aggregates. In Chapter 6, the effective aggregate input method has been researched. The second is to find a solution to the speed problem. Usually, aiming at a model regulates parameters time and again to carry out numerical tests so that a computational example needs several days and even several weeks' time until the time cannot be accepted. Although division scale theory is used to simplify the problem, the computation speed problem should be solved fundamentally for the parallel computable selection of three-dimensional mesonumerical analysis.

PFEPG is the finite element program that automatically generates the additional system network edition IFEPG subsystem. Based on the PFEPG system, the parallel program generated is almost together with the serial program to use the same element document (PDE/VDE document and algorithm document) so that the compiled concrete numerical simulation test FEPG document in Section 15.3 is very easy to generate the corresponding parallel program. However, this parallel program can only be used under Unix or Linux. Thus far, all the computation data prepared should correspond to Unix or under Linux data. The structure diagram of the compiled and written to generate parallel finite element program is shown in Fig. 15.61, whose structure diagram has not much difference from that in Fig. 8.1b. The corresponding main progressing and subordinate progressing program functions are the same. It is only to consider the material nonlinearity in the loading process, with more renewing iteration circulation of material parameter increased.

15.8.1 MESOMODEL SELECTION AND PARALLEL PROGRAM GENERATION

With 150 mm × 150 mm × 550 mm standard wet-sieved test specimen as an example, the three-element point loading method is adopted. Taking test specimen span mid-width of 150 mm zone serving as the mesomodel, the rest of the parts are treated in terms of macroconcrete materials: (1) aggregate model – in terms of grade 2 aggregate proportion, and within 150 mm × 150 mm × 150 mm cube, there are 30 mid-aggregate grains with particle size of 30 mm, 438 small aggregate grains with particle size of 15 mm, with aggregate model shown in Fig. 15.62. The proportion of coarse aggregate mass is about 47%. (2) Unit profile separation – the zone profile separation with test specimen mid-width of 150 mm can be the hexahedron unit with side length of 3.75 mm, being connected with the macroprofile separation zone in tetrahedron unit transforming from denseness to

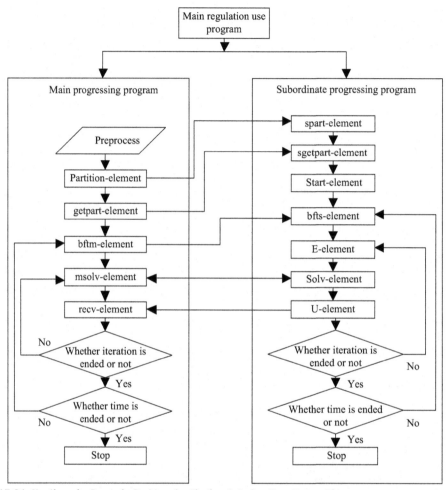

FIGURE 15.61 Nonlinearity Dynamic Problem for Finding Solution Flow Frame Diagram

looseness. A hexahedron unit with larger sizes is adopted for the outside of the supports as shown in Fig. 15.63. Figure 15.63b left end support middle point three directions are fixed. The rest of the points x, z directions are fixed; y direction is free. Figure 15.63c is the model spatial grid view. The profile along the test specimen's three symmetries surfaces is shown in Fig. 15.63d. The midpart position in Fig. 15.63 is the mesoprofile separation zone, and the concrete mesoaggregate model can be seen vaguely. There are 76,810 nodes in total in the whole profile separation zone. There are 64,000 nodes in mesohexahedron unit of the test specimen middle part, of which there are 14,889 nodes in solidification cement mortar unit, as shown in Fig. 15.63e. There are 9571 nodes in aggregate unit, as shown in Fig. 15.63f. There are 47,788 nodes in macrotetrahedron unit, and there are 200 nodes in macrohexahedron unit beyond support outside.

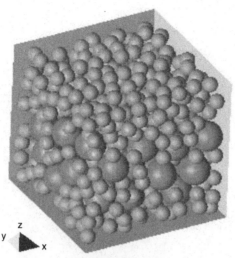

FIGURE 15.62 Three-Dimensional Random Aggregate Distribution

Material mechanics property parameter of concrete mesoeach-phase component takes value in accordance with Table 15.10. Concrete damage and enhanced parameters are seen in Table 15.11. Static loading step length takes 0.25 kN; the loading speed rate of impacting loading is 150 kN/s; loading time step takes 0.001 s.

Based on FEGEN, GID finite element prior and postprocessing software should be subdivided into units, and the generated random aggregate model can be projected into the meshes. In terms of aggregate model positions, aggregates, interfaces, and solidification cement mortar should be endowed with unit attributes.

In order to make the unit code data blocks in the test specimen span middle separated from other zones so as to put aggregate models into the zone, they are endowed with mesophysical parameters. Every zone corresponds to different unit types and stdy MDI document can be revised as follows:

```
3dxy y

    #a 3 3 u v w                          \physical field name is a; 3 unknown quantities

    VDE stdya c8g2 1c8g2 w4               \from stdya. PDE generates aec8g2.for、 aelc84g2.for and aew4.for

    # b 0 7 sxx syy szz syz sxz sxy did   \physical field name is b;0 initial value,7 unknown quantities
    VDE stdyb c8g2 1c8g2 w4               \\from stdya.PDE generates bec8g2.for、 belc84g2.for and bew4.for

        #
```

At the same time, the VDE document is slightly revised, making it able to generate a three-dimensional unit computation program. Then the MDI command is implemented from stdya. PDE generates aec8g2.for, aelc8e.for, and aew4.for, from stdyb. PDE generates bec8g2.for, belc84g2.for, aew4.for, and so on. Unit subprograms: After a single-computer serial program is generated, the parallel program for that problem is generated in terms of PFEPG again.

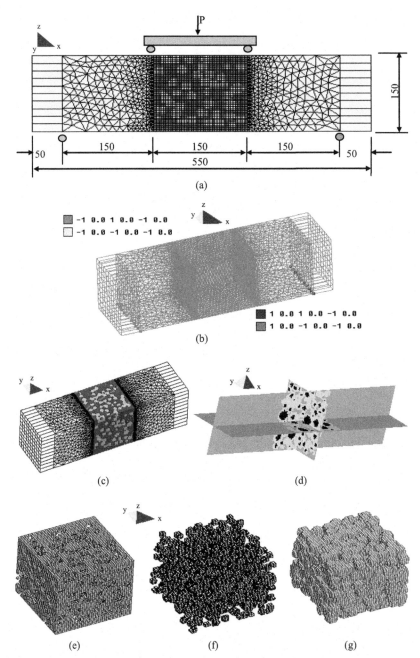

FIGURE 15.63 Wet-Sieved Concrete Flexural Tensile Three-Dimensional Numerical Model

(a) Three-dimensional numerical model plane diagram (unit: mm), (b) Constrain condition, (c) Special mesh or grid, (d) Symmetric surface grid cutting surface, (e) Solidification cement mortar unit, (f) Aggregate unit, (g) Interface unit.

Table 15.10 Material Mechanics Property Parameters of Each Concrete Phase Component

Materials	Elastic Modulus		Poisson's Ratio	Tensile Strength		Bulk Density (kN·m^{-3})
	Mean Value (GPa)	Variability Coefficient		Mean Value (MPa)	Variability Coefficient	
Aggregates	50.00	0.10	0.20	6.00	0.10	27.00
Solidification	25.00	0.10	0.20	3.50	0.10	21.00
Viscous interface	21.00	0.10	0.20	2.70	0.10	24.00
Concrete	28.00	0.10	0.20	3.00	0.10	24.00

Table 15.11 Concrete Damage and Enhanced Parameters

Residual Strength Coefficient	Residual Strain Coefficient	Limit Strain Coefficient	Strength Enhanced Parameters			Elastic Modulus Enhanced Parameters		
λ	η	ξ	A_t	B_t	C_t	A_E	B_E	C_E
0.1	3.0	10.0	0.2	6.0	2.0	0.14	6.0	1.5

After the parallel program is generated, the parallel solver provided by PFEPG can be used. PFEPG adopts the Krylov subspace interaction algorithm program bank provided by Aztec, including conjugate gradient (CG) method, general minimum residual measure method (GMres), and biconjugate gradient method (Bidgstab). The parallel solver and preprocessing program are in realignment. The PFEPG system lacking a parallel solver is the AZ-solver. The part of displacement field computation selects the stable biconjugate gradient solver (AZ-Bicgstab). The preprocessing method adopts AZ-sym-Gs (Symmetric Gauss–Siedel). Stress field and damage parameter adopt the least squares method smoothing.

15.8.2 THREE-DIMENSIONAL MESOMODEL PARALLEL COMPUTATION ANALYSIS UNDER PRESTATIC LOADING ACTION

Based on the PFEPG platform, a three-dimensional concrete mesomechanics static–dynamic analysis parallel program has been completed. In order to reflect the nonuniformity of concrete mesostructure mechanics parameters, the concrete mesomechanics model requires that mesoeach-phase unit mechanics property parameters should be in terms of logarithmic normal distribution. In fact, the number of material types in the VDE document is the individual number of all the units in the computation zone. Accordingly, after FEPG.GIO is used to complete the prior treatment of the finite element, the material parameters must be revised, that is, each field put-element coded and material parameter documents elem0 and elemb0, the corresponding material parameters are replaced as the unit material parameters generated in terms of random aggregates and random parameters, and then the parallel computation is carried out. This work carries out mesosubdivisions of models, whose major procedures are as follows:

1. The finite element nodes of mesozone generated by FEPG.GID are coded, and the unit coded data blocks are sorted out, which serve as the input data of the mesosubdivision program.
2. The program is generated through aggregate model so as to generate aggregate distribution.
3. It is necessary to project the aggregate distribution model onto the finite element meshes of mesozones through mesosubdivision program. The mesosubdivision program should be put under operation so as to obtain mesounit attribute, that is, aggregate unit, solidification cement mortar unit, and interface unit attribute and to make the statistics of numbers of units of various types, and then, the parameters of mechanics properties of elements or units of various kinds in terms of Table 15.9. Based on Monte Carlo method, it is necessary to determine the elastic modulus and tensile strength of every unit, and at this time, one unit corresponds to the material parameter of one kind.
4. The newly generated unit material parameters are used to cover the unit codes and material parameters of GID-generated elem0 and elemb0 in mesozones.

Up to now, all the prior treated data are already complete. In computation, the coordinate partition algorithm, corresponding to parallel computation nodes, is used to divide the algorithm zone into four blocks; partitions dat document are as follows:

4   2   \4 partitions, 2 expresses the partitions by following coordinate partitions;
4   2   1   \x direction subdivision number is 2; y direction subdivision number is 2; z
direction subdivision number is 1.

The partition diagram is shown in Fig. 15.64

The parallel computation analysis is made of the wet-sieved concrete three-dimensional mesonumerical model under the different prestatic action in Fig. 15.65. Figure 15.65 is the dynamic enhanced coefficient obtained in terms of stress–displacement curves in Fig. 15.66; that is, the relative flexural tensile strength is the varying situation along with static load level prior to applying impact load. The figures indicate that the dynamic enhanced coefficient along with prestatic loading level varying trend

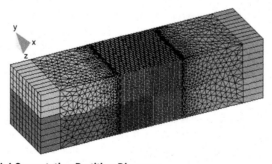

FIGURE 15.64 Model Parallel Computation Partition Diagram

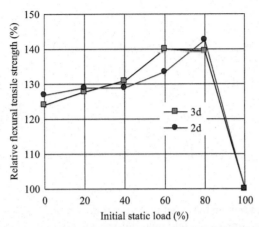

FIGURE 15.65 Effect of Prestatic Loading Upon Static–Dynamic Comprehensive Flexural Tensile Strength

is found to be in basic agreement with the result of the corresponding two-dimensional model computation (Fig. 15.65).

The hardware environment of the mentioned three-dimensional model parallel computation is as follows: CPU frequency is 3 GHz, internal memory is 1 GB 4 PC nodes, and Giga Ethernet is connecting parallel computer group. The maximum iteration times are 15 times. Static computation time-consumption 76,230 s is about 15.6 h, and the prestatic loading situation is in-between. This model uses the compiled serial program to have the static loading computation time-consumption of 355.5 h being about 15 days on a Sun6800 (CPU frequency 1.2 GHz). In comparison, parallel computation has raised computation efficiency by a great margin.

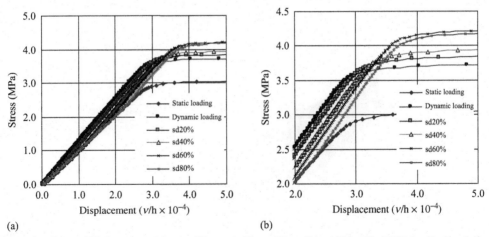

FIGURE 15.66 Stress–Displacement Curves of Concrete Three-Dimensional Meso-Model Under the Prestatic Loading Action

(a) Stress–displacement curves, (b) Stress–displacement local magnified curves.

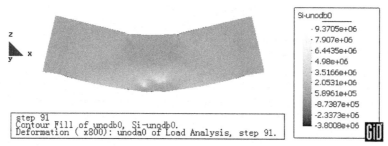

FIGURE 15.67 The First Main Stress Cloud Map in Static Limit Stress

In addition, under the mentioned computation environment, the wet-sieved test specimen in each zone of Fig. 15.59 takes the macroconcrete material parameters to carry out static computation consuming 10 h, which is less than 5.6 h in mesostatic computation. This is obviously due to mesomodel parameters increasing to enlarge communication volume on the one hand, and to the enlargement of internal memory holding capacity to cause it on the other hand.

15.8.3 POSTPROCESSING OF COMPUTATION RESULTS

PFEPG has generated the computation result document of the finite element prior and postprocessing software GID interface, including displacement field, stress field, and unit damage parameters. The GID software platform can output the cloud chap atlas, iso-line map, vector diagram, deformation diagram, cures, cartoons, and so on, of the computation results.

Figure 15.67 gives out the ninety-first static loading step, that is, corresponding to the stress cloud atlas and deformation conditions of the first main stress when stress–displacement curve limit stress is 2.97 MPa, as shown in Fig. 15.67, as well as to the damage variable cloud atlas in the case of limit loading as shown in Fig. 15.68. Their results are similar to the computation results by the two-dimensional

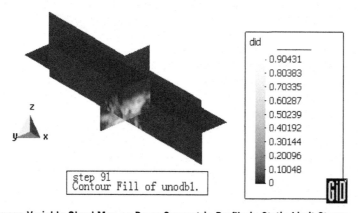

FIGURE 15.68 Damage Variable Cloud Map on Beam Symmetric Profile in Static Limit Stress

Table 15.12 Parallel Computation Efficiency

Number of Computation Node	Computation Time (s)	Communication Time (s)	Parallel Computation Efficiency (%)
1 (2CPU)	3139	140	100
2 (4CPU)	1613	150	97
3 (6CPU)	1199	150	87
4 (8CPU)	986	150	80
5 (10CPU)	858	160	73

model in Chapter 9. When loading is applied to a certain extent, the inner part near the lower verge begins to the boundary surface unit damage. When the loading is going on, the damaged units on the boundary surface are increasing, and the damages are extending downward the lower verge, and then upward. It can be seen from the first main stress cloud atlas varying process that stresses on the two lower verges are loose and tensile main stress peak value is enlarging and moving upward until it is out of stability.

15.8.4 PARALLEL COMPUTATION EFFICIENCY

Investigation into the mentioned parallel program computation efficiency is carried out on the Association & Deep Dump 1800 computer group of the Aseismic Center of Chinese Water Resources and Electric Power Academy (Huaifa et al., 2007). Apart from the main control nodes, this system has five computation nodes; each node has two main frequencies of 3.0 GHz CPU, sharing 2 GB internal memory. Billi Ethernet switching system in communication is adopted. The software environment is RedHat Advanced Intel MPI Library 2.0 Parallel Compiling Program environment and Intel Fortran 9.0/C Compiler 9.0.

As to the prior five loading steps in the static loading process, one computation node, two computation nodes, till five computation nodes are adopted to carry out their computation program. Tested results are indicated in Table 15.12. It can be seen from Table 15.12 that with an increase in the number of computation nodes, computation time and efficiency decrease, but communication time is prolonged. But five nodes with 10 CPU parallel computation efficiency still remain over 70%, whereby illustrating that the parallel computation program is effective.

References

Abrahamson, N.A., Silva, W.J., 2007. NGA ground motion relations for the geometric mean horizontal component of peak and spectral ground motion parameters. Pacific Earthquake Engineering Research Center, University of California, Berkeley, pp. 1–378.

Ahmad, S.H., Shah, S.P., 1985. Behavior of hoop confined concrete under high strain rates. ACI J. (82), 634–647.

Alliche, A., Francois, D., 1986. Fatigue behavior of hardened cement paste. Cement Concrete Res. 16 (2), 199–206.

Atchley, B.L., Furr, H.L., 1967. Strength and energy absorption capabilities of plain concrete under dynamic and static loadings. ACI J. 64 (11), 745–756.

Bai, C., Yu, X., et al., 1996. Fuzzy identification of data distribution types in rock engineering environment. J. Shenyang Inst. Gold Technol. 15 (4), 324–329.

Baokun, L., Yijiang, P., 2001. Strength and size effect analyses of meso-level damage fracture for RCC specimens. J. North China Univ. Water Resour. Elec. Power 22 (3), 50–53.

Bazant, Z.P., Tabbara, M.R., Kazemi, M.T., Pijaudier-Cabot, G., 1990. Random particle models for fracture of aggregate or fiber composites. ASCE J. Eng. Mech. 116 (8), 1686–1705.

Belyschko, T., et al., 1981. Nonlinear Finite Elements for Continua and Structures. John Wiley & Sons, Ltd, New York.

Biggs, D., 2006. Reclamation Nations: The US Bureau of Reclamation's Role in Water Management and Nation Building in the Mekong Valley, 1945–1975. Comparative Technology Transfer and Society 4 (3), 225–246.

Bing, C., Wu, Y., Dong, Z., Keru, W., 2001. Study on the AE characteristics of fracture process of concrete beams. J. Building Mater. 4 (4), 332–338.

Bischoff, P.H., Perry, S.H., 1991. Compressive behavior of concrete at high strain rates. Mater. Struct. 24 (6), 425–450.

Bischoff, P.H., Perry, S.H., 1995. Impact behavior of plain concrete loaded in uniaxial compression. J. Eng. Mech. 121 (6), 685–693.

Bofang, Z., 1998. Finite Element Method Theory and Applications, second ed. China Water Power Press, Beijing.

Boore, D.M., Atkinson, G.M., 2007. Boore-Atkinson NGA Ground Motion Relations for the Geometric Mean Horizontal Component of Peak and Spectral Ground Motion Parameters. Pacific Earthquake Engineering Research Center University of California, Berkeley, pp. 1–234.

Boyan, Z., Houqun, C., et al., 2002. Effect of transverse slots of concrete arch dam on abutment seismic stability. Water Resources Hydropower Eng. 33 (6).

Boyan, Z., Houqun, C., Jian, T., 2004. Improved FEM based on dynamic contact force method for analyzing the stability of arch dam abutment. J. Hydraulic Eng. 35 (10).

Brace, W.F., Joncs, A.H., 1971. Comparison of uniaxial deformation in shock and static loading of three rocks. Geophys. Res. 76 (20), 4913–4921.

Brune, J.N., 1983. Tectonic stress and the spectra of seismic shear waves from earthquakes. Geophys. Res. 75, 4997–5009.

Campbell, K.W., Bozorgnia, Y., 2007. Campbell-Bozorgnia NGA Ground Motion Relations for the Geometric Mean Horizontal Component of Peak and Spectral Ground Motion Parameters. Pacific Earthquake Engineering Research Center University of California, Berkeley, pp. 1–204.

Chao, S., Tiantang, Y., Hongdao, J., 2002. FEM-based dynamic optimum design method for high arch dams and its application. J. Hehai Univ. 30 (1), 1–5.

Chen, H.L., Cheng, C.T., Chen, S.E., 1992. Determination of fracture parameters of mortar and concrete beams by using acoustic emission. Mater. Eval. (7), 888–894.

Chen, S., Ning, B., et al., 2007. A damage constitutive model of cemented soil on meso-fracture process testing. Rock Soil Mech. 28 (1), 93–96.

Chengkui, H., Guofan, Z., Renjee, S., Yupu, S., 1997. Research on strength and deformation behavior under dynamic loading and its test methods. Design Hydroelec. Power Station 13 (1), 17–22.

Chengqiu, Y., Zheng, W., 2000. Experimental study on basic mechanical properties of full mix concrete. Adv. Sci. Technol. Water Resources 20 (3), 29–32.

Chenguang, B., Yiming, W., Jianming, Z., 1996. The relation between dimension of defects of rock material and its damage evolution. Min. Metall. 5 (2), 17–19.

Chengzhi, Q., Qihu, Q., 2003. Physical mechanism of dependence of material strength on strain rate for rock-like material. Chinese J. Rock Mech. Eng. 22 (2), 177–181.

Chiou, B.S.J., Youngs, R.R., 2007. Chiou and Youngs PEER-NGA Empirical Ground Motion Model for the Average Horizontal Component of Peak Acceleration and Pseudo-Spectral Acceleration for Spectral Periods of 0.01 to 10 Seconds. Pacific Earthquake Engineering Research Center University of California, Berkeley, pp. 1–219.

Cho, S.H., Ogata, Y., Kaneko, K., 2003. Strain-rate dependency of the dynamic tensile strength of rock. Int. J. Rock Mech. Min. Sci. 40, 763–777.

Chuhan, Z., Feng, J., Pekou, O.A., 1995. Time domain procedure of FE-BE-IBE coupling for seismic interaction of arch dam and canyons. Earthquake Eng. Struct. Dynam. 24, 1651–1666.

Chuhan, Z., 2005. High dam-key science & technology problem in hydropower station construction. Guizhou Water Power 19 (2), 1–4.

Chunan, T., Wanchen, Z., 2003. Concrete Breakage & Rupture-Numerical Simulation Tests. Science Press, Beijing.

Cowell, W.L., 1966. Dynamic Properties Of Plain Portland Cement Concrete (No. NCEL-TR-447). Naval Civil Engineering Lab, Port Hueneme, Calif.

CSMIP, Report OSMS 94-04 [M], Strong-motion data from the Northridge/San Fernando valley earthquake of January 17, 1944.

Cuiran, Z., Houqun, C., 2007. Study on evolutionary power spectra of non-stationary earthquake accelerograms. J. Hydraulic Eng. 38 (12), 1475–1481.

Cuiran, Z., Houqun, C., 2008. Prediction of non-stationary earthquake accelerograms compatible with design evolutionary spectra. Earthquake Eng. Eng. Vibration 28 (3), 24–32.

Cuiran, Z., Houqun, C., Min, L., 2007. Earthquake accelerogram simulation with statistical law of evolutionary power spectrum. Acta Seismologica Sinica 29 (4), 409–428.

Cunzhong, Z., 1991. Earthquake Dictionary. Shanghai Wordbook Press, Shanghai.

Dai, R., Li, Z., Liu, H., 2006. Research and implementation serial parallel explicit Lagrangian finite difference method. Chinese J. Rock Mechan. Eng. 2006 (8), 83–89.

Dan, Z., Qingbin, L., 2004. Dynamic constitutive equations for concrete based on meso-cracks and a static constitutive model. J. Tsinghua Univ. 44 (3), 410–412.

Daorong, W., Shisheng, H., 2002. Influence of aggregate on the compression properties of concrete under impact. J. Exp. Mech. 17 (1), 23–27.

Dasgupta, G.A., 1982. Finite element formulation for unbounded homogeneous continua. J. Appl. Mech. 49, 136–140.

Dhir, R.K., Sangha, C.M., 1972. A study of relationships between time, strength, deformation and fracture of plain concrete. Magazine Concrete Res. 24 (81), 197–208.

Dilger, W.H., Koch, R., Kowalczyk, R., 1984. Ductility of plain and confined concrete under different strain rates. ACI J. 81 (1), 73–81.

Ding, M., Hvid, I., 2000. Quantification of age-related changes in the structure model type and trabecular thickness of human tibial cancellous bone. Bone 26 (3), 291–295.

Dominguez J, et al. Model for the seismic analysis of arch dams including interaction effects [C]. Proc. 10th WCEI, Madrid. 1992.

Dong, Y., Xie, H., Zhao, P., 1997. Experimental study and constitutive model on concrete under compression with different strain rate. Journal of Hydraulic Engineering 7, 72–77.

Eibl, J., Curbach, M., 1989. An attempt to explain strength increase due to high loading rates. Nuclear Engineering and Design 112, 45–50.

Elfahal, M.M., Krauthammer, T., Ohno, T., Beppu, M., Mindess, S., 2005. Size effect for normal strength concrete cylinders subjected to axial impact. Int. J. Impact Eng. 31 (4), 461–481.

Eshelby, J.D., 1957. The determination of the elastic field of an ellipsoidal inclusion and related problems. Proc. R. Soc. London Ser A 240, 367–396.

Eshelby, J.D., 1959. The elastic field outside an ellipsoidal inclusion. Proc. R. Soc. London A252, 561–569.

Evans.R.H., 1968. Effect of rate of loading on some mechanical properties of concrete [A]. Proc. of conference on mechanical properties of non-metallic brittle materials [C]. London, 175–192.

Faning, D., 2005. Damage-fracture evolution theory of rock and soil(I): damage-fracture space. Rock Soil Mech. 26 (4), 513–519.

Faning, D., 2005. Damage-fracture evolution theory of rock and soil (II): physical state indexes and divisional damage-fracture theory. Rock Soil Mech. 26 (5), 673–679.

Faning, D., Xiaotao, Y., et al., 2005. Subarea breakage constitutive model of rock mass based on CT test. Chin. J. Rock Mech. Eng. 24 (22), 4003–4009.

FEMA, Federal Guidelines for Dam Safety, Earthquake Analyses and Design of Dams, May 2005.

Feng, M.Q., Kim, Y.J., Park, K., 2006. Real-time and hand-held microwave NDE technology for inspection of FRP-wrapped concrete structures. FRP Int. 3 (2), 2–5.

Flangan, D.P., Belytschko, T., 1981. A uniform strain hexahedron and quadrilateral with orthogonal hourglass control. Int. J. Numer. Methods Eng. 17, 679–706.

Gao, Z., Liu, G., 2003. Two-dimensional random aggregate structure for concrete. Journal of Tsinghua University 43 (5), 710–714.

Gary, G., Bailly, P., 1998. Behavior of quasi-brittle material at high strain rate experiment and modeling. Eur. J. Mech. 17 (3), 403–420.

Ge Xiurun, Ren Jianzi, Pu Yibin, et al., 1999. A real-in-time CT triaxial testing study of meso-damage evolution law of coal. Chinese Journal of Rock Mechanics and Engineering 18 (5), 497–502.

Gengshe, Y., Dingyi, X., Changqing, Z., et al., 1996. CT identification of rock damage properties. Chinese J. Rock Mech. Eng. 15 (1), 48–54.

Gengshe, Y., Dingyi, X., Changqing, Z., 1998. The quantitative analysis of distribution regulation of CT values of rock damage. Chinese J. Rock Mech. Eng. 17 (3), 279–285.

Genshi, Y., Changqing, Z., 1998. Rock Mass Damage and Detection. Shaanxi Sci-Tech Press, Xian.

Grote, D.L., Park, S.W., Zhou, M., 2001. Experimental characterization of the dynamic failure behavior of mortar under impact loading. J. Appl. Phys., 2115–2123.

Guangting, L., Zhenggou, G., 2003. Random 3-D aggregate structure for concrete. J. Tsinghua Univ. 43 (8), 1120–1123.

Guangting, L., Zongmin, W., 1996. Numerical simulation study of fracture of concrete materials using random aggregate model. J. Tsinghua Univ. 36 (1), 84–89.

Gueiming, L., 2006. Non-Destructive Detection Technology. National Defense Press, Beijing.

Guoping, L., 1990. Finite element program generator and finite element language. Adv. Mech. 20 (2), 199–204.

Guoping, L., 2009. Finite Element Language. Science Press, Beijing.

Guoxin, W., 2001. A study on attenuation of strong ground motion. Institute of Engineering Mechanics. China Earthquake Administration, Harbin.

Haifei, Y., Shijiao, Y., Sheng, Z., 2008. Study on calculating method for reliability of geotechnical engineering. J. Water Resour. Archit. Eng. 6 (2).

Hanks, T.C., Kanamori, H., 1979. A moment-magnitude scale. Geophys. Res. 84, 2348–2350.

Hao, J., Weishen, Z., Shucai, L., Yibin, P., et al., 2002. Real-time CT Test of hydraulic fracture process for jointed rock-masses. Chinese J. Rock Mech. Eng. 21 (11), 1655–1662.

Harris, D.W., Mohorovic, C.E., Dolen, T.P., 2000. Dynamic properties of mass concrete obtained from dam cores. ACI Mater. J. 97 (3), 290–296.

Harris, David W., Mohorovic, Caroline E., and Dolen, Timothy P., (2000). Dynamic properties of mass concrete obtained from dam cores, Title No. 97-M35, ACI Mater. J., May-June.

Hatano, T., Tsutsumi, H., 1960. Dynamical compressive deformation and failure of concrete under earthquake load. Proc Second World Conference on Earthquake Engineering. Tokyo: Science Council of Japan. pp. 1963–1978.

Hatano, T., 1960. Dynamical Behaviors of Concrete Under Impulsive Tensile Load. Technical Laboratory, Central Research Institute of Electric Power Industry.

Hatano T. (1960). Relations between strength of failure, strain ability, elastic modulus and failure time of concretes. Technical Report C-6001, Technical Laboratory of the Central Research Institute of Electric Power Industry.

He, S., Plesha, M.E., Rowlands, R.E., Bažant, Z.P., 1992. Fracture energy tests of dam concrete with rate and size effects. Dam Engineering 3 (2), 139–159.

Hejin, T., Biao, W., Xin, Z., Jie, Z., 1998. An experimental unit for the dynamic mechanical properties of rock. J. Xían Shiyou Univ. 13 (1), 19–21.

Heping, X., 1990. Rock Concrete Damage Mechanics. China Metallurgical University Press, Jiangsu.

Heping, X., 1996. Fractal-Concrete Mechanics Introduction Theory. Science Press, Beijing.

Heping, X., Yang, J., 1997. Fractal characteristics of meso/micro damage and fracture of concrete. J. China Coal Soc. 22 (6), 586–590.

Higdon, R.L., 1991. Absorbing boundary conditions for elastic waves. Geophysics 56 (2), 231–241.

Hongguang, J., Tianshen, Z., Meifeng, C., et al., 2000. Experimental study on concrete damage by dynamic measurement of acoustic emission. Chinese J. Rock Mech. Eng. 19 (2), 65–168.

Hordijk, D.A., 1992. Tensile and tensile fatigue behavior of concrete: experiments, modeling and analyses. HERON 37 (1), 5–45.

Horsch, T., Wittmann, F.H., 2001. Three-dimensional numerical concrete applied to investigate effective properties of composite materials. Fracture mechanism of concrete structures, 57–64.

Houqun, C., 2006. Discussion on seismic input mechanism at dam-site. J. Hydraulic Eng. 37 (12), 1417–1423.

Houqun C., Deyu L., Analysis of dynamics of gravity dams and research on earthquake resistance design [m] «gravity dam design for 20 years» Contributed by Zhou Jianping, Niu Xinqiang, Jia Jinsheng, China Electric Power Press, 2008.

Houqun, C., Huaifa, M., Yuncheng, L., 2007. Influence of random aggregate shapes on flexural strength of dam concrete. J. China Inst. Water Resources Hydropower Res. 5 (4), 1–6.

Houqun, C., Zeping, X., Min, L., 2008. Wenchuan earthquake and seismic safety of large dams. J. Hydraul. Eng. 10 (19), 1158–1167.

Hua, C., Rui, F., Youming, Z., Fachun, Z., 2009. Stability analysis of rock high side-slope based on finite element strength reduction method. Highway (10), 83–86.

Huaifa, M., Houqun, C., 2008. Research On Concrete Dynamic Damage And Destruction Mechanism Of Full-Gradation Large Dams And Its Meso-Mechanics Analysis Method. China Electric Power Press, Beijing.

Huaifa, M., Hou-qun, C., Bao-kun, L., 2004. Meso-structure numerical simulation of concrete specimens. J. Hydraul. Eng. (10), 27–35.

Huaifa, M., Houqun, C., Baokun, L., 2004. Review on micro-mechanics studies of concrete. J. China Inst. Water Resour. Hydropower Res. 2 (2), 124–130.

Huaifa, M., Houqun, C., Baokun, L., 2005. Influence of meso-structure heterogeneity on dynamic bending strength of concrete. J. Hydraul. Eng. 36 (7), 846–852.

Huaifa, M., Houqun, C., Baokun, L., 2005. The concrete dynamical flexure bending strength under static pre-loading. J. China Inst. Water Resour. Hydropower Res. 3 (3), 168–172.

Huaifa, M., Houqun, C., Baokun, L., 2005. Influence of strain rate effect on dynamic bending strength of concrete. J. Hydraul. Eng. 36 (1), 69–76.

Huaifa, M., Shuzhen, M., Houqun, C., 2006. A generating approach of random convex polygon aggregate model. J. China Inst. Water Resour. Hydropower Res. 4 (3), 196–201.

Huaifa, M., Houqun, C., Yongfa, Z., Baokun, L., 2007. Parallel computation study on 3D meso-mechanics of dam concrete. Eng. Mech. 24 (10), 74–79.

Huaifa, M., Houqun, C., Jianping, W., Baokun, L., 2008. Study on numerical algorithm of 3D meso-mechanics model of dam concrete. Chinese J. Comput. Mech. 25 (2), 241–247.

Huaifa, M., Litao, W., Hou-qun, C., Deyu, L., 2010. Mechanism of dynamic damage delay characteristic of concrete. J. Hydraul. Eng. 41 (6), 659–664.

Huaifa, M., Houqun, C., Shufeng, X., 2012. Review on the advance of seismic response analysis of high concrete dam systems under strong earthquake. J. China Inst. Water Resour. Hydropower Res. 10 (1), 1–7.

Hughes, B.P., Gregory, R., 1972. Concrete subjected to high rates of loading in compression. Magazine Concrete Res. 24 (78), 25–36.

Hughes, B.P., Watson, A.J., 1978. Compressive strength and ultimate strain of concrete under impact loading. Magazine Concrete Res. 30 (105), 188–199.

Idriss, I.M., 2007. Empirical Model for Estimating the Average Horizontal Values of Pseudo -absolute Spectral Accelerations generated by Crustal Earthquakes. Pacific Earthquake Engineering Research Center University of California, Berkeley, pp. 1–76.

Ji, C.L., Xia, S.Y., Deng, D.S., 1995. Dynamic Analysis of Asphalt Concrete Core Rockfill Dams on Deep Overburden. Journal Of Hohai University (Natural Sciences) 23 (5), 74–80.

Jianfei, L., Shisheng, H., Daorong, W., 2001. A novel experimental method of Hopkinson pressure bar system for brittle material. J. Exp. Mech. 16 (3), 283–290.

Jianfei, Z., 2004. Parallel Boundary Element Method on Cluster and its Application in Hydraulic Structure Analysis. Hohai University, Nanjing.

Jiangrong, J., 1998. Study on the material properties of high-strength mass concrete. Water Power (3), 17–20.

Jianxi, R., Xiurun, G., et al., 2000. Real- time CT test on the meso- damage evolution of rock under uniaxial compression. China Civil Eng. J. 33 (6), 99–104.

Jiazeng, P., 2007. Water Power and China: Practical Cases Of China Large Dam Technology Advanced Level And Engineering. China Electric Power Press, Beijing.

Jin, T., Shunzai, H., Houqun, C., 2000. Influence of longitudinal joint on seismic response of gravity dam. J. Hydraul. Eng. (12), 53–58.

Jinchen, X., 1981. Point-Set Topography Lecture Sheets. Higher Education Press, Beijing.

Jinfu, Z., Xin, Q., 1991. An efficient parallel algorithm for multi-transputer system—ABC method. J. Nanjing Univ. Aeronautics Astronautics 23 (2), 55–62.

Jinsheng, H., Degao, T., Xiangxin, C., Zaosheng, Z., 2003. Method of enhancing experimental precision for big radial size SHPB equipment. J. PLA Univ. Sci. Technol. (2), 71–74.

Jinsheng, J., Yulan, Y., Zhongli, M., 2007. A Brief Description of Chinese Large Dam Statistics and an Introduction to the Conditions of the World Large Dams. The Practical Examples of Chinese Large Dam Technology Development and Levels. China Electric Power Press, Beijing.

Jinting, W., Xiuli, D., Chuhan, Z., 2002. Numerical stability of explicit finite element schemes for dynamic system with Rayleigh damping. Earthquake Eng. Eng. Vibration 22 (6).

Jin-xian, Wang, 1991. The parallel direct method on MIMD computer for finite element structure analysis. Mathematica Numerica Sinica (4), 433–438.

Kaplan, S.A., 1980. Factors affecting the relationship between rate of loading and measured compressive strength of concrete. Magazine of Concrete Research 32 (111), 79–88.

Katsuta, T., 1943. On the elastic and plastic properties of concrete in compression tests with high deformation velocity, Part 1. Trans Inst Jap Arch 29, 268–274.

Kawakata, H., Cho, A., Kiyama, T., et al., 1999. Three-dimensional observations of faulting process in Westerly granite under uniaxial conditions by CT X-ray scan. Tectonophysics (313), 293–305.

Kennedy, G., Ferranti, L., 2002. Dynamic high-strain-rate mechanical behavior of microstructurally biased two-phase TIB2+AL$_2$O$_3$ ceramics. J. Appl. Phys., 1921–1927.

Khair, A.W., 1995. Stress rate Effect on a.e. Acoustic Emission/Micro-seismic Activity in Geological Structure and Materials, In: Hardy Jr., H.R. (Ed.). Proceedings of the 5th Conference, pp. 29–43.

Krajcinovic, D., Fonseka, G.U., 1981. The continuous damage theory of brittle materials. Part 1: general theory. J. Appl. Mech. 48, 809–815.

Kvirikadze, O.P., 1977. Determination of the ultimate strength and modulus of deformation of concrete at different rates of loading. In: Proc. RILEM International Symposium on Testing *In-Situ* Concrete Structures. Budapest. pp. 9–17.

Labuz, J.F., Cattane, S., Chen, L.-H., 2001. Acoustic emission at failure in quasi-brittle materials. Constr. Building Mater. 15 (5), 225–233.

Landis, E.N., 1999. Micro-macro fracture relationships and acoustic emissions in concrete. Constr.Building Mater. 13 (1), 65–72.

Lei, Z., Hongdao, J., Hai-lin, P., 2006. Building, testing, and application of high performance computer clusters. Adv. Sci. Technol. Water Resour. (4), 65–69.

Lei, Z., Hongdao, J., Jianfei, Z., 2007. Numerical computation based on campus grid. Microelectron. Comput. 24 (9), 42–44.

Li, P., Liu, G., Gao, H., Chen, F., 2001. Study on experimental method for determining autogenous volume deformation and its application. Journal of Tsinghua University (Science and Technology) 41 (11), 114–117.

Li, Z., Kulkarni, S.M., Shah, S.P., 1993. New test method for obtaining softening response of un-notched concrete specimen under uniaxial tension. Exp. Mech. 33 (3), 181–188.

Li, Z., Shah, S.P., 1994. Localization of microcracking in concrete under uniaxial tension. ACI Materials Journal 91 (4), 372–381.

Litao, W., Houqun, C., Huaifa, M., 2009. Computation of dynamical contact problems with artificial boundary conditions based on FEPG. J. Hydroelectric Eng. 28 (5), 179–182.

Liu, G., Gao, Z., 2003. Random 3-d aggregate structure for concrete. Journal of Tsinghua University 43 (8), 1120–1123.

Liu, G., Wang, Z., 1996. Numerical simulation study of fracture of concrete materials using random aggregate model. Qinghua Daxue Xuebao/Journal of Tsinghua University 36 (1), 84–89.

Liu, Q., Chen, X., Gindy, N., 2006. Investigation of acoustic emission signals under a simulative environment of grinding burn. International Journal of Machine Tools and Manufacture 46 (3), 284–292.

Lu, X., 1995. Application of identification methodology to shaking table tests on reinforced concrete columns. Engineering Structures 17 (7), 505–511.

Lu, Y.F., Aoyagi, Y., 1995. Acoustic emission in laser surface cleaning for real-time monitoring. Japanese journal of applied physics 34 (11B), L1557.

Lv, P.Y., 2001. Experimental study on dynamic strength and deformation of concrete under uniaxial and biaxial action. Dalian University of Technology, Dalian.

Lysmer, J., Kuhlemeyer, R.L., 1969. Finite dynamic model for infinite media. J. Eng. Mechan. Div. ASCE 95 (4), 859–878.

Malvar, L.J., Ross, C.A., 1998. Review of strain rate effects for concrete in tension. ACI Mater. J. 95 (6), 735–739.

McCabe, W.M., Koerner, R.M., Lord, Jr., A.E., 1976. Acoustic emission behavior of concrete laboratory specimens. J. Am. Concrete Inst. 73 (7), 367–371.

Mellinger F. M., Birkimer D. L., 1966. Measurement of stress and strain on cylindrical test specimens of rock and concrete under impact loading. Technical Report 4–46, U.S. Army Corps of Engineers, Ohio River Division Laboratories, Cincinnati, Ohio.

Mihashi, H., Nomura, N., Niiseki, S., 1991. Influence of aggregate size on fracture process zone of concrete detected with three dimensional acoustic emission technique. Cement Concrete Res. 21 (5), 737–744.

Ming, Z., Zhongkui, L., Xia, S., 2005. Probabilistic volume element modeling in elastic damage analysis of quasi-brittle materials. Chinese J. Rock Mech. Eng. 24 (23), 4282–4288.

Motazedian, D., Atkinson, G., 2005. Dynamic corner frequency: a new concept in stochastic finite fault modeling, reviewing. BSSA 28 (14), 2531–2545.

Muravin, G.B., Lezvinskaya, L.M., 1982. Investigation of the spectral density of acoustic-emission signals. Defektoskopiya 7, 10–15.

Muravin, G.B., Likhod ko, A.D., Lezvinskaya, L.M., 1982. Investigation of the dynamic strength of concrete and of its acoustic emission. Strength Mater. 6, 883–886.

Naihua, R., Zhonghua, J., 1995. Large Dam Incidents and Their Safety·Arch Dams. China Water Resources & Electric Power Press, Beijing.

Naiyang, D., Yingjie, T., 2004. New Method in Data Digging – Support Vector Machine. Science Press, Beijing.

Nakayama, T., Fujiwara, H., Komatsu, S., Sumida, N., 1994. Non-stationary response and reliability of linear systems under seismic loadings. In: Schueller, Shinozuka, Yao (Eds.), Structural Safety & Reliability, Balkema, Rotterdam, pp. 2179–2186.

Ohtsu, M., 1996. The history and development of acoustic emission in concrete engineering. Mag. Concrete Res. 48 (177), 321–328.

Ohtsu, M., Okamoto, T., Yuyama, S., 1998. Moment tensor analysis of acoustic emission for cracking mechanisms concrete. ACI Struct. J. 95 (2), 87–95.

Paulmann, K., Steinert, J., 1982. Concrete under very short-term loading. Beton (6), 225–228.

Paulmann, K., Steinert, J., 1983. Verhalten der Biegedruckzone nach schneller, hoher und kurzzeitiger Belastung.

Peiyin, L., Yupu, S., 2002. Dynamic compressive test of concrete and its constitutive model. Ocean Eng. 2, 43–48.

Peiyin, L., Yupu, S., Zhimin, W., 2001. Strength and deformation characteristics of concrete subjected to different loading rates combined with confined stress. J. Dalian Univ. Technol. 6, 716–720.

Priestley, M.B., 1965. Evolutionary spectra and non-stationary processes. J. R. Stat. Soc., Ser B Methodol. 27 (2), 204–237.

Proulx, J., Darbre, G.R., Kamileri, N., Aug 1–6, 2004. Analytical and experimental investigation of damping in arch dams based on recorded earthquakes. Proc 13th World Conference Earthquake Engineering. Vancouver, Canada. , Paper No. 68.

Qi, C., Qian, Q., 2003. Physical mechanism of dependence of material strength on strain rate for rock-like material. Chinese Journal of Rock Mechanics & Engineering 22 (2), 177–181.

Qiang, Y., Jicheng, R., Hao, Z., 2002. Numerical simulation of the pullout of anchor in rock. J. Hydraul. Eng. 12, 68–73.

Qiang, Y., Hao, Z., Weiyuan, Z., 2002. Lattice model for simulating failure process of rock. J. Hydraul. Eng. 4, 46–50.

Qiang, Y., Yonggang, C., Hao, Z., 2003. Simulation of cracking processes of rock materials by lattice model. Eng. Mech. 20 (1), 117–120.

Qiuling, L., Jue, Z., 2002. Vector analysis for collision detection of 3-D models with triangulated mesh surface. Comput. Aided Eng. 3 (1), 69–72.

Quyang, C.S., Landis, E., Shah, S.P., 1991. Damage assessment in concrete using quantitative acoustic emission. J. Eng. Mech. Div. 117 (11), 2681–2698.

Raphael, J.M., 1984. Tensile strength of concrete. ACI J. 81 (2), 158–165.

Reinhardt, H.W., Rossi, P., Van Mier, J.G.M., 1990. Joint investigation of concrete at high rates of loading. Mater. Struct. (23), 213–216.

Rome, J.I., 2002. Experimental Characterization and Micromechanical Modeling of the Dynamic Response and Failure Modes of Concrete. University of California, San Diego.

Rome, J., Isaacs, J., Nemat-Nasser, S., 2002. Hopkinson techniques for dynamic triaxial compression tests. Recent Advances in Experimental Mechanics. Springer, Netherlands, 3–12.

Rong, D., Zhongkui, L., Hui, L., 2006. Research and implementation of serial-parallel explicit Lagrangian finite difference method. Chinese J. Rock Mech. Eng. 8, 83–89.

Ross, C.A., Tedesco, J.W., Kuennen, S.T., 1995. Effects of strain rate on concrete strength. ACI Mater. J. 92 (1), 37–47.

Ross, C.A., Jerome, D.M., Tedesco, J.W., Hughes, M.L., 1996. Moisture and strain rate effects on concrete strength. ACI Mater. J. 93 (3), 293–300.

Rossi, P., 1991. A physical phenomenon which can explain mechanical behavior of concrete under high strain rates. Mater. Struct. 24 (6), 422–424.

Rossi, P., 1991. Influence of cracking in the presence of free water on the mechanical behavior of concrete. Mag. Concrete Res. 43 (154), 53–57.

Rossi, P., 1997. Strain rate effects in concrete structures: the LCPC experience. Mater. Struct. 3, 54–62.

Rossi, P., Boulay, C., 1990. Influence of free water in concrete on the cracking process. Mater. Struct. 42 (152), 142–146.

Rossi, P., Miex, V., 1994. Effect of loading rate on the strength of concrete subjected to uniaxial tension. Mater. Struct. 27, 260–264.

Rossi, P., Toutlemonde, F., 1996. Effect of loading rate on the tensile behavior of concrete: description of the physical mechanisms. Mater. Struct. 29, 116–118.

Rossi, P., Ulm, F.J., Hachi, F., 1996. Compressive behavior of concrete: physical mechanisms and modeling. Journal of Engineering Mechanics 122 (11), 1038–1043.

Rossi, P., Wu, X., 1992. A probabilistic model for material behavior analysis and appraisement of the concrete structures. Mag. Concrete Res. 44 (161), 271–280.

Rossi, P., Robert, J.L., Gervais, J.P., Bruhat, D., 1989. Identification of physical mechanisms underlying acoustic emission during the cracking of concrete. Mater. Struct. 22 (129), 194–198.

Rossi, P., Robert, J.L., Gervais, J.P., Bruhat, D., 1990. The use of acoustic emission in fracture mechanics applied to concrete. Engineering Fracture Mechanics 35 (4-5), 751–763.

Rossi, P., Van Mier, J.G.M., Boulay, C., Le Maou, F., 1992. The dynamic behavior of concrete: influence of free water. Mater. Struct. 25, 509–514.

Rossi, P., Van Mier, J.G., Toutlemonde, F., Le Maou, F., Boulay, C., 1994. Effect of loading rate on the strength of concrete subjected to uniaxial tension. Materials and structures 27 (5), 260–264.

Ruqing, Z., 1993. Parallel Computation Structure Mechanics. Chongqin University Press, Chongqin, China.

Sadowska-Boczar, E., 1983. Application of acoustic emission to investigate the mechanical strength of ceramic materials. Sec. Ceram. 12 (8), 639–642.

Sagaidak, A.I., Elizarov, S.V., 2007. Acoustic emission parameters correlated with fracture and deformation processes of concrete members. Constr. Building Mater. 21 (3), 477–482.

Schlangen, E., Garbocai, E.J., 1997. Fracture simulations of concrete using lattice models: computational aspects. Eng. Frac. Mech. 57 (2/3), 319–332.

Schlangen, E., van Mier, J.G.M., Sept. 1991. Lattice model for numerical simulation of concrete fracture. International conference on dam fracture (Vol. 527). Denver, Colorado, USA. pp. 513–527.

Shang, R.J., 1994. Studying of the dynamic constitutive behavior of concrete. Ph.D. Thesis, Dalian University of Technology, Dalian, China.

Shaochiu, S., Lili, W., 2000. Passive confined pressure SHPB test method for materials under quasi-one dimensional strain state. J. Exp. Mech. 15 (4), 377–384.

Shengxing, W., Shunxiang, Z., Dejian, S., 2008. An experimental study on Kaiser effect of acoustic emission in concrete under uniaxial tension loading. China Civil Eng. J. 48 (4), 31–39.

Shi, X.Q., Pang, H.L.J., Zhou, W., Wang, Z.P., 2000. Low cycle fatigue analysis of temperature and frequency effects in eutectic solder alloy. International Journal of Fatigue 22 (3), 217–228.

Shisheng, H., Daorong, W., 2002. Dynamic constitutive relation of concrete under impact. Explosion Shock Waves 22 (3), 242–246.

Shisheng, H., Daorong, W., Jianfei, L., 2001. Experimental study of dynamic mechanical behavior of concrete. Eng. Mech. 10, 115–126.

Shiyou, W., Liuqing, H., Xibing, L., 2000. The fractal study on the damage of joint rock mass. Nonferrous Metals Sci. Eng. 14 (3), 14–16.

Shiyun, X., Gao, L., Zhe, W., Jingzhou, L., 2001. Effects of strain rate on dynamic behavior of concrete in tension. J. Dalian Univ. Technol. 41 (6), 721–725.

Shiyun, X., Gao, L., Zhe, W., Jingzhou, L., 2002. Effect of strain rate on dynamic behavior of concrete in compression. J. Harbin Univ. Civil Eng. Arch. 35 (5), 35–39.

Shuquan, Z., Weitai, L., 1993. Parallel direct integration algorithm in dynamic analysis of structures on YH-1 computer. Chinese J. Comput. Phys. 3, 6–16.

Smith, W.D., 1975. The application of finite element analysis to body wave propagation problem. Geophys. J. 4, 747–768.

Sadowska-Boczar, E., 1983. Application of acoustic emission to investigate the mechanical strength of ceramic materials. Science of Ceramics 12, 639.

Somerville, P., Irikula, K., Graves, R., Sawada, S., Wald, D., Abrahamson, N., Iwasaki, Y., Kagawa, T., Smith, N., Kowada, A., 1999. Characterizing crustal earthquake slip models for the prediction of strong ground motion. Seism. Res. Lett. 70 (1), 59–80.

Soroushian, P., Choi, K.-B., Alhamad, A., 1986. Dynamic constitutive behavior of concrete. ACI J. 83 (2), 251–259.

Sparks, P.R., Menzies, J.B., 1973. The effect of loading upon the static and fatigue strength of plain concrete in compression. Mag. Concrete Res. 25 (83), 73–80.

Spooner, D.C., 1972. Stress–stain–time relationships for concrete. Mag. Concrete Res. 24 (81), 197–208.

Suzuki, T., Ohtsu, M., 2004. Quantitative damage evaluation of structural concrete by a compression test based on AE rate process analysis. Constr. Building Mater. 18, 197–202.

Takashi, I., Koichi, K., Yuichiro, M., et al., 1983. Preliminary study on application of X-ray CT scanner to measurement of void fractions in steady state two-phase flows. J. Nuclear Sci. Technol. 20 (1), 1–12.

Takeda, J., Tachikawa, H., 1962. The mechanical properties of several kinds of concrete at compressive, tensile, and flexural tests in high rates of loading. Trans. Arch. Inst. Japan 77, 1–6.

Takeda, J., Tachikawa, H., 1972. Deformation and fracture of concrete subjected to dynamic load. Proc International Conference on Mechanical Behavior of Materials. Kyoto, Japan. pp. 267–277.

Takeda, J., Tachikawa, H., Fujimoto, K., 1974. Mechanical behavior of concrete under higher rate loading than in static test. Mech. Behav. Mater. Kyoto Soc. Mater. Sci, 1–10.

Takeda, S., 1971. U.S. Patent No. 3,597,292. U.S. Patent and Trademark Office, Washington, DC.

Tang, J., 2006. Composite grid method based on FEPG and the Lagrange multipliers methods of contact problems. University of Electric Technology of Guilin. MD Dissertation (in Chinese).

Teng, J.G., Yao, J., 2000. Self-weight buckling of FRP tubes filled with wet concrete. Thin-walled structures 38 (4), 337–353.

Tenshanbangjiu. Application of Acoustic Emission (AE) Technology (Feng Xiating, Trans.). Metallurgical Industrial Press, Beijing; 1997.

Terrien, M., 1980. Emission Acoustique et comportement mécanique post-critique d'un.béton sollicité en traction. Bull liaison Ponts Chaussées 105, 65–72.

Tham, L.G., Liu, H., Tang, C.A., Lee, P.K.K., Tsui, Y., 2005. On tension failure of 2-crack specimens and associated acoustic emission. Rock Mech. Rock Eng. 38 (1), 1–19.

Tiantang, Y., Hongdao, J., 1999. Elasto-plastic substructure parallel finite element method with multi-front parallel processing. Chinese J. Comput. Mech. 16 (4), 493–496.

Tiantang, Y., Hongdao, J., 1999. A parallel solution of dynamic response of structure. J. Hehai Univ. 27 (3), 75–78.

Tingjie, Z., Deyao, Z., Xu, D., 1994. Dynamic mechanical behavior of materials under plate impact loading. Rare Metal Mater. Eng. (8), 1–6.

Ueta, K., Tani, K., Kato, T., 2000. Computerized X-ray topography analysis of three-dimensional fault geometries in basement-induced wrench faulting. Eng. Geol. (56), 197–210.

Van Mier, J.G., Chiaia, B.M., Vervuurt, A., 1997. Numerical simulation of chaotic and self-organizing damage in brittle disordered materials. Computer methods in applied mechanics and engineering 142 (1), 189–201.

Van Mier, J.G.M., 1991. Fracture Processes of Concrete-Assessment of Material Parameters for Fracture Models. CRC Press, New York.

Van Mier, J.G.M., Shah, S.P., Arnaud, M., Balayssac, J.P., Bascoul, A., Choi, S., et al., 1997. Strain-softening of concrete in uniaxial compression. Materials and Structures 30 (4), 195–209.

Van Mier, J.G.M., Vervuurt, A., Van Vliet, M.R.A., 1999. Materials engineering of cement-based composites using lattice models. In: Catpinteri, A., Aliabadi, M. (Eds.), Computational Fracture Mechanics in Concrete Technology. WIT Press, Boston, Southampton, pp. 1–32.

Walraven, J.C., 1991. High-strength concrete. Concrete Precasting Plant and Technology (BFT) 6, 45–52.

Walraven, J.C., Reinhardt, H.W., 1991. Theory and experiments on the mechanical behavior of cracks in plain and reinforced concrete subjected to shear loading. Heron 26 (1A), 26–35.

Wancheng, Z., Chunan, T., et al., 2002. Numerical simulation on the fracture process of concrete specimen under static loading. Eng. Mech. 19 (6), 148–153.

Wang, J., 1991. The parallel direct method on MIMD computer for finite element structure analysis. Chinese J. Comput. Math. 1991 (4), 433–438.

Wang, J.-T., Du, X.-L., Zhang, C.-H., 2002. Numerical stability of explicit finite element schemes for dynamic system with Rayleigh damping. Earthquake Eng. Eng. Vibrat. 22 (6), 18–24.

Wang, L., Shen, W., Xie, H., Neelamkavil, J., Pardasani, A., 2002. Collaborative conceptual design—state of the art and future trends. Computer-Aided Design 34 (13), 981–996.

Wang, W.M., 1997. Stationary and propagative instabilities in metals – a computational point of view. TU Delft, Delft University of Technology.

Wang, X., Min, S., 2003. Finite Element Method. Tsinghua University Press, Tsinghua.

Wang, Z., 2000. Numerical simulation of strain softening and localization for concrete materials. Journal of Bioscience & Engineering.

Watstein, D., 1953. Effect of straining rate on the compressive strength and elastic properties of concrete. ACI J. (8), 729–744.

Weibull, W., 1939. The phenomenon of rupture of solids. I.V.A. Prov., No. 153, Stockholm.

Weihua, D., Yanqin, W., Yinbin, P., et al., 2003a. X-ray CT approach on rock-interior crack evolution under low strain rate. Chinese J. Rock Mech. Eng. 22 (11), 1793–1797.

Weihua, D., Yanqin, W., Yinbin, P., et al., 2003b. Measurement of crack width in rock interior based on X-ray CT. Chinese J. Rock Mech. Eng. 22 (9), 1421–1425.

Weihua, D., Houqun, C., Faning, D., et al., 2009. Manufacture of portable material test device forming a complete set with medical X-ray CT and its application in concrete damage study. J. Exp. Mech. 24 (3), 207–214.

Wells, D.L., Coppersmith, K.J., 1994. New empirical relationships among magnitude, rupture length, rupture width, rupture area, and surface displacement. Bull. Seism. Soc. Am. 84 (4), 974–1002.

Wen, S.-Y., Hu, L.-G., et al., 2000. The fractal study on the damage of joint rock mass. Jiangxi Nonferrous Metals 14 (3), 14–16.

Wenda, X., 2004. Reliability analysis of slope based on Monte Carlo – finite element method. J. Fuzhou Univ. 32 (1), 73–77.

Wu, K., Chen, B., Yao, W., 2000. Study on the AE characteristics of fracture process of mortar, concrete and steel-fiber-reinforced concrete beams. Cem. Conc. Res. 30 (10), 1495–1500.

Wu, S.X., Zhang, S.X., Shen, D.J., 2008. An experimental study on Kaiser effect of acoustic emission in concrete under uniaxial tension loading. China Civil Engineering Journal 41 (4), 31–39.

Wu, Z., 2002. Image segmentation based on fractal theory. Nanjing University Of Aeronautics and Astronautics. Master's Degree Thesis.

Wu, X., Fan, Y.M., Ying, S.H., 1997. Study on high strength concrete mixed with composite micro-aggregates. Journal of South China University of Technology 25 (11), 108–113.

Xiao, S.Y., 2002. Rate-dependent constitutive model of concrete and its application to dynamic response of arch dams. Ph.D. Dissertation, Dalian University of Technology, Dalian, China.

Xiao, Y., Yun, H.W., 2002. Experimental studies on full-scale high-strength concrete columns. ACI Structural Journal 99 (2), 199–207.

Xiaojun, L., Zhenpeng, L., Xiuli, D., 1992. An explicit finite difference method for viscoelastic dynamic problem. J. Earthquake Eng. Eng. Vibration 12 (4), 74–80.

Xin, Y., Ming, Y., Min, W., 2005. MPI based concurrent calculation and study on runoff and sediment mathematical model of the Yellow River. Yellow River 27 (3), 49–53.

Xiuren, G., Jianxi, R., Yibin, P., et al., 2004. Research on Macro-Meso Tests of Rock Soil Damage Mechanics. Science Press, Beijing.

Xiurun, G., Jianxi, R., Yibin, P., et al., 2000. CT real-time tests of rock-meso-damage expansion laws. Scientia Sinica Technologica (Series E) 30 (2), 104–111.

Xuefeng, Z., Yuanming, X., 2001. Development of material testing apparatus for intermediate strain rate test. J. Exp. Mech. 16 (1), 13–18.

Yan, D., Lin, G., 2006. Dynamic properties of concrete in direct tension. Cement and concrete research 36 (7), 1371–1378.

Yang, G., Xie, D., 1996. CT identification of rock damage properties. Chin. J. Rock Mech. Eng. 15 (1), 48–54.

Yang, G., Xie, D., et al., 1998. The quantitative analysis of distribution regulation of CT values of rock damage. Chin. J. Rock Mech. Eng. 17 (3), 279–285.

Yanwen, L., 2007. Study on the Mechanics Behavior of the Concrete Through CT Technique. Xian University of Technology, Xi'an.

Yingren, Z., Shangyi, Z., 2004. Application of strength reduction fem in soil and rock slope. Chinese J. Rock Mech. Eng. 23 (19), 3381–3388.

Yinhui, Z., 2003. A study on coal or rock acoustic emission mechanism and relevant experiments. J. Hunan Univ. Sci. Technol. 18 (4), 18–21.

Yiyan, L., Shaoxi, H., 1995. Experimental research on the complete stress-strain curve with hoop tension for SFRC. Eng. J. Wuhan Univ. 28 (3), 285–292.

Yon, J.-H., Hawkins, N.M., Kobayashi, A.S., 1992. Strain-rate sensitivity of concrete mechanical properties. ACI Mater. J. 89, 146–153.

Yon, J.H., Hawkins, N.M., Kobayashi, A.S., 1992. Strain-rate sensitivity of concrete mechanical properties. ACI Mater. J. 89 (2), 146–153.

Yong, Y., 2004. Controlling Early Cracks of Concrete Structure. Science Press, Beijing.

Youyun, L., 2005. The multi-scale analysis for the elastics mechanics performance of the composite material with rectangle/ cuboid periodicity. J. Hunan City Univ. 14 (4), 30–33.

Yuli, D., Heping, X., Peng, Z., 1997. Experimental study and constitutive model on concrete under compression with different strain rate. J. Hydraul. Eng. (7), 72–77.

Yuncheng, L., Huaifa, M., Houqun, C., Xiao, H., 2006. Approach to generation of random convex polyhedral aggregate model and plotting for concrete meso-mechanics. J. Hydraul. Eng. 37 (5), 588–592.

Yushan, N., 1997. Fractal analysis of micro-structural. J. Dalian Univ. Technol. 37 (S1), 72–76.

Yuxian, H., 2006. Earthquake Engineering, second ed. Seismological Press, Beijing.

Zaihua, L., Zhenghan, C., Yibin, P., 2002. A CT study on the crack evolution of expansive soil during drying and wetting cycles. Rock Soil Mech. 23 (4), 417–422.

Zaihua, L., Zhenghan, C., et al., 2002. Study on damage evolution of natural expansive soil with computerized tomography during triaxial shear test. Shuili Xuebao 6, 106–112.

Zaishun, H., Jinyu, L., Jianguo, C., et al., 2002. Dynamic tests and study on the fully-graded concrete for high arch dam. Water Power (1), 51–54.

Zengyan, C., Jianhua, Q., 1994. Preliminary study of concrete dynamic behaviors. Hydropower Design 10 (4), 81–85.

Zhaixia L., Jicheng T., Yinhe Y., Progress in dynamic damage mechanics, J. Hehai Univ., 22 (Special Edition of Calculation Mechanics), 1994, 112–118.

Zhengguo, G., Guangting, L., 2003. Two-dimensional random aggregate structure for concrete. J. Tsinghua Univ. 43 (5), 710–714.

Zhengpeng, L., 2002. Introductory Engineering Wave Motion Theory, second ed. Science Press, Beijing.

Zhigang, Y., Zhenkai, X., Chunwen, W., 2004. The applications of improved Monte Carlo method in hydro-concrete's reliability analysis. J. Nanchang Univ. 26 (2), 66–69.

Zhou, S., et al., 1993. Parallel direct integration algorithm in dynamic analysis of structures on YH-1 computer. Chinese J. Comput. Phys. 1993 (3), 6–16.

Zhu, D., Mobasher, B., Rajan, S.D., 2010. Dynamic tensile testing of Kevlar 49 fabrics. Journal of Materials in Civil Engineering 23 (3), 230–239.

Zhu, H.B., Sun, Z.Y., 2010. Quality Control and Monitoring for Abutment Groove Excavation of Xiluodu Arch Dam. Journal of Water Resources and Architectural Engineering 4, 010.

Zhu, J., Xin, Q., 1991. An efficient parallel algorithm for multitransputer system – ABC method. J. Nanjing Univ. Aeronaut. Astronaut. 23 (2), 55–62.

Zhujiang, S., 2003. Breakage mechanics in geological materials: an ideal brittle elastoplastic model. Chinese J. Geotech. Eng. 25 (3), 253–257.

Zhurkov, V.S., Pechennikova, E.V., Fel'Dt, E.G., Lkh, G., Tkhiem, T.K., 1979. Analysis of the chromosome aberrations in the bone marrow cells of white rats after inhalational ozone exposure. Gigiena I Sanitariia (9), 12–14.

Zielinski, J., Reinhardt, H.W., Kormeling, H.A., 1981. Experiments on concrete under uniaxial impact tensile loading. Mater. Struct. 14, 103–112.

Zietlow, W.K., Labuz, J.F., 1998. Measurement of the intrinsic process zone in rock using acoustic emission. Int. J. Rock Mech. Min. Sci. 35 (3), 291–299.

Zixing, L., Changjin, T., Mingbao, H., Ren, W., 1995. Experimental investigation of the impact mechanical properties of polyurethane rigid (PUR) foam plastics. Polymer Mater. Sci. Eng. (11), 76–81.

Zongjing, M., Pingren, D., 1995. Problems Concerning Present-Day Crust Motion. Seismic Press.

Zongmin, Wang., 2000. Numerical simulation of strain softening and localization for concrete materials. J. Basic Sci. Eng. 8 (2), 187–194.

Index

A

Abrahamson–Silva (ASOT) attenuation relation, 65, 71, 72
  acceleration response spectrum attenuation relation
      formula, 74
Abscissa cycle, 59
Absolute acceleration response time process, 59
Abutment aseismic stability, 139
  safety coefficient, 141
Acceleration response spectrum, 77
  attenuation relation of, 72
Acceleration time process, 56, 98
Acceleration time process generation, random finite fault
    method
  centro-NS seismic motion, 85
  dam-site seismic motion, 96
  evolutionary power spectrum, 85
    sketch example for earthquake magnitudes and
      epicenter distance fitting, 87
  fault face and weight distribution
    subsource classification on, 96
  fault fracture face and seismic source basic parameters, 95
  real measured seismic evolutionary power spectrum
    fitting time process curves in terms of, 89
  steps to, 92–96
    fault face geometric characteristics, determination of, 92
    real examples of engineering application, 95–96
    seismic motion time process generated by subsource
      fracture in dam-site, 93–94
    site earthquake motion time process synthesis, 94–95
    subfault face, classification of, 92–93
  time variation power spectrum empiric model
    parameters, 86
  zero direction seismic motion time process correlative
    coefficients, 89
Acoustic emission (AE) technique, 321, 322
  accumulative curve, 329, 331, 332, 383
  accumulative number, 361, 363
    curves, 368, 375
  acquisition system
    preamplifier (PAC-2/4/6) of, 329
    setting, 377
  activity, 363, 367, 369
  aggregates, analysis of, 357
    axial stress and acoustic emission, 360
    axial tension, loading rate of, 363
    granite aggregate specimens, 357
    hit rise time and amplitude distribution analysis, 361
    waveform spectrum analysis, 362

  amplitude, time–history curves, 379
  analysis of, 329, 338, 344
    characteristic parameters, 347
    impact dynamic load, 338
    luffing triangle wave dynamic load, 339
    source location, 347
    waveform spectrum under impact dynamic load, 341
  axial stress and cumulative number, 377
  characteristics
    mechanical properties, 341
    parameters, 379
  of concrete materials
    exhaustive method algorithm, three-dimensional source
      location, 326
    experimental study of correlation, 322
    source location technology, 324
    wave velocity determination, 324
  count, time-history curves of, 386, 388
  dam concrete, damage process of
    dynamic and static flexural failure, 321, 327
    settings of acquisition system, 327
  under different loading rate test, 371
  duration time
    in concrete uniaxial tensile failure process, 390
    time–history curves, 380
  events, 352, 391
  failure mechanism identification, 335–338
  feature parameters of, 322
    three-graded concrete, 337
    wet sieving concrete, 337
  frequency, 381, 382
    centroid of concrete under uniaxial tensile load, 391
  frequency–amplitude distribution
    time history three-dimensional histogram of, 353
  full-graded/wet-sieving arch dam concrete specimens
    dynamic flexural test, 338
    static bending test of, 329
  full waveform, in flexural tensile failure process, 343
  hit aggregate time–history curve
    of concrete specimen CO_2 spectrum
      characteristics, 360
    of concrete specimen CO_1 spectrum characteristics, 359
  hit cumulative number, 378, 385
    tensile stress, curves of, 384
  hit distribution contour maps of, 355
  hit load curve, 332
  hit number, 372
    normalized cumulative time–history curves, 382, 383

Acoustic emission (AE) technique *(cont.)*
   hit wave power spectrum diagram, 358
   location, analysis of, 338
   measurement system, 352
   monitoring, 246
      data acquisition systems of, 351
   monotonic uniaxial load
      analysis of, 377
      characteristic parameters and wave spectrum
         analysis, 379
      waveform spectrum, characteristic parameters, 381
   mortar specimens, analysis, 367
      axial tensile loading rates, 369
      mechanical properties, 367
      rise time and amplitude analysis, 368
      waveform spectrum analysis, 369
   parameters, 381
   peak frequency, time–history curves, 380
   rate
      peak, 364
      variation, 347
   ringing accumulative, axial stress, 360
   ring number, 385
   sensors, 377
      layout plan of, 329
      types of, 321
   signals, 363
      basic parameters of, 322
      energy, 362
      hit advantage, 376
      hit-count figure (HC) of, 335
   softening stage, cyclic loading, 383
      analysis of, 383
      Kaiser effect, 385
      source location, analysis, 390
      uniaxial tensile stress–strain, 385
      wave spectrum analysis, 387
   software, search method of exhaustion, 326
   source localization techniques, 324, 347, 351
      spatial distribution of, 328
      three-dimensional projection, 347, 350
   source location projections, three-dimensional, 340
   testing, 357
      device, 378
   time-history scatter diagrams of, 387
   uniaxial tensile failure process of concrete containing
      softening stage, 377
   waveform, 377
      spectrum, 333
         under impact dynamic load, 341
   wet-sieved concrete, impact dynamic bending
      test, 344
      arch dams with different loading rates, 344

Action of inertia, 250
Active faults, 35
Aggregates
   axial tensile damage, 367
   dynamic tensile strength of, 315
   model, 567
   mortar interface specimens, 371
      acoustic emission
         activity, analysis of, 371
         hit waveform, 373
         rise time and amplitude analysis, 372
         rise time statistics, 372
         waveforms, spectrum analysis of, 373
      axial tension
         damage distribution contour plot, 373
         experiment, 373, 374
      static uniaxial tension failure process, 372
   mortar junction interview, 372
   particle size distribution, 501
      grade 2 gradation aggregate, 501
      grade 3 gradation sphere random aggregate model
         finite element grids, 503
   random distribution static–dynamic comprehensive
      flexural tensile strength, 544
   specimen acoustic emission hit, 362
      number, percentage of, 361
   specimen axial stress, 361
   specimens, static axial tensile damage of, 363
   statistics window containing, 563
   strain rates, comparison of, 315
Algebraic group coefficient matrix, 199
Angle frequency, 94
ANSYS software, 301
Approximate method, 104
Arch crown beam, displacement response curses of, 213
Arch dam
   aseismic stability analysis concept, 144
   body in Switzerland
      locations of three strong seismic observation and
         measurement stations of, 17
   body-near regional foundation finite element mesh
      model, 121
   seismic response analysis, 167
   shoulder seismic resistance, 20
   system finite element model, homology boundary zone
      mesh diagram of, 122
Arch dam abutment
   arch end residual displacement, 149
   aseismic stability analysis behaviors
      of, 137
   consequence river displacement values changing with
      overloading coefficient
      relation curve of, 152

consequence river sliding move quantity with changes in
overloading coefficient
relative curve of, 151
Dagangshan arch dam, 150
system finite element model, 146
dam body part nodal point displacement time process, 148
dam foundation face sliding quantity changing with
overloading coefficient, 149
earthquake disaster enlightenment, aseismic stability
analysis behaviors of, 137–139
engineering applications, practical examples of, 144–153
of high arch dam, 137–153
Jinping grade I arch dam system finite element model, 147
Xiluodu arch dam system finite element model, 146
Arch dam contact nonlinear problem, numerical simulation of,
127–134
body-near regional foundation finite element mesh
model, 121
boundary constraint conditions and current states of
seeking solution method to, 128–129
contact interstice and upper and lower sides, 129
3-D contact problem, tangential sliding direction of, 133
finite element model of, 120
sliding fracture block body and dam body in foundation
relative position schematic diagram of, 121
static and dynamic combination computation method
of, 134
system finite element model, homology boundary zone
mesh diagram of, 122
transverse interstice motion contact face force model,
129–134
contact point displacement, decomposition of, 129–130
3-D motion contact problem numerical computation,
realization of, 132–134
normal contact force and caused nodal point
displacement, computation of, 130–131
tangential friction force and caused displacement,
computation of, 131–132
Arch dam transverse interstice motion contact face force
model, 129–134
contact point displacement, decomposition
of, 129–130
3-D motion contact problem numerical computation,
realization of, 132–134
normal contact force and caused nodal point displacement,
computation of, 130–131
tangential friction force and caused displacement,
computation of, 131–132
Architecture Aseismic Design Code, 73, 74
Arch support rock block stability, 137–142
current arch dam aseismic stability analysis method,
problems of, 137–142
of high arch dam, 137–153

rock block instability safety coefficient of arch dam, basic
concept of, 137–141
Artificial boundary processing, 210
operational condition I, 210
operational condition II, 210
operational condition III, 210
Artificial boundary realization in FEPG, 170–192
artificial homology/transmission boundary in FEPG,
188–192
realization of artificial viscoelastic boundary, 180–188
algorithms document, 183–185
partial differential equation, descriptive document of,
180–183
seeking solution stream and physical coupling
document, 185–187
viscoelastic boundary conditions, computation example
test of, 187–188
viscoelastic boundary input formula of seismic wave in,
170–180
artificial boundary surface loading, 173–178
viscoelastic boundary virtual work forms, 179
Artificial excitation source positioning error, 327
Artificial homology
boundary realization method, 121, 188, 200, 203
transmission boundary in FEPG, 188–192
FBC document, 190
NFE document, 191–192
PDE document, 189–190
Artificial nodal point current scattering wave value, 127
Artificial seismic displacement waves, 210
Artificial viscoelastic boundary, 215
surface force, 174
Artificial wave velocity, 126
Aseismic Design and Recheck Guidelines, 40
Aseismic Design Code
for Architecture Structure, 47
of Hydraulic Structures, 73
Aseismic design prevention level/standard framework
guiding thinking for working out, 42–46
corresponding to performance objectives, 43–44
performance objectives, concrete quantification of,
45–46
staging design and prevention and multiperformance
objectives, adoption of, 44–45
Aseismic Engineering program, 62
Aseismic safety analysis, 144
Aseismic stability safety coefficient, 150
ASOT. *See* Abrahamson-Silva (ASOT) attenuation relation
Association & Deep Dump 1800 computer group, 210, 215,
233, 234
parallel computers, 216
Association high performance parallel computer group
system, 233

Attenuation coefficient, 397
Attenuation exponent, 71
Attenuation relations, 65
  sets of, 63
Automatic genesis finite element program, 160
Average response spectrum, 56
Axial tensile damage, 366
Axial tensile failure process
  corresponding power spectrum, 356
Axial tension damage process, 376
Axial tension test, of concrete, 367
  acoustic emission
    activity, analysis of, 353
    collection system settings, 351
    rise time/amplitude, analysis of, 354
    waveforms, spectrum analysis of, 355
  concrete/components, comparative analysis of, 375
    acoustic emission hit accumulative number curves, 375
    acoustic emission hit rise time, 376
    acoustic emission wave spectrum, 376

B

Baihetan hydropower station, 224, 225
Baihetan left bank side slope sliding blocks, 223–234
  computation basic data, 224–229
  computation model, material parameters of, 226
  dynamic stability computation, 229–232
  efficiency, analysis of, 232–234
  engineering general conditions, 224
  mesh division of sliding block computation zone of, 225
  parallel and serial stability analysis, time comparison, 233
Balance force system, 512
Beam fracture process
  dynamic loading action, 551
  static loading action, 550
Beam symmetric profile, in static limit stress
  damage variable cloud map, 574
Bedrock sliding cracking face shear strength, 151
  store safety coefficient, 148
Bedrock slipping crack faces, 129
Beihetan arch dam, 3
Bending strength ratio
  on dynamic and static, 293
Bending tensile failure process
  acoustic emission full waveform, 346
Biconjugate gradient method (Bidgstab), 571
Binding effect, 403
Bonded steel plates
  uniaxial tensile specimens, 313
BOOLEAN operation, 474
Boundary conditions, 559
Boundary constraint

conditions and current states of seeking solution method to, 128–129
Boundary element method, 102
Boundary normal direction cosine vector, 105
Branch Commission of Seismic Engineering and Aseismic Design Norm, 41
British EMI Company, 398
Broadband sensor, 321, 353
Brune ω^2 model, 93
Bureau of Reclamation in United States, 251

C

California Energy Resource Commission, 62
Campbell–Bozorgnia (CBO7) attenuation relation, 65
Capacity safety coefficient, 139
Cardio matrix parallel machine (CPM), 158
Casting concrete, equipment for, 262
Cement mortar
  acoustic emission hit accumulative number curve, 375
  axial tension study, 313
  dynamic tensile strength, 304
  elastic modulus, 306, 307
  failure photos, 304
  low strain rate, 304
  strain rates, comparison of, 315
  tensile strength of, 305
Cement stone mesofracture process, 461
Central difference method, 122, 123
Centro-N seismic motion acceleration time process, 85
Changchun Municipal Zhaoyang Testing Instrument Co Ltd., 401
Channel threshold, 321
Chap developing process, 423
Chap evolution process, 459
Chendu Survey and Design Academy
  of Sino-hydro Consultation Group, 7
Chengdu Engineering Corporation Limited, 251
Chengdu Investigation Survey and Design Academy, 219
China
  aseismic design and prevention level framework in suggestions on revision of, 46–49
  construction of big dams and reservoirs, 2
  earthquake intensity zoning map, 32
  earthquake motion parameter zonation map, 33, 53, 97
  earthquake motion response spectrum performance cycle zonation map, 73
  East Survey and Design Academy, 224
  hydropower technology in, 1
  natural disasters in, 3
  per capita water resources in, 1
  Seismic Bureau, 3
  seismic intensity, 29

seismic safety of hydraulic concrete structures in, 22

Shapai rolled concrete arch dam in, 7, 16

social development in, 1

water energy resources in, 1

Chinese Institute of Water Resources and Hydropower Research, 251

Chinese National Development and Reform Committee, 90

Chiou–Youngs (CYO7) attenuation relation, 65

Classification Standards of Architecture Seismic Damage Grades, 47

Clayton–Engquist artificial boundary conditions, 102

CME. *See* Controlling maximum earthquake (CME)

Coal-fired electric power, 1

Coarse aggregate

    meso-zone containing part, 562

    minimum particle sizes, 564

Code for Architectural Aseismic Design, 45

Code for Design and Safety Assessment of Dams and Hydraulic Structures, 40

Code for Hydraulic Structure Earthquake Resistance Design, 47, 60

Coefficient matrix, 199

Coloration theory, application of, 158

Communication device, 193

Complete gradation concrete

    aggregate, 544

        grain distribution, 553

    beam stress–strain curves, 546

    test, 547, 548

        computation analysis of, 551

        specimen dynamic strength-growing coefficient of, 549

        specimen mesonumerical computation of, 547

Complex amplitude function, 85

Compression process, 459

Compressive stress, 217

Computation analysis, 215

    efficiency, parallel, 575

    of structure displacement, 216

Computation equation, 162

Computation process, 131

    successive loading processes, 203

Computation system, 398

Computed tomography (CT) number

    average value, 435

        curves varying with stress of, 421, 422

    magnitude of, 425

    variance, 418

    variation, 472

Computed tomography (CT) scan technology, of concrete

    advantage of, 398

    application in concrete material tests, 395–400

        equipment, brief introduction to, 398–400

        principle, description of, 396–397

CONC-10 concrete test specimen, stress–strain relation of, 465

CONC-13-2-2 cross-section, 413

    CT number average value curves, 414

CONC-044 stress–strain curve, 415

CONC-13 test specimen

    CT number average value with stress variation, curves of, 413

    loading time process curves, 411

CONC-17 test specimen, 405, 408, 409

CONC-044 test specimen

    statistic zone, 417

CONC-055 test specimen

    loading and displacement time curves, 419

    stress–strain curves, 419

    test statistic zone, 421

cone-shaped beam CT scanning, working principle of, 399

cycle loading load–time process curve, schematic diagram of, 410

cylinder test specimen, 426

damage and fracture field intercept joint, 432

damage evolution equation and constitutive relation based on, 460–474

    aggregate contact unit diagram, 475

    average CT number damage evolution equation, 464–468

    damage variable and strain relations, 464

    geotechnical damage equation and constitutive model based on, 460–462

    scanning cross-section, damage variable statistics of, 471

    statistic damage evolution equation, 462–464

    statistic damage variable of, 463

different cross-section changing with stress, 412

electro-hydraulic servo value, 402

fractal dimension computation and analysis of, 452–460

    concrete fractal characteristics under static and dynamic actions, comparison of, 459–460

    concrete uniaxial compression CT images, 454–457

        under dynamic loading action, 457–459

    dimensional computation method, 452–454

    fractal theory, 452–454

    image fractal dimensionality–strain relation curves, 456

    results, 456, 458

    stress relation curves of, 458

fracture dimensionality damage evolution equation, 468

image analysis, 406

    analysis of average numbers of each cross-section, 407

    chap morphology, 406, 410

    cross-section average number analysis, 412

    difference value image analysis, 412

inner setting-in, scheme consideration photo of, 400

Computed tomography (CT) scan technology, of concrete *(cont.)*
 instrument pressure silo, 402
 large dam concrete dynamic-static damage and failure,
  testing research on, 395–489
 loading damage process, three-dimensional cartoon
  demonstration of, 486–489
 m(d$_\lambda$) and stress curve, relation between, 431
 medical-use CT machine
  portable type dynamic test loading equipment in
   realignment with, 400–404
 mesostatistics damage model, 470
 number average value, curves varying with stress of,
  421, 422
 number variance, 418
 operating platform of, 403
 perfect degree and fracture degree, distribution diagram
  of, 427
 process of
  dynamic loading, 411
  static loading, 406
 quantitative analysis of, 408
 results and initial analysis, 404–425
  dynamic and static tensile and compression crack
   process, 423–425
  uniaxial dynamic and static compression destructive
   process, 405–413
  uniaxial static, dynamic tensile failure process,
   413–423
 sample λ-level nonperfect field and λ-level damage and
  fracture field, 429
  curve varying with, 430
 samples, intercept joints of, 433
 scanning mode of, 398
 specimen installation, types of, 404
 stage and stress–strain curve, 405
 static compressed test specimen, 424
 static tensile test specimen destructed surface
  shape, 424
 statistics zones, schematic diagram of, 407
 stress–strain curves
  of three test specimens, 415
 subzone breaking theory, 425–444
 support vector machine for concrete images, classification
  of, 445–452
 test specimen centering, schematic diagram of, 404
 three-dimensional mesomechanics analysis,
  474–486
 two-way dynamic servo oil cylinder, 402
 used portable dynamic loading equipment, structure
  schematic diagram of, 402
 working principle of, 396
Computed tomography (CT) tests, 460
Computer memory capacity, limitation of, 547

CONC-10 concrete test specimen, stress–strain relation
  of, 465
CONC-13-2-2 cross-section, 413
 CT number average value curves, 414
Concentration mass method, 166
Concrete
 AE activity characteristics, 354
 AE rise time statistics, 355
 aggregate grain distribution, 548
 average CT number damage evolution equation,
  464–468
  establishing damage variable, 466
  establishment of, 466–468
 axial tensile test of, 351
 complete curve test, dynamic direct tension of, 316
 components, comparative analysis of, 375, 537
  acoustic emission hit accumulative number
   curves, 375
  acoustic emission hit rise time, 376
  acoustic emission wave spectrum, 376
  component medium uniaxial AE spectrum
   categories, 376
  parameter analysis of the acoustic emission
   signal, 376
 CT test analysis, 426
 CT test specimens, 403
 damage, 542, 551, 552
  enhanced parameters, 537, 566, 571
  and fracture field intercept joint, 432
  process flow chart of mechanism recognition program
   of, 336
 damage evolution equation and constitutive relation based
  on, 460–474
  concrete aggregate contact unit diagram, 475
  concrete average CT number damage evolution
   equation, 464–468
  concrete statistic damage evolution equation,
   462–464
  damage variable and strain relations, 464
  geotechnical damage equation and constitutive model
   based on, 460–462
  scanning cross-section, damage variable statistics
   of, 471
  statistic damage variable of, 463
 deformation, mechanics properties of, 470
 dynamic and static tensile and compression crack process,
  423–425
  in dynamic compression tests, 424–425
  in dynamic tensile tests, 425
  static compression loading, under the action of, 423
  in static tensile tests, 423–424
 dynamic loading, stress–strain curves of, 317
 failure process of, 324

flexural tensile meso-numerical tests, full gradation, 566
flexural tensile test specimens, 561
mesodamage evolution process, 395
mesomechanics
    concentrates concrete materials, 507
    method, 492
mesostatistics damage model, 470
mesostructures, 540, 550
    perfect degree and fracture degree, distribution diagram of, 427
    phase component materials, 541
        mechanics property parameters, 552, 566, 571
    sample λ-level nonperfect field and λ-level damage and fracture field, 429
        curve varying with, 430
    samples, intercept joints of, 433
    specimens
        time-domain waveform of, 356
        uniaxial tension failure process of, 357, 362
    static–dynamic comprehensive flexural tensile strength, 542
    strength, 539
    stress–strain softening stage, characteristics of, 387
    tensile behavior of, 317
    tensile strength, relation curves of, 509
Concrete dams
    actual performance behaviors of, 22
    characteristics of, 251–255
        dynamic elastic modulus of, 254–255
        dynamic tensile strength of, 252–254
        main features of, 251–252
    designed code for, 290
    dynamic flexural strength, strain rate effects, 529
        computation model, 529
        numerical computation model, 529
        parameter taking values, 529
    dynamic flexural tests of, 257
    dynamic/static damage failure, 321
Concrete fracture dimensionality damage evolution equation, 468
    constitutive relation, establishment of, 472–474
    evolution equation, establishment of, 472
    mesostatistics damage mode, establishment of, 468–470
    mesostatistics damage model, determination of parameter, 470
    variables, establishment of, 471–472
Concrete loading damage process, three-dimensional cartoon demonstration of, 486–489
    cartoon path of cartoon setting, 488

    cartoon setting, 488
    material quality regulation, 488
    model preparation, 487
    output synthesis, 489
    prestage data processing, 487
    three-dimensional images
        of dynamic loading, damage chaps of, 489
        of static loading, damage chaps of, 489
Concrete material nonuniformity effects upon dynamic flexural strength, 539
    aggregate morphology effect upon concrete flexural tensile strength, 551
    concrete gradation effect upon dynamic flexural tensile strength, 545
    concrete test specimen static–dynamic comprehensive flexural tensile mesodamage mechanism, 547
    different gradation concrete mesomodels, numerical test comparison of, 551
    mechanics parameter discreteness, affecting dynamic flexural tensile strength, 541
    mesoprofile separation zones upon computation results, 547
    model parameters taking values, 551
    random aggregate parameter model (RARM) 1, 539
    wet-sieved concrete test specimens, mesoanalysis of, 543
Concrete materials, 442
    dynamic uniaxial, 298
        data acquisition, 303
        loading modes, 303
        specimen clamping modes, 302
        specimen processing, 303
        specimen shape, 301
    enhancement parameters, physical significance of, 536
    exhaustive method, algorithm of three-dimensional source location, 326
    experimental study of correlation, 322
    fractal characteristics of, 457
    mechanics, 436
    safety zone set, 442
    source location technology, 324
    tests application in, 395–400
        equipment, brief introduction to, 398–400
        principle, description of, 396–397
    wave velocity determination, 324
Concrete results and initial analysis, 404–425
    dynamic and static tensile and compression crack process, 423–425
    uniaxial dynamic and static compression destructive process, 405–413
    uniaxial static, dynamic tensile failure process, 413–423

Concrete subzone breaking theory, 425–444
  basic assumption of, 426
  constitutive relation, subzone description of, 442–443
  damage and fracture
    criterions for, 440
    generated positions, 439–440
    space, constitutive theory of, 443–444
    zone lower limitation and upper limitation boundary
      surface, 441–442
  degree of failure, 426–428
  degree of perfect, 426–428
  λ_1-λ_2 cutoff or intercept joint, 431–436
  λ-level damage and fracture ratio and λ-level damage and
    fracture rate, 437–438
  λ-level perfect field and λ-level damage and fracture field,
    428–429
  λ-level perfect field and λ-level damage and fracture field
    measure, 429–431
  λ_1-λ_2 intercept joint ratio and intercept joint rate, 438
  perfect space and damage and fracture, 436
  relation among concrete λ-level damage and fracture and
    CT number and density, 438–439
  research, existing conditions and problems of, 425–426
  safety zone, damaging and fracturing zone, and damaged
    and fractured zone, 440–441
  summary, 444
  weakening–strengthening norms, 442
Concrete test specimens
  cross-section stress distribution, 484
  damage and failure numerical simulation finite element
    equation, 507
    strain rate effect enhancement relations, 507
  damage and fracture developing stage, 467
  failure stage, 468
  linear elastic stage, 467
  MesoModel, grade 3 gradation, 561
  softening stage after the peak, 467
  vertical stress distribution of, 484
Concrete three-dimensional mesomechanics analysis, 474–486
  concrete damage process numerical simulation, damage
    constitutive relation, 478–486
    computation conditions, 478–479
    numerical simulation results, comparison of,
      484–486
    stress and displacement computation results, analysis
      of, 480–484
  concrete test specimen three-phase material finite element
    grids, 480
  cylinder test specimen
    cross-section diagram of, 481
    damage and fracture diagram, 481
    horizontal displacement cross-section of, 482
    horizontal stress cross-section of, 483

  unit damage longitudinal cross-section diagram of, 485
    vertical displacement cross-section of, 482
  different numerical test specimens
    loading–displacement curves contrast of, 486
  finite element three-dimensional rebuilding
    grids and computation results, 479
    model, 478
  stochastic aggregate digital computation model, 474–478
    contact surface digital computation model, 474–476
    rebuilding digital computation model, 476–478
Concrete three-dimensional stochastic aggregate digital
  computation model, 474
Concrete uniaxial dynamic and static compression destructive
  process, 405–413
  grade I concrete
    uniaxial dynamic compression online CT test,
      409–413
    uniaxial static compress, online CT test of,
      405–409
Concrete uniaxial static, dynamic tensile failure process,
  413–423
  grade I concrete uniaxial dynamic tensile, online CT tests
    of, 419–423
  grade I concrete uniaxial static tensile, online CT tests
    of, 414–419
CONC-13 test specimen
  CT number average value with stress variation, curves
    of, 413
  loading time process curves, 411
CONC-17 test specimen, 405, 408, 409
CONC-055 test specimen
  loading and displacement time curves, 419
  stress–strain curves, 419
  test statistic zone, 421
CONC-044 test specimen, statistic zone, 417
Cone-shaped beam CT scanning, working principle
  of, 399
Conjugate gradient (CG) method, 571
Contact element method, 128
Contact node relative displacement curve, 230, 231
Contact nonlinear problem, 117
Continuous particle size random aggregate mode, 498
Controlling maximum earthquake (CME), 41
Control node three-direction displacement curved diagram,
  226, 227
Conventional earthquake hazard evaluation, 91
Conventional measuring technology, 246
Convex aggregate algorithm, 498
Convex polygon aggregate model, 492
Convex polyhedron aggregate model
  generating program diagram, 499
  mesofinite element profile separation of, 503
Convolution reverse projection method, 397

Correlation design seismic motion parameters on dam site
average response spectrum, 56
design response spectrum, 58–73
determination of, 53–81
earthquake magnitudes
distances, and fault inclination upon hanging side
effect, influence of, 70
ground surface seismic record-distance distribution in
NGA data bank, 63
epicenter distances and seismic magnitudes EP/PC, 57
fault facture types with influence of attenuation relation,
comparison of, 68
hanging side effect upon regularized response spectrum β
influence, 71
NGA attenuation relation
comparison, five sets of, 64
curves comparison, five sets of, 66
peak ground acceleration, design of, 53–58
response spectrum attenuation, 73–81
relation, 58–73
site correlation design response spectrum, 78–81
standard response spectrum, 73–76
uniform probability response spectrums, 76–78
response spectrum comparison before and after reducing
PGA, 55
seismic motion acceleration response spectrum after
accounting hanging side effect
increasing rate of, 69
Coulomb's law, 131
Coupling agent, 322
CPM. *See* Cardio matrix parallel machine (CPM)
Crack, without penetrating, 319
CRAY-MP machines, 158
CT scan. *See* Computed tomography (CT) scan
technology
Current arch dam aseismic stability analysis, 141
Current hydraulic engineering
aseismic specifications in, 141
Current nodal point motion equation, 103
Cycle loading load-time process, 409
curve, schematic diagram of, 410
Cyclic amplitude fatigue effect, 292
Cylinder test specimen, 426
Cylindrical constructed specimens
with aggregate, 312

D

Dagang mountain high arch dam, 3
engineering, 91
Dagangshan arch dam
wet-sieved concrete
frequency centroid distribution statistics in, 335

Dagangshan arch dams, 48, 149, 150, 338, 339
concrete, 333
cyclic loads, 265
designed mix proportions for, 264
dynamic and static flexural strength
of four-graded concrete for, 268
of wet-sieved concrete for, 268
dynamic and static flexural strength of four-graded
concrete for, 268
dynamic and static flexural strength of wet-sieved concrete
for, 268
engineering project, 264
four-graded concrete, 265, 278
full-graded concrete for, 292
loading modes on dynamic increase factor of flexural
strength, 269
mixing proportions for, 264
wet-sieved concrete, 265, 278, 286,
334, 335
Dagangshan gradation I concrete dynamic
and static tensile CT test, 422
Dagangshan hydropower engineering, 95
Dam abutment aseismic stability research, 138
Damage
constitutive equation, 472
constitutive relation, 470
evolution equation, 461
fracture evolution relations, 526
and fracture field concept, 428
and fracture space concept, 444
influence factor, 460
morphology, 425
probability, 468
variable concept, 469, 471
Dam body–bedrock system, 146, 150
static stability, 150
Dam body finite element model, 119
Dam body–foundation system, 134, 210
Dam body structure analysis model, 101
Dam body whole time process
stress enveloping cloud diagram, 213
Dam concrete
acoustic emission (AE) technique, 321
damage and failure process, 297
damage process of
dynamic and static flexural failure, 327
settings of acquisition system, 327
nonlinear finite element analysis, 297
tensile stress–strain curves. *See* Tensile stress-strain
curves
Dam crest nodes stabilizing course under action of
dam body gravity, 217
Dam–foundation system, 143

Damping, 538
  matrix, 103, 122, 124
  ratio, 53, 58, 59
Dam-site earthquake
  hazard evaluation, 42
    procedures of, 37
  motion displacement spectrum, 94
Dam-site seismic motion
  acceleration time processes, 95
  input mechanism, 97–109
    dam body and near regional foundation substructure
        boundary model, 105
    free-field incident seismic motion input mechanism,
        100–106
      artificial homological boundary method,
          102–104
      substructure method based on dynamics,
          104–106
    relation b/w computation point and finite element
        dispersion nodal point, 103
    seismic motion input mode, 99–100
      problems need to be clarified and discussed in,
          106–108
    seismic wave radiation paths, 98
    site design seismic motion peak ground acceleration
        input
      basic concept of, 97–99
    suggestions, 108–109
  spectrum displacement, 93
Dam site seismic peak acceleration
  transcendental probability curves of, 38
Deformation
  controlled modes, 304
  strain rate
    dynamic increase factor, 311
    stress–strain curves, 312
    tensile elastic modulus, 310
Deforming structure effects method, 139
Deji double-curvature arch dam, 6, 17
Density function, 37
Design acceleration time process, 83–96
  generation, random finite fault method, 85–96
    random finite fault method behaviors, basic thinking
        ways and effects, 91–92
    steps to, 92–96
  nonsmooth acceleration time process, amplitude and
      frequency, 83–90
    artificial fitting amplitude and frequency genesis,
        87–90
    evolutionary power spectrum, seeking solution to,
        84–85
    objective evolutionary power spectrum empiric model
        fitting, 85–87

Design Codes for Architectural Earthquake
    Resistance, 67
Design response spectrum method, 58–73, 78
  method and procedures from, 78
  by set earthquake, 79
3-D finite element model, 119
Diagonal matrix, 205
Differential box accounting method, 454
Direct tensile behavior
  stress-strain curves for, 318
Direct tensile tests, 298
  interface, on dynamic mechanical behavior, 312
  mortar and aggregate, 312
Displacement
  boundary conditions, 514
  input time process, 229
  time process, 209
    of artificial seismic wave, 220
    curves of dam upstream surface arch, 222
  time processing curves, 188
Displacement–time curve, 401
Distribution probability density function, 37
Domestic parallel computation method, 159
Double-scale analysis method
  for two-phase materials, 558
Double-scale compound material, 557
Drop hammer impacting tests, 246
Drop hammer system, 245
Duhamel integral, 59
Dynamic amplification effect, 60
  of abutment rock mass seismic motion, 138
Dynamic analysis methods, 7
Dynamic combination computation method, 134
Dynamic contact problem handling method, 160–169
Dynamic explicit computation format, 160–169
  finite element in structure seismic response
      analysis, 160–161
  Lagrange's method of multipliers of motion contact
      nonlinearity, 166–168
  off-diagonal additional mass matrix, point by point
      multiplier method for, 168–169
Dynamic flexural
  experimental researches, for dam concrete,
      257–295
    Dagangshan arch dams
      designed mix proportions for, 264
      dynamic and static flexural strength of four-graded
          concrete for, 268
      dynamic and static flexural strength of wet-sieved
          concrete for, 268
      loading modes on dynamic increase factor of
          flexural strength, 269
    discussions on experimental results, 288–295

cubic compressive and splitting tensile strength,
  ratio of static flexural tensile strength to,
  291–292
dam concrete strength, influence of age on,
  293–294
dynamic flexural strength, effect of initial loads,
  292–293
flexural deformation and value of dynamic elastic
  modulus, analysis of characteristics,
  294–295
flexural strength and effects of dynamic loading
  mode, strain rate effects on, 292
full graded concrete to wet-sieved concrete, ratio
  for static strength, 288–291
dry/wet-sieved concrete specimens, 263
dynamic and static bending strength ratio, initial static
  load on, 293
elastic modulus
  changing of, 273
  dynamic and static flexural, 272, 282
flexural initial cracking stress and strain, 273
flexural region
  dynamic deformation of, 280
  normal section in, dynamic deformation of, 275
  normal section in, static deformation of, 275
  static deformation of, 280
flexural strain value under different loading
  modes, 282
flexural stress–strain curves, 274, 276–278
  for combined initial static loads and impact loads,
    270, 271
  four-graded concrete for Dagangshan arch dam
    under dynamic impact loading, 281
  four-graded concrete for Dagangshan arch dam
    under static loading, 281
  wet-sieved concrete for Dagangshan arch dam,
    282–284
full-graded concrete, 263
  cubic compressive strength of, 289, 290
hydraulic servo loading system, 259
initial cracking strain, schematic figure for, 262
Poisson's ratio, 272, 282
strain gauges, bonded locations for, 261
testing methods for, 257–263
  acquisition of experimental data, 260–262
  casting of dam concrete specimen, 262
  dynamic loading mode, 258–260
  instruments, fixtures, and the requirements, 258
testing on dynamic mechanical characteristics of,
  263–288
  compressive strength tests, 265
  experimental materials, 263–264
  experimental results, 265–288

experiments for age effects on flexural
  characteristics, 286–288
flexural characteristics, experiments on strain rate
  effects, 278–286
flexural deformation characteristics, 265–278
flexural strength tests, 265
loading scheme, 264–265
splitting tensile strength tests, 265
three-graded concrete
  under loading modes, maximum flexural strain
    of, 274
  under triangle wave, dynamic tensile elastic
    modulus of, 272
three points beam loading method, 258
triangle loading wave, 260
wet-sieved concrete
  cubic compressive strength of, 287–289
  cubic concrete specimens, splitting tensile strength
    for, 266
  dynamic and static flexural strength of, 287
  elastic modulus of, 279, 284
  flexural peak chord modulus of, 286
  flexural strain of, 279, 285
  flexural strength and dynamic increase factor
    of, 285
  maximum flexural strain value of, 286
  Poisson's ratio of, 286
  tangent modulus at, 285
Xiaowan arch dams
  designed mix proportions for, 263
  loading modes on dynamic increase factor (DIF),
    influence of, 267, 268
  strength rate sensitivity, effects of, 535
Dynamic flexural tensile strength, 529, 533, 545
Dynamic flexural tests, 258
Dynamic friction coefficient, 131
Dynamic mechanical behavior
  characteristic study, key problem and technical way in,
    255–256
  dam concrete, characteristics of, 251–255
    dynamic elastic modulus of, 254–255
    dynamic tensile strength of, 252–254
    main features of, 251–252
  of high arch dam concrete, research progress
    on, 239–256
  of normal concrete, 239–250
    experimental techniques for, 245–247
    mesomechanical numerical analysis of, 247–249
    strain rate effect, discussion on mechanisms,
      249–250
    strength, deformation, and its influencing factors
      of, 239–245
Dynamic resistance matrix, 104

Dynamic tensile online CT test, 419
Dynamic tensile strength
    of dam concrete, 252–254
        determination in dam design, 252–253
        with initial static loads, 254
        strain rate effects for, 253–254
Dynamic tensile tests, 255
Dynamic uniaxial tensile stress–strain curves
    of concrete materials, 298
Dynamometers, 13

E

Earthquake design ground motion (EDGM), 39
Earthquake hazard evaluation, 34–38, 42, 79, 80, 106
    basic concept of, 34–35
    dam site seismic motion parameters
        determination of transcendental probability of, 37
    latent seismic area, classifications of, 35
    seismic activity parameters in latent seismic source areas
        determination of, 35–37
Earthquakes, 4, 11, 15, 27, 138
    analysis of dams, 40
    cyclic reaction, 257
    in-depth analysis of, 5
    intensity of, 3
    magnitudes, 28, 37, 54, 63, 77, 79, 93
        definitions of, 27, 28
        distances, and fault inclination upon hanging side
            effect, influence of, 70
        ground surface seismic record-distance distribution
            in NGA data bank, 63
    motion acceleration time process, 94
    motion hazard evaluation, 56
    resistance safety, 47
    response spectrum peak cycle, 54
EDGM. *See* Earthquake design ground motion (EDGM)
E element, 201
    program, 199
Effective peak acceleration (EPA), 33, 53
Effective peak velocity (EPV), 33
Elastic damage constitutive relation, 474
Elasticity analysis, 158
Elastic modulus, 272, 273, 282, 306, 307
    dynamic, 207, 254, 530
    enhancement parameter, 530
    rate sensitivity
        upon dynamic bendin, 536
    sensitivity, 535
    strain rate equations, 536
Elastic three-dimensional finite element analysis, 302
Electrohydraulic servo loading system, 245, 246, 257
    value, 402

Element square matrix, 123
ELXSI-6400 shared storage type MIMD system, 158
Energy
    absorption boundary, 107
    dissipation mechanisms, 249
    resources, 1
Engineering real example analysis, 207–234
Entity model, discreteness of, 162
EPA. *See* Effective peak acceleration (EPA)
EPV. *See* Effective peak velocity (EPV)
Equation discrete method, 161
Euro-Asia earthquake belt, 3
European Code, 242
European Concrete Commission-International Pre-stress
    Association (CEB-FIP), 45
Evolutionary power spectrum, 83, 87, 88, 90
    amplitude and phase angle of, 88
Explicit algorithm format, 168
Explicit computation method, 161
    advantages of, 161
    stability of, 165
Explicit finite element method, 129
Explicit integrated computation format
    numerical stability, influence factors on, 161–166
        nonminimum side length affecting stability in
            nonequilateral length unit, 166
        unit inner integration mode affecting stability, 162–166
Explicit solution FEPG element program, 189

F

Fault
    distance, 27
    facture types with influence of attenuation relation,
        comparison of, 68
FBC document, 183
Federal Government Department, 41
FEM. *See* Finite element machine (FEM)
FEPG. *See* Finite element program generator (FEPG)
Finite calculation model, 102
Finite element analysis, 124
Finite element application framework, 159
Finite element computation program, 157, 169
    BFT element program, 169
    E element program, 169
    SOLV solver, 169
    START program, 169
    U element program, 169
Finite element explicit computation format, 200
Finite element expression formula, 195
Finite element grid
    mesh of grade 3 gradation concrete test specimens, 537
Finite element machine (FEM), 157

Finite element method, 115, 122, 157, 160, 476
Finite element model, 120
Finite element parallel computation program, 157, 223
    development and existing conditions of, 157–160
Finite element program generator (FEPG)
    artificial boundary realization in, 170–192
    in carrying out finite element computing general
        process, 172
    finite element document and generated FORTRAN sources
        document
        logic corresponding relationship of, 172
    finite element language, 517
    finite unit element method on, 170
    grammar rules, 521
    scenario documents, 187
    system, 161, 169, 170, 180, 186
        element program, 187
        existing element program in, 200
        program structure performances, 169–170
        tools, 179
Finite element program solution flowchart framework, 171
Finite element time process, 151
Finite unit element method, 115
Flexural region
    dynamic deformation of, 280
    normal section in
        dynamic deformation of, 275
        static deformation of, 275
    static deformation of, 280
Flexural strain value, under different loading modes, 282
Flexural stress–strain curves, 274, 276–278, 294
    for combined initial static loads and impact
        loads, 270, 271
    four-graded concrete for Dagangshan arch dam under
        dynamic impact loading, 281
    four-graded concrete for Dagangshan arch dam under
        static loading, 281
    wet-sieved concrete for Dagangshan arch dam, 282–284
Flexural tensile failure process
    mechanical properties and acoustic emission activity
        characteristics, 347
Flexural tensile strain, 272
Flexural tensile strength
    transverse and longitudinal magnitudes, 541
Flexural tensile test
    schematic diagram of, 530
Flexural tests, 294
Forewarning phenomena, 4
FORTRAN source program, 169, 180, 182, 191
    documents, 170
Foundation deformation
    under the action of hill body self-weight, 211
Foundation variation cloud diagram, 211

Four-graded concrete
    accumulative curve of acoustic emission hits, 342
    in Dagangshan arch dam, 343
Fractal dimensionality, 454, 458, 470
Fractal dimension computation and analysis, 452–460
    concrete fractal characteristics under static and dynamic
        actions, comparison of, 459–460
    concrete uniaxial compression CT images, 454–457
        under dynamic loading action, 457–459
    dimensional computation method, 452–454
    fractal theory, 452–454
    image fractal dimensionality–strain relation
        curves, 456
    results, 456, 458
    stress relation curves of, 458
Fractal theory, 452
Fracture
    face shear dislocation, 92
    model, 510
    rate definition, 439
    section of specimen, 319
Fractured specimens, part of, 313
Freedom site, 98
Free-field incident seismic displacement wave, 103
Free-field stress, 105
Free-vibration cycle, 53
Frequency centroid distribution statistics, 333
Frictionless model, 118
Friction model, 118
Fulei three-dimensional aggregate gradation curves, 540
Fulget coefficient, 115
Fuller gradation curve, 248
Full gradation large dam concrete dynamic behaviors
    concrete constitutive relation, 508
    concrete dam dynamic flexural strength, strain rate
        effects, 529
        computation model, 529
        numerical computation model, 529
        parameter taking values, 529
    concrete material enhancement parameters, physical
        significance of, 536
    concrete material nonuniformity effects upon dynamic
        flexural strength, 539
        aggregate morphology effect upon concrete flexural
            tensile strength, 551
        concrete gradation effect upon dynamic flexural tensile
            strength, 545
        concrete test specimen static–dynamic comprehensive
            flexural tensile mesodamage mechanism, 547
        different gradation concrete mesomodels, numerical
            test comparison of, 551
        mechanics parameter discreteness, affecting dynamic
            flexural tensile strength, 541

Full gradation large dam concrete dynamic behaviors *(cont.)*
  mesoprofile separation zones upon computation results, 547
  model parameters taking values, 551
  random aggregate parameter model (RARM) 1, 539
  wet-sieved concrete test specimens, mesoanalysis of, 543
  concrete test specimens, damage and failure numerical simulation finite element equation, 507
  strain rate effect enhancement relations, 507
  convex polyhedron aggregate model, mesofinite element profile separation of, 503
  damage evolution model, 508
  inner-structure damage and failure mechanism, 491
  mesomechanism numerical method, 491
    dimensional random aggregate model, 492
    placeholder project method, 493
    random convex polyhedron aggregate model, 494
  multiscale algorithm to predict concrete material parameter method, 557
    basic theory, 557
    complete gradation concrete mesoparameter prediction, 561
    complete gradation concrete test specimen, mesoflexural tensile numerical simulation, 564
    on double-scale method, 560
    grade 3 concrete test specimens, mesoflexural tensile numerical tests, 565
    small aggregate melting, 564
  numerical computation analysis, 530
    dynamic flexural tensile strength
      concrete strain rate sensitivity, 533
      initial prestatic loading, 531
      loading rate effects, 530
  numerical simulation test FEPG document, 515
    algorithms document, 517
    displacement algorithm stdya NFE document, 517
    GCN command document, 516
    GID command stream document, 516
    stress algorithm stdyb NEF document, 520
  parallel finite element program generator (PFEPG), concrete mesoanalysis/parallel computation, 567
    computation results, postprocessing of, 574
    mesomodel selection/parallel program generation, 567
    parallel computation efficiency, 575
    three-dimensional mesomodel parallel computation analysis, under prestatic loading action, 571

  prestatic loading dynamic equation, 513
    finite element increment equation, virtual work of, 514
    increment equation, 513
  random spherical aggregate model, 493
  scenario document, 521
    unit displacement computation, 521
    unit stress computation VDE document, 526
  spherical aggregate model, finite element grid profile separation program, 502
  static–dynamic system, virtual work equation, 511
    static balance equation, 512
    static balance equation, increment, 511
Full-graded concrete, 263
  cubic compressive strength of, 289, 290
Fuzzy concept, 142

G

Gauss' integral, 165
  accuracy, 165
  integration integral, 521
  mode, 176
Gauss–Saider iteration method, 168, 169
GCN document, 185
GCN system, 183
General minimum residual measure method (GMres), 571
Geo-mechanic model test, 140
Geometric diffusion function, 94
GID software
  document, 516
  interface, prior and postprocessing software, 574
  platform posterior processing environment, 506
GIO document, 186, 516
Gneissic quartz diorite, 7
Gouss radial base kernel function, 449
Grade particle size aggregate proportion, 498
Grain-enhanced material mechanics parameter, 560
Grains distribution
  with same particle sizes, 556
Granite
  acoustic emission hit rate, 364, 365
  aggregate specimens
    axial tension test, 357
    failure photos, 310
    peak tensile strain of, 311
    ultimate tensile strength of, 310
Grid profile separation program, 504
Ground surface reflection wave, 176
Gudenbao–Liket seismic magnitude–frequency relation formula, 80
Gudengbao and Richter logarithm, 35

Guidelines for Selecting Seismic Parameters for Large
Dams, 39
Guizhou provinces, 3

H

Hausdorff–Besicovitch dimensional number, 452
Hausdorff dimensionality, 453
mathematic expression equation, 453
Higdon-keys artificial boundary condition, 102
High arch concrete dams in China
arch dam body in Switzerland
locations of three strong seismic observation and
measurement stations of, 17
construction and seismic safety of, 1–4
general conditions of high concrete arch dam
construction in our country, 2–4
seismic general conditions in dam site of, 3–4
contact nonlinear seismic response parallel finite element
program, 202
structure, 201
local broken rock blocks below left bank gravity pier, 13
Pacoima arch dam
acceleration observation and measurement stations
layout figure, 14
measurement values of, 15
acceleration observation and measurement stations,
layout figure, 14
cracks on gravity pier of the left dam shoulder of, 12
crown section, 10
plane figure, 9
research progress on, 239–256
resisting force body side slopes, 8
seismic safety evaluation, 4–23
basic concept of, 4–5
of seismic risk, 6–23
Shapai arch dam after earthquake, birds-eye-view of, 8
High arch dam body–foundation–reservoir water systems
research and development (R&D) project of, 142
High arch dam body–foundation system
artificial homology boundary implementation method,
124–134
arch dam contact nonlinear problem, numerical
simulation of, 127–134
computation method of boundary nodal point
displacement, 126–127
concrete computation steps when seismic wave is
incident, 127
homology formula based on finite element mesh nodal
point, 124–126
contact problem handling/solving method, 117–119
normal contact condition, 118
tangential contact conditions, 118–119

engineering seismic response analysis in strong seismic
areas performance and engineering background of,
115–117
seismic response analysis method, 119–124
dynamic equation discretion, 122–124
high arch dam body–reservoir water–foundation
system numerical model, 119–122
three-dimensional contact nonlinear dynamic analysis
method, 115–134
High arch dam systems, 119
aseismic gradation design and prevention level and
determination of corresponding performance
objective of, 38–49
of foreign dams, 38–42
guiding thinking for working out, 42–46
aseismic safety evaluation of, 144
concept of seismic safety of, 6
deformation performance of, 143
foundation system seismic response analysis
method, 134
seismic cases of, 7
seismic design of, 257
seismic problem of, 4
seismic resistance of, 5
complexity of, 5
optimum design of, 5
seismic safety, research on, 5
structure dynamic parallel computation stream,
203–205
contact nonlinear seismic response parallel finite
element program structure, 201
displacement time processing curves, 188
finite element program generator (FEPG), FORTRAN
sources document
logic corresponding relationship of, 172
finite element program generator (FEPG), in carrying
out finite element computing general
process, 172
finite element program solution flowchart
framework, 171
loading process simulation, 203–204
maximum stable time step length obtained from
practical calculation by adopting different
units, 166
PFEG system working model schematic
diagram, 195
PFHPG system, linear dynamic problem solution
program stream frame work in, 198
principal project documents, 204–205
real computation *vs.* theoretic prediction, 164
speed time processing, 188
viscoelastic boundary, physical model
of, 173

High arch dam systems *(cont.)*
 structure seismic motion response
  dynamic contact problem handling method, 160–169
  dynamic explicit computation format, 160–169
   finite element in structure seismic response analysis, 160–161
    Lagrange's method of multipliers of motion contact nonlinearity, 166–168
    off-diagonal additional mass matrix, point by point multiplier method for, 168–169
   finite element parallel computation, development and existing conditions of, 157–160
   finite element program generator (FEPG)
    artificial boundary realization in, 170–192
    finite unit element method on, 170
    system program structure performances, 169–170
   large-scale structure response, research and significance of, 155–157
    high performances parallel computation developing situation in hydraulic structure field, 155–156
    seismic motion parallel computation significance of research on, 156
   PEFPG system, parallel computation program development based on, 192–205
    dynamic equation, explicit format parallel program development based on, 200–203
    high arch dam structure dynamic parallel computation stream, 203–205
    message passing programming interface, 192–194
    structure and work mode, 194–199
   research on parallel computation of, 155–205
High-frequency component, 28
High-order calculation accuracy method, 166
High-performance campus grid platform, 155
High performance computation application, 159
High-performance parallel computation technology, 155, 156
High-performance parallel computer group system, 156
High-precision loading sensor, 260
High-pressure water jet, 11
Hilbert–Schmidt principle, 448
Hill mass gravity function, 215
Homogeneous material uniaxial compression test, 439
"Homogenization" functions, 18
Homology boundary method, 203
Horizontal earthquake motion acceleration, 58
Horizontal wave, 27
Hydraulic engineering, 155
Hydraulic servo loading system, 259
Hydrodynamic pressure, 168, 221
 in analysis, 166

I
ICOLD. *See* International Committee of Large Dams (ICOLD)
Impact loading rate, 551
Imperial Valley earthquake, 85
Implicit computation method, 160
Incident free-field seismic wave amplitude, 106
In-depth analysis, 16
Industrial CT machines, 399–400
Inertia force, 538, 539
Infinite regional model, 102
Initial cracking strain
 determination of, 262
 schematic figure for, 262
Initial failure process, 423
In situ vibration measurement, 19
In situ vibration tests, 17
Instability safety coefficient, 139
Installing test specimen, 403
Integral constants, 58
Intercept joint concept, 432, 434
Intercept theory, 433
Interfaces
 dynamic tensile strength of, 315
 strain rates, comparison of, 315
 strength discreteness upon static–dynamic comprehensive flexural tensile strength, 549
 variability coefficient, 549
 wave motion, 550
International Committee of Large Dams (ICOLD), 38
 guidelines, 39
Interpolation method, 119
Irregular structure engineering analysis, 158
Isotropic plastic damage mechanics
 constitutive relation of, 510

J
Jacobi block iteration method, 158
Jinping grade 1 concrete dam, 3, 151
 double curvature arch dam, 2

K
Kaiser effect, 385
Kernel functions, 448, 449
Kilomaga Ethernet switching system, 210
Kuhn–Tucker condition, 446
Kuming Investigation Survey and Design Academy, 207
Kunming Hydropower Investigation Design & Research Institute, 254

L
Lagrange's explicit difference serial program, 155
Lagrange's method of multipliers, 117, 167, 446
 number, 167

Lamete strain equivalent principle, 510
Lanczos method, 158
Land surface environment vibration, 19
Large dam aseismic design and prevention level frameworks
  principles for the establishment of, 46
Large dam concrete dynamic-static damage and failure,
  testing research on, 395–489
Large dam design and prevention levels
  amendment of, 49
Large dam engineering earthquake resistance, 97
Large dam–foundation system
  nonlinear static and dynamic analysis system
    of, 153
Large dam visco-elasticity boundary, 119
Large-scale high arch dam engineering
  aseismic safety analysis and evaluation comparison
    of, 144
Large-scale structure response, research and significance
    of, 155–157
  high performances parallel computation developing
    situation in hydraulic structure field,
      155–156
  seismic motion parallel computation significance of
    research on, 156
Latent seismic source, 78
Latent sliding blocks, 137
Lattice system, 247
Lemaitre strain equivalence principle, 472
Lepu wave, 27
Life and death unit method, 474
Limit tensile strain, 511
Linear complementary equation, 129
Load balance problem, 155
Load–displacement relationship, 300
Loading
  displacement curve, 401
  equipment noise levels, 353
  equipment technical indexes, 401
  monotonic, 304
  numerical test analysis, 553
  process, 414, 440
    simulation, 203–204
    week's time, 557
  strain rates, 312
  system, 377
    analysis testing on stiffness, 301
    stiffness influence, 299
Loading scheme
  dynamic flexural experimental researches, for dam
    concrete, 264–265
    cyclic loads with variable triangle wave, 265
    impact wave loading, 264–265
    static load, 264

Load-time process curve
  for specimens, 318
Local coordinate system, 167
Local earthquake magnitude, 27
Low-cycle fatigue effects, 257

M

Marconi M8000 screw CT scanning instrument, 405
Mass absorption coefficient, 396
Material damage process, 478
Mathematic measure theory, 436
MATLAB language, 454
Matrix earthquake magnitude, 92
Matrix seismic magnitude, 72
Maximum convincible earthquake (MCE), 38
  seismic motion acceleration time process, 91
Maximum design earthquake (MDE), 38
  design and prevention levels of, 47
MCE. *See* Maximum convincible earthquake (MCE)
MDE. *See* Maximum design earthquake (MDE)
MDI document, 516
Mechanics test concrete specimens
  loading–displacement curves of, 486
Medical-use CT machine, 398–400, 423
  portable type dynamic test loading equipment in
    realignment with, 400–404
    specimen installation and load, 403–404
    technical requirement and structure of test loading
      equipment, 400–401
  screw-type CT machine, fifth generation of, 399
Mercalli (MM) intensity table, 40
MesoAggregate model computation
  of complete gradation concrete, 553
Mesodamage evolution process, 395
Mesoeach-phase
  material elastic modulus, 545
  unit elastic modulus, 547
Meso-earthquakes, 45
Mesomechanics angle, 400
Mesomechanism numerical method, 491
  dimensional random aggregate model, 492
  placeholder project method, 493
  random convex polyhedron aggregate model, 494
MesoNonuniformity effects
  static–dynamic comprehensive pull and bend strength, 543
Mesostatistic damage mechanics, 466
Mesostratum
  finite element profile separation of, 567
MesoZone numerical computation result, 547
Message passing interface (MPI), 155
  communication devices, 193
  parallel program, 193
  standardization, 192

Microcracks, 347
Microwave CT, 398
Model parallel computation partition diagram, 572
Mohr–Coulomb criterion, 248
Mohr–Coulomb friction model, 118
Mohr–Coulomb's law, 128
Mohr–Coulomb theory, 52
Monotonic uniaxial load
    analysis of, 377
    characteristic parameters and wave spectrum analysis, 379
    waveform spectrum, characteristic parameters, 381
Monte Carlo method, 248, 474, 492, 493
Mortar specimen
    acoustic emission activity of, 367
    analysis, 367
        axial tensile loading rates, 369
        mechanical properties, 367
        rise time and amplitude analysis, 368
        waveform spectrum analysis, 369
    axial tension damage of, 368, 370
    axial tension failure process, 369
    dynamic tensile strength of, 315
    static axial tensile damage, 370
Motion balance equation, 123
Motion contact force model, 129
Motion differential equation, 125
Motion state function, 130
Mountain body topography, 18
MPI. *See* Message passing interface (MPI)
MTS322 electrical hydroservo machine, 298, 304, 309
Multihomology formula, 126
Multiple wave filter method, 88
Multiscale algorithm to predict concrete material parameter
        method, 557
    basic theory, 557
    complete gradation concrete
        mesoparameter prediction, 561
        test specimen, mesoflexural tensile numerical
            simulation, 564
    on double-scale method, 560
    grade 3 concrete test specimens, mesoflexural tensile
        numerical tests, 565
    small aggregate melting, 564

N

NASA. *See* National Aeronautics and Space Administration
    (NASA)
National Aeronautics and Space Administration (NASA), 157
National Natural Science Foundation, 400
National Seismic Department, 35
Near-field foundation boundary, 215
New Zealand Society of Large Dams (NZSOLD), 41

Next generation attenuation (NGA) research, 62
    attenuation relation
        comparison, five sets of, 64
        curves comparison, five sets of, 66
    attenuation relation models, 62
NFE algorithm, 183
    bank, algorithm documents, 184
    document, 517
    document nomenclature
        stipulations for, 184
    scenario document, 515
NGA. *See* Next generation attenuation (NGA) research
Ningxia, Haiyuan earthquake in, 3
NMR-CT. *See* Nuclear magnetic resonance CT (NMR-CT)
Nodes, spatial convex, 496
Noise interference, 322
Noncontact measuring system, 246
    high speed camera system, 247
    laser extensometer, 246
    nondestructive microwave monitoring technique, 247
Nondestructive detection technology, 399
Nondestructive microwave monitoring techniques, 246
Nondiagonal matrix, 124
Nonlinear attenuation function, 94
Nonlinear dynamic analysis, 83, 143
Nonlinear finite element, 183
Nonlinearity dynamic problem
    for finding solution flow frame diagram, 568
Nonlinear Lagrange's method of multipliers, 189, 200
Nonpenetrability, definition of, 118
Nonsmooth random earthquake motion acceleration time
        process, 83, 87
Normal concrete, 239–250
    age, 245
    coarse aggregate, 244
    curing and humidity conditions, 244
    dynamic loading system, 245
    dynamic mechanical behavior of, 239–250, 255
        experimental techniques for, 245–247
            dynamic loading instruments for, 245–246
            dynamic measuring technique for, 246–247
            X-ray CT technique, application of, 247
        mesomechanical numerical analysis of, 247–249
            material mechanical characteristics for medium
                in, 249
            mesomechanical mathematical model, 247–248
    strain rate effect, discussion on mechanisms,
        249–250
        problems in mechanisms for, 250
        revealed by existing experimental results,
            preliminary consensus of, 249–250
    strength, deformation, and its influencing factors of,
        239–245

deformation properties of, 242–243
    effect of strain rate on increase, 240, 241
    influencing factor of dynamic strength, 243–245
    strength characteristics of, 239–242
  elastic modulus of, 242
  experimental techniques for, 245–247
  failure condition for pure flexural region in specimen, 259
  free water viscosity, influence of, 250
  hardening process of, 251
  loading equipment for specimens, 259
  mesomechanical numerical analysis of, 247–249, 255
  multiaxial strength of, 241
  peak strain of, 242
  Poisson's ratio of, 243
  size effect, 245
  static strength of, 243
  strain rate effect, 249
    discussion on mechanisms, 249–250
  strength, deformation, and its influencing factors of, 239–245
  temperature, 245
  testing equipment of tensile testing, 239
  uniaxial and flexural strength of, 239
  water–cement ratio, 244
Normal direction contact force, 118
Normalization response spectrum, 80
  probability of, 81
Northridge earthquake, 13
Nuclear energy, 1
Nuclear magnetic resonance CT (NMR-CT), 398
Nuclear Power Aseismic Design Code, 73
Numerical instability, 162
Numerical stability, 166
NZSOLD. *See* New Zealand Society of Large Dams (NZSOLD)

O

OBE. *See* Operating basic earthquake (OBE)
Occupied zone reject method, 493
Operating basic earthquake (OBE), 38
Operational safety seismicity, 101
Orthogonal increment process, 84
Overloading safety coefficient, 139, 140, 146, 149

P

Pacific Gas and Electric Power Company, 62
Pacoima arch dam, 16, 18, 20, 21
  acceleration observation and measurement stations
    layout figure, 14
    measurement values of, 15
  acceleration observation and measurement stations, layout
    figure, 14

  cracks on gravity pier of the left dam shoulder
    of, 12
  crown section, 10
  plane figure, 9
PAC-R6α sensors, 323
PAC-WD broadband transducer, 327
Parallel boundary element method, 155, 157
Parallel computation analysis, 205
  arch crown beam, displacement response curses
    of, 213
  Baihetan left bank side slope sliding blocks, 223–234
    computation basic data, 224–229
    computation model, material parameters
      of, 226
    dynamic stability computation, 229–232
    efficiency, analysis of, 232–234
    engineering general conditions, 224
    mesh division of sliding block computation zone of,
      225
    parallel and serial stability analysis, time comparison,
      233
  contact node relative displacement curve, 230, 231
  control node three-direction displacement curved diagram,
    226, 227
  dam body whole time process stress enveloping cloud
    diagram, 213
  dam crest nodes stabilizing course under action of dam
    body gravity, 217
  dam upstream surface arch crown nodes under action of
    seismic waves
    displacement time process curves of, 222
  efficiency, 211
  efficiency analysis, 215–216
  engineering real example analysis of, 207–234
  foundation deformation under the action of hill body
    self-weight, 211
  load balance problem in, 155
  program structure seismic response analysis, 204
  relation curve
    between strength reduction factor and sliding move
      quantity, 228
    between strength relation factor and sliding move
      quantity, 232
  scheme, 232
  static–dynamic loading whole time process
    stress envelope cloud diagram for, 218
  static loading
    dam body and foundation deformation under action
      of, 212
    dam body main stress cloud diagram under, 212
  technology, application of, 216
  typical transverse crack, opening time process
    of, 214

Parallel computation analysis *(cont.)*
  Xiaowan arch dam
    computation model, foundation rock mass material parameters of, 208
    input seismic displacement waves, 209
    results of seismic dynamic response of, 216–217
    seismic dynamic analysis, and serial computation of conditions and time consumption, 216
    of seismic motion responses, 207–217
      under artificial viscoelastic boundary conditions, 215–217
      computation model, 207–210
      parallel computation under condition of transmitting boundary, 210–214
    system 3-D discrete mesh, 209
  Xiluodu arch dam
    body seismic loading, maximum tensile stress distribution under action of, 223
    computation input, artificial seismic wave time process of, 221
    computation model, foundation rock mass material parameters for, 219
    seismic motion dynamic response of, 217–223
      basic data for computation, 219–220
      parallel computation of, 220–223
    seismic response analysis, three-dimensional finite element discrete mesh for, 220
    serial computation and parallel computation, conditions and time consumption of, 222
Parallel computer group system, 157
Parallel finite element computation programs, 194
Parallel finite element generator (PFEG)
  system working model schematic diagram, 195
Parallel finite element program generator (PFEPG), 160
  concrete mesoanalysis/parallel computation, 567
    computation results, postprocessing of, 574
    mesomodel selection/parallel program generation, 567
    parallel computation efficiency, 575
    three-dimensional mesomodel parallel computation analysis, under prestatic loading action, 571
  platform, 491
  structure and work mode, 194–199
    data structure, 194
    generate parallel program structure, 197–199
    parallel program, generation and operation of, 196–197
    system working model, 194–195
  system, 194, 567, 571
    parallel computation program development based on, 192–205
      dynamic equation, explicit format parallel program development based on, 200–203
      high arch dam structure dynamic parallel computation stream, 203–205

message passing programming interface, 192–194
  structure and work mode, 194–199
steps for applying, 196
working mode of, 195
Parallel realization platform, 192
Parallel solving process, 158
Partial differential equation method, 170
Particle size continuous distribution
  random aggregate model, 498
Parzen window function, 88
Past International Dam Commission, 49
PDE document, 179, 180
  sections, 180
Peak frequency distribution statistics
  dagangshan arch dam wet-sieved and four-grade concrete under flexural load, 334
Peak ground acceleration (PGA), 53, 79, 95
  design of, 53–58
Peak tensile strain, 307, 308
Peleg Model, 453
Performance-based seismic design concept, 43
Performance response spectrum, 60
PET. *See* Positron emission CT (PET)
PFEPG. *See* Parallel finite element program generator (PFEPG)
PFHPG system
  linear dynamic problem solution program stream frame work in, 198
PGA. *See* Peak ground acceleration (PGA)
Philips 16-Row Screw CT machine, 413
Physical field coupling GIO document, 186
Piezometers, 11
Point-by-point/stepwise multiplier method, 168
Point CT magnitude δ neighboring zone magnitude, 441
Poisson's ratio, 163, 260, 272, 274, 282, 507, 510, 561, 562
  of normal concrete, 243
Polygon aggregate model, 553
Polynomial kernel function, 449
Portable dynamic loading equipment, structure schematic diagram of, 402
Portable type dynamic loading equipment, 401, 405
Portland cement, 304
Positron emission CT (PET), 398
Postulation process, 177
Power spectrum
  in band, 362
Preprocessing partition module, 197, 199
Prestatic loading
  action of concrete three-dimensional meso-model, 573
  dynamic equation, 513
    finite element increment equation, virtual work of, 514
    increment equation, 513
  stress–displacement curves, 531

Principal project documents, 204–205
Probability implication, 80
Probability method, 77, 90
Profile division technique, 474
Program automatic genesis technology, 160
Projection model method, 474
Punt-Dal-Gall arch dam, 16, 107

Q

Quality-free rock mass, 115

R

Radiation damping effects, 17, 100
Random aggregate
    distribution, three-dimensional, 569
    and mechanics characteristics model, 248
Random aggregate model, 248
    profile separation program frame diagram of, 505
    three-dimensional, 555
Random aggregate parameter model (RARM) 1, 539
Random convex aggregate model
    gradation 3, 500
    mesounit profile separation of, 505
Random distribution, 559
    compound materials, with multiscale grains, 560
    variable, 468
Random finite fault method, 91, 95
    advantages, 91
    basic thinking ways, 91
    behaviors, 91
    effects, 92
    problems, 92
Random grain distribution, 558
Random packing algorithm
    for three-dimensional aggregate, 248
Random spherical aggregate model, 493
Rayleigh damping, 161
    conditions, 123
    medium finite element discrete model, 161
    ratio, 163
    structure damping, 539
Rayleigh wave, 27
Realization, concrete procedures for, 87
Real-time scanning, 396
RedHat Advanced Intel MPI Library 2.0 Parallel Compiling
        Program, 575
Red Hat Advanced Server V4.0 operating system, 210
Refracted waves, 53
Regional decomposition method, 155, 197
Regional foundation free boundary dynamic resistance, 105
Regression fitting empiric model, 86
Relation curve

between strength reduction factor and sliding move
        quantity, 228
    between strength relation factor and sliding move quantity,
        232
Reluctant force, 137
Renewable energy, 1
Research frontier, 43
Research Project of Lifeline System, 62
Reservoir earthquakes, 49–52. *See also* Earthquakes
    genesis mechanism of, 50
    safety evaluation, 52
    structure type of, 51
    structure-types of, 52
Reservoir-induced earthquake (RIE), 38
Reservoir triggered earthquake (RTE), 38
Reservoir water
    function of, 50
    hydro-geological conditions for, 51
Residual displacement coefficient, 148
Residual strength coefficient, 510
Resistance force, 83
Resisting force
    body side slopes, 8
Resonance phenomena, 119
Response spectrum, 58
    approximate formula, 83
    comparison, before and after reducing PGA, 55
    curves, 59
    performance cycle, 73
Response spectrum attenuation
    design response spectrum, 73–81
        site correlation design response spectrum, 78–81
        standard response spectrum, 73–76
        uniform probability response spectrums, 76–78
    relation, 58–73, 80
Response time process, 229
RIE. *See* Reservoir earthquake; Reservoir-induced
        earthquake (RIE)
Rigidity limit balance method, 116, 141
Rigid pan body, 99
Rock mass resistance force, 139
Rotary parallel X-ray scanning, 399
RTE. *See* Reservoir triggered earthquake (RTE)

S

Safe shut down earthquake (SSE), 44
Safety coefficient, 139
    of stability, 7
Safety evaluation earthquake (SEE), 39
Safety limit earthquake
    damping ratio of, 101
Saint Philando earthquake, 21. *See also* Earthquakes

SAMOS™ Series 16-channel acoustic emission, 321
Scanning system, 398
Scattering field, 18
Sedimentation altitude, 207
Sediment dynamic loading functions, 168
SEE. *See* Safety evaluation earthquake (SEE)
Seismic active faults, 35
Seismic catastrophes, 90
Seismic danger analysis
   of engineering worksite, 27–38
     bases for design and prevention, 27–34
     earthquake hazard evaluation, general description of,
       34–38
Seismic Design Code of Hydraulic Structures, 34, 207, 219
Seismic disasters, 3
   changes, 48
Seismic dynamic magnification effects, 20
Seismic fortification and hazard analysis at dam site
   aseismic design and prevention level framework in China
     suggestions on revision of, 46–49
   dam site seismic peak acceleration
     transcendental probability curves of, 38
   high arch dams, aseismic gradation design and prevention
     level and determination of corresponding
     performance objective of, 38–49
     of foreign dams, 38–42
     guiding thinking for working out, 42–46
   outline of bases of, 27–52
   reservoir earthquake, 49–52
   seismic danger analysis of engineering worksite, 27–38
     bases for design and prevention, 27–34
     earthquake hazard evaluation, general description of,
       34–38
   standard design acceleration response spectrum, 33
   Xiaowan arch dam, dam site latent seismic source area
     division of
     schematic diagram of, 36
Seismic function effects, 45
Seismic hazard evaluation, 40
Seismic inertia force, 137
Seismic intensity, 3, 28, 32
   attenuation, 61
   model, 61
Seismic magnitudes, saturation phenomena of, 28
Seismic motion
   acceleration, 98
   attenuation relations, 61
   behaviors, 60
   designing, basic concept of, 99
   input mechanism, 97, 115
   response spectrum, 61
Seismic motion acceleration response spectrum, 80
   after accounting hanging side effect
     increasing rate of, 69

Seismic motion acceleration time process, 53, 83, 84, 143
Seismic motion prediction model, 61
Seismic motion time process, 94
Seismic radiation model, 93
Seismic resistance
   dynamics behaviors of, 6
   stability, 18
Seismic response analysis, 99, 116
   performances of, 117
Seismic response spectrum, 53
Seismic safety evaluation, 4–23
   arch dam aseismic stability with deformation as core and
     aseismic safety evaluation, 142–143
   basic concept of, 4–5
     paying attention to practical test, 5
     protruding engineering viewpoint, 5
     stressing comprehensive evaluation, 5
   of high arch dam, 137–153
   of seismic risk, 6–23
     analysis and enlightenment of, 18–23
     typical seismic cases of, 6–17
Seismic source, 27
   depth, 27
   distance, 27
Seismic wave motion response analysis, 122, 134
Seismic waves, 98, 170, 210
   dam upstream surface arch crown nodes under action of
     displacement time process curves of, 222
   transmission, 34
     approaches, 4
Seismic zone, 35
Seismo-station orientation effect, 93
Self-shock cycle, 165
Self-vibration cycle, 33
Sensors
   layout scheme, 377
   types of, 321
Shapai arch dam, 7
   after earthquake, birds-eye-view of, 8
   spectrum for, 7
Shapai rolled concrete arch dam, 7
Shear chap, 423
Shear dislocation, 93
Shear index, 143
SHPB. *See* Split Hopkinson pressure bar (SHPB) technique
Side slope dynamic safety coefficient, 231, 232
Sigmoid kernel function, 449
Signal analysis, 322
Signal-to-noise ratio, 322
Signal waveform, 323
Simulated cracking extension process, 248
Single computer serial program, 221
Single-degree freedom system, 58
Single-freedom linear elastic system, 33

Single mass point system, 53
Single-orientation wave transmission theory, 102
Single photon emission CT (SPECT), 398
Site seismic motion time process, 92
Slave process program, 199
SLE. *See* Strength level earthquake (SLE)
Sliding fracture block body
     and dam body in foundation, relative position schematic
          diagram of, 121
Sliding friction stress, 118
Sliding instability, 140
Softening stage
     cyclic loading, 383
          analysis of, 383
          Kaiser effect, 385
          source location, analysis, 390
          uniaxial tensile stress–strain, 385
          wave spectrum analysis, 387
     stress–strain relationship, 393
Solidified cement mortar body units, 543
Solid mechanics electric magnetic field, 170
solv element, 201
SOR parallel iteration solving process, 158
Spatial convex polyhedron, 497
Spatial correlation, 51
Spatial distribution model, 248
Spatial mesh profile separation, 556
Spatial vector mixture product principle, 494
Special interface unit, judgment of, 504
Specimen C150, artificial excitation source positioning
     error, 327
Specimen clamping modes, 302
Specimen damaging and fracturing zone, 440
Specimen safety zone, 440
SPECT. *See* Single photon emission CT (SPECT)
Spectrum attenuation formula, 61
Speed time processing, 187, 188
Sphere aggregate input procedure fame diagram, 495
Sphere random aggregate model-generating program, 494
Spherical aggregate model, 551
Split Hopkinson pressure bar (SHPB) technique, 244, 245
     instruments, 241
SSE. *See* Safe shut down earthquake (SSE)
Standard component test
     axial tensile device, 299
Standard design acceleration response spectrum, 33
Standard response spectrum, 73
Standard wave motion equation, 97
Static analysis method, 141
Static and dynamic compressed tests, 424
Static axial tensile damage of aggregate–mortar interface, 374
Static bending test
     flexural failure process of, 333
Static compressed test specimen, 424

Static computation, 215
Static cubic compressive strength, 288
Static/dynamic combination computation method, 134
Static–dynamic comprehensive flexural tensile strength, 531,
     534, 550
     concrete gradation, 546
     prestatic loading, effect of, 573
Static–dynamic loading whole time process
     stress envelope cloud diagram for, 218
Static flexural tensile strength, 555
Static friction coefficient, 131
Static limit stress, cloud map, 574
Static loading computation, 513
     dam body and foundation deformation under action
          of, 212
     dam body main stress cloud diagram under, 212
Static motion sliding friction, 128
Static tensile tests, 424
     specimen destructed surface shape, 424
Static uniaxial tension failure process
     AE rise time statistics of, 368
Statistical regression analysis, 85
Statistics mean elastic modulus
     different times, 563
Statistics mean Poisson ratio
     different times, 563
Statistics zones, schematic diagram of, 407
Stdy MDI document, 569
Steel cap
     influence of thickness, 303
Steel standard specimens, 299
Stefan effect, 250, 539
Straight-down earthquake, 32
Strain
     coefficients, 443
     expression formulas, 443
     gauges, bonded locations for, 261
     Kaiser effect, 321
     rate effect, 256, 507, 539
     rate effects, 240, 250, 293
          mechanisms for, 250
     time-history curve of, 388
Strength criterion, 248
Strength decreasing coefficient, 148, 151
Strength level earthquake (SLE), 40
Strength reduction factor, 229
Strength reduction method, 229, 232
Strength safety coefficient, 144
Stress
     axial tension failure process, 354
     curve of acoustic emission hit rate varying, 349
Stress–displacement curves, 536, 543, 556
     of concrete three-dimensional meso-model, 573
Stress–displacement process, 529

Stress–strain curves, 272, 297, 379, 401, 414, 464, 473
AE hit numbers, 391
of concrete, 300
deformation, strain rate, 312
different sample test specimens, 565
different strain rates, 309
dynamic and static, 308
nonlinearity, 478
random distribution test specimens, 564
of specimens, 298
strain gauges and extensometer, 394
of three test specimens, 415
Stress–strain expression formula, 409
Structural Engineers Association of California, 45
Subspace iteration method, 158
Substructure boundary, 108
Substructure solving process parallel computation, 158
Successive damage theory, 468
Sun6800, 573
Super-high arch dams, 115, 116
Super-high-speed network, 158
Supersonic CT, 398
Support vector machine (SVM)
classification, 451
result, 452
classification method of, 448–449
concrete component, classification of, 449–452
for concrete CT images, classification of, 445–452
broad sense optional classification face, 447–448
optimum classification faces, 445–447
for concrete images, classification of, 445–452
image classification, 448
learning algorithm, 445
learning efficiency of, 449
one pair multimethod, 449
one-to-one method, 449
optimum classification face support vector, 445
result analysis of, 450
specimen zone, section of, 450
theory, 445
Surface seismic acceleration, 53
Surface tensile stress distribution, 223
SVM. *See* Support vector machine (SVM)
Swiss Guidelines on Assessment of Earthquake Behavior
of Dam, 41
Synthesis acceleration time process, 94
System deformation process, 143
System inherent circular frequency, 58

T

Tangential friction force, 131
Tangential revised coefficients, 105

Tensile–compressive cyclic action, 258
Tensile force, 12
Tensile safety coefficients, 7
Tensile strength, 547
dynamic increase factor, 314
of interface under different strain rates, 314
Tensile stress, 116, 211
distribution, of cylindrical and prism specimen, 301
dynamic strength, 306
time–history curves, 379, 380
time history curves of, 392
time-history curves of, 386, 390
Tensile stress–strain curves. *See also* Stress–strain curves
cement mortar, dynamic tensile characteristics for, 304
elastic modulus, 306
peak tensile strain, 307
static stress–strain curves, 308
of concrete, 316
dynamic direct tensile characteristics, experimental
research on, 309
dynamic tensile strength
interface, mortar, and aggregate, 315
dynamic uniaxial, concrete materials, 298
data acquisition, 303
loading modes, 303
specimen clamping modes, 302
specimen processing, 303
specimen shape, 301
loading system, stiffness of
experimental research, 298
quasibrittle materials, stable fracture conditions for, 298
Tensile test, 413
Test Code for Hydraulic Concrete, 251, 257, 258
Test specimen tensile, 401
Thermal activation and macroviscous mechanisms, 250
Three-dimensional finite element model, 141, 477
Three-dimensional mesomechanical numerical analysis, 247
Three-dimensional rebuilding method, 476
ANSYS generating connection or node according to
layers, 477
connecting the corresponding nodes to generate element, 477
CT image format conversion, reading, and writing, 476
generating element connection or node, 477
image enhancement, 477
making discrimination of, 477
read-in CT images into MATLAB, 477
Three-dimensional space, intermonitoring, 497
Three-graded concretes
under loading modes, maximum flexural strain of, 274
under triangle wave, dynamic tensile elastic modulus of, 272
Three Grade Project hydropower station, 217
Three points beam loading method, 258
Topologic space concept, 436

Total deformation energy, 298
Traditional rigid limit balance method, 7
Transcendental probability, 37, 38, 80, 95
Transcendental probability curves, 76
Transmission process, 34, 53, 60, 98
Transputer chip distribution pattern MIMD system, 158
Triangle loading wave, 260
Two-way dynamic servo oil cylinder, 402
Typical transverse crack, opening time process of, 214

U

U element, 201
Uniaxial tensile failure process of concrete containing
    softening stage, 377
Uniaxial tensile strength
    strain rate sensitivity of, 240
Uniaxial tension
    failure process, 323
    load, 387
    test, 352
Unified Standard for Reliability Design of Hydraulic
    Engineering Structures, 252
Uniform probability response spectrums, 76, 77, 78
United States
    architecture code (UBC97), 67
    Bureau of Reclamation, 251, 253, 254, 291, 292, 294
    California Communication Department, 62
    ENWEI company, 158
    Pacoima arch dam, 7, 16
        purpose of, 7
        Seismic disasters of, 138
United States Committee of Large Dams (USCOLD), 39
Unit multiprofile division technique, 474
USCOLD. *See* United States Committee of Large Dams
    (USCOLD)

V

Variable bandwidth storage method, 158
Vaseline, coupling agent, 322
VDE document, 181, 196, 528, 571
Vector machine, 158
Vibration system, 58
Virtual work equation, 183
Viscoelastic boundary, 104
    condition, 184
    conditions, 191
    physical model of, 173
    processing, 203
Viscous interface tensile strength, 548

W

Walraven formula, 492, 540
Water and energy resources, 1

Water attenuation coefficient, 397
Water–cement ratio effect, 244
Waveform
    frequency centroid, 381
    spectrum, 333
Wave-front method, 158
Wave motion matrix, 99
Wave motion process, 124
Wave velocity test
    error of source localization, 325
    exhaustive method, diagram of, 326
    first group of specimens, 325
    second group of specimens, 325
Weibull distribution, 468
Weight coefficient vector, 446
Wet-sieved concretes, 273
    accumulative curve of acoustic emission hits,
        341, 342
    acoustic emission characteristics parameters,
        347, 350
    AE features parameter, 337
    bending tensile failure process, 344, 345
    cubic compressive strength of, 287–289
    cubic concrete specimens, splitting tensile strength
        for, 266
    in Dagangshan arch dam, 342
    dynamic flexural strength, 287, 290
    dynamic flexural tensile strength growth
        coefficient, 545
    elastic modulus of, 279, 284
    flexural peak chord modulus of, 286
    flexural strain of, 279, 285
    flexural tensile
        three-dimensional numerical model, 570
    flexural tests of, 327
    impact dynamic bending test, 344
        arch dams with different loading rates, 344
    maximum flexural strain value of, 286
    mechanical property/AE characteristics, 330
    MesoAggregate model computation, 554
    MesoFinite element model, 554
    mesofinite element model, 543
    Poisson's ratio of, 286
    splitting tensile tests, 291
    static flexural strength value of, 287, 290, 291
    tangent modulus at, 285
    tests
        specimen flexural tensile damage, 554
        three-dimensional mesoflexural tensile numerical
            test, 557
    three-dimensional AE source location
        projections, 339
Wet sieving method, 251

X

Xianshui River fracture zone, 95
Xian University of Technology, 401
Xiaowan arch dams, 3, 263, 330, 344
    computation model, foundation rock mass material
        parameters of, 208
    computation stream sequence, 205
    concrete
        acoustic emission peak frequency statistic histogram
           of, 333
    concrete double-curvature arch dam, 2
    cyclic loads, 265
    dam site latent seismic source area division of
        schematic diagram of, 36
    designed mix proportions for, 263
    dynamic and static flexural strength of
        three-graded concrete for, 266
        wet-sieved concrete for, 267
    dynamic computation, 220
    engineering project, 264
    full-graded concrete for, 285
    input seismic displacement waves, 209
    loading modes on dynamic increase factor (DIF), influence
        of, 267, 268
    mixing proportions for, 263
    results of seismic dynamic response of, 216–217
    seismic dynamic analysis, 216
    seismic dynamic analysis, and serial computation of
        conditions and time consumption, 216
    of seismic motion responses, 207–217
        under artificial viscoelastic boundary conditions,
           215–217
        parallel computation under condition of transmitting
           boundary, 210–214
        Xiaowan arch dam computation model, 207–210
    system, 209

system 3-D discrete mesh, 209
three-graded concrete, 265, 267
wet-sieved concrete, 265, 278
    dynamic and static flexural elastic modulus, 285
    dynamic and static Poisson's ratio of, 286
    flexural strength of, 285
    maximum flexural strain value, 286
Xiaowan engineering project, 2
Xiaowan hydropower station, 207
Xilou ferry arch dam, 3
Xiluodu arch dam
    body seismic loading, maximum tensile stress distribution
        under action of, 223
    computation input, artificial seismic wave time process
        of, 221
    computation model, foundation rock mass material
        parameters for, 219
    damage process of, 146
    overloading safety coefficient, 146
    seismic motion dynamic response of, 217–223
        basic data for computation, 219–220
        parallel computation of, 220–223
    seismic response analysis, 234
        three-dimensional finite element discrete mesh
           for, 220
    serial computation and parallel computation, conditions
        and time consumption of, 222
Xiluodu cascade hydropower station, 217, 224
X-ray, 396
    absorption coefficient, 438
    computer-aided tomography, 427
    CT technology, 247, 255, 395
        application of, 247, 395
    line absorption coefficient, 439
    scanning machine, 398
    tube, 398

Printed in the United States
By Bookmasters